Soil Biology

Volume 13

Series Editor
Ajit Varma, Amity Institute of Microbial Sciences,
Noida, UP, India

Volumes published in the series

Applied Bioremediation and Phytoremediation (Vol. 1)
A. Singh, O.P. Ward (Eds.)

Biodegradation and Bioremediation (Vol. 2)
A. Singh, O.P. Ward (Eds.)

Microorganisms in Soils: Roles in Genesis and Functions (Vol. 3)
F. Buscot, A. Varma (Eds.)

In Vitro Culture of Mycorrhizas (Vol. 4)
S. Declerck, D.-G. Strullu, J.A. Fortin (Eds.)

Manual for Soil Analysis – Monitoring and Assessing Soil
Bioremediation (Vol. 5)
R. Margesin, F. Schinner (Eds.)

Intestinal Microorganisms of Termites and Other Invertebrates (Vol. 6)
H. König, A. Varma (Eds.)

Microbial Activity in the Rhizosphere (Vol. 7)
K.G. Mukerji, C. Manoharachary, J. Singh (Eds.)

Nucleic Acids and Proteins in Soil (Vol. 8)
P. Nannipieri, K. Smalla (Eds.)

Microbial Root Endophytes (Vol. 9)
B.J.E. Schulz, C.J.C. Boyle, T.N. Sieber (Eds.)

Nutrient Cycling in Terrestrial Ecosystems (Vol. 10)
P. Marschner, Z. Rengel (Eds.)

Advanced Techniques in Soil Microbiology (Vol. 11)
A. Varma, R. Oelmüller (Eds.)

Microbial Siderophores (Vol. 12)
A. Varma, S. Chicholkar (Eds.)

Patrice Dion • Chandra Shekhar Nautiyal
Editors

Microbiology of Extreme Soils

Foreword by John D. Rummel

Professor Dr. Patrice Dion
Département de phytologie
Pavillon Charles-Eugène Marchand
1030, avenue de la Médecine
Université Laval
Québec (Québec) G1V 0A6
Canada
E-mail: patrice.dion@plg.ulaval.ca

Professor Dr. Chandra Shekhar Nautiyal
National Botanical Research Institute
Rana Pratap Marg
Lucknow 226001
India
E-mail: csn@nbri.res.in

ISBN 978-3-540-74230-2   e-ISBN 978-3-540-74231-9

Soil Biology ISSN: 1613–3382

Library of Congress Control Number: 2007934829

© 2008 Springer-Verlag Berlin Heidelberg

This work is subject to copyright. All rights are reserved, whether the whole or part of the material is concerned, specifically the rights of translation, reprinting, reuse of illustrations, recitation, broadcasting, reproduction on microfilm or in any other way, and storage in data banks. Duplication of this publication or parts thereof is permitted only under the provisions of the German Copyright Law of September 9, 1965, in its current version, and permission for use must always be obtained from Springer. Violations are liable to prosecution under the German Copyright Law.

The use of general descriptive names, registered names, trademarks, etc. in this publication does not imply, even in the absence of a specific statement, that such names are exempt from the relevant protective laws and regulations and therefore free for general use.

Cover design: WMXDesign GmbH, Heidelberg, Germany

Printed on acid-free paper   5 4 3 2 1 0

springer.com

Foreword

My auxiliaries are the dews and rains which water this dry soil, and what fertility is in the soil itself, which for the most part is lean and effete.

– Henry David Thoreau, Walden Pond

The concerns that Thoreau had about his beans were nothing to those that would face a similarly conscientious gardener in the Atacama Desert or on the planet Mars, where dews are rare, or frozen, and rains are extremely rare – or absent altogether. Yet we live in a time when an appreciation of the differences and similarities among soils (or regolith: no organics detected on Mars, as yet!) can provide a perspective on life at its most fundamental level: that of microbiology.

Microbes are the Earth's finest chemists, and most prodigious chemical engineers. Beyond pure chemistry, they know tricks with electrons that would make any Silicon Valley chip designer blush with pride. And yet their size and association with human food (good) and diseases (bad) has for more than a century obscured their essential place in making the Earth a habitable planet for humans. One of the most interesting facets of this book is that we are shown those chemists at work in one of their most important habitats. Soils comprise both a pervasive environment on our planet and one of the most important (even most fruitful!) of habitats with respect to human survival. What the chapters of this book make clear is that extreme soils have lessons that reflect on our understanding of all natural soil processes, and that, as a site for scientific exploration, these soils have unique attributes that are worth of study in their own right.

This volume provides an excellent introduction to the study of extreme soil microbiology, and a variety of the challenging and fascinating environments that Earth-bound microbes face. Some are natural, and some are the result of human activity, and all of them have lessons to teach us about life's adaptations within the "extreme" horizons of terrestrial soils. What's more, each of these chapters (including the chapter on the soils of Mars by Ronald L. Crawford and David A. Newcombe) can give us insights into strategies that may make life possible beyond the safe confines of our present-day biosphere, to other worlds in this solar system and beyond.

John D. Rummel
Senior Scientist for Astrobiology
NASA Headquarters
Washington, DC 20546
U.S.A

Preface

Two founding intuitions inspired this book. First comes the suggestion that "All soils are extreme, but some are more extreme than others". Indeed, as is discussed amply in the chapters that follow, heterogeneities and discontinuities in the physico-chemical environment are a hallmark of soil systems, with the obligation for soil microbes to constantly reinvent themselves through metabolic control and adaptation. In addition to this tell-tale variable character, some soils, designated here as "extreme", harbor a constant, defining parameter that limits colonization by most organisms found elsewhere. The second inspiration for the present book is a sense of awe in view of the virtually limitless inventiveness of the microbial world. Investigating this diversity and how it developed has become a task fundamental to our understanding of the world. In no other province of human activity do more stories await to be told: each new day brings its microbiological marvels, and not a few of them come from the examination of soil micro-organisms. Hence, the present book proposes the concept of "extreme soil microbiology" to researchers and advanced students, as a means for organizing current knowledge and stimulating further developments.

Technological advances and sociological changes are such that scientific mores keep evolving. If Charles Darwin was living today, who knows if he would not choose to publish his ideas as a one-page article in a top journal? Nevertheless, even today, there remain people to write and read scientific books. We believe that one reason for this is that broadening one's view, through the examination of a text of wide and extensive coverage, nurtures one's capacity for learning and reflection. Whereas scientific ideas seem to arise in the most unexpected manner, typically while jumping out of bed or at the movies, they are often elicited by a largely unconscious process involving scrambling and relating a vast corpus of notions. To serve in this process, those notions have to be timely and accurate. They also must be presented clearly, so as to trigger curiosity and stimulate the imagination. We hope that the reader will find the current book to fulfill these expectations.

The book is organized into three sections. The first section, "Principles of Extreme Soil Microbiology" presents a conceptual framework for further comparisons and coincident analyses of the various extreme soil systems. Chapter 1 provides an overview and places the other book chapters into perspective. Chapter 2 focuses on microbial communities from various types of extreme soils, presenting elements that contribute to their structure and diversity. Chapter 3 considers evolutionary mechanisms taking place in extreme soils. That extreme soils ecosystems are exquisitely amenable to scientific exploration is eloquently advocated in Chapter 4. A recurrent theme in this first part of the book is the significance of extreme soils as representations and models for other terrestrial or extraterrestrial environments.

In the second section, soils that owe their extreme character to natural processes are considered, with a description of the corresponding microbial communities, either observed or sometimes deduced or hypothesized. Chapter 5 shows that salinity is superimposed on other soil characters to shape selective pressure for colonizing organisms, which, in response, have developed traits that offer potential for biotechnology and bioremediation. Chapter 6 deals with organisms finding unlikely havens in areas where rain remains an abstract, literally evaporated notion. Permafrost soils are particularly fragile and, accordingly, generate preoccupations that are echoed in various sections of the book; their microbiology is the object of particular scrutiny in Chapter 7. Curious and unexpected phenomena are presented in Chapter 8, relating to the distribution of thermophilic *Bacillus* that colonize geothermal soils of the Antarctic. Peatlands, that provide a unique refuge to microbes driving peculiar biogeochemical cycles, are covered in Chapter 9. Chapters 10 and 11 deal with life as it occurs or may occur well outside our current human range. Convincing evidence is provided for microbial functioning in an iron-rich subterranean environment, whereas some Martian environments protected from the deadly Sun may well meet life's requirements.

The third section examines the results of human action on soil microbes, with an eye on future biological consequences as we continue remodelling our environments. Many lessons are to be learned from how micro-organisms deal with hydrocarbon contamination in extremely cold Antarctic soils, as is discussed in Chapter 12. While offering quite a contrast, the hot Kuwaiti desert is also home to oil-degrading microbial communities, which Chapter 13 discusses. Fires, whether they be of short or long duration, present soil micro-organisms with particular challenges, as is discussed in Chapter 14. Whereas hyperaccumulating plants have attracted much attention as colonizers of heavy metal-contaminated soils, Chapter 15 stresses that such plants owe much of their properties to rhizospheric and endophytic micro-organisms. Finally, Chapter 16 examines fungal responses to radionuclide contamination of soil, particularly in view of the Chernobyl accident. It should be remarked that the book subdivisions remain somewhat artificial, as some phenomena, particularly fires and presence of heavy metals, may be of natural or else anthropogenic origin. So the placement of chapters reflects the editors' prejudice and, perhaps more deeply, a human propensity to self-criticism.

Except for residing on a small desert island with a single coconut tree and really nothing much to do, every human process entails some form of compromise. Preparing this book proved to be no exception, as a balance was sought between size and completeness. So it will perhaps be felt that some important topics are not covered, although every effort was made to provide a comprehensive coverage of the most prevalent extreme soil systems. For this and other failings of the book, the editors take the entire responsibility.

Preface

We have been privileged as book editors to count on the collaboration of dedicated and expert authors. We should like to thank them here for their professionalism and enthusiasm, and also for their patience in dealing with our frequent and sometimes almost harassing demands. Our gratitude also goes to Professor Ajit Varma, for having believed in this project and for his continued support as the Series Editor, and to Dr. Jutta Lindenborn, Life Sciences Editorial at Springer, for her expert guidance throughout the book preparation process. We also wish to thank Cécile Gauthier, for helpful suggestions and help in the editing process. As this project reaches to a close, we are persuaded that the results will prove well worth the effort, and that the reader will share our amazement as, page after page, life thriving in seemingly inhospitable soils reveals its secrets.

Patrice Dion	Québec City
Chandra Shekhar Nautiyal	Lucknow
	July, 2007

Contents

Part I Principles of Extreme Soil Microbiology

1. **The Microbiological Promises of Extreme Soils** 3
   Patrice Dion

2. **Microbial Diversity, Life Strategies, and Adaptation to Life in Extreme Soils** 15
   Vigdis Torsvik and Lise Øvreås

3. **Extreme Views on Prokaryote Evolution** 45
   Patrice Dion

4. **Biodiversity: Extracting Lessons from Extreme Soils** 71
   Diana H. Wall

Part II Natural Extreme Soils

5. **Halophilic and Halotolerant Micro-Organisms from Soils** 87
   Antonio Ventosa, Encarnacion Mellado, Cristina Sanchez-Porro, and M. Carmen Marquez

6. **Atacama Desert Soil Microbiology** 117
   Benito Gómez-Silva, Fred A. Rainey, Kimberley A. Warren-Rhodes, Christopher P. McKay, and Rafael Navarro-González

7. **Microbial Communities and Processes in Arctic Permafrost Environments** 133
   Dirk Wagner

8. **Aerobic, Endospore-Forming Bacteria from Antarctic Geothermal Soils** 155
   Niall A. Logan and Raymond N. Allan

9 **Peatland Microbiology**... 177
Shwet Kamal and Ajit Varma

10 **Subsurface Geomicrobiology of the Iberian Pyritic Belt**........... 205
Ricardo Amils, David Fernández-Remolar, Felipe Gómez,
Elena González-Toril, Nuria Rodríguez, Carlos Briones,
Olga Prieto-Ballesteros, José Luis Sanz, Emiliano Díaz,
Todd O. Stevens, Carol R. Stoker, the MARTE Team

11 **The Potential for Extant Life in the Soils of Mars**................ 225
Ronald L. Crawford and David A. Newcombe

Part III Anthropogenic Extreme Soils

12 **Bacteriology of Extremely Cold Soils Exposed
to Hydrocarbon Pollution** 247
Lucas A.M. Ruberto, Susana C. Vazquez, and Walter P. Mac Cormack

13 **Microbiology of Oil-Contaminated Desert Soils and Coastal
Areas in the Arabian Gulf Region** 275
Samir Radwan

14 **Microbial Communities in Fire-Affected Soils**.................... 299
Christopher Janzen and Tammy Tobin-Janzen

15 **Endophytes and Rhizosphere Bacteria of Plants Growing
in Heavy Metal-Containing Soils** 317
Angela Sessitsch and Markus Puschenreiter

16 **Interactions of Fungi and Radionuclides in Soil**.................. 333
John Dighton, Tatyana Tugay, and Nelli Zhdanova

Index 357

Contributors

Aguilera, Angeles
Centro de Astrobiología (INTA-CSIC), Torrejón de Ardoz, Spain

Allan, Raymond N.
Department of Biological and Biomedical Sciences, Glasgow Caledonian University, Cowcaddens Road, Glasgow G4 0BA, U.K. R.Allan@gcal.ac.uk

Amils, Ricardo
Centro de Astrobiología (INTA-CSIC), 28850 Torrejón de Ardoz, Madrid, Spain, Centro de Biología Molecular (UAM-CSIC), Cantoblanco, 28049 Madrid, Spain, ramils@cbm.uam.es

Briones, Carlos
Centro de Astrobiología (INTA-CSIC), 28850 Torrejón de Ardoz, Madrid, Spain, brioneslc@inta.es

Cannon, Howard
NASA Ames Research Center, Mountain View, CA, USA

Crawford, Ronald L.
Environmental Biotechnology Institute, University of Idaho, Moscow, ID 83844, USA, crawford@uidaho.edu

Davila, Fidel
Centro de Astrobiología (INTA-CSIC), Torrejón de Ardoz, Spain

Díaz, Emiliano
Centro de Biología Molecular (UAM-CSIC), Cantoblanco, 28049 Madrid, Spain, eediaz@cbm.uam.es

Dighton, John
Rutgers University Pinelands Field Station, New Lisbon, NJ 08064, USA, dighton@camden.rutgers.edu

Dion, Patrice
Département de phytologie, Pavillon Charles-Eugène-Marchand, 1030, avenue de la Médecine, Université Laval, Québec (Québec) G1V 0A6, Canada, patrice.dion@plg.ulaval.ca

Dunagan, Steven
NASA Ames Research Center, Mountain View, CA, USA

Fairén, Alberto G.
Centro de Biología Molecular (UAM-CSIC), U. Autónoma de Madrid, Madrid, Spain

Fernández-Remolar, David
Centro de Astrobiología (INTA-CSIC), 28850 Torrejón de Ardoz, Madrid, Spain, fernandezrd@inta.es

Glass, Brian
NASA Ames Research Center, Mountain View, CA, USA

Gómez, Felipe
Centro de Astrobiología (INTA-CSIC), 28850 Torrejón de Ardoz, Madrid, Spain, gomezgf@inta.es

Gómez-Elvira, Javier
Centro de Astrobiología (INTA-CSIC), Torrejón de Ardoz, Spain

Gómez-Silva, Benito
Instituto del Desierto y Unidad de Bioquímica, Facultad Ciencias de la Salud, Universidad de Antofagasta, Casilla 170, Antofagasta, Chile, bgomez@uantof.cl

González-Toril, Elena
Centro de Astrobiología (INTA-CSIC), 28850 Torrejón de Ardoz, Madrid, Spain, etoril@cbm.uam.es

Janzen, Christopher
Chemistry Department, Susquehanna University, Selinsgrove, PA 17870, USA, janzen@susqu.edu

Kamal, Shwet
Amity Institute of Microbial Sciences, Amity University Uttar Pradesh, Noida 201303, India, skamal@amity.edu

Lemke, Lawrence G.
NASA Ames Research Center, Mountain View, CA, USA

Logan, Niall A.
Department of Biological and Biomedical Sciences, Glasgow Caledonian University, Cowcaddens Road, Glasgow G4 0BA, UK, nalo@gcal.ac.uk

Lynch, Kennda
NASA Johnson Space Center, Houston, TX, USA

Mac Cormack, Walter P.
Instituto Antártico Argentino, Departamento de Biología. Cerrito 1248, C1010AAZ, Buenos Aires, Argentina, wmac@huemul.ffyb.uba.ar

Marquez, M. Carmen
Department of Microbiology and Parasitology, Faculty of Pharmacy, University
of Sevilla, 41012 Sevilla, Spain, cmarquez@us.es

McKay, Christopher P.
Space Science Division, NASA Ames Research Center, Moffett Field, CA
94035-1000, USA, cmckay@mail.arc.nasa.gov

Mellado, Encarnacion
Department of Microbiology and Parasitology, Faculty of Pharmacy, University
of Sevilla, 41012 Sevilla, Spain, emellado@us.es

Navarro-González, Rafael
Laboratorio de Química de Plasmas y Estudios Planetarios,
Instituto de Ciencias Nucleares, Universidad Nacional
Autónoma de México, Apartado Postal 70-543, México D.F. 04510,
México, navarro@nucleares.unam.mx

Newcombe, David
Environmental Science Program, University of Idaho, 721 Lochsa Street, Suite 3,
Post Falls, ID 83854, USA, dnewcombe@uidaho.edu

Øvreås, Lise
Department of Biology, University of Bergen, P. Box 7800, Jahnebakken 5,
N-5020 Bergen, Norway, lise.ovreas@bio.uib.no

Parro, Victor
Centro de Astrobiología (INTA-CSIC), Torrejón de Ardoz, Spain

Prieto-Ballesteros, Olga
Centro de Astrobiología (INTA-CSIC), 28850 Torrejón de Ardoz, Madrid, Spain,
prietobo@inta.es

Puschenreiter, Markus
Department of Forest and Soil Sciences, University of Applied
Life Sciences and Natural Resources, A-1190 Vienna, Austria,
markus.puschenreiter@boku.ac.at

Radwan, Samir
Department of Biological Sciences, Faculty of Science, Kuwait University,
P.O. Box 5969, Safat 13060, Kuwait, radwan@kuc01.kuniv.edu.kw

Rainey, Fred A
Department of Biological Sciences, 202 Life Sciences Building, Louisiana State
University, Baton Rouge, LA 70803, USA, frainey@lsu.edu

Rodríguez, Nuria
Centro de Astrobiología (INTA-CSIC), 28850 Torrejón de Ardoz, Madrid, Spain,
nrodriguez@cbm.uam.es

Ruberto, Lucas A.M.
Cátedra de Microbiología Industrial y Biotecnología, Facultad de Farmacia y
Bioquímica, Universidad de Buenos Aires. Junin 956, C1113AAD, Buenos Aires,
Argentina, lruberto@ffyb.uba.ar

Sanchez-Porro, Cristina
Department of Microbiology and Parasitology, Faculty of Pharmacy, University
of Sevilla, 41012 Sevilla, Spain, sanpor@us.es

Sanz, José Luis
Centro de Biología Molecular (UAM-CSIC), Cantoblanco, 28049 Madrid, Spain,
joseluis.sanz@uam.es

Sessitsch, Angela
Austrian Research Centers GmbH, Department of Bioresources, A-2444
Seibersdorf, Austria, angela.sessitsch@arcs.ac.at

Souza-Egipsy, Virginia
Centro de Astrobiología (INTA-CSIC), Torrejón de Ardoz, Spain

Stevens, Todd O.
Department of Biology, Portland State University, Portland, OR, USA,
tstevens@gorge.net

Stoker, Carol R.
NASA Ames Research Center, Mountain View, CA, USA, cstoker@mail.arc.nasa.gov

Tobin-Janzen, Tammy
Biology Department, Susquehanna University, Selinsgrove, PA 17870, USA,
tobinjan@susqu.edu

Torsvik, Vigdis
Department of Biology, University of Bergen, P. Box 7800, Jahnebakken 5,
N-5020 Bergen, Norway, vigdis.torsvik@bio.uib.no

Tugay, Tatyana
Department of Physiology and Taxonomy of Micromycetes, Institute
of Microbiology and Virology, Ukrainian National Academy of Sciences,
154 Zabolotny Street, Kiev 252143, Ukraine, tatyanatugay@rambler.ru

Varma, Ajit
Amity Institute of Microbial Sciences, Amity University Uttar Pradesh,
Noida 201303, India, ajitvarma@aihmr.amity.edu

Vazquez, Susana C.
Concejo Nacional de Investigaciones Científicas y Técnicas (CONICET),
Buenos Aires, Argentina, svazquez@ffyb.uba.ar

Ventosa, Antonio
Department of Microbiology and Parasitology, Faculty of Pharmacy, University
of Sevilla, 41012 Sevilla, Spain, ventosa@us.es

Wagner, Dirk
Alfred Wegener Institute for Polar and Marine Research, Research Unit Potsdam,
Telegrafenberg A45, 14473 Potsdam, Germany, Dirk.Wagner@awi.de

Wall, Diana H.
Department of Biology and Natural Resource Ecology Laboratory, Colorado State
University, Fort Collins, CO 80523-1499, USA, Diana@nrel.colostate.edu

Warren-Rhodes, Kimberley A.
Space Science Division, NASA Ames Research Center, Moffett Field, CA
94035-1000, USA, kwarren-rhodes@mail.arc.nasa.gov

Zavaleta, Jhony
NASA Ames Research Center, Mountain View, CA, USA

Zhdanova, Nelli
Department of Physiology and Taxonomy of Micromycetes, Institute of
Microbiology and Virology, Ukrainian National Academy of Sciences,
154 Zabolotnoy Street, Kiev 252143, Ukraine, zhdanova_imv_ua@rambler.ru

Part I
Principles of Extreme Soil Microbiology

Part 1

Chapter 1
The Microbiological Promises of Extreme Soils

Patrice Dion

1.1 Introduction

Whereas the notion of extreme environment has received much attention from microbiologists, this generalization does not systematically include extreme soils. There may be at least two reasons for this. First, any soil may be considered as extreme for the colonizing microbes constantly facing starvation, desiccation, predation, and other attacks. In this sense, the notion of "extreme soil" would appear pleonastic. A second reason for the uncommon use of the term "extreme soil" might be the opinion that there is little to be gained from it, inasmuch as every soil has its particularities and, in its very nature, is refractory to human efforts at unification and simplification. In this sense, any grouping of soils from, say, the Antarctic or hot deserts into a common category designated as "extreme" would appear futile, if not detrimental to a precise understanding of soils and their microbial populations. However, one might take the stance that, although it certainly serves to be aware of these difficulties, there is still much to be learned from running into them. Hence, it is hoped that the present book will be a demonstration of the usefulness of the extreme soil concept to microbiologists.

Various extreme soils have been the topic of numerous and fruitful studies dealing with the characterization of microbial communities and processes. These studies have done much to enrich our understanding of microbial diversity and of biogeochemical and other biological mechanisms. They allow us to grasp microbial adaptability and to envision practical applications. Reading the enclosed collection of chapters will make it clear that unifying these studies under the general theme of "extreme soils," and associating this concept with the broader notion of "extreme environments" leads to essential theoretical and practical advances.

We qualify a soil as "extreme" when it supports colonization by organisms presenting a specific and common adaptation. The specifying character of an extreme soil may be physical in nature, and correspond to extreme values of temperature, or

---

Patrice Dion
Département de phytologie, Pavillon Charles-Eugène Marchand, 1030, avenue de la Médecine, Université Laval, Québec (Québec), Canada G1V 0A6
e-mail: patrice.dion@plg.ulaval.ca

to exposure to radiation or intense heat. Alternatively, it may also be defined in chemical terms, with the salt content or the presence of toxic pollutants exerting a preponderant influence on microbial processes. The extreme character may be conferred on soils by climatic, geological, or other environmental factors, or else by human activities. One might observe here that the distinction between nature-driven and anthropogenic processes is becoming increasingly blurred, as a result of our improved capacity to relate effects to their cause, and also as human activities exert an ever stronger influence on an expanding scale.

The chapters included in this book provide a careful description of microbial communities exposed to various soil-borne challenges. Bringing together analyses of a wide range of soil systems invites comparative assessments and well-founded extrapolations. Thus, it is hoped that, in addition to providing timely knowledge about extreme soil microbial dwellers, the book will pay tribute to a vast and largely unexplored territory wide open to microbiological enquiry. Indeed, extreme soils promise crucial progress in our understanding of microbial activities and adaptation processes, as well as in our ability to rationally influence ecosystems upon which terrestrial life depends.

1.2 Extreme Soils and Microbial Community Structure and Evolution

Following an unprecedented search for marine microbial sequences, an immense diversity of marine bacteria was revealed, with at least 25,000 different types of micro-organisms being estimated to exist per litre of seawater. The methods used in the study made it possible to relate intraribotype genetic variation to environmental factors. Species may be organized into subtypes, and the corresponding variation results from physical barriers, short-term stochastic effects, and functional differentiation acting in combination (Rusch et al. 2007). Comparison of marine and terrestrial organisms showed that 68% of the nearly 7,000 examined protein domains varied between the two classes of micro-organisms, this variation being in part the result of different metabolic requirements for marine and terrestrial life (Yooseph et al. 2007). These results are a testimony to adaptability of microbial life.

Extreme soils offer us an opportunity to understand microbial diversity as an adaptive response that both reflects and multiplies environmental diversity. Indeed, the soil can be hot or cold, acidic or basic, wet or dry, saline, radioactive, polluted with heavy metals or hydrocarbons, or located on Earth or some other celestial body. Soil microbes pay tribute to this multiplicity of characters by undergoing and maintaining diversification (see Chapter 2).

Extreme environments present peculiar evolutionary challenges to prokaryotes, to which might correspond peculiar evolutionary responses. Identification of these specificities may reveal hitherto unnoticed aspects of evolutionary processes. In particular, extreme conditions might force evolution of traits that are otherwise invariant. More generally, studies on extremophile evolution may help shed light on the roles of environmental pressure in driving evolution. Indeed, competition between individuals may play a larger role in nonextreme environments, whereas environmental pressures would be determinant under extreme conditions. Soils are physically constituted by the orderly collection of sizable and interacting structures. Physicochemical parameters are superimposed on this primary framework. Microbial niches and corresponding diversity are defined by a series of combinatorial operations that the various organizational levels of the soils render possible (see Chapter 3).

In comparing microbial communities of nonextreme and extreme soils, it becomes apparent that extreme soil communities reach unique equilibria, corresponding to under- or overrepresentation of certain community components. The degree to which community member exclusions and inclusions occur, and the nature of these processes, vary in different extreme soils. Attempting a synthesis of these particular responses may lead to identification of crucial microbial mechanisms for survival, growth, dissemination, and adaptation. Such a synthesis may also provide insights on the forces at play to shape and structure biological communities in extreme as well as nonextreme soils (see Chapter 4). Biogeochemical cycles, as they operate under extreme conditions, bring into sharp focus important aspects of microbial community functioning and bear direct relevance to global equilibria (see Chapters 7, 9, and 14).

1.3 Extreme Soils and Microbial Physiology

Comparisons of microbial adaptations in extreme soils and other extreme environments suggest commonalities in physiological processes and cell adaptations. For example, patterns of adaptation to heat and cold, through adjustments in protein thermal stability and membrane composition (see Chapters 2, 3, and 8), are similar in extreme soils as in other environments. Also, compatible solutes contribute to maintain osmotic balance in halophilic organisms from saline soils and other saline environments (see Chapter 5). Such a pattern of common adaptations, superimposed on additional and specific adaptations to soil, water, or other environments, is suggestive of modularity in microbial evolutionary processes. Modules can be defined as "building blocks of interacting elements that operate in an integrated and relatively autonomous manner" (Schlosser 2004). Modules may be thought of as structures, but also as processes, that would be articulated according to three principles. These principles are: (1) connectivity, meaning that they are triggered in a switchlike fashion by a variety of inputs; (2) hierarchization, implying that modules may be spatiotemporally embedded in higher-order modules, or overlap by sharing common elements; and (3) multiple instantiation, which contributes to delimitate a module from others of the same type by its independent perturbability during development. Specifically, modules of evolution are units of integrated and context-insensitive evolutionary changes (Schlosser 2004). From these considerations, it appears that adaptive processes in extreme soil bacteria may be modular, in the sense that adaptations to the soil and extreme components of the environment may occur somewhat independently, while influencing each other through epistasis (see Chapter 3).

It might be worth mentioning in this respect that relatively little is known on physiology of life in arid environments and the adaptations involved. In most studies, water stress refers to external solute excess, rather than dehydration. Whereas salt and water stress often come together (Wierzchos et al. 2006), and may be both dealt with through osmotic adjustment (see Chapter 2), it is possible to distinguish between tolerance to salt and to water in plants (Munns 2002), and also perhaps in micro-organisms (Zahran 1999). Most studies where dehydration is considered refer to food (Grant 2004). Studies on arid soils deal mainly with community structure (McKay et al. 2003; see also Chapter 6) and little is known about microbial physiology and adaptations, as many of the dominant organisms in arid soils yet remain to be isolated (Drees et al. 2006). One aspect that is being studied is the physiology of photosynthesis in cyanobacteria colonizing arid soils (Lüttge et al. 1995; Ohad et al. 2005). It is also known that, in *Deinococcus radiodurans*, radiation and desiccation resistance are correlated (Mattimore and Battista 1996), and related at least in part to a remarkable capacity for reassembly of broken chromosomes (Zahradka et al. 2006).

1.4 Microbial Functions in Extreme Soils

Modularity, which occurs at the level of cell adaptation to the extreme and soil environmental components, may also be identified as a defining factor of soil microbial communities, where it arises as a consequence of redundancy and complementarity of soil functions (Nannipieri et al. 2003). Redundancy allows microbial groups to be considered as equivalent and interchangeable with respect to function, thus buffering biogeochemical cycling and other ecosystem processes against restricting changes, such as selection and metabolic tradeoffs.

The extreme character acts on redundancy, diminishing it without altering ecosystem processes (see Chapter 4). This results in a strengthening of the correspondence between identity and function, and leads to a better illustration of microbial function as it more directly relates to particular microbial types. Clues to this simplifying, looking-glass effect are proposed in this volume, and are provided, for example, by observations on dissimilatory sulfate reduction by certain halophiles in saline soils and their role in ecosystem functioning (see Chapter 5), hydrocarbon degradation as it is performed by permafrost microbial communities (see Chapter 12), and involvement of endo- and ecto-mycorrhizae in radionuclide uptake and transfer to plants (see Chapter 16). It may well be that further studies on these and other systems will highlight the potential of extreme soils as simplified objects for study of microbial processes.

1.5 Practical Value of Extreme Soils

Like a cat, the Earth cleans itself continuously, although there is growing concern that autogenous processes may not suffice given the extent of human-inflicted damage. Soil contamination occurs worldwide, and it is striking that bacteria from pristine Arctic soils express *mer* (or mercury resistance) genes (Poulain et al. 2007).

One might consider the degradation or elimination of pollutants as yet another manifestation of overall soil productivity, and then be led to suggest that the intrinsic cleanup potential will be less in extreme soils, as compared to nonextreme soils. This suggestion arises from the conjunction of two observations, which are, first, that microbial diversity is often lower in extreme soils than in nonextreme soils (see Chapter 2), and, second, that ecosystem productivity increases with biodiversity (Tilman 1999). On the other hand, extreme soils might be more responsive than nonextreme soils to application of exogenous microbial inoculant with remediating activity, as lower diversity might make communities more susceptible to invasions (Tilman 1999).

The relationship between diversity and stability of an ecosystem is the object of some controversy and may be influenced by species composition (Bezemer and van der Putten 2007; Tilman et al. 2007). This may explain why attempts at accelerating the degradation of oil pollutants in Antarctic soils through bioaugmentation are often unsuccessful (see Chapter 12).

Conceptually, two situations may arise, with the extreme soil parameter being the direct object of the bioremediation process or simply acting as an intrinsic confounding factor. Attempts to use bacteria to reduce the impact of salts (Bacilio et al. 2004; Ashraf et al. 2006) or heavy metals (see Chapter 15), belong to the first category, and, in this case, the extreme character of the soil will become attenuated following successful treatment. Bioremediation efforts of hydrocarbon-contaminated cold (see Chapter 12) or arid (see Chapter 13) soils fall into the second class of bioremediation interventions.

Extremophiles offer an ever-expanding domain to biotechnological applications (Podar and Reysenbach 2006), and extreme soils represent a rich and still relatively unexplored source of useful microbes. For example, halophiles colonizing saline soils produce a wealth of macromolecules and small compounds of potential use (see Chapter 5).

1.6 Extreme Soils and the Boundaries of Life

Although there are sound and totally unemotional reasons to send humans into space instead of strictly relying on robots (Crawford 2004), the fascination exerted by human expansion through the Earth and beyond points to additional motivations. In yielding to our relentless drive to expand, we are doing little more than applying virtually unlimited imagination and technical skills to satisfy our innate desire at self-perpetuation. Along the process, and especially since the end of the Cold War, we are conducted to establish the pacifying image of humanity as a constant creator of knowledge and settings. Not surprisingly then, life as it is envisioned elsewhere in the cosmos is often portrayed as akin to Earthly biological processes (Pace 2001), which is certainly a reasonable view as long as we restrict our exploratory range to our immediate planetary neighbors. Looking farther beyond, a variety of other chemistries based on liquids other than water may be envisioned (Bains 2004; Benner et al. 2004).

Following its formation 4.6 billion years ago and early bombardment, Mars is believed to have been through three geologic ages, the first of which, termed the phyllosian era, coincided with a nonacidic aqueous alteration of planetary material. The planet might have been habitable during this period, which ended 3.5 billion years ago, when surface water became increasingly acidic during the theiikian era, and then disappeared to initiate the siderikian era, that is still going on today (Bibring et al. 2006). However, liquid water might have resided very close to the surface throughout Mars' history, as is indicated by the observation of gullies filled during the last decade with a liquid that might be water (Malin et al. 2006). These observations leave open the possibility that Martian life, if indeed it was initiated during the Earth-like phyllosian era, might have been maintained in isolated subterranean oases up to this day (see Chapter 11).

The eventual demonstration and examination of extraterrestrial life will have an unimaginable impact on human thinking. It appears, however, that the mere search for this life is already bearing fruit. It leads us to systematically consider how life can be recognized and studied (see Chapter 11). It provides an incentive to investigate Mars-like ecosystems of our planet, such as sun-bathed deserts (see Chapter 6) or rocks where metabolic energy is extracted from metal chemistry (see Chapter 10). It also leads us to reflect upon and better comprehend ecological processes upon which our lives immediately depend (Wilkinson 2003). Ultimately, we are presented with pictures of our past and our future (see Sections 1.7 and 1.8) and, consequently, our very nature.

While reflecting on the nature of life, Erwin Schrödinger wrote: "Living matter, while not eluding the 'laws of physics' as established up to date, is likely to involve 'other laws of physics' hitherto unknown, which, however, once they have been revealed, will form just as integral a part of this science as the former" (Schrödinger 1944). Although this view has been amply commented and criticized (Sarkar 1991), it may retain some intuitive value, in the sense that we have entrenched the notion that our current physical knowledge does not fully account for life, even as we know it in its most current manifestations. The history of biological enquiry may be viewed as an endless tentative to restrict and even contradict this notion of life as escaping common physical laws.

Today, we maintain a desire to define life from the outside, that is, to determine where life can and cannot exist and then identify absolute differences between these two classes of environments. Certainly, extremely arid (see Chapters 6 and 13) or cold (see Chapters 7 and 12) land may be viewed as a patchwork of life-prone and life-hostile zones, as does extraterrestrial soil (see Chapter 11). For its part, isolation might not impede life development itself, but rather narrow its manifestations (see Chapter 8). In gaining such a subtractive vision of life, in the sense of determining where life cannot be and what factors restrict its range, we demonstrate the limitations of biological processes and adaptations. Thus, a frontier is drawn, within which life is thought to remain. Within this frontier composed of water and carbon, every known life-related phenomenon exists, from the citric acid cycle to consciousness. However, lingering questions remain: can this frontier be pushed just a bit further? Is there really nothing living beyond it?

1.7 Extreme Soils and Our Past

Inspired by Beijerinck, Bass Becking has said that "Everything is everywhere, but the environment selects," and this statement has been the object of numerous comments (de Wit and Bouvier 2006). Certainly, some observations reported in this book, for example, on the presence of thermophilic endospore formers in cold soils and even Arctic ice (see Chapter 8), or of *Rhodococcus* strains with a capacity for hydrocarbon degradation in pristine Antarctic soils (see Chapter 12) seem to corroborate the notion of a universally shared microbial substratum for environmental selection. However, upon further consideration of this notion, two questions arise, which concern the nature of, first, the environmental selection implied in the statement from Bass Becking, and, second, the mechanisms that have allowed "everything" to have become present "everywhere."

With respect to the first question, bacterial dispersal is likely to set the table for what has been termed "postimmigration evolution" (Novak 2007), that implies that selection of invasive species is accompanied by adaptive genotypic modifications. Furthermore, it appears that these two recognized components of evolutionary processes, mutation and selection, are functionally related and environmentally determined, albeit to various extents. The mutation rate is influenced by various environmental components in addition to classical mutagens, this influence being exerted in the absence of any directionality to mutation events (see Chapter 3). Bacteria take different avenues towards genetic changes, that include horizontal gene transfer in addition to mutation. The ability to acquire foreign genes may itself be influenced by environmental parameters. This may occur indirectly, through the creation of hotspots of high bacterial density where plasmid transfer occurs at increased frequency (Sørensen et al. 2005), or even directly, by SOS-mediated enhancement of expression of genes involved in DNA transfer (Beaber et al. 2004).

Dispersal, that may occur over considerable distances, comes immediately to mind as a possible explanation for the apparent ubiquity of bacteria. However, the notion of bacterial ubiquity must be put in perspective. Some bacterial species, including various plant and animal pathogens, are distributed worldwide, whereas others are endemic. Within cosmopolitan species, some particular genotypes may in fact be endemic. The ability for dispersal would be associated with the presence of particular genes (Ramette and Tiedje 2007a). Reflecting this variation in distribution range, bacterial populations exhibit clear spatial patterns that arise both from environmental heterogeneity and from spatial distance taken as a measure of past historical events and disturbances. Within-species diversity may be influenced by spatial distance as well as by environmental heterogeneity, whereas species abundance and the composition of communities are more influenced by local environmental interactions. However, a considerable proportion of bacterial diversity remains unexplained by the currently measured variables (Ramette and Tiedje 2007b). Because they represent well-defined and localized environments, extreme soils may prove valuable objects of study in bacterial biogeography. For example, suitable and unsuitable habitats for extant soil communities in the Antarctic Dry Valleys are defined by a combination of factors that include current and historical elements. Among the latter are past climates, which influence the level of lakes and glaciers. Also of relevance are legacies of productivity from past ecosystems that influence the chemistry of contemporary soils and their organic matter (Virginia and Wall 1999; see also Chapter 4).

1.8 Extreme Soils and Our Future

Very few areas of our planet now escape human influence, and the global environment may be described as the "human-made context of our lives" (Dalby 2002). Nearly half of the land surface has been transformed by direct human action (Steffen et al. 2004), and sometimes humans may cause catastrophic soil changes that extend over considerable time and space scales. The great Centralia coal fire and other similar events are a case in point, and have considerable impact on soil microflora and other soil properties (see Chapter 14). Indirect human action, particularly through global climate change and what has been termed "global distillation" (or the spread of persistent organic pollutants) may be even more pervasive, with the result that virtually no pristine land is said to remain (Dalby 2002). Thus, we may be experiencing a global extremization of the soils (see Chapter 4), which may affect interactions between humans and a variety of soil-dependent or soil-residing organisms, would these particular organisms be sources of food or disease.

Throughout its modern history, humankind has been able to escape the consequences of its actions by exporting, either its waste or its populations themselves, some less fortunate human groups occasionally suffering in the process as unwilling recipients. Now we have reached a point where these movements appear useless, with a radical transformation of air, water, and soil threatening the self-maintained balance upon which all life forms depend. Particularly worrisome is the spectre of a spiralling uncontrollable process, whereby environmental change becomes the trigger of ever greater and more damaging changes. Such self-sustained processes may eventually occur in peatlands, which have been acting as net carbon sinks and methane sources since the early Holocene. Desiccation of Siberian peatlands would elevate CO_2 concentrations through peat oxidation but would reduce CH_4 emissions, with the outcome on net radiative balance being difficult to estimate. However, the balance may well tilt towards an intensification of greenhouse warming (Smith et al. 2004). This is one reason why microbial communities and processes of peatlands deserve special scrutiny (see Chapter 9). Permafrost soils, where carbon storage and methanogenesis also exist, are similarly preoccupying (see Chapter 7).

Industrialized societies have been especially good at minimizing the impact of microbial pathogens, through a combination of sanitation and hygiene measures, large-scale vaccination strategies, and heavy antibiotic use. Now there are indications that this relatively disease-free era may be coming to an end. Global changes are expected to play a role in this resurgence of microbial pathogenicity, with the poorest nations being hit first and strongest. Soils are well-known reservoirs of microbial human pathogens (Santamaría and Toranzos 2003), so that soil changes may well be associated with modifications in transmissibility and incidences of various diseases.

Many of these changes as they relate to disease transmission concern dispersal. A striking example is provided by Vozrozhdeniye Island, which was once in the middle of the now drying Aral Sea, and hosted the main Soviet center for testing biological weapons, antidotes, and vaccines. Upon shutdown of the Soviet program in 1988, slurries of anthrax spores and other pathogens were buried on the island. Vozrozhdeniye has now grown from a 33-km island to a 145-km peninsula, and, with some of the anthrax spores having remained viable (Whish-Wilson 2002), has become the object of intense surveillance and decontamination programs. In addition to persisting anthrax, a military-grade, antibiotic-resistant strain of plague bacterium may also have survived among the rodent population in the testing range (Pala 2005). There is a possibility of short-term survival of *Yersinia pestis* in soils, embedded in flea feces or tissues of dead animals, and even of its persistence in soils in dormant form or as an intracellular parasite of soil protozoa (Gage and Kosoy 2005).

The recent reconstruction of the 1918 Spanish flu epidemic virus, from the body of a flu victim buried in Alaskan permafrost (Tumpey et al. 2005), should also provide us with an opportunity to ponder the infectious potential of currently inaccessible or inert material. Although not documented still, the possibility exists that seemingly vanished pathogens, including the smallpox virus (Stone 2002), remain preserved in frozen state in permafrost burial sites. Hence, it appears that the study of permafrost-inhabiting microbial communities (see Chapters 7 and 12), while providing a remarkable insight into microbial adaptations and processes, may take unexpected relevance and offer twist endings. The possibility that some Archaea may be as of yet unrecognized human pathogens (Eckburg et al. 2003) also places soil microbial reservoirs and soil changes into a new perspective.

Cioran tells us of Socrates, who, the day before he would die, was learning a tune on the flute. Being asked what good it would do to him, he answered "To know this tune before I die." Following the philosopher's example, we will continue learning about extreme soils and their microbial inhabitants. This will help us monitor the changes ahead. We will also be in a better position to cope with these changes, which it is hoped will not prove as dreadful as a cup of hemlock tea.

References

Ashraf M, Hasnain S, Berge O (2006) Effect of exo-polysaccharides producing bacterial inoculation on growth of roots of wheat (*Triticum aestivum* L.) plants grown in a salt-affected soil. *Int J Environ Sci Tech* 3:43–51

Bacilio M, Rodriguez H, Moreno M, Hernandez J-P, Bashan Y (2004) Mitigation of salt stress in wheat seedlings by a *gfp*-tagged *Azospirillum lipoferum*. *Biol Fertil Soils* 40:188–193

Bains W (2004) Many chemistries could be used to build living systems. *Astrobiology* 4:137–167

Beaber JW, Hochhut B, Waldor MK (2004) SOS response promotes horizontal dissemination of antibiotic resistance genes. *Nature* 427:72–74

Benner SA, Ricardo A, Carrigan MA (2004) Is there a common chemical model for life in the universe? *Curr Opin Chem Biol* 8:672–689

Bezemer TM, van der Putten WH (2007) Diversity and stability in plant communities. *Nature* 446: E6–E7

Bibring J-P et al. (2006) Global mineralogical and aqueous Mars history derived from OMEGA/Mars Express data. *Science* 312:400–404

Crawford I (2004) Human exploration of the Moon and Mars: Implications for Aurora. *Astron Geophys* 45: 2.28–2.29

Dalby S (2002) Security and ecology in the age of globalization. ECSP Rep 8:95–108

de Wit R, Bouvier T (2006) 'Everything is everywhere, but, the environment selects'; what did Baas Becking and Beijerinck really say? *Environ Microbiol* 8:755–758

Drees KP et al. (2006) Bacterial community structure in the hyperarid core of the Atacama desert, Chile. *Appl Environ Microbiol* 72:7902–7908

Eckburg PB, Lepp PW, Relman DA (2003) Archaea and their potential role in human disease. *Infect Immun* 71:591–596

Gage KL, Kosoy MY (2005) Natural history of plague: Perspectives from more than a century of research. *Annu Rev Entomol* 50:505–528

Grant W (2004) Life at low water activity. *Phil Trans R Soc B* 359:1249–1267

Lüttge U, Büdel B, Ball E, Strube F, Weber P (1995) Photosynthesis of terrestrial cyanobacteria under light and desiccation stress as expressed by chlorophyll fluorescence and gas exchange. *J Exp Bot* 46:309–319

Malin MC, Edgett KS, Posiolova LV, McColley SM, Dobrea EZN (2006) Present-day impact cratering rate and contemporary gully activity on Mars. *Science* 314:1573–1577

Mattimore V, Battista J (1996) Radioresistance of *Deinococcus radiodurans*: Functions necessary to survive ionizing radiation are also necessary to survive prolonged desiccation. *J Bacteriol* 178:633–637

McKay CP, Friedmann EI, Gómez-Silva B, Cáceres-Villanueva L, Andersen DT, Landheim R (2003) Temperature and moisture conditions for life in the extreme arid region of the Atacama desert: Four years of observations including the El Niño of 1997–1998. *Astrobiology* 3:393–406

Munns R (2002) Comparative physiology of salt and water stress. *Plant Cell Environ* 25:239–250

Nannipieri P, Ascher J, Ceccherini MT, Landi L, Pietramellara G, Renella G (2003) Microbial diversity and soil functions. *Eur J Soil Sci* 54:655–670

Novak SJ (2007) The role of evolution in the invasion process. *Proc Natl Acad Sci USA* 104:3671–3672

Ohad I et al. (2005) Inactivation of photosynthetic electron flow during desiccation of desert biological sand crusts and *Microcoleus* sp.-enriched isolates. *Photochem Photobiol Sci* 4:977–982

Pace NR (2001) The universal nature of biochemistry. *Proc Natl Acad Sci USA* 98:805–808

Pala C (2005) To save a vanishing sea. *Science* 307:1032–1034

Podar M, Reysenbach A-L (2006) New opportunities revealed by biotechnological explorations of extremophiles. *Curr Opin Biotechnol* 17:250–255

Poulain AJ et al. (2007) Potential for mercury reduction by microbes in the High Arctic. *Appl Environ Microbiol* 73:2230–2238

Ramette A, Tiedje JM (2007a) Biogeography: An emerging cornerstone for understanding prokaryotic diversity, ecology, and evolution. *Microb Ecol* 53:197–207

Ramette A, Tiedje JM (2007b) Multiscale responses of microbial life to spatial distance and environmental heterogeneity in a patchy ecosystem. *Proc Natl Acad Sci USA* 104:2761–2766

Rusch DB et al. (2007) The Sorcerer II Global Ocean Sampling expedition: Northwest Atlantic through Eastern tropical Pacific. *PLoS Biology* 5:e77

Santamaría J, Toranzos GA (2003) Enteric pathogens and soil: A short review. *Int Microbiol* 6:5–9
Sarkar S (1991) What is life? Revisited. *BioScience* 41:631–634
Schlosser G (2004) The role of modules in development and function. In: Schlosser G, Wagner GP (eds) *Modularity in Development and Evolution*. The University of Chicago Press, Chicago, pp 519–582
Schrödinger E (1944) What is life? The book page (http://home.att.net/~p.caimi/schrodinger.html)
Smith LC et al. (2004) Siberian peatlands a net carbon sink and global methane source since the early holocene. *Science* 303:353–356
Sørensen SJ, Bailey M, Hansen LH, Kroer N, Wuertz S (2005) Studying plasmid horizontal transfer *in situ*: A critical review. *Nature Rev Microbiol* 3:700–710
Steffen W et al. (2004) *Global Change and the Earth System: A Planet Under Pressure*. Springer-Verlag, Berlin (Executive summary available at http://www.igbp.net/documents/IGBP_ExecSummary.pdf)
Stone R (2002) Is live smallpox lurking in the Arctic? *Science* 295:2002
Tilman D (1999) The ecological consequences of changes in biodiversity: A search for general principles. *Ecology* 80:1455–1474
Tilman D, Reich PB, Knops J (2007) Diversity and stability in plant communities (Reply). *Nature* 446:E7–E8
Tumpey TM et al. (2005) Characterization of the reconstructed 1918 Spanish influenza pandemic virus. *Science* 310:77–80
Virginia RA, Wall DH (1999) How soils structure communities in the Antarctic Dry Valleys. *BioScience* 49:973–983
Whish-Wilson (2002) The Aral Sea environmental health crisis. *J Rural Remote Environ Health* 1:29–34
Wierzchos J, Ascaso C, McKay CP (2006) Endolithic cyanobacteria in halite rocks from the hyperarid core of the Atacama desert. *Astrobiology* 6:415–422
Wilkinson DM (2003) The fundamental processes in ecology: A thought experiment on extraterrestrial biospheres. *Biol Rev* 78:171–179
Yooseph S et al. (2007) The Sorcerer II Global Ocean Sampling expedition: Expanding the universe of protein families. *PLoS Biology* 5:e16
Zahradka K et al. (2006) Reassembly of shattered chromosomes in *Deinococcus radiodurans*. *Nature* 443:569–573
Zahran HH (1999) *Rhizobium*-legume symbiosis and nitrogen fixation under severe conditions and in an arid climate. *Microbiol Mol Biol Rev* 63:968–989

Chapter 2
Microbial Diversity, Life Strategies, and Adaptation to Life in Extreme Soils

Vigdis Torsvik(✉) and Lise Øvreås

2.1 Introduction: What Is an Extreme Environment?

There is no general consensus on how to define an extreme environment. From an anthropocentric point of view, physicochemical conditions supporting mammalian life appear as normal, and conditions deviating from these are considered as extreme. However, what is extreme and what is normal for microbes remains debatable, and the concept "extreme" as we use it may not necessarily be appropriate for micro-organisms (Gorbushina and Krumbein 1999). Micro-organisms dwell in virtually all types of soil habitats. These range from extremely dry and cold deserts in the Antarctic and deep into permafrost soils to geothermal and humid soils in volcanic areas, from extremely acid mines with sulfuric acid to high alkaline areas. Microbial life can also exist in salt crystals, under extremely low water activity, and low nutrient concentrations. As a group, micro-organisms have the highest ability of all life forms to adapt to extreme and stressful environments. This includes new types of habitats created by anthropogenic activities, such as those polluted with heavy metals, radionuclides, and high concentrations of toxic xenobiotic compounds (e.g., polychlorinated biphenyls, hydrocarbons, and pesticides).

Environments which we consider extreme can be inhabited by well-adapted microbiota, and if the environment is stable the resident micro-organisms may not experience any stress but metabolize and grow successfully under strictly limiting conditions, which appear normal to them. During the approximately 3.8 billion years that micro-organisms inhabited Earth, dramatic changes in physicochemical conditions of Earth's surface have periodically occurred. Thus, conditions that can be considered as "normal" for life have also changed.

An alternative view is that any stable environment can be regarded as "normal," and that an extreme environment is one with highly fluctuating conditions where the organisms experience episodic or periodic dramatic environmental changes. In unstable and extreme environments, the metabolic costs to survive stress may be

---

Vigdis Torsvik
Department of Biology, Centre for Geobiology, University of Bergen, P.Box 7800, Jahnebakken 5, N-5020 Bergen, Norway
e-mail: vigdis.torsvik@bio.uib.no

high, and most organisms will probably die. However, some micro-organisms have very high physiological and ecological plasticity, which makes them well adapted to survive even in environments where the conditions may change suddenly and dramatically. Organisms that can tolerate considerable environmental stress caused by fluctuating conditions have been termed poikilotrophic or poikilophilic (*poikilo* = various; Gorbushina and Krumbein 1999). Poikilo-environments have prevailing hostile conditions for life (extremely low water potential, extreme temperatures, low nutrients, high levels of toxic substances), but the conditions may occasionally and sporadically change and become suitable for microbial activity and growth. The best examples of poikilotrophic organisms are rock-dwelling prokaryotes and fungi. Organisms living in deserts or arid fell-field soils with extremely low nutrients, precipitation, and highly variable temperatures can also be regarded as poikilotrophic.

A more objective view of extreme environments is one based on the fact that there are specific physical and chemical limitations to cellular processes. These limitations are related to the characteristics of biomolecules and biochemical reactions and set the boundaries for cellular life. Extreme conditions can, according to this view, be defined as those near the limits for cell functioning, that is, limiting for enzyme activities or damaging to biomolecules (Rothschild and Mancinelli 2001; Marion et al. 2003). The best example of a physical limit to life is the presence of liquid water. Life is not possible without water, because it is the solvent necessary for all biochemical reactions. Other constraints are extreme conditions typical at the end of gradients, such as low and high temperature, low and high pH and E_h, high salinity, high radiation doses, high concentrations of toxic compounds, and extremely low nutrient concentrations. Despite the physicochemical limits to biochemical processes and stability of biomolecules, the evolving micro-organisms have extended the boundaries for their life processes.

Organisms living under extreme conditions are divided into different categories according to the nature of their adaptation (Table 2.1). They are classified as thermo- (high temperature), psychro- (low temperature), halo- (high salt), acido- or alkali- (extreme low or high pH), and xero- (low water activity). The suffix -phile is used for those that require the extreme condition for growth, and -troph or -tolerant for those that tolerate the extreme condition. These designations are not exclusive, because two or more factors can be extreme in the same environment, as independent or interrelated conditions. The organisms may therefore belong to multiple categories, and, for example, be considered as both psychro- and xerophile.

The extremophiles are adapted to and limited by very narrow sets of environmental conditions, and they thrive in or require the extreme conditions. Extremotrophic or extremotolerant organisms can survive and proliferate under a wider set of environmental conditions. They tolerate extreme environments but normally grow better at moderate conditions.

In ecology, the predominant growth strategy of organisms is often described as r- and K-strategy (Panikov 1999; van Elsas et al. 2006). Organisms with a predominant K-strategy can live near the carrying capacity of the environment. They have relatively low growth rates, but compensate this by competitive advantages such as

Table 2.1 Micro-organisms in extreme soil

Environmental Conditions	Types of Extreme Micro-organisms	Definitions	Adaptation Mechanisms	Examples
Temperature (low)	Psychrophile	Growth ≤0–20°C; $T_{opt} <20°C$	Production of unsaturated fatty acids to counteract decreased membrane fluidity. Reduced cell size. Increased fraction of ordered cellular water	*Psychrobacter*
	Psychrotroph	Growth ≤5–30°C; $T_{opt} >20°C$		
Temperature (high)	Thermophile	Growth 60–80°C	Heat-stable biomolecules (proteins, nucleic acids, lipids). Production of long-chained, saturated, branched fatty acids and cyclic lipids	*Picrophilus sp.*; *Thermobaculum terrenum*; *Pyrolobus fumarii*
	Hyperthermophile	Growth >80°C		
Radiation	Radiation resistant/tolerant	Tolerate high levels of ionizing radiation (e.g., UV, γ-radiation)	Strong pigmentation (melanoids, carotenoids, etc). Efficient DNA repair mechanisms	*Deinococcus radiodurans*; *Rubrobacter sp.*
Desiccation	Xerophile/xerotolerant	Water potential MPa: −4 to −10 (bacteria) −10 to −70 (actinomycetes and fungi)	Accumulation of inorganic or organic osmolytes; Production of extracellular polysaccharides; Differentiation into desiccation resistant cells (spores)	Actinobacteria: "group-1 Rubrobacteria" and members of Actinomycetales; Cyanobacteria, fungi
Salinity	Halophile/halotolerant Osmophile	2–5 M NaCl; water potential −1.5 to −40 Mpa	Accumulating osmolytes. Salt-tolerant (and salt-dependent) enzymes	*Haloarcula japonica*; Cyanobacteria
pH (low)	Acidophile	$pH_{opt} ≤2–3$	Proton exclusion, efficient efflux system. Acid stable membrane. Proton-driven secondary transporters	*Acidithiobacillus*; *Picrophilus sp.*; *Ferroplasma sp*
pH (high)	Alcaliphile	$pH_{opt} ≥9$	Membrane impermeability for OH-ions. Efficient proton uptake mediated by membrane antiporters. Negatively charged cell wall polymers	*Natronobacterium*; *Bacillus firmus OF4*
Heavy metals	Metalo-resistant/tolerant	Tolerate high metal concentrations	Efflux pump. Sequestering and/or detoxification of metals (reduction, alkylation)	*Ferroplasma acidarmanus*
Toxins and xenobiotics	Toxin-resistant/tolerant; Xenobiotic decomposers	Tolerate high concentrations of toxic compounds	Efficient efflux pump. Detoxification or decomposition of toxins and xenobiotics	*Ralstonia sp.*; *Pseudomonas putida*

high affinity for substrates, low maintenance energy, the ability to uncouple growth from transport, and accumulation of storage polymers. In contrast, r-selected organisms have the potential of rapid proliferation and fast response to abundant and readily available substrates. A specific life strategy, designated L-strategy, is used to characterise organisms that are selected under unfavorable conditions and are highly tolerant or resistant to stress (Panikov 1999). This also includes micro-organisms with specific adaptation to manmade stress conditions. Thus, soils may be characterized as extreme which support the growth of micro-organisms that can tolerate anthropogenic disturbance and adverse conditions caused by high concentrations of pollutants and toxic compounds.

2.2 Physicochemical Factors Limiting to Life

2.2.1 *Water*

Soil water is either adsorbed onto surfaces or present as free water in pores or films between soil particles. The soil water status is described by the water potential, which is a measure of the energy and forces that hold and move water in the soil. It relates to water activity (a_w) and is the difference in free energy (in pascals or Pa; energy per unit mass) between pure water and soil water. The main components of soil water potential are the matric potential (the energy with which soil water binds to solid surfaces or is retained in pores) and the osmotic potential (a function of dissolved salt concentrations; Vetterlein and Jahn 2004). Because these components reduce the free energy of water, the water potential is negative. This indicates the energy that organisms must exert to withdraw water from soil, and as the water potential decreases, water becomes less available and the stress level of organisms increases.

Extreme soil water stress occurs periodically in most soils, even in climates with ample precipitation, where the water availability depends on soil composition, rainfall drainage, and plant cover. Soil prokaryotes live in water films surrounding particles or inside water-filled pores, and are therefore very susceptible to water depletion. Soil fungi are normally more tolerant to water stress than prokaryotes. Furthermore, they can tolerate drought due to hyphal growth, which allows them to cross dried pores and obtain water from smaller pores where the water remains for longer periods (Killham 1994). Prokaryotic cells have a turgor pressure, with a concentration of solutes inside the cell being slightly higher than that outside. Changes in water activity in the environment are rapidly followed by a water flux across the semipermeable cell membrane from high to low water potential. This can cause swelling and lysis of the cells under hypotonic conditions or dehydration under hypertonic conditions (Kempf and Bremer 1998). Therefore the cell must maintain an intracellular water potential similar to that existing outside the cells.

To maintain cell integrity at nonextreme temperatures (10–40°C) usually requires soil water potential above −4 MPa (0.95 a_w) for most bacteria, and above −22 MPa (0.86 a_w) for actinomycetes and most fungi. It is generally considered that the lower limit of water potential for life is −70 MPa (0.60 a_w; Zvyagintsev et al. 2005), but recently it has been demonstrated that spore germination and elongation of some actinomycetes can occur at water potential of −96 MPa (0.50 a_w; Doroshenko et al. 2005; Zvyagintsev et al. 2005). Some organisms, for example lichens, can even survive on water vapor rather than liquid water. Inasmuch as cell damage cannot be repaired during desiccation, these organisms must exhibit an efficient repair system upon rehydration.

Xerotolerant or xerophile micro-organisms are able to withstand water and salt stress because they can counterbalance a low water potential in the environment by accumulating highly soluble small molecules in the cytoplasm (Kempf and Bremer 1998). The solutes can be inorganic salts or organic molecules (amino acids, polyols, carbohydrates, quaternary ammonium compounds). The accumulation of solutes results in decreased internal water potential. These molecules can influence and modulate specific enzyme activities, but do not inhibit the overall metabolism of the cells, and are therefore termed compatible solutes or osmoprotectants (see Section 2.2.2). Some compatible solutes are constitutively produced, whereas others are induced. Osmoregulation is an energy-requiring process, but seems to be a general mechanism enabling soil micro-organisms to preserve the activity of intracellular enzymes under long-term and severe water stress.

Other strategies that protect prokaryotes from desiccation are the production of extracellular polysaccharides which retain water (Wright et al. 2005). Formation of microaggregates of cells where elevated water activities are retained may further protect micro-organisms from desiccation. Actinomycetes are particularly well osmoregulated, as their cell membranes have restricted permeability and keep salt ions out and organic solutes inside the cells. Like fungi, they can differentiate into dormant cells that are resistant to drying (Dose et al. 2001).

2.2.2 Salinity

Salt or osmotic stress is closely related to water stress because solutes strongly affect the water activity. In contrast to water stress which occurs frequently in most terrestrial habitats, high salinity typically occurs in restricted habitats. Soils with high salinity are often characterized by highly uneven temporal and spatial water distribution (Brown 1976). Such fluctuations cause special stress for the microbes and reduce their ability to survive, because they need to respond rapidly to desiccation and adapt to high salt concentrations (see Chapter 5).

Saline terrestrial habitats are typical for naturally arid regions with high evaporation rates. They may also be the result of pollution from mining activities or chemical and metallurgic industries. Most of the micro-organisms that inhabit saline soils are salt tolerant (halotolerant), but also halophilic micro-organisms that require salt for maintaining their membrane integrity and enzyme stability and activity are present. Extremely halophilic prokaryotes can tolerate very low water potential, and grow well at −40 Mpa (0.75a_w, the value of saturated NaCl solution). This limit is determined by the solubility of salt rather than by the physiology of the cells (Brown 1976). As with the xerophile and xerotolerant organisms, the halophiles and halotolerant organisms accumulate a variety of small molecules in the cytoplasm (osmolytes or compatible solutes; see Section 2.2.2) to counteract the external osmotic pressure (Kempf and Bremer 1998; Roberts 2005). Generally, bacteria produce and accumulate compatible organic solutes that are zwitterions (e.g., proline, glycine betaine, ectoine, methylamines, and derivatives) or nonionic (such as polyols, carbohydrates, neutral peptides, and amino acids and derivatives).

In archaea the osmolytes are often inorganic cations that are taken up by passive diffusion or by selective ion transport across the membrane. Many archaea have evolved negatively charged acid polypeptides that require cation counterions such as K^+ for proper protein folding and activity. Organic osmolytes that accumulate in archaea belong to the same types as for bacteria, but the majority of the solutes in archaea are anionic (due to negatively charged groups such as carboxyl, phosphate, and sulfate groups; Martin et al. 1999). In addition to their functions as osmotically active substances, the compatible solutes may function as chemical chaperones that protect proteins from denaturation and increase their activity.

2.2.3 Temperature

The temperature limits of life are related to the boiling and freezing points of water. However, many micro-organisms have developed mechanisms to extend the temperature ranges beyond the values for pure water and atmospheric pressure. At present, the temperature limitations for microbial activity are regarded as ranging from approximately −40°C to +130°C (Kashefi and Lovley 2003; Price and Sowers 2004). Arctic micro-organisms are well adapted to an extremely cold climate and several authors have reported microbial activities at temperatures as low as −10 to −20°C (Panikov and Dedysh 2000; Bakermans et al. 2003; Jakosky et al. 2003; Callaghan et al. 2004; Gilichinsky et al. 2005; see Chapter 7). At subzero temperatures there can still be liquid water present in soils, as adsorbed water forms a thin liquid film on the surface of soil particles (hygroscopic water; Rivkina et al. 2000; Steven et al. 2006).

Growth at low temperatures requires significant membrane alterations in order to maintain the fluidity necessary for nutrient transport across the membrane. The low temperature modifications involve less saturated and less branched membrane fatty acids. Below the minimum growth temperature the membrane becomes solid and transmembrane transportation stops. Life at subzero temperatures is also facilitated by accumulation of antifreeze compounds (high concentrations of salts, hydrocarbons, or amino acids) in the cytoplasm. Archaea have many of the same mechanisms for adaptations to low temperatures as bacteria; these involve altered membrane composition (cold-adapted lipids) as well as cold-active proteins involved in fundamental cell functions (e.g., protein synthesis; Cavicchioli et al. 2000).

In psychrophiles, the specific cold adaptation implies such drastic changes in the chemical composition of the cell that life outside of cold environments becomes impossible. For example, micro-organisms adapted to low temperatures have enzymes and ribosomes becoming unstable at temperatures 1–2°C above their optimum temperatures. Accordingly, the psychrophiles have optimum temperatures at or below 15°C and maximum temperatures below 20°C. The psychrotrophic organisms can also grow at temperatures close to or even below 0°C, but their optimum temperature is above 15°C, and their maximum temperature can be as high as 30 to 40°C. Price and Sowers (2004) studied the temperature dependence of metabolic rates in different environments including permafrost. These authors distinguished three categories of metabolic rates: first, rates sufficiently high to allow growth; second, intermediate rates sufficient for maintenance of functions, but too low for growth; and third, basal rates sufficient for survival of cells and repair of damaged macromolecules, but otherwise permitting only cell dormancy. They did not observe any minimum temperature for metabolism, but at low temperatures the metabolic rate was extremely low. At elevated temperatures, micro-organisms from permafrost showed metabolic rates similar to those found in temperate soils.

At the other extreme of the temperature range for life are the thermophiles and the thermotrophs. Thermophilic organisms cannot grow at temperatures below 50°C, whereas the thermotrophs have a lower temperature limit (20–30°C). At their upper temperature limit, cells undergo instability and irreversible denaturation of their proteins and nucleic acids, and therefore the ability of these molecules to perform their functions is lost. Thermal soils, with temperatures above 50°C, can be natural or manmade (see Chapters 8 and 14). Thermophilic micro-organisms have been isolated from natural thermal soils such as decomposing litter, volcanic, geothermal and tropical desert soils, and from manmade thermal soils such as compost piles and coal refuse piles (Botero et al. 2004).

Micro-organisms adapted to high temperatures have mechanisms for protecting their proteins and nucleic acids from irreversible denaturation. Biomolecules from such organisms are thermostable and remain active at temperatures that generally inactivate proteins, lipids, and nucleic acids in mesophilic organisms (Rothschild and Mancinelli 2001). In some proteins, the stabilization is caused by changes in amino acid residues that make the proteins more hydrophobic and increase the stability of subunit interactions (Singleton and Amelunxen 1973). The nucleic acids are also thermostabilized, for example as the result of interactions with histonelike proteins. At high temperatures, the membrane fatty acids acquire longer chains, and they become more saturated and more branched. Such changes in the membrane structure and composition lead to decreased membrane fluidity and consequently better thermostability (Pakchung et al. 2006).

2.2.4 pH

Taken as a group, prokaryotes can live in environments with pH values ranging from below 0 to 13 (Marion et al. 2003), although most prokaryotes grow at relatively narrow pH ranges close to neutrality. Extreme acidophiles grow in the pH range 0 to 3. It seems that a general adaptation to extreme pH is to regulate the intracellular pH and keep it close to neutral. The bacterium *Acidithiobacillus ferrooxidans* (previously *Thiobacillus ferrooxidans*) lives in acidic environments with a pH of about 1 (see Chapter 10). However, its intracellular pH is around 5.5, which indicates an active mechanism for excluding protons.

Whereas all known cytoplasmic enzymes have pH optima from pH 5 to pH 8, some enzymes found in the bacterial outer membrane tend to have low pH optima. Among archaea there are acidophiles that can grow at pH 0. Members of the cell wall-lacking archaeon *Ferroplasma* have been isolated from highly acidic environments associated with sulfide ores, solfatara fields, and the like (Golyshina and Timmis 2005). They can mobilize metals from sulfide ores and tolerate high concentrations of heavy metals (Edwards et al. 2000; Baker-Austin et al. 2005). The extremely thermoacidophilic archaea *Picrophilus torridus* and *P. oshimae* were first isolated from solfataric Japanese soils. They have optimal growth rates at pH 0.7 and 60°C (Schleper et al. 1995). In contrast to most other acidophilic micro-organisms, the intracellular pH is very low (pH 4.6), and *P. torridus* cannot grow above pH 4.0. A common feature of acidophile and acidotolerant micro-organisms is the presence of acidophilic lipids with cyclic rings and alkyl side chains in their fatty acids. Adaptation to acid environments is also enhanced by unusual tetraether lipids in the cell membrane (in *Ferroplasma* and *Picrophilus*) or lipopolysaccharides with a low content of fatty acids (in *Acidithiobacillus*). The tetraether lipids render the membrane of acidophilic archaea impermeable to protons, and a proton pump expels protons from the cytoplasm. High ratios of genes for proton-driven secondary transporters over ATP-consuming primary transporters indicate that the cells utilize the proton gradient for solute transport into the cell (Futterer et al. 2004).

Extreme alkaliphilic micro-organisms have pH optimum between 9 and 11, and do not grow near neutral pH. However, the cytoplasmic pH is at least two units lower than the external pH (Krulwich et al. 1998), which indicates that the cell membrane presents an efficient barrier to fluxes of OH^- ions and that there is an efficient inward proton translocation system (e.g., Na^+/H^+ or K^+/H^+ antiporters, proton-translocating ATP synthase). pH in bulk soil is generally between 4.0 and 8.5 (Lynch 1979). Decomposition of plant litter tends to reduce the pH, and soil with accumulated organic matter is normally acid. Redox reactions also influence the soil pH. For example, oxidation of NH_4^+, S, and FeS_2 results in production of mineral acids such as HNO_3 and H_2SO_4, and thereby decreases the pH. Thus, it is apparent that prokaryotes exert a profound influence on their own environment. As a result of metabolic activity, the pH in microbial microhabitats can be at least 3–4 pH units lower than in the bulk soil. High pH is common in soils containing alkaline minerals, or occurs temporally in restricted zones, for instance due to the presence of animal excretions.

Anaerobic reactions, such as the reduction of nitrate to N_2 and of sulfate to sulfide, also increase the pH. Recently alkaliphilic psychrotolerant bacteria were isolated from permafrost soil in the Qinghai-Tibet Plateau (Zhang et al. 2007). The colony-forming units of alkaliphilic bacteria in these soils varied between 10^2 and 10^5 cells g^{-1} of dry soil. The isolates could grow at pH 6.5–10.5 with optimum pH of 9.0–9.5, and optimum growth temperatures of 10–15°C.

2.2.5 Radiation

High doses of ionizing radiation and ultraviolet (UV) light are lethal to most microbes, although some of them can tolerate surprisingly high radiation doses (see Chapter 16). Normally high correlations are observed between tolerance towards radiation, desiccation, and DNA damaging chemicals (Shukla et al. 2007). *Deinococcus radiodurans* is the most radiation-resistant organism known and can survive doses of $1,000 J/m^2$ UV-light, and more than 20 kGy of γ-radiation (1 Gy = 100 Rad), which is approximately 4,000 times the dose that will kill a human (Battista 1997; Marion et al. 2003; Rainey et al. 2005). This micro-organism is remarkably well adapted to extreme conditions as it can survive drought and lack of nutrients, in addition to extremely high radiation dosages (Battista 1997). This red spherical bacterium was discovered in 1957, in a can of ground meat that was spoiled despite having been sterilized by radiation. The bacterium is widely distributed and has been found in a variety of soil environments as well as in granite in Antarctic dry valleys.

Dehydration and radiation cause very similar types of DNA damage. Resistance is conferred on *Deinococcus* by a particularly efficient system for DNA repair. Desiccation or high doses of radiation lead to massive double-strand breaks in DNA. *Deinococcus* has 4–10 copies of its chromosome (Battista 1997) and repairs the fragments by intrachromosomal recombination that reconstitute an intact chromosome in just a few hours (Minton and Daly 1995; Sale 2007). Prokaryotes and algae on surfaces of barren polar soil have adaptive strategies that allow them to avoid, or at least minimize UV injury. In this case, substances such as pigments and amino acids (e.g., melanoids, carotenoids, scytenomin, mycosporine-like amino acids) protect against the excessive light radiations and oxidative damage (Ehling-Schulz et al. 1997; Bowker et al. 2002; Wright et al. 2005).

2.2.6 Low Nutrients

Environments characterized by extreme physical or chemical conditions, such as desert and permafrost soils, are often poor in organic and inorganic nutrients. Most of the indigenous micro-organisms in such environments are probably oligotrophs, which means they are adapted to low nutrient supply rates. However, high numbers of oligotrophic prokaryotes may also be found in bogs and other soils with high amounts of organic matter. In such soils the major organic compounds are humic matter that is recalcitrant and not readily decomposable (Koch 2001; Fierer et al. 2007). Bacteria in the phylum Acidobacteria are generally regarded as oligotrophic. They are especially abundant in soils with low resource availability and their abundance decreases after amendment with a readily available carbon source (Fierer et al. 2007). Prokaryotes that are adapted to grow in oligotrophic environments are normally K-selected (Bernard et al. 2007). They have low growth rates, but very efficient uptake systems with low half-saturation constants (down to nM levels) for uptake of organic substrates. This adaptation often results in their inability to grow under high nutrient levels.

Other mechanisms for adaptation to low nutrient levels are the ability to use many different substrates simultaneously (Eichorst et al. 2007). In environments where nutrient supplies fluctuate, prokaryotes can store nutrients as intracellular polymers (e.g., polysaccharides, poly-β-hydroxybutyrate, polyphosphate). However, in constant oligotrophic environments, especially cold environments, the nutrient supply probably is too low to support any intracellular storage. It has been suggested that the organisms' affinity for substrates decreases at low temperature due to loss of membrane fluidity that impedes active transport, and that the minimum substrate concentration needed for growth therefore increases near the organisms' lower temperature limits (Wiebe et al. 1992). If liquid water is present, growth limitation by decreased temperature may be the result of reduced active uptake of nutrients, which eventually becomes so low that the cell's minimum maintenance requirements is no longer met (Nedwell 1999).

2.2.7 Pollution

Acid deposition has affected soil microbial communities and activities for some decades. This pollution is caused by acid precipitation, the result of nitrogen oxide (NO_x) and sulfur dioxide (SO_2) emitted into the atmosphere and oxidised to SO_4^{2-} and NO_3^-. Despite the effort to reduce the primary sources of acid input, the effect is still apparent in many regions. The effect of acid deposition on soil ecosystems depends on the concentration of SO_4^{2-} and NO_3^-, the amount of precipitation, and the buffering capacity of the soils (the cation exchange capacity through bases). Nitrogen and sulfur provided by acid rain may stimulate growth of some soil micro-organisms. On the other hand, even low-level but prolonged acid rain will result in soil acidification that may have adverse effects on soil bacteria, whereas the effect on fungi seems to be minor (Pennanen et al. 1998a; Bååth and Anderson 2003).

The effect of acid deposition can be direct or indirect. The lower pH and reduced concentrations of divalent cations (Ca^{2+}, Mg^{2+}) can lead to mobilization and increased bioavailability of heavy metals and other toxic compounds (Francis 1986). Acidification of soils may also reduce the solubility of organic matter and thereby reduce the substrate availability for microbes. Increased soil acidity does not seem to affect prokaryotic biomass to any extent, but rather to reduce prokaryotic growth rates and activity (Francis 1986; Pennanen et al. 1998b). Reduced activity of a number of soil enzymes, such as dehydrogenases, ureases, and phosphatases have been observed at significant pH reductions (Killham et al. 1983). The reduced microbial growth observed with increased acidity may indicate that more metabolic energy is used for maintenance rather than for biosynthesis of cell materials. It has been suggested that increased metabolic quotient (qCO_2, the ratio of basal respiration to microbial biomass) indicates a shift in energy use from growth to maintenance, and that increased energy demand is a sensitive indicator of physiological adaptation to environmental stress (Post and Beeby 1996; Liao and Xie 2007).

Soil can have naturally high concentrations of heavy metals as the result of weathering of parental material with high amounts of heavy metal minerals (e.g., mineral sulfides). Other sources are contaminations associated with mines and metal smelters, which have led to increased soil concentrations of heavy metals such as zinc, cadmium, copper, and lead. Sewage sludge may also contain heavy metals, and it has been demonstrated that long-term application of heavy metal containing sewage sludge to agricultural soils can have profound effects on the microbial diversity and community composition (Sandaa et al. 1999; Gans et al. 2005). The effect of heavy metal toxicity depends on soil abiotic factors such as organic matter and clay content, divalent cation concentrations (cation exchange capacity), and pH (Giller et al. 1998). These factors influence complex formation and immobilization of heavy metals.

Irrespectively of soil types, however, the relative toxicity of different metals seems to be the same, namely Cd > Cu > Zn > Pb (Bååth 1989). In soil contaminated for 40 years with high concentrations of Cr and Pb, the microbial biomass and activity was reduced and soil organic carbon accumulated (Shi et al. 2002). These results indicated that Pb presented greater stress to soil microbes than Cr. Soil micro-organisms vary widely in their tolerance to heavy metal contamination, and the proportion of culturable resistant micro-organisms can range from 10% to nearly 100%. The activity of enzymes in soil may serve as indicators for heavy metal contamination as there are generally high correlations between reduced enzyme activities (of, e.g., dehydrogenases, acid phosphatases, and ureases) and increased heavy metal contamination (Bååth 1989). It has been reported that heavy metal contamination has a different effect on soil bacteria and fungi (Rajapaksha et al. 2004). Metal addition decreased bacterial activity whereas it increased fungal activity, and the increased fungal activity was found to persist in contaminated as compared to control soil after 35 days. The different effect of heavy metals was also demonstrated by an increase in the relative fungal/bacterial ratio (estimated using phospholipids fatty acid analysis) with increased metal concentrations.

Mechanisms for metal resistance include stable complex binding (chelation) with organic ligands (extracellular or intracellular sequestering), transportation out of the cells, and biotransformation of the ions to less bioavailable or less toxic metal species. Genes for metal resistance (e.g., mercury resistance) are often harbored on plasmids and can easily be disseminated through a population or a community in response to selection pressure associated with toxic metal exposure (Drønen et al. 1998).

Hydrocarbon contamination of soils caused by human activities increasingly occurs in all parts of the world. Petroleum is a rich carbon source and most of the hydrocarbon components are biodegradable by micro-organisms. The rate of degradation is normally rather low, because crude oil has low concentrations of phosphorus and nitrogen, which do not allow extensive growth of indigenous hydrocarbon-degrading micro-organisms in petroleum-contaminated soils. However, growth can be stimulated by addition of phosphorus and nitrogen fertilizers. In many extreme environments, hydrocarbon-polluted areas are found (Margesin and Schinner 2001). The success of bioremediation in such environments depends on the presence of biodegrading microbes that are adapted to the prevailing environmental conditions.

Pesticides are classified according to their primary target organisms, that is, herbicides, fungicides, and insecticides (Johnsen et al. 2001). Normally the pesticides are very specific and restricted to a narrow range of target organisms. However, they can be modified in the environment and become toxic also to nontarget organisms. For instance, triazines, which normally target photosynthetic enzymes in C3 plants, may be chlorinated in the triazine ring and thus become toxic to a wide range of organisms. The effect of pesticides on soil microbes depends on their bioavailability, which in turn is influenced by the crop being grown, as well as soil properties affecting the sorption and leaching of pesticides. The micro-organisms can develop resistance to the pesticides through their ability to decompose or transform them to less toxic compounds.

2.3 Soil as Habitat for Micro-Organisms

Soil has been defined as the upper weathered layer of the earth's crust, with a complex mixture of particulate materials derived from abiotic parent minerals, living biota, and particulate organic detritus and humic substances (Odum 1971). Formation of soil is the result of climate (temperature and moisture), parental material, time, topography, and organisms (Jenny 1994), and involves complex interactions of physical, chemical, and biological processes. Soil texture (the relative proportion of particles with different sizes) and mineral constituents depend on the parent material (rocks), and transportation by water, ice, and wind. Soil structure is the distribution of pores of various sizes that occur between soil particles. The pore sizes depend on the level of aggregation of soil particulate material, and the pores contain gases and water.

The vegetation and soil biota affect soil development by weathering and controlling organic matter accumulation and mineralization. The recognition of close interactions between soils and vegetation is reflected in the division of soils into major types, which are associated with climatic vegetation zones. Micro-organisms are able to modify and shape their physical and chemical environment. They dissolve and alter minerals derived from the parental material, contribute to and mineralize soil organic matter, and recycle nutrients. Microbes produce biopolymers (polysaccharides) as cell envelopes. Such polymers facilitate formation and stabilization of soil aggregates, and thereby improve the soil water-holding capacity. Together with colloid clay particles and humus, the polymers create complex structures with extensive surfaces, which adsorb minerals and organic molecules. Adsorption of proteins and nucleic acids to surfaces protects them from biodegradation and denaturation. Adsorbed DNA remains available for horizontal gene transfer by transformation of competent cells (Lorenz and Wackernagel 1994). The activity of extracellular enzymes is maintained or even increased by adsorption on minerals, whereas adsorption to humic substances can either maintain or decrease their activity (Nannipieri et al. 1990; Allison 2006). The adsorption to soil colloids may strongly reduce the availability of organic molecules as nutrients for micro-organisms, and contribute to soils being oligotrophic environments.

Surfaces of soil minerals, especially clay colloids, can serve as catalyst for abiotic chemical reactions. Clay particles are coated with metal oxides and hydroxides and have net electronegative charges. They can mediate electron transfer reactions and catalyse oxidation of phenols and polyphenols. They also contribute to humus formation by catalysing reactions such as deamination, polymerization, and condensation of organic molecules. It has been suggested that microbial processes such as decomposition and mineralization of organic substances prevail under moderate conditions, whereas abiotic reactions are more dominant under harsh conditions where microbial activities are hampered (Huang 1990; Ruggiero et al. 1996).

Soils are among the habitats that have been shown to support the highest abundance and diversity of micro-organisms. Soil habitats are distinguished from aquatic habitats by being much more complex and spatially heterogeneous. A characteristic feature is the wide range of steep physicochemical gradients (e.g., of substrate concentrations, redox potential, pH, available water) which may occur across short distances approaching the size of a soil aggregate. Thus, even an aggregate of a few mm can offer many different microenvironments that would collectively be colonized by different types of micro-organisms (Standing and Killham 2006). The size scale of microhabitats is typically a few µm for unicellular prokaryotes, but may be much larger for filamentous actinomycetes and fungi. Microhabitats for prokaryotes exist either within or between aggregates. Intra-aggregate habitats have typically small pores that are often water-filled and anaerobic, whereas inter-aggregate habitats are more frequently aerobic. However, the living conditions in these habitats can undergo considerable changes both in space and time, therefore soils are highly dynamic systems.

The distribution, activity, and interactions (e.g., predation) of soil biota depend on soil properties such as texture, structure, and available nutrients and water. The growth conditions are normally most favorable on surfaces, and most (80–90%) of the soil micro-organisms are attached to surfaces (Hattori et al. 1997), often aided by extracellular biopolymers which stick to particles. However, surfaces also expose micro-organisms to the highest risks of desiccation and predation. Specific soil habitats such as organic litter aggregates, biofilms, rhizosphere, and animal droppings, are rich in readily available organic nutrients and can support very high microbial activities. The bulk soil on the other hand often contains low levels of easily decomposable substrates, and most of the organic matter is refractory. Thus the distribution of biomass and activities of soil microbes is generally very patchy, and the space that is occupied by micro-organisms may be less than 5% of the overall space in soil (Nannipieri et al. 2003).

Moderate soils with no stress factors are characterized by a high microbial abundance (10^9-10^{10} prokaryotes g^{-1} soil dry weight) and high genetic, phylogenetic, and functional diversity of microbial communities (Giller et al. 1997; Torsvik and Øvreås 2006). Micro-organisms are by far the most active and functionally diverse component of the soil biota. It has been estimated that 80–90% of the soil processes are mediated by the microbiota, including prokaryotes and fungi. Generally, about one third of the organic carbon added to temperate soils is transformed to humus and microbial biomass, whereas about two thirds of the carbon is respired to CO_2 by micro-organisms (Stotzky 1997).

Interestingly, the deep subsurface terrestrial environments, which can extend for several hundred meters below the soil surface, have been proven to sustain ample microbial biomasses. Although the cell numbers are much lower than in the surface soil, a variety of micro-organisms, primarily prokaryotes, is present in deep subsurface soils. For example, in samples collected aseptically from bore holes drilled down to 300m, a diverse array of micro-organisms has been found. These organisms most likely have access to organic nutrients present in the groundwater percolating down the subsurface material and flowing through their habitat. Studies on the microbial ecology of deep basalt aquifers have shown that both chemoorganotrophic and chemolithotrophic prokaryotes are present, but that the chemolithotrophs are dominating in these environments (Stevens and McKinley 1995).

2.4 Extreme Soils

Extreme soil microbiology deals with micro-organisms adapted to extreme or stressful soil conditions. Soil properties are determined by the parental material (geological properties), climate, and biota, and are influenced by anthropogenic activities. Odum (1971) divided soils in two categories, those which are mainly controlled by climate and vegetation types, and those which are mainly controlled by parent materials or other pedological or environmental factors (topography, drainage, pollution, etc.). These controlling factors can strengthen each other. In regions with extreme climate, minor differences in edaphic factors may create large differences in the structure and activity of soil microbiota.

In many temperate climate zones, soil water and/or temperature stress occurs periodically. Also nutrient stress occurs periodically in many soils, and this will influence the soil microbiota so that organisms adapted to periodic stress become dominant. In some areas, wide fluctuations in environmental conditions occur. It has been reported that in Antarctic desert soil the temperature could change from −15°C to nearly 30°C in three hours (Cowan and Tow 2004). For the micro-organisms to survive freeze–thaw cycles and sudden differences of more than 40°C, very specialized adaptations are required. Thus, in soil where a stress situation is maintained over an extended time period, the microbiota will develop specialized adaptations and life strategies that differentiate them from microbiota in nonextreme environments. However, such specialization may correspond to a tradeoff between life under adverse and harsh situations and loss of adaptability. Indeed, extremophiles are often not able to adapt to less extreme conditions, and do not compete effectively under moderate conditions.

Soil microbial communities under nonextreme and relatively stable environmental conditions are characterized as functionally redundant. Moderate environmental stress and perturbations seem to have little impact on overall soil processes such as respiration and mineralisation, although the microbial community structure can be profoundly changed. This is explained by the insurance hypothesis (Yachi and Loreau 1999), which states that in an ecosystem there are many different populations which can perform the same function, so that when some micro-organisms disappear others proliferate and take over the function (Giller et al. 1997). Microbial communities in extreme environments, especially those with fluctuating conditions, often comprise some numerically dominant species. In such environments, ecological processes may be more sensitive to changes in diversity imposed by additional stress factors. Lack of functional redundancy in extreme soils is illustrated by the observation that, in these environments, microbiota plays an increasing role at all trophic levels. For example, cyanobacteria and microalgae contribute significantly to primary production when the conditions become so harsh that higher plants can no longer grow.

Two types of extreme soils are dominant on Earth, namely desert and tundra soils. Typical of these biomes is that vegetation is sparse and consists mainly of low vegetation, with no trees being present.

2.4.1 Desert Soils

Water is an overall limiting factor in terrestrial ecosystems, and within a specific climate zone the annual net primary production correlates well with annual precipitation. Deserts occur in regions having less than 250 mm of rainfall per year (Odum 1971). Arid areas, that can be either extremely hot or cold, cover more than 30% of Earth's terrestrial surface (Rainey et al. 2005). Dry soil ecosystems are characterized by spatial patterns and high spatial heterogeneity. In temperate deserts, spatial variability is strongly influenced by vegetation, whereas in polar deserts, which often lack vascular plants, physical processes control the spatial variation in soil properties. An example is the formation of frost fissure patterns.

The micro-organisms living in such environments have to deal with unfavorable life conditions such as absence of water, high or low temperatures, and lack of nutrients. The most extreme deserts are found in Antarctica (Ross Desert, Dry Valleys), in northern Chile (Atacama Desert ; see Chapter 6), and in central Sahara where there is virtually no rainfall. The low precipitation is caused by high subtropical atmospheric pressure (Sahara), position in rain shadow areas (Chile), high altitude (Tibet and Gobi), or latitude (Antarctic Dry Valley). In hot deserts, most of the matric water evaporates during the day and micro-organisms obtain moisture by absorbing dew water during cool nights. The cold deserts in Antarctica suffer from extreme temperature in addition to extreme water stress, although in some areas and during restricted periods water can come from melting snow. The air temperature is −10°C on average during the summer and down to −55°C during winter, and there are often strong winds, which are responsible for high sublimation rates.

In the past, doubts were raised if any organism could proliferate under such climatic conditions, but micro-organisms have been isolated from even the harshest desert environments. In deserts, the microbiota often inhabit pores in sandstones or they tend to form biological soil crusts. In the Ross Desert, an Antarctic cold desert, cryptoendolithic micro-organisms grow in the near-surface layer of porous sandstone rocks, where the microclimate is less hostile. They transform and mobilize iron compounds, and depend on the unsteady interactions between biological and environmental factors for survival. If the balance between these factors changes and becomes unfavorable, they will die but leave behind trace fossils and a characteristic iron-leaching pattern caused by their activity (Friedmann and Weed 1987). In the most extreme cold deserts, conditions suitable for microbial metabolism may occur only 2–10 days per year.

The crust communities are composed of prokaryotes, fungi, microalgae, and lichens. They are extremely important in desert ecosystems as they form stable soil aggregates with increased water retention responsible for functions such as primary production, nitrogen fixation, and nutrient cycling. Lichens are especially well adapted to extreme conditions, as they can withstand desiccation for long periods. Under cold conditions, the lichen algae keep the water in their cytoplasm in liquid form by producing and accumulating polyol intracellular solutes. The desiccation tolerance characteristic for desert micro-organisms is often correlated with salinity tolerance, extreme oligotrophy, and radiotolerance. Chanal et al. (2006) analysed microbial diversity in the Tataouine sand dunes in south Tunisia. The climate at this site is arid with a high seasonal variation in precipitation. The mean annual rainfall is 115 mm, and there is almost no precipitation in summer. Despite these unfavorable conditions, an unexpectedly high diversity of micro-organisms was revealed. The community contained a broad spectrum of micro-organisms, with 16S rRNA sequences affiliated with 11 bacterial divisions and some archaeal lineages. After irradiation of this soil with 15 kGy, radiotolerant organisms affiliated with *Bacillus, Deinococcus/ Thermus*, and the Alphaproteobacteria could be isolated. In fact, many of the environments from which radioresistant organisms have been isolated are extremely dry, and many of these isolates are also desiccation-resistant (Rainey et al. 2005; Chanal et al. 2006).

In deserts with the most extreme dry habitats, micro-organisms can be totally dried out, and in arctic deserts they may actually be freeze-dried. The best strategy for micro-organisms to survive in such extreme environments may be to completely abolish their metabolism during the most unfavorable time period, and switch into a dormant state until the conditions improve. Therefore many of them have resting stages or spores (Barak et al. 2005).

2.4.2 Tundra Soils

Climate has an overriding effect on species diversity on a global scale and biodiversity generally decreases with increased latitudes and altitudes. For eukaryotes, this trend is seen in polar regions both in terms of number of species and growth forms. However, on smaller spatial scales, biodiversity may not be any lower in Arctic tundra than in temperate soils. Permafrost represents approximately 26% of terrestrial soil ecosystems and can extend hundreds or even thousands of meters down into the subsurface (Steven et al. 2006). The permafrost environment is considered extreme because indigenous micro-organisms must survive long-term exposure to subzero temperatures and withstand background radiation. Low temperature and a short growing season (approximately 60 days) characterize extreme tundra and high altitude fell-field soils. Here such soils are considered and described together as extreme environments.

On a global scale, most of the tundra consists of Arctic wetlands covered by vegetation. However, the most extreme high Arctic tundra offers bare soil without vegetation except for sparse areas of lichens, sedges, and grasses. In the high Arctic we find permafrost soils, where the ground is permanently frozen except for a few dm of active layer during the growth season (see Chapters 7 and 12). Characteristic in permafrost soils are the ice-wedge polygon structures. These are topographic features formed by a network of ice-wedges, with either a depressed central area caused by thawing of the ice-rich permafrost in the centre, or a relatively elevated central area due to melting of the surrounding ice-wedges (see Chapter 7). Alpine fell-field tundra occurs in high mountains in temperate zones, and such tundra soils do not have permafrost.

In some areas such as the Antarctic deserts, several harsh environmental factors interact, such as low temperature, low annual precipitation, and strong desiccating winds. The Antarctic Dry Valleys are regarded as the coldest and driest place on Earth. The precipitation is only a few millimetres a year and occurs mainly as snow. As most of this snow is blown away, the potential evaporation exceeds precipitation. These regions are further characterised by a long period of winter darkness and low temperatures, followed by a very short summer with 24 h daily light for a few weeks, and even then the temperature rarely exceeds 0°C. Organisms in these environments must therefore tolerate long periods of desiccation and dormancy, and a common opinion has been that the microbial biomass is very low in these soils (Horowitz et al. 1972; Virginia and Wall 1999; Smith et al. 2006). Recent investigations suggest that the biomass is several orders of magnitude higher than previously recognized (Cowan et al. 2002).

In polar tundra areas there may also be profound microclimate differences. One factor which exerts a major influence on soil temperature is the snow depth. In Siberia and Alaska, it has been observed that, whereas exposed areas with low snow cover had soil temperatures down to −30 to −40°C, soils under the snow cover had temperatures around −5 to −10°C. During the active summer season the diurnal temperature fluctuations in the upper soil layers can vary considerably over short time periods. The amplitude of such fluctuations is influenced by the soil water content and vegetation cover, but in dry barren mineral soils temperatures can vary by more than 20°C, sometimes by nearly 40°C, with minima below 0°C. As a result of such abrupt temperature changes, freeze–thaw cycles occur that can be lethal to soil organisms. The organisms have therefore developed mechanisms that allow them to survive repeated freeze–thaw cycles.

Survival of adapted microbes depends on their hydration state, their compatible solute content, and their ability to switch metabolism to cryoprotectant synthesis. In some arid mineral soils, the micro-organisms are also subjected to osmotic stress due to accumulated salts. However, the presence of salt may result in water remaining liquid in cold environments, and active microbes can exist in thin films of liquid water present in permafrost or in permafrost brine lenses, called cryopegs, at below freezing temperatures (Gilichinsky et al. 2003). Cryopegs are layers of unfrozen ground that are perennially cryotic (forming part of the permafrost), but in which freezing is prevented by freezing-point depression due to high concentrations of dissolved substances in the pore water. An unfrozen cryopeg is entirely surrounded by frozen ground (Gilichinsky et al. 2005). Such habitats allow for microbial growth at −10°C and metabolic activity at −20°C and even lower (Bakermans et al. 2003).

2.5 Microbial Diversity and Community Structure in Extreme Soils

The term microbial diversity describes different aspects of complexity and variability within microbial populations and communities. This comprises genetic variability within taxons (species), variability in community composition, complexity of interactions, trophic levels, and number of guilds, this latter parameter defining the functional diversity. Diversity is expressed in different ways: as inventories of taxonomic groups or as single numbers (diversity indices), which are based on the number of taxons or OTUs (operational taxonomic units). Diversity may also be represented as phylogenetic trees, or appreciated from the number of functional guilds.

In moderate and stable environments, soil microbial communities will normally develop into complex systems with high phylogenetic and functional diversities. Therefore, such communities are among the most difficult to characterize phenotypically and genetically. In addition, huge and coherent discrepancies between the total and cultivable cell numbers in natural environments has led to the introduction of "the great plate count anomaly" concept (Staley and Konopka 1985). This means that diversity measurements based on cultured micro-organisms are restricted to a subset of 1% or less of the community members (Torsvik et al. 1990; Ward et al. 1990) and applying culture-dependent methods will only reveal information about the very small fraction of micro-organisms able to grow under the given conditions (Sørheim et al. 1989).

Molecular methods and direct in situ studies circumvent the selective and biased culturing step and allow both cultured and noncultured members of a community to be surveyed (Pace et al. 1986; Torsvik et al. 1998). Some molecular methods allow for an in situ detection of prokaryotes in more or less intact soil samples whereas other methods require effective separation of cells from soil particles. Analysis of total DNA extracted directly from a community generates information derived from all the community members, and provides estimates of the microbial diversity and a comprehensive picture of soil microbial community composition (Fig. 2.1). The information contained in nucleic acids can be used to address diversity at different levels from the entire microbial community and populations to within species levels. A schematic overview of various methods used to obtain such information is given in Fig. 2.2 and in Øvreås (2000).

The total genetic diversity can be estimated by measuring the reassociation rate of community DNA (Britten and Kohne 1968; Torsvik et al. 1990), which is a low-resolution method that allows analyses of broad-scale differences in microbial communities (Torsvik and Øvreås 2006).

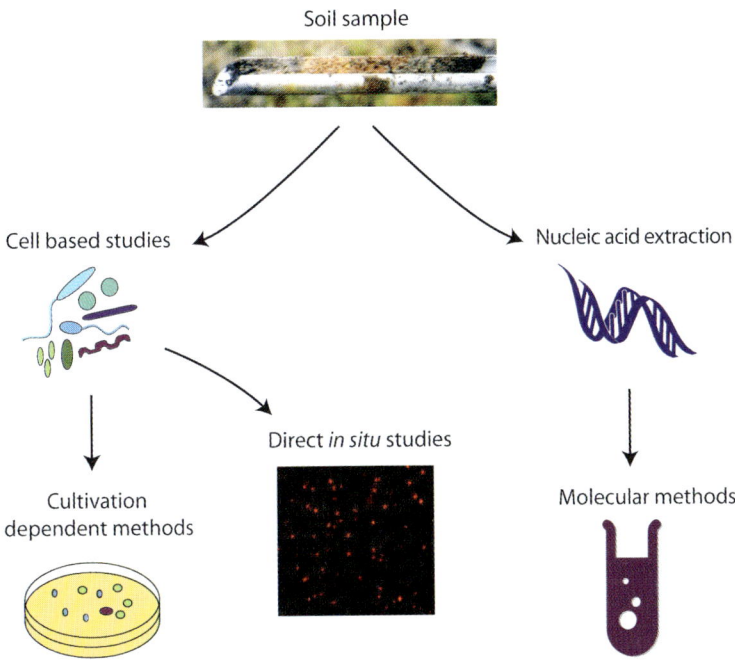

Fig. 2.1 Schematic drawing showing the overall approaches available for measuring bacterial diversity in soil

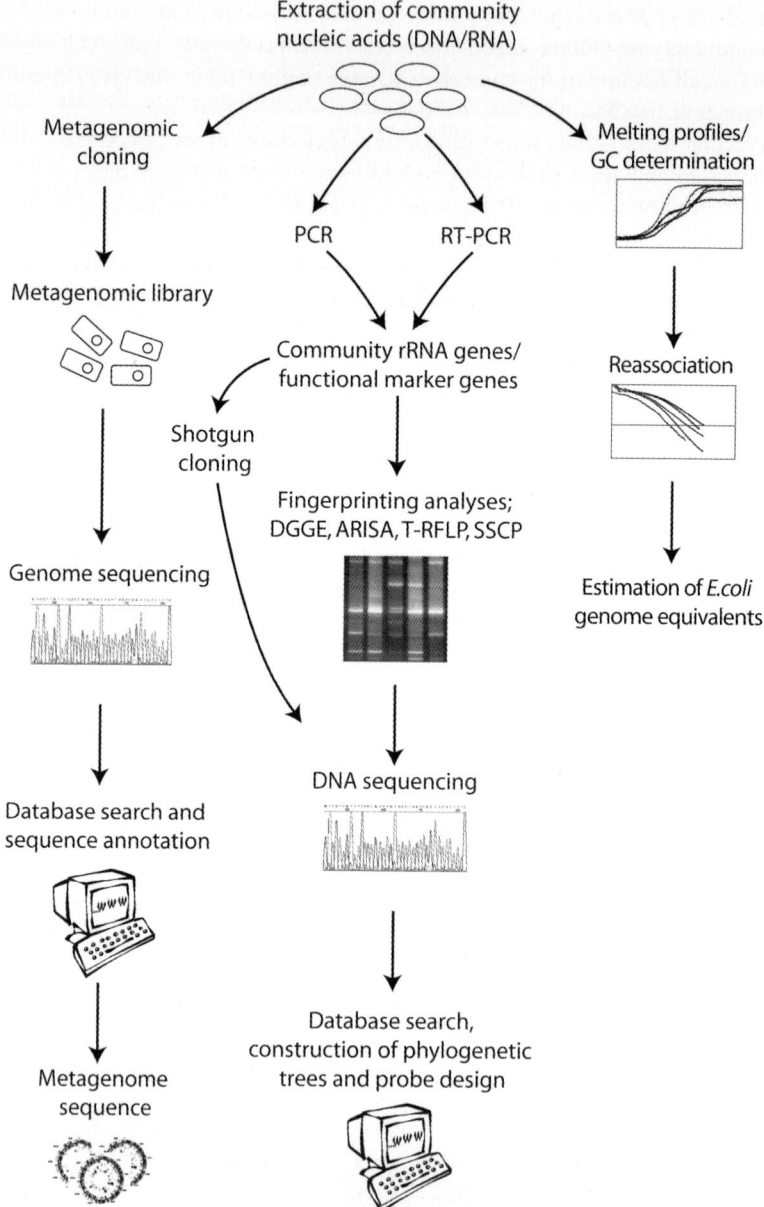

Fig. 2.2 Schematic drawing of the basic principles of some molecular methods

The metagenome can be regarded as the sum of all the microbial genomes in a given sample. Therefore, the metagenome approach represents a different whole community DNA-based analysis. It circumvents the cultivation anomaly as well as the PCR biases by cloning and sequencing genes directly from the environmental

DNA. This method involves construction of complex community libraries by direct cloning of large genomic DNA fragments (40–80 kb) from environmental samples into fosmid or BAC (bacterial artificial chromosome) vectors. The challenge of this application lies in the ability to extract DNA of high molecular weight and high purity. The metagenome approach can be used to generate information on the potential functioning of individual microbial species in soil environments in order to study the broader role of micro-organisms in the ecosystem (Rondon et al. 2000; Tringe et al. 2005).

The most common approach for assessing microbial diversity is to use polymerase chain reaction (PCR) to amplify 16S rRNA genes (rDNA) from the community (Pace et al. 1986). The amplified genes can then be cloned, and clones can be identified by DNA sequencing, which are then amenable to comparative analyses. Cloning and sequencing approaches are time-consuming and labor-intensive for routine analysis of large sample sets. To screen for changes in time and space, community fingerprinting techniques such as denaturing gradient gel electrophoresis (DGGE), terminal restriction fragment length polymorphism (T-RFLP), automated ribosomal intergenic spacer region analysis (ARISA), and single strand conformational polymorphism analysis (SSCP) of PCR amplified 16S rRNA genes are frequently used (Muyzer et al. 1993; Macnaughton et al. 1999a,b; Kozdroj and van Elsas 2001; Kuske et al. 2002; Hoj et al. 2005, 2006; Nakatsu et al. 2005; Neufeld and Mohn 2005; Becker et al. 2006; Joynt et al. 2006; Loisel et al. 2006; Hewson et al. 2007).

All these methods provide information about the numerically dominant community members, and the motivation for choosing one particular method instead of another lies in the expertise and equipment available in various laboratories. Furthermore, the phylogenetic affiliation of the numerically dominant organisms can be assessed by subsequent sequence analysis of, for instance, DGGE-separated PCR products. Such rDNA-based fingerprinting and cloning approaches offer higher resolution than DNA reassociation analysis, and have led to the discovery of several new prokaryotic taxa, even some entirely new divisions (Hugenholtz et al. 1998, 2001). Many studies have showed that most of the 16S rRNA sequences obtained from a given soil sample are unique (Hugenholtz et al. 2001; Smith et al. 2006). Due to the complexity of soil communities and the efforts required for the cloning and sequencing-based approach, only a limited number of soil environments have been surveyed using such methods, and our understanding of the extent of microbial diversity in soils is still very limited. More detailed descriptions of molecular methods for analysing microbial communities are provided in recent review papers (Johnsen et al. 2001; Prosser 2002; Lynch et al. 2004; van Elsas et al. 2006).

Regardless of their limitations, the culture-independent molecular methods have greatly expanded our view of extreme soils as habitats for micro-organisms. Several investigations have addressed the effect of extreme conditions created by human activities and pollution on microbial communities. Microbial communities in soils treated for many years with heavy metal-contaminated sewage sludge were investigated with respect to diversity and composition (Sandaa et al. 1999).

The control soil was amended with "uncontaminated" sewage sludge, whereas the contaminated soils received sewage sludge with two different levels of heavy metal concentrations (resulting in low and high levels of metal contamination). The total genetic diversity in microbial communities in unpolluted soil was high. In this case, the complexity of the community genome corresponded to approximately 9,800 different bacterial genomes with an average *E. coli* genome size. The diversity of the metal-polluted soils was reduced and depended on the level of pollution. The complexity of DNA isolated from the soils with low and high levels of metal pollution corresponded to a diversity of approximately 4,600 and 1,500 *E. coli* genomes, respectively. Thus, it seems that the genetic diversity can be an indicator of environmental stress caused by pollution.

It was further observed that environmental stress induces profound changes in the community structure. Pollution and perturbation lead both to reduced species richness and evenness, as some species become numerically dominant. It can be concluded that, under extreme stress conditions or strong pollution, microbial diversity may be reduced, and microbial community structure changed to the extent that functioning of this community is altered (Giller et al. 1997; Griffiths et al. 2004).

Both DNA reassociation and clone library analysis suggest that the overall prokaryotic diversity in pristine Arctic tundra soils can be very high (see Chapter 7), even higher than in soils from temperate regions (Øvreås et al. 2004; Neufeld and Mohn 2005). The prokaryotic communities in tundra soils have representatives of the same phylogenetic divisions as found in soils at lower latitudes (Cowan et al. 2002; Smith et al. 2006). They also carry out the same microbial processes as in temperate soils although at slower rates.

The increased human activities in polar regions depend on petroleum hydrocarbon for power generation and transportation. As a consequence of this and the exploitation of oil field reservoirs in the Arctic, increased oil pollution has become a significant problem in these cold environments (Aislabie et al. 2006; see Chapter 12). Several studies of hydrocarbon-contaminated polar soils indicate that hydrocarbon degraders are widely distributed in polar soils and that oil spill will result in marked enrichment of these micro-organisms (Atlas 1986; Aislabie et al. 2004, 2006; Sunde 2005). In comparative studies of pristine and oil-polluted Arctic and Antarctic tundra, significant shifts in the microbial diversity and community composition have been observed as a result of oil contamination. Disturbed tundra soil had lower microbial diversity than pristine soils, and in the polluted soils some populations were very predominant (Neufeld and Mohn 2005; Saul et al. 2005; Smith et al. 2006).

An investigation from Arctic tundra at Svalbard, Norway, showed that the proportion of clones with sequence similarities to cultured bacteria was much higher in polluted (36%) than in pristine soil (6%). Even then, the phylogenetic groups that were most abundant in pristine tundra soil have so far only been represented in the databases by a limited number of sequences from cultured organisms (Yndestad 2004). These, as well as clone library data from Antarctic pristine tundra soils, indicate that most of the sequences are derived from unknown and uncultured micro-organisms, and may represent new and undescribed taxa. Enrichment cultures with three different oil types incubated at 4°C demonstrated that different oils promoted the establishment of different communities. In these experiments, the efficient oil-degrading organisms showed phylogenetic affiliation to well-known hydrocarbon-degrading organisms within the Proteobacteria and the Gram positive bacteria (Sunde 2005). In oil-contaminated soil, biodegradation of petroleum hydrocarbons by indigenous cold-adapted microbial populations at low temperatures has been observed (Whyte et al. 1999, 2001; Rike et al. 2003; Sunde 2005), but the in situ rates of degradation were low. Therefore, the activity of the indigenous hydrocarbon-degrading microbes is limited in cold soil, most likely by a combination of unfavorable conditions including low temperature and moisture, nutrient limitation, alkalinity, and potentially inhibitory hydrocarbons.

2.6 Conclusions: The Significance of Studying Extremophiles in Soil

Extreme soils have highly selective physicochemical properties and many of them have low microbial diversity relative to nonextreme soils. Therefore, they can serve as model systems for exploring fundamental ecological principles such as the relationships between diversity and activity of micro-organisms and soil environmental conditions (Smith et al. 2006). Furthermore, studies of microbial community composition and functions in extreme soils may be of great value for applications in environmental cleanup, pollution prevention, or energy production.

Improved knowledge about extreme ecosystems will lead to important advances in the understanding of microbial adaptation mechanisms, and facilitate the design of biotechnological applications for enzymes and other compounds adapted to function under extreme physicochemical conditions. In addition, extreme prokaryotes may be interesting for bioprospecting, as they can be expected to contain a number of bioactive compounds potentially useful in medicine as well as in the pharmaceutical and environmental industries.

Defining the limiting conditions for life on our planet can aid us in speculation on comparable limits in the universe. Extreme environments on Earth may resemble those that exist on other planets and moons. Thus, investigation of the most challenging environments on Earth, can give us some clues as to under which conditions we can expect to find life on other planets (see Chapters 6, 7, and 10). It can also provide some hints of what to search for when looking for signs of extraterrestrial life (see Chapter 11). The study of microbes in extreme soils is therefore highly relevant for astrobiology. It will advance our understanding, at the molecular and physiological levels, of specializations and adaptations required for the maintenance and proliferation of remote and as yet unrecognized forms of life.

Acknowledgements We thank Beate Helle for technical assistance with the figures.

References

Aislabie J, Saul DJ, Foght JM (2006) Bioremediation of hydrocarbon-contaminated polar soils. *Extremophiles* 10:171–179

Aislabie JM, Balks MR, Foght JM, Waterhouse EJ (2004) Hydrocarbon spills on Antarctic soils: Effects and management. *Environ Sci Technol* 38:1265–1274

Allison SD (2006) Soil minerals and humic acids alter enzyme stability: Implications for ecosystem processes. *Biogeochem* 81:361–373

Atlas RM (1986) Fate of petroleum pollutants in Arctic ecosystems. *Water Sci Tech* 18:59–67

Baker-Austin C, Dopson M, Wexler M, Sawers RG, Bond PL (2005) Molecular insight into extreme copper resistance in the extremophilic archaeon 'derromicsk acidiphilum' Fer1. *Microbiol* 151:2637–2646

Bakermans C, Tsapin AI, Souza-Egipsy V, Gilichinsky DA, Nealson KH (2003) Reproduction and metabolism at −10 degrees C of bacteria isolated from Siberian permafrost. *Environ Microbiol* 5:321–326

Barak I, Ricca E, Cutting SM (2005) From fundamental studies of sporulation to applied spore research. *Mol Microbiol* 55:330–338

Battista JR (1997) Against all odds: The survival strategies of *Deinococcus radiodurans*. *Ann Rev Microbiol* 51:203–224

Becker JM, Parkin T, Nakatsu CH, Wilbur JD, Konopka A (2006) Bacterial activity, community structure, and centimeter-scale spatial heterogeneity in contaminated soil. *Microb Ecol* 51:220–231

Bernard L, Mougel C, Maron P-A, Nowak V, Leveque J, Henault C, Haichar FeZ, Berge O, Marol C, Balesdent J, Gibiat F, Lemanceau P, Ranjard L (2007) Dynamics and identification of soil microbial populations actively assimilating carbon from ^{13}C-labelled wheat residue as estimated by DNA- and RNA-SIP techniques. *Environ Microbiol* 9:752–764

Botero LM, Brown KB, Brumefield S, Burr M, Castenholz RW, Young M, McDermott TR (2004) *Thermobaculum terrenum* gen. nov., sp. nov.: A non-phototrophic gram-positive thermophile representing an environmental clone group related to the Chloroflexi (green non-sulfur bacteria) and Thermomicrobia. *Arch Microbiol* 181:269–277

Bowker MA, Reed SC, Belnap J, Phillips SL (2002) Temporal variation in community composition, pigmentation, and fv/fm of desert Cyanobacterial soil crusts. *Microb Ecol* 43:13–25

Britten RJ, Kohne DE (1968) Repeated sequences in DNA. *Science* 161:529–540

Brown AD (1976) Microbial water stress. *Bact Rev* 40:803–846

Bååth E (1989) Effects of heavy metals in soil on microbial processes and populations (a review). *Water Air Soil Poll* 47:335–379

Bååth E, Anderson TH (2003) Comparison of soil fungal/bacterial ratios in a pH gradient using physiological and PLFA-based techniques. *Soil Biol Biochem* 35:955–963

Callaghan TV, Björn LO, Chernov Y, Chapin T, Christensen TR, Huntley B, Ims RA, Johansson M, Jolly D, Jonasson S, Matveyeva N, Panikov N, Oechel W, Shaver G, Elster J, Henttonen H, Laine K, Taulavuori K, Taulavuori E, Zöckler C (2004) Climate change and UV-B impacts on arctic tundra and polar desert ecosystems; Biodiversity, distributions and adaptations of arctic species in the context of environmental change. *Ambio* 33:404–417

Cavicchioli R, Thomas T, Curmi PMG (2000) Cold stress response in Archaea. *Extremophiles* 4:321–331

Chanal A, Chapon V, Benzerara K, Barakat M, Christen R, Achouak W, Barras F, Heulin T (2006) The desert of Tataouine: An extreme environment that hosts a wide diversity of microorganisms and radiotolerant bacteria. *Environ Microbiol* 8:514–525

Cowan D, Russell N, Mamais A, Sheppard D (2002) Antarctic Dry Valley mineral soils contain unexpectedly high levels of microbial biomass. *Extremophiles* 6:431–436

Cowan DA, Tow LA (2004) Endangered Antarctic environments. *Ann Rev Microbiol* 58:649–690

Doroshenko EA, Zenova GM, Zvyagintsev DG, Sudnitsyn II (2005) Spore germination and mycelial growth of streptomycetes at different humidity levels. *Microbiol* 74:690–694

Dose K, Bieger-Dose A, Ernst B, Feister U, Gomez-Silva B, Klein A, Risi S, Stridde C (2001) Survival of microorganisms under the extreme conditions of the Atacama desert. *Ori Life Evol Biosph* 31:287–303

Drønen AK, Torsvik V, Goksøyr J, Top EM (1998) Effect of mercury addition on plasmid incidence and gene mobilizing capacity in bulk soil. *FEMS Microbiol Ecol* 27:381–394

Edwards KJ, Bond PL, Gihring TM, Banfield JF (2000) An archaeal iron-oxidizing extreme acidophile important in acid mine drainage. *Science* 287:1796–1799

Ehling-Schulz M, Bilger W, Scherer S (1997) UV-B-induced synthesis of photoprotective pigments and extracellular polysaccharides in the terrestrial cyanobacterium *Nostoc commune*. *J Bacteriol* 179:1940–1945

Eichorst SA, Breznak JA, Schmidt TM (2007) Isolation and characterization of bacteria from soil that define *Terriglobus gen. nov.*, in the phylum *Acidobacteria*. *Appl Environ Microbiol* 73:2708–2717

Fierer N, Bradford M, Jackson R (2007) Towards an ecological classification of soil bacteria. *Ecology* 88:1354–1364

Francis AJ (1986) Acid rain effects on soil and aquatic microbial processes. *Cell Molec Life Sci (CMLS)* 42:455–465

Friedmann EI, Weed R (1987) Microbial trace-fossil formation, biogenous, and abiotic weathering in the Antarctic cold desert. *Science* 236:703–705

Futterer O, Angelov A, Liesegang H, Gottschalk G, Schleper C, Schepers B, Dock C, Antranikian G, Liebl W (2004) Genome sequence of *Picrophilus torridus* and its implications for life around pH 0. *Proc Natl Acad Sci USA* 101:9091–9096

Gans J, Wolinsky M, Dunbar J (2005) Computational improvements reveal great bacterial diversity and high metal toxicity in soil. *Science* 309:1387–1390

Gilichinsky D, Rivkina E, Bakermans C, Shcherbakova V, Petrovskaya L, Ozerskaya S, Ivanushkina N, Kochkina G, Laurinavichuis K, Pecheritsina S (2005) Biodiversity of cryopegs in permafrost. *FEMS Microbiol Ecol* 53:117–128

Gilichinsky D, Rivkina E, Shcherbakova V, Laurinavichuis K, Tiedje J (2003) Supercooled water brines within permafrost - An unknown ecological niche for microorganisms: A model for astrobiology. *Astrobiology* 3:331–341

Giller KE, Beare MH, Lavelle P, Izac AMN, Swift MJ (1997) Agricultural intensification, soil biodiversity and agroecosystem function. *Appl Soil Ecol* 6:3–16

Giller KE, Witter E, McGrath SP (1998) Toxicity of heavy metals to microorganisms and microbial processes in agricultural soils: A review. *Soil Biol Biochem* 30:1389–1414

Golyshina OV, Timmis KN (2005) *Ferroplasma* and relatives, recently discovered cell wall-lacking archaea making a living in extremely acid, heavy metal-rich environments. *Environ Microbiol* 7:1277–1288

Gorbushina AA, Krumbein WE (1999) The poikilotrophic micro-organism and its environment. In: Seckbach J (ed) *Enigmatic Microorganisms and Life in Extreme Environments*. Kluwer Academic, Dordrecht, pp 177–185

Griffiths BS, Kuan HL, Ritz K, Glover LA, McCaig AE, Fenwick C (2004) The relationship between microbial community structure and functional stability, tested experimentally in an upland pasture soil. *Microb Ecol* 47:104–113

Hattori T, Mitsui H, Haga H, Wakao N, Shikano S, Gorlach K, Kasahara Y, El BA, Hattori R (1997) Advances in soil microbial ecology and the biodiversity. *Ant v Leeuw* 72:21–28

Hewson I, Jacobson-Meyers ME, Fuhrman JA (2007) Diversity and biogeography of bacterial assemblages in surface sediments across the San Pedro Basin, Southern California Borderlands. *Environ Microbiol* 9:923–933

Hoj L, Olsen RA, Torsvik VL (2005) Archaeal communities in High Arctic wetlands at Spitsbergen, Norway (78[deg]N) as characterized by 16S rRNA gene fingerprinting. *FEMS Microbiol Ecol* 53:89–101

Hoj L, Rusten M, Haugen LE, Olsen RA, Torsvik VL (2006) Effects of water regime on archaeal community composition in Arctic soils. *Environ Microbiol* 8:984–996

Horowitz NH, Cameron RE, Hubbard JS (1972) Microbiology of the Dry Valleys of Antarctica. *Science* 176:242–245

Huang PM (1990) Role of soil minerals in transformation of natural organics and xenobiotics in soil. In: Bollag J-M, Stotzky G (eds) *Soil Biochemistry*. Marcel Dekker, New York, pp 29–115

Hugenholtz P, Goebel BM, Pace NR (1998) Impact of culture-independent studies on the emerging phylogenetic view of bacterial diversity. *J Bacteriol* 180:4765–4774

Hugenholtz P, Tyson GW, Webb RI, Wagner AM, Blackall LL (2001) Investigation of candidate division TM7, a recently recognized major lineage of the domain bacteria with no known pure-culture representatives. *Appl Environ Microbiol* 67:411–419

Jakosky BM, Nealson KH, Bakermans C, Ley RE, Mellon MT (2003) Subfreezing activity of microorganisms and the potential habitability of Mars' polar regions. *Astrobiology* 3:343–350

Jenny H (1994) *Factors of Soil Formation. A System of Quantitative Pedology*. Dover Press, New York. (Reprint, with Foreword by R. Amundson, of the 1941 McGraw-Hill publication). Reprint McGraw-Hill, New York

Johnsen K, Jacobsen C, Torsvik V, Sørensen J (2001) Pesticide effects on bacterial diversity in agricultural soils – A review. *Biol Fertil Soils* 33:443–453

Joynt J, Bischoff M, Turco R, Konopka A, Nakatsu CH (2006) Microbial community analysis of soils contaminated with lead, chromium and petroleum hydrocarbons. *Microb Ecol* 51:209–219

Kashefi K, Lovley DR (2003) Extending the upper temperature limit for life. *Science* 301:934

Kempf B, Bremer E (1998) Uptake and synthesis of compatible solutes as microbial stress responses to high-osmolality environments. *Arch Microbiol* 170:319–330

Killham K (1994) *Soil Ecology*. Cambridge University Press, Cambridge

Killham K, Firestone MK, JGM (1983) Acid rain and soil microbial activity: Effects and their mechanisms. *J Environ Qual* 12:133–137

Koch AL (2001) Oligotrophs versus copiotrophs. *BioEssays* 23:657–661

Kozdroj J, van Elsas JD (2001) Structural diversity of microbial communities in arable soils of a heavily industrialised area determined by PCR-DGGE fingerprinting and FAME profiling. *Appl Soil Ecol* 17:31–42

Krulwich TA, Ito M, Hicks DB, Gilmour R, Guffanti AA (1998) pH homeostasis and ATP synthesis: studies of two processes that necessitate inward proton translocation in extremely alkaliphilic *Bacillus* species. *Extremophiles* 2:217–222

Kuske CR, Ticknor LO, Miller ME, Dunbar JM, Davis JA, Barns SM, Belnap J (2002) Comparison of soil bacterial communities in rhizospheres of three plant species and the interspaces in an arid grassland. *Appl Environ Microbiol* 68:1854–1863

Liao M, Xie XM (2007) Effect of heavy metals on substrate utilization pattern, biomass, and activity of microbial communities in a reclaimed mining wasteland of red soil area. *Ecotox Environ Safety* 66:217–223

Loisel P, Harmand J, Zemb O, Latrille E, Lobry C, Delgenes JP, Godon JJ (2006) Denaturing gradient electrophoresis (DGE) and single-strand conformation polymorphism (SSCP) molecular fingerprintings revisited by simulation and used as a tool to measure microbial diversity. *Environ Microbiol* 8:720–731

Lorenz MG, Wackernagel W (1994) Bacterial gene transfer by natural genetic transformation in the environment. *Microbiol Mol Biol Rev* 58:563–602

Lynch JM (1979) The terrestrial environment. In: Lynch JM, Poole NJ (eds) *Microbial Ecology: A Conceptual Approach*. Blackwell Scientific, Oxford, pp 67–91

Lynch JM, Benedetti A, Insam H, Nuti MP, Smalla K, Torsvik V, Nannipieri P (2004) Microbial diversity in soil: Ecological theories, the contribution of molecular techniques and the impact of transgenic plants and transgenic microorganisms. *Biol Fertil Soils* 40: 363–385

Macnaughton S, Stephen JR, Chang YJ, Peacock A, Flemming CA, Leung K, White DC (1999a) Characterization of metal-resistant soil eubacteria by polymerase chain reaction - Denaturing gradient gel electrophoresis with isolation of resistant strains. *Can J Microbiol* 45:116–124

Macnaughton SJ, Stephen JR, Venosa AD, Davis GA, Chang YJ, White DC (1999b) Microbial population changes during bioremediation of an experimental oil spill. *Appl Environ Microbiol* 65:3566–3574

Margesin R, Schinner F (2001) Biodegradation and bioremediation of hydrocarbons in extreme environments. *Appl Microbiol Biotechnol* V56:650–663

Marion GM, Fritsen CH, Eicken H, Payne MC (2003) The search for life on Europa: Limiting environmental factors, potential habitats, and Earth analogues. *Astrobiology* 3:785–811

Martin DD, Ciulla RA, Roberts MF (1999) Osmoadaptation in *Archaea*. *Appl Environ Microbiol* 65:1815–1825

Minton KW, Daly MJ (1995) A model for repair of radiation-induced DNA double-strand breaks in the extreme radiophile *Deinococcus radiodurans*. *BioEssays* 17:457–464

Muyzer G, Dewaal EC, Uitterlinden AG (1993) Profiling of complex microbial-populations by denaturing gradient gel-electrophoresis analysis of polymerase chain reaction amplified genes coding for 16S ribosomal-RNA. *Appl Environ Microbiol* 59:695–700

Nakatsu CH, Carmosini N, Baldwin B, Beasley F, Kourtev P, Konopka A (2005) Soil microbial community responses to additions of organic carbon substrates and heavy metals (Pb and Cr). *Appl Biochem Microbiol* 71:7679–7689

Nannipieri P, Ascher J, Ceccherini MT, Landi L, Pietramellara G, Renella G (2003) Microbial diversity and soil functions. *Eur J Soil Sci* 54:655–670

Nannipieri P, Grego S, Ceccanti B (1990) Ecological significance of the biological activity in soil. In: Bollag J-M, Stotzky G (eds) *Soil Biochemistry*. Marcel Dekker, New York, pp 293–355

Nedwell DB (1999) Effect of low temperature on microbial growth: Lowered affinity for substrates limits growth at low temperature. *FEMS Microbiol Ecol* 30:101–111

Neufeld JD, Mohn WW (2005) Unexpectedly high bacterial diversity in Arctic tundra relative to boreal forest soils, revealed by serial analysis of ribosomal sequence tags. *Appl Environ Microbiol* 71:5710–5718

Odum EP (1971) *Fundamentals of Ecology*, 3rd edn. W.B. Saunders, Philadelphia

Øvreås L (2000) Population and community level approaches for analysing microbial diversity in natural environments. *Ecol Lett* 3:236–251

Øvreås L, Daae FL, Yndestad S, Jørgensen SL, Torsvik V, Brandvik PJ (2004) Microbial community analysis in pristine and polluted environments from Arctic. In: *10th International Symposium for Microbial Ecology (ISME10)*, Cancún, Mexico

Pace NR, Stahl DA, Lane DJ, Olsen GJ (1986) The analysis of natural microbial populations by ribosomal-RNA sequences. *Adv Microb Ecol* 9:1–55

Pakchung AAH, Simpson PJL, Codd R (2006) Life on earth. Extremophiles continue to move the goal posts. *Environ Chem* 3:77–93

Panikov NS (1999) Understanding and prediction of soil microbial community dynamics under global change. *Appl Soil Ecol* 11:161–176

Panikov NS, Dedysh SN (2000) Cold season CH_4 and CO_2 emission from boreal peat bogs (West Siberia): Winter fluxes and thaw activation dynamics. *Global Biogeochem Cycles* 14:1071–1080

Pennanen T, Fritze H, Vanhala P, Kiikkila O, Neuvonen S, Baath E (1998a) Structure of a microbial community in soil after prolonged addition of low levels of simulated acid rain. *Appl Environ Microbiol* 64:2173–2180

Pennanen T, Perkiomaki J, Kiikkila O, Vanhala P, Neuvonen S, Fritze H (1998b) Prolonged, simulated acid rain and heavy metal deposition: Separated and combined effects on forest soil microbial community structure. *FEMS Microbiol Ecol* 27:291–300

Post RD, Beeby AN (1996) Activity of the microbial decomposer community in metal-contaminated roadside soils. *J Appl Ecol* 33:703–709

Price PB, Sowers T (2004) Temperature dependence of metabolic rates for microbial growth, maintenance, and survival. *Proc Natl Acad Sci USA* 101:4631–4636

Prosser JI (2002) Molecular and functional diversity in soil micro-organisms. *Plant Soil* 244:9–17

Rainey FA, Ray K, Ferreira M, Gatz BZ, Nobre MF, Bagaley D, Rash BA, Park MJ, Earl AM, Shank NC, Small AM, Henk MC, Battista JR, Kampfer P, da Costa MS (2005) Extensive diversity of ionizing-radiation-resistant bacteria recovered from Sonoran Desert soil and description of nine new species of the genus *Deinococcus* obtained from a single soil sample. *Appl Environ Microbiol* 71:5225–5235

Rajapaksha RMCP, Tobor-Kaplon MA, Bååth E (2004) Metal toxicity affects fungal and bacterial activities in soil differently. *Appl Environ Microbiol* 70:2966–2973

Rike AG, Haugen KB, Borresen M, Engene B, Kolstad P (2003) In situ biodegradation of petroleum hydrocarbons in frozen arctic soils. *Cold Reg Sci Tech* 37:97–120

Rivkina EM, Friedmann EI, McKay CP, Gilichinsky DA (2000) Metabolic activity of permafrost bacteria below the freezing point. *Appl Environ Microbiol* 66:3230–3233

Roberts MF (2005) Organic compatible salutes of halotolerant and halophilic microorganisms. *Sal Syst* 1:5–30

Rondon MR, August PR, Bettermann AD, Brady SF, Grossman TH, Liles MR, Loiacono KA, Lynch BA, MacNeil IA, Minor C, Tiong CL, Gilman M, Osburne MS, Clardy J, Handelsman J, Goodman RM (2000) Cloning the soil metagenome: a strategy for accessing the genetic and functional diversity of uncultured microorganisms. *Appl Environ Microbiol* 66:2541–2547

Rothschild LJ, Mancinelli RL (2001) Life in extreme environments. *Nature* 409:1092–1101

Ruggiero P, Dec J, Bollag J-M (1996) Soil as a catalytic system. In: Bollag J-M, Stotzky G (eds) *Soil Biochemistry*. Marcel Dekker, New York, pp 79–122

Sale JE (2007) Radiation resistance: Resurrection by recombination. *Curr Biol* 17:R12–R14

Sandaa R-A, Torsvik V, Enger O, Daae FL, Castberg T, Hahn D (1999) Analysis of bacterial communities in heavy metal-contaminated soils at different levels of resolution. *FEMS Microbiol Ecol* 30:237–251

Saul DJ, Aislabie JM, Brown CE, Harris L, Foght JM (2005) Hydrocarbon contamination changes the bacterial diversity of soil from around Scott Base, Antarctica. *FEMS Microbiol Ecol* 53:141–155

Schleper C, Puehler G, Holz I, Gambacorta A, Janekovic D, Santarius U, Klenk HP, Zillig W (1995) *Picrophilus gen. nov., fam. nov*.: A novel aerobic, heterotrophic, thermoacidophilic genus and family comprising archaea capable of growth around pH 0. *J Bacteriol* 177:7050–7059

Shi W, Bischoff M, Turco R, Konopka A (2002) Long-term effects of chromium and lead upon the activity of soil microbial communities. *Appl Soil Ecol* 21:169–177

Shukla M, Chaturvedi R, Tamhane D, Vyas P, Archana G, Apte S, Bandekar J, Desai A (2007) Multiple-stress tolerance of ionizing radiation-resistant bacterial isolates obtained from various habitats: correlation between stresses. *Curr Microbiol* 54:142–148

Singleton JR, Amelunxen RE (1973) Proteins from thermophilic microorganisms. *Bacteriol Rev* 37:320–342

Smith JJ, Tow LA, Stafford W, Cary C, Cowan DA (2006) Bacterial diversity in three different Antarctic cold desert mineral soils. *Microb Ecol* 51:413–421

Staley JT, Konopka A (1985) Measurement of *in situ* activities of nonphotosynthetic microorganisms in aquatic and terrestrial habitats. *Ann Rev Microbiol* 39:321–346

Standing D, Killham K (2006) The soil environment. In: van Elsas JD, Jansson JK, Trevors JT (eds) *Modern Soil Microbiology*, 2nd edn. CRC Press, Taylor and Francis, Boca Raton, FL, pp 1–22

Steven B, Léveillé R, Pollard WH, Whyte LG (2006) Microbial ecology and biodiversity in permafrost. *Extremophiles* 10:259–267

Stevens TO, McKinley JP (1995) Lithoautotrophic microbial ecosystems in deep basalt aquifers. *Science* 270:450–454

Stotzky G (1997) Soil as an environment for microbial life. In: van Elsas JD, Trevors JT, Wellington EMH (eds) *Modern Soil Microbiology*, 1st edn. Marcel Dekker, New York, pp 1–20

Sunde IR (2005) Enrichment, isolation and phylogenetic characterisation of hydrocarbon-degrading bacteria from oil-contaminated tundra. In: *Department of Biology*. University of Bergen, Bergen, p 103

Sørheim R, Torsvik VL, Goksoyr J (1989) Phenotypical divergences between populations of soil bacteria isolated on different media. *Microb Ecol* 17:181–192

Torsvik V, Daae FL, Sandaa RA, Øvreas L (1998) Novel techniques for analysing microbial diversity in natural and perturbed environments. *J Biotechnol* 64:53–62

Torsvik V, Goksoyr J, Daae FL (1990) High diversity in DNA of soil bacteria. *Appl Environ Microbiol* 56:782–787

Torsvik V, Øvreås L (2006) Microbial phylogeny and diversity in soil. In: van Elsas JD, Jansson JK, Trevors JT (eds) *Modern Soil Microbiology*, 2nd edn. CRC Press, Taylor and Francis Group, Boca Raton, FL, pp 23–54

Tringe SG, von Mering C, Kobayashi A, Salamov AA, Chen K, Chang HW, Podar M, Short JM, Mathur EJ, Detter JC, Bork P, Hugenholtz P, Rubin EM (2005) Comparative metagenomics of microbial communities. *Science* 308:554–557

van Elsas JD, Torsvik V, Hartmann A, Øvreås L, Jansson JK (2006) The bacteria and archaea in soil. In: van Elsas JD, Jansson JK, Trevors JT (eds) *Modern Soil Microbiology*, 2nd edn. CRC Press, Taylor and Francis, Boca Raton, FL pp 83–105

Vetterlein D, Jahn R (2004) Combination of micro suction cups and time-domain reflectometry to measure osmotic potential gradients between bulk soil and rhizosphere at high resolution in time and space. *Eur J Soil Sci* 55:497–504

Virginia RA, Wall DH (1999) How soils structure communities in the Antarctic Dry Valleys *BioScience* 49:974–983

Ward DM, Weller R, Bateson MM (1990) 16S ribosomal-RNA sequences reveal numerous uncultured microorganisms in a natural community. *Nature* 345:63–65

Whyte LG, Slagman SJ, Pietrantonio F, Bourbonniere L, Koval SF, Lawrence JR, Inniss WE, Greer CW (1999) Physiological adaptations involved in alkane assimilation at a low temperature by *Rhodococcus sp* strain Q15. *Appl Environ Microbiol* 65:2961–2968

Whyte LG, Goalen B, Hawari J, Labbe D, Greer CW, Nahir M (2001) Bioremediation treatability assessment of hydrocarbon-contaminated soils from Eureka, Nunayut. *Cold Reg Sci Tech* 32:121–132

Wiebe WJ, Sheldon WM, Jr., Pomeroy LR (1992) Bacterial growth in the cold: Evidence for an enhanced substrate requirement. *Appl Environ Microbiol* 58:359–364

Wright DJ, Smith SC, Joardar V, Scherer S, Jervis J, Warren A, Helm RF, Potts M (2005) UV irradiation and desiccation modulate the three-dimensional extracellular matrix of *Nostoc commune* (Cyanobacteria). *J Biol Chem* 280:40271–40281

Yachi S, Loreau M (1999) Biodiversity and ecosystem productivity in a fluctuating environment: The insurance hypothesis. *Proc Natl Acad Sci USA* 96:1463–1468

Yndestad S (2004) Microbial diversity in oil polluted and pristine tundra on Svalbard (Norwegian). In: *Department of Biology*. University of Bergen, Norway, Bergen

Zhang G, Ma X, Niu F, Dong M, Feng H, An L, Cheng G (2007) Diversity and distribution of alkaliphilic psychrotolerant bacteria in the Qinghai-Tibet Plateau permafrost region. *Extremophiles* 11:415–424

Zvyagintsev DG, Zenova GM, Sudnizin II, Doroshenko EA (2005) The ability of soil *Actinomycetes* to develop at an extremely low humidity. *Doklady Biol Sci* 405:461–463

Chapter 3
Extreme Views on Prokaryote Evolution

Patrice Dion

Wilt thou not view, then, the truth, in my mirror so clearly depicted?

– Goethe

3.1 Introduction

Not two soils are identical. Not two bacteria are identical either, with perhaps even not two daughter cells as they emerge from symmetric division being rigorously the same (Stewart et al. 2005). However, it might be easier to draw generalizations from observations of *Escherichia coli* or *Caulobacter crescentus* than from soil studies. This might explain why the bacterial cell has been deemed a more popular object of theoretical reflections than has the soil environment, although remarkable concepts on soil biology have emerged (Wardle et al. 2004; Young and Crawford 2004). This being said, extreme soils may appear as a particularly fruitful ground for those of us who feel seduced by the "idea of the soil" and who wish to venture on such a (admittedly slippery) terrain.

Life constantly reinvents itself, under new aspects that owe much to ancient ones, to the point that nothing, at least microbial, can be said to have ever disappeared: the author is aware that this opinion may hold only if emphasis is placed on the modular construction of prokaryotic cells. In this sense, the last universal common ancestor or rather perhaps a set of basic processes acting on communally evolving primitive cells (Woese 2002), is still among us in a transformed and diversified form, and having seeded among its progeny many clues of its ancient organization. This apparent persistence of functional attributes over evolutionary time has been termed the "fecundity of function principle" (Staley 1997). Anecdotally, it might also be remarked that even the smallpox virus is still preserved somewhere behind well-locked doors.

Life on Earth has been proposed to have originated in a hot (Di Giulio 2003; Schwartzman and Lineweaver 2004) or a cold (Price 2007) environment. As a reconciling alternative to these opposing possibilities, it has also been suggested that

---

Patrice Dion
Département de phytologie, Pavillon Charles-Eugène Marchand, 1030, avenue de la Médecine, Université Laval, Québec (Québec), Canada G1V 0A6
e-mail: patrice.dion@plg.ulaval.ca

the earliest organisms might have been mesophiles, soon to be replaced by thermophiles which would have been advantaged, in particular, by their greater resistance to various stresses (Islas et al. 2003). Indeed, various aspects of primitive environments might have been rude by human standards. Now we see that climates and environments are changing, with soils becoming hotter, colder, dryer, more saline or toxic, the risk of nuclear accidents or fire notwithstanding. So the examination of microbial life reactions to these changes (Staley 1997) is timely.

3.2 Scope and Limitations of This Review

This review attempts to describe extreme soils as two-component environments, and to examine the corresponding bacterial response. One must obviously be guarded against generalizations such as those that are made here, as the genetic response to selection varies among different bacterial species and even among different regions of the genome of a given organism (Van Belkum et al. 1998). As a second word of caution, it must be stressed that not considered in this review is the role of horizontal gene transfer in microbial adaptation to extreme soil conditions, and especially the possibility that mutators would cooperate with nonmutator variants to generate adapted strains with preserved essential genes (Loewe 2004). In fact, horizontal gene transfer plays a major role in bacterial evolution (Goldenfeld and Woese 2007), and its contribution to the life and metamorphoses of extreme soil micro-organisms would be well worth investigating. Finally, the author is aware that soils are fluctuating environments, where stochastic phenotypic changes (Kussell and Leibler 2005) and phase variations (Thattai and van Oudenaarden 2004) might occur, and where persisters (Kussell et al. 2005) might arise, influencing evolutionary responses. However, a static point of view is adopted here, in the hope of designing a general framework sufficiently broad and solid to later accommodate evolutionary changes accompanying environmental fluctuations and the horizontal transmission of discrete, perhaps modular, characters.

Prokaryotic evolution has given rise to deep divisions reflecting ancient divergences in cell structure and function (Woese 1987), but it has also produced comparatively minute differences within a species, subspecies, or clone (Schloter et al. 2000). The relationships between the mechanisms of macroevolution and microevolution are not elucidated, and it has been suggested that macroevolutionary events cannot be deduced from a simple extrapolation of microevolutionary changes (Erwin 2000). However, macroevolution and microevolution are both based on environmental and clade constraints (Moore and Woods 2006), implying that there need not be mechanistic differences between macroevolutionary and microevolutionary changes. Specifically, punctuated evolution may occur in clonal bacterial populations by simple mechanisms such as selection of rare mutations with a large effect (Elena et al. 1996). Many of these mutations are likely to be pleiotropic, altering global regulators (Zinser and Kolter 2004). Hence, it appears that, for the sake of simplicity at least, major evolutionary changes and minor adjustments, which would both be expected to occur in micro-organisms adapting to extreme soil conditions, can be accommodated within the same evolutionary framework.

3.3 Selection as It Occurs in the Laboratory and in the Real World

Experiments on bacterial evolution are conducted under laboratory conditions that differ markedly from those that prevail in extreme soils or other environments. Evolutionary changes are provoked in the laboratory by applying unique and discriminating selective pressures. Typically, bacteria are experimentally challenged for the utilization of a particular carbon source, or for phage or antibiotic resistance. In most cases, the experimental design calls for an all or none response, such as life or death, growth or no growth. However, some experimental systems have led to the observation of small fitness gains and to an elucidation of the mutations involved (e.g., see Cooper et al. 2001).

A successful response to experimental selective conditions generally involves the modification of a single cell component, such as a catabolic enzyme or a phage receptor (Cairns et al. 1988). On the contrary, environmental challenges to bacteria are multiple and often associate to build global physiological constraints. Bacteria may gain advantage from the utilization of a substrate, not as the sole, but as an additional carbon source. They may benefit from resistance to a particular bacteriophage, while remaining sensitive to other viruses and prey to protozoa. Even a slight increase in growth rate, for example resulting from adaptation to physical conditions or chemical resistance, may contribute usefully to environmental success. Thus, the extent to which the results of laboratory studies on bacterial evolution can be extrapolated to natural systems calls for careful evaluation.

At least two aspects of this problem ought to be considered. The first, alluded to above, refers to differences between artificial and natural modes of discrimination between wild-type and mutant cells. The second aspect concerns environmental effects on genome stability. Just as natural environments may reveal more complex and graded phenotypic changes than those that are commonly considered in the laboratory, nature may also act on genome stability in subtle ways, our experimental manipulations of this stability appearing crude in comparison. Current experimental models that appear to best mimic natural evolution are based on the study of bacteria in stationary phase. Indeed, stationary-phase populations are highly dynamic and generate extensive genetic diversity (Finkel and Kolter 1999).

Approaches for the understanding of natural adaptation may also resort to comparative biology. One possible strategy would involve comparisons in the bacterial dimension. For example, differences may exist in the adaptive mode of an enteric bacterium, such as *E. coli*, and that of an isolate from Antarctic ice or a hydrothermal vent. A second, perhaps complementary, strategy, would be to evaluate adaptive responses along an environmental axis, graded, for example, from cold to hot, or from alkaline to acidic. In this manner, an understanding of bacteria as variable and responding components of their environments may become achievable.

Along those lines, the present review tentatively proposes to consider extreme soils as applying a composite selection pressure, comprised of a complex and heterogeneous soil component, and of a unique and defined extreme component. In this light, extreme soils may be perceived as combining the characters of both natural and experimental environments. The natural component would be represented by the soil matrix, with its complex environmental heterogeneity and spatial distance factors (Ramette and Tiedje 2007). For its part, the defining extreme parameter of extreme soils would introduce selection pressure akin to that produced in well-characterized experimental systems. Thus, an internal standard would become available, in the form of an adaptive response to a defining extreme parameter, to evaluate and compare evolutionary responses as they occur in different soils and in different organisms.

The general forces shaping soil microbial communities are discussed in the next section. Then the particular case of extreme soils is presented, with an examination of the selective pressures at play and of the prokaryotic adaptive response to these pressures. Finally, these various elements are combined in the description of a prototypal fitness landscape that would be characteristic of extreme soils.

3.4 Bacteria in Soils, a Substrate for Evolutionary Change

Heterogeneity, functional complementarity, and intense communication are characters shared by both the soil and immune cell systems. These two systems have been constituted by a process of evolutionary layer accretion, corresponding, first, to the maintenance of environments along a geological or evolutionary scale, second, to the sustained coexistence and interaction of organisms or cells having appeared in succession in these various environments, and, third, to the perpetuation of mechanisms involved in symbiotic relationships (Dion 2008).

3.4.1 The Soil Environment

Soil heterogeneity is a multifactorial character (Giri et al. 2005). It results from the interplay of spatiotemporal, physical, chemical and nutritional variables delineating spheres of influence (Beare et al. 1995) that may separate bacteria with respect to location, physiology, or genetics. Heterogeneity contributes to diversification in populations (Kassen 2002), and this applies to the particular case of soils (Korona et al. 1994; Bardgett 2002; Torsvik et al. 2002). Conceivably, genotypic diversity might represent local adaptation to very specific niches, the continuous dispersal of locally nonadapted genotypes, or simply neutral variation (Vos and Velicer 2006). However, it seems well established that genetic patterns among soil microbial populations are influenced by habitat (McArthur et al. 1988; Noguez et al. 2005), and particularly soil type (Gelsomino et al. 1999).

The making of a soil bacterium involves both global genomic adaptations, such as adjustments in codon usage preference (Willenbrock et al. 2006), and fine tuning. An illustration of fine tuning is provided by the local adaptation of soil bacteria to trophic conditions found at a particular site (Belotte et al. 2003). Extensive adaptation to soil conditions depends on expression of many properties, such as survival and stress responses, substrate utilization, chemotaxis, surface adhesion, interaction with plants and other microbes, and signalling pathways (Wipat and Harwood 1999).

Adaptation and selection occur at very small scales in the soil heterogeneous substrate, as suggested by the observation of extensive diversity at the centimetric, millimetric, or even submillimetric scale (Grundmann and Normand 2000; Torsvik and Øvreås 2002; Vogel et al. 2003), coupled with the demonstrated capacity of particular clones to spread and become distributed in soils (Grundmann 2004). Microscale analysis indicates quantitative and qualitative heterogeneity within soil bacterial or archaeal communities (Nunan et al. 2002; Nicol et al. 2003).

Soil colonization patterns are influenced by soil structure, as microbial communities established outside and inside aggregates show quantitative and structural differences. Most bacteria are located inside the aggregates or associated with the dispersible clay fraction (Ranjard et al. 2000). Clay and organic carbon-based colloids may facilitate microbial survival and growth by providing nutrients and protection against predation. Sand fractions appear to be preferentially colonized by fungi and by bacteria that resist grazing and that are adapted to nutrient limitations (Van Gestel et al. 1996; Ranjard et al. 2000; Sessitsch et al. 2001).

In soil, bacteria are distributed in patches, the location of which may correspond to local substrate deposits stimulating growth and creating cell-density gradients (Nunan et al. 2003). This patchy distribution of growing cells may have a regulatory effect on bacterial activity, as a result of diffusion of substrates and metabolites (Grundmann et al. 2001). Metabolic interactions also contribute to structure soil microbial communities. For example, populations of nitrifying bacteria establish spatial associations of two bacterial types carrying out two successive biochemical transformations (Grundmann and Debouzie 2000). Bacteria collaborating in the stepwise oxidation of ammonia are located together in randomly distributed and preferentially colonized patches within the soil matrix; they are distributed nonrandomly within these preferentially colonized patches (Grundmann et al. 2001). Hence, it is apparent that bacterial distribution is fine-tuned at the microscale, to reflect preferential colonization abilities and thus maximize adaptation. These small-scale spatial relationships are themselves influenced by various soil parameters, such as management practices (Webster et al. 2002).

3.4.2 Bacterial Adaptations to Soil and the Role of the Stress Response

Soil microbial selection is the result of a complex interplay of biotic interactions, involving competition for nutrients, antibiosis, and predation (Lockwood 1988; Hoitink and Boehm 1999), and of abiotic factors, the latter including lack of water or oxygen, nutrient starvation, exposure to toxic pH or metal cations, and high osmotic or matric tension (Van Veen et al. 1997). These selection components help establish the heterogeneous nature of the soil as a microbial habitat. They apply external forces on the microbial community, shaping it while at the same time eliciting a biological response that is expressed as a capacity for self-organization (Young and Crawford 2004). Self-organization primarily arises as a byproduct of stress responses. Indeed, stress induced by environmental factors provides both a mechanism and a substrate for selection, because it both reduces fitness and elicits a controlled genetic instability leading to increased variability and the emergence of better adapted clones. In this manner, stress provides a dynamic and self-responding mechanism for shaping bacterial communities (Saint-Ruf and Matic 2006).

A global stress response, resulting in the induction of more than 70 genes in reaction to a variety of physical and chemical stresses (Hengge-Aronis 2002), is under the control of a specialized sigma factor, such as RpoS in *E. coli* (Hengge-Aronis 1999) or SigB in *Bacillus subtilis* (Price et al. 2001). RpoS interferes with substrate utilization, leading cells to balance their capacity for substrate utilization and stress resistance through mutation of the *rpoS* gene (King et al. 2006). High intracellular concentrations of RpoS increase mutation rates by downregulating the mismatch repair system and the *dinB* gene, coding for the Pol IV translesion synthesis DNA polymerase (Saint-Ruf and Matic 2006). In addition to the general stress response, *E. coli* also exhibits a mutagenic SOS response in reaction to extensive DNA damage and inhibition of DNA synthesis (Janion 2001; Michel 2005). This response, under the control of the LexA repressor and the RecA coprotease, involves the production of error-prone DNA polymerases II, IV, and V responsible for translesion synthesis (Napolitano et al. 2000). It has been suggested that a variety of environmental stimuli, such as starvation, could elicit the SOS response by endogenous induction of DNA damage (Aertsen and Michiels 2006). Antibiotic-induced defective cell wall synthesis has also been identified as an inducer of the SOS response in *E. coli* (Miller et al. 2004).

Microbial populations may reconfigure themselves through mutation and natural selection, so as to reduce stress. Mutations improve the correspondence between environmental factors and physiological requirements, through adjustment of either the mutating cell itself or else its surroundings, the latter effect resulting from bacteria-induced microenvironmental changes (Banas et al. 2007). In a second-order response, the ability to reduce stress through mutation is itself a selectable trait (Earl and Deem 2004), with a general evolutionary trend towards a reduction of the standard mutation rate, and allowing higher mutation rates under specific circumstances (Drake et al. 1998). This results in the acquisition of stress-induced and transiently expressed "evolution-accelerating systems" (Kivisaar 2003).

3.4.3 The Stress Response as a Mechanism and an Object of Evolutionary Change

The existence of an evolved genetic instability response to environmental stress is suggested by three sets of observations. The first concerns apparently divergent responses of an organism to stress, with stress-induced mechanisms for both increased or decreased genetic instability having been identified in *E. coli* and a variety of environmental bacteria (see below, this section). A second observation is that orthologous stress-related proteins have divergent effects on gene stability in different organisms. The most notable example is RecA, whose action promotes hypermutability in *E. coli*, but prevents it in *Thermus thermophilus* (Castán et al. 2003). Perhaps in keeping with these differences in function, *recA* is diversely regulated in *E. coli* and other organisms (Mazón et al. 2006). A third observation suggesting evolution of genetic instability in bacteria is that differences exist in intrinsic genetic stability in different organisms. For example, it has been suggested that the hyperthermophile crenarchaeon *Pyrobaculum aerophilum* is deficient in mismatch repair and has developed the ability to survive as a permanent mutator (Fitz-Gibbon et al. 2002). By contrast, spontaneous mutations occur at low rates in the thermoacidophilic archaeon *Sulfolobulus acidoalcarius* (Grogan et al. 2001). Thus, it appears that the mutagenic response to stress is subject to second-order selection (Bjedov et al. 2003) and intense diversifying evolution (Saint-Ruf and Matic 2006).

As a result of this second-order evolutionary process, bacteria have acquired the capacity to switch phenotypically between high and low mutation rates depending on environmental conditions. Several such switches have been postulated to exist in bacteria, and the functioning of some of these may involve the RpoS-mediated general stress response (Bjedov et al. 2003). They allow acquisition of a high mutation rate, followed by a return to genetic stability. In addition to occurrences where they hitchhike beneficial traits to fixation in small asexual populations (Sniegowski et al. 2000), mutator phenotypes will be selected when gene diversity limits adaptation. This may occur, for example, following the introduction of a strong stress-generating parameter. However, mutator phenotypes are counterselected once adaptation is achieved, as a result of continuous production of deleterious mutations (Denamur and Matic 2006). On broad terms, phenotypic switching and return to low mutation rates may be rare or intense. Rare switching occurs when the mutator phenotype is stably expressed as a result of a mutation in an antimutator gene, and in this case reduction of the mutation rate may be achieved by reversion, secondary mutation, or reacquisition of the wild-type antimutator allele by horizontal gene transfer (Denamur and Matic 2006).

Intense phenotypic switching occurs when the mutator phenotype is transiently expressed through physiological adjustments. Indeed, a variety of physiological and genetic mechanisms contribute to modulate the level of genetic stability as a function of environmental factors (Table 3.1).

An examination of the structure of soil microbial communities (see Section 3.4.1) suggests that, in soils, stresses are spatially and temporally heterogeneous, and that, correspondingly, the bacterial stress response is patchy and localized. Hence, soil microbial communities can be viewed as an assemblage of individuals responding to various levels of stress by modulating genetic stability. Genetic instability becomes advantageous when maladaptation increases the potential benefit of mutation. In turn, expression of a newly produced beneficial mutation will result in alleviation of stress, which can then be viewed as a physiological relay between the environment and cell determinants of genetic stability.

Table 3.1 Operators of genetic change or stability in bacteria

Description of Operator	Mechanism Involved	Conditions for Observation	Impact	Organisms	References
Operators of Genetic Change					
General hypermutation	Amplification of *dinB*, encoding error-prone DNA polymerase IV	Starvation	Possibility of widespread loss of information	*Escherichia coli*	(Torkelson et al. 1997; Slechta et al. 2003)
General hypermutation	Deletion or inactivation of genes for DNA repair (mutator mutation)	Nutrient limitation	Possibility of widespread loss of information; expected low frequency of phenotype reversal	*E. coli*	(Notley-McRobb et al. 2002; Denamur and Matic 2006)
General hypermutation	SOS induction and synthesis of Pol II, IV and V DNA polymerases	Starvation	Possibility of widespread loss of information	*E. coli*	(Taddei et al. 1995; Yeiser et al. 2002; Nohmi 2006)
General hypermutation	Mutagenesis cassette containing genes for error-prone DNA polymerases	LexA regulation	Possibility of widespread loss of information	Found across the Bacteria domain	(Erill et al. 2006)
General hypermutation	Plasmid-borne *umuDC*, encoding error-prone DNA polymerase V	Variety of environmental stresses	Possibility of widespread loss of information	Variety of environmental bacteria	(Tark et al. 2005)
General hypermutation	RpoS-dependent gene expression	Variety of environmental stresses	Possibility of widespread loss of information	*E. coli*	(Saint-Ruf and Matic 2006)
Hypermutation localized to contingency genes	Discriminate mutational mechanisms occurring preferentially at loci characterized by certain nucleotide arrangements	Various environments (frequently host)	No or localized loss of information	Variety of bacteria (frequently symbiotic)	(Moxon et al. 1994; Van Belkum et al. 1998)

(continued)

Table 3.1 (continued)

Description of Operator	Mechanism Involved	Conditions for Observation	Impact	Organisms	References
Chromosomal rearrangements	Insertion elements or DNA repeats providing substrate for increased homologous recombination	Various environmental stresses (frequently host-related)	No or localized loss of information	Various bacteria	(Van Belkum et al. 1998; Kresse et al. 2003; Rocha 2004; Kang et al. 2006)
Translational stress-induced mutagenesis	Increased frequency of errors in translation of genes coding for DNA polymerases	Amino acid starvation	Modulation of replication fidelity	*E. coli*	(Al Mamun et al. 2006)
Transcription-associated mutagenesis	Increased mutagenesis associated with transcriptional derepression	Amino acid starvation	May potentiate mutation at alleles under direct selection	*Bacillus subtilis*	(Ross et al. 2006)
Transient hypermutability	Errors in transcription and/or translation that generate faulty proteins	Occurs as an intrinsic process unrelated to environmental stimulation	Production of mutants with multiple mutations	*E. coli* and other organisms	(Drake et al. 2005)
Transient gene amplification	Increase in amount of DNA substrate for mutation; RecA-dependent	Starvation	May lead to structural instability or hypermutability; contradicts tendency to compactness of genome	*E. coli*	(Andersson et al. 1998; Rocha 2004; Roth et al. 2006)
Genetic diversification in biofilms	RecA-dependent genetic modification of various cellular functions	Growth in biofilms	Production of specialist cells better adapted to particular niches within the biofilm; such specialists are less prone to further adaptation	*Pseudomonas aeruginosa*	(Buckling et al. 2003; Boles et al. 2004; Kirisits et al. 2005)

(continued)

Table 3.1 (continued)

Description of Operator	Mechanism Involved	Conditions for Observation	Impact	Organisms	References
Induction of transformation competence	Ability of cells to take up exogenous DNA	Constitutive or various regulatory modes (quorum sensing, nutritional controls, starvation)	Gain of function through homologous recombination	Variety of environmental and pathogenic bacteria	(Solomon and Grossman 1996)
Stimulation of horizontal gene transfer	SOS induction (at least in some cases)	DNA damage	Gain or sharing of adaptation; possibility of transfer of toxic genes; when the transferred sequence is adaptive, loss of competitive advantage of the donor strain	*Vibrio cholerae* (for SOS induction of transfer); environmental bacteria (for evolutionary consequences of transfer)	(Berg and Kurland 2002; Kurland et al. 2003; Beaber et al. 2004)
Activation of mobile genetic elements	Involves multiple signal transduction networks and, in some cases at least, RpoS function	Starvation	Transposition-related mutations and gene fusions	*E. coli*, *Pseudomonas putida*, other organisms	(Ilves et al. 2001; Shapiro 2005)
Operators of Genetic Stability					
Increase in DNA repair	Overexpression of DNA glycosylase Mug and other DNA-processing enzymes	Starvation	Mutation avoidance	*E. coli*	(Mokkapati et al. 2001)
Prevention of hypermutational phenotype	RecA function	Thermophilic conditions	Increased gene stability at high temperature	*Thermus thermophilus*	(Castán et al. 2003)
Dormancy	Production of endospores, myxospores, other types of resting cells	Starvation, quorum sensing	Reversible stabilization of the genome	Variety of multicellular and unicellular bacteria	(Ishihama 1997; Lazazzera 2000; Yamamoto 2000)

(continued)

Table 3.1 (continued)

Description of Operator	Mechanism Involved	Conditions for Observation	Impact	Organisms	References
Programmed cell death	SOS-dependent synthesis of SulA, an inhibitor of cell division	DNA damage	Destruction of cells with altered genome	*E. coli*	(Lewis 2000)
Definition of exchange community boundaries	Horizontal transfer mechanisms	Coexistence of bacteria differing in genome size, G+C content, carbon metabolism or oxygen tolerance	Preferential transfer among bacteria with similar characters	Variety of environmental bacteria	(Jain et al. 2003)
Polyploidy	Control of DNA replication	Phases of growth or of life cycle	Multiple genomic copies serve as buffer to deleterious mutation	Variety of environmental bacteria	(Bendich and Drlica 2000; Poole et al. 2003)
Polyploidy	Reassembly of shattered chromosomes	Resistance to desiccation and radiation	Chromosomal fragments with overlapping homologies are used both as primers and as templates for massive synthesis of complementary single strands	*Deinococcus radiodurans*	(Zahradka et al. 2006)

Within the highly structured and heterogeneous soil environment, a defining extreme parameter would act as a cross-gradient factor. Thus, application of this extreme parameter will attenuate the heterogeneous character of the soil environment, as a result of cross-protection (Jenkins et al. 1988; Hengge-Aronis 2002) that is provided by response to one particular stress (such as that caused by the defining extreme parameter) against unrelated stresses (such as soil-associated stresses). It may also be envisioned that there exists a maximal level of induction of the stress responses through the RpoS and RecA/LexA regulation, and that superimposition of a defining extreme parameter will result in constant induction of these responses at maximal or near-maximal levels. In this manner, the heterogeneous character of the soil, which was originally expressed as variations in stress levels according to space and time, will be tamed.

3.5 The Context of Extreme Soils

Initial land colonists might have been confronted with harsh conditions, characterized by intense radiation and low moisture. Whether early soil colonization was effected by thermophiles is debatable. It has been proposed that, along the history of the Earth, biotic evolution was tightly coupled with that of climate (Schwartzman and Lineweaver 2005). Climate might have been as warm or warmer in the Archean than today, and possibly very hot (Des Marais 1998; Kharecha et al. 2005). Surface Archean temperatures were estimated to have been ~55–80°C (Knauth and Lowe 2003), and the oceans might have been at a temperature of 70°C in the Precambrian, for much of the time between 3.5 and 2.5 billion years ago (Ga; Knauth 2005; Robert and Chaussidon 2006). Hence, it may be that initial land colonizers were thermophiles. At later times, major glaciation events, occurring at about 2.4 Ga and between 0.8 and 0.6 Ga (Kirschvink et al. 2000), may have resulted in massive psychrophilic adaptations (see Chapter 2).

In addition to global changes, local perturbations, such as the initiation of geothermal or volcanic activity (see Chapter 8) or else a change in the course of an acidic or heavy metal-rich stream, may superimpose an extreme character on a soil. The advent of a defining extreme parameter can have an abrasive or thinning effect on soil evolutionary layers, decreasing the number of species or individuals represented within a particular evolutionary layer (see Chapter 4), or removing some layers altogether when a particular metabolism becomes impossible. This would be the case, for example, of obligate photoautotrophy, as photosynthetic activity is inhibited above 73°C (Brock 1985; Rothschild and Mancinelli 2001). At the same time, this extreme parameter will have a founding effect, as the process of evolutionary layer accretion will be reinitiated on the basis of novel mutual dependences.

In extreme soils, a single defining parameter is superimposing a strong selective pressure on the soil system. The nature of this defining extreme component varies, and correspondingly the physiological response to this component also varies. The reader is referred to other chapters in this volume (see Chapters 2, 5, 8, 10, and 16) and various reviews, some examples of which are given below, for descriptions of physiological adaptations to heat (Stetter 2001), cold (D'Amico et al. 2002; Scherer and Neuhaus 2006), radiation (Battista et al. 1999; Baumstark-Khan and Facius 2001), extreme pH (Baker-Austin and Dopson 2007), and salt (Kunte et al. 2001; Oren 2006).

Given the complexity and sophisticated character of the adaptive responses to various extreme conditions (Cleaves and Chalmers 2004), it may be debatable whether such adaptations appear *de novo* in a soil newly exposed to extreme conditions. Certainly, extremophiles may pre-exist in soils in advance of the superimposition of extreme conditions (Norris et al. 2002), and some micro-organisms can be transported over considerable distances (Griffin et al. 2001; Kellogg and Griffin 2006). However, there are numerous examples to show that dispersal does not impede microbial diversification and the appearance of endemic populations (Whitaker and Banfield 2005). More specifically, it has been proposed that the biogeographic distribution of micro-organisms results from a balance between origination and extinction processes, with the environment being in part responsible for the spatial variation in microbial diversity (Martiny et al. 2006).

A study of the diversity and biogeographic distribution of soil bacterial communities across the American continent led to the conclusion that distribution is influenced primarily by edaphic variables, with soil pH playing a major role in the nonextreme soils that were examined (Fierer and Jackson 2006). Thus, the changes that are considered here do not correspond exclusively to the genetic creation of heterogeneity. The response to a recently introduced soil extreme component would be defined as a balance between selection of pre-existing colonists and generation of newly adapted forms. In discussing the respective contributions of dispersal and adaptation in establishment of biogeographic diversity patterns, it must be remembered that combinatorial processes occurring in mosaic genome pools enhance the adaptive potential of populations beyond that of individuals that compose them (Allen et al. 2007).

Although physiological adaptations to the various environmental extremes appear to be vastly different, commonalities become apparent when basic mechanisms and initial responses are considered. At the physiological level, changes in plasma membrane potential and/or ion flux modifications are amongst the earliest cellular events in response to a great variety of stresses and biological or mechanical stimulations (Shabala et al. 2006). At the genetic level, the stress response is a shared trait, universally expressed in prokaryotes and eukaryotes. Archaea share some components of the stress response with bacteria and others with eukaryotes (Macario et al. 1999). The universal stress response involves a core of common genes, and a minimal stress response proteome comprising 44 proteins, including the bacterial RecA or its eukaryotic homolog Rad51, has been postulated to exist in all cellular organisms (Kültz 2005). Of course, other components that complement this minimal protein set vary between organisms, and the modes of action of common components also vary. For example, the activity of RecA orthologs differ, as mentioned earlier (see Section 3.4.3).

The generalizations that can be drawn at the level of the basic microbial responses suggest that commonalities also exist between evolutionary processes. In particular, adaptation to edaphic and extreme environmental factors may involve similar mechanisms and influence each other, exerting a combined influence on resident bacteria. This compound influence is also noticeable at the level of community structure, as both the soil (Cho and Tiedje 2000) and extreme components (Whitaker and Banfield 2005) of the environment contribute to isolation of the resident populations. Indeed, isolation favors the creation of island dynamics, introducing into populations geographically restricted prokaryotes, or geovars (Staley and Gosink 1999), carrying environment-specific genes (Tringe et al. 2005).

The combined action on stress-responsive mechanisms of the adapting cell remains decomposable into its two constitutive elements. This implies that, although influencing the same target processes, the soil and defining extreme parameters may act synergistically or antagonistically on adaptative change (see also the discussion on epistasis below, Section 3.6.1).

Evolution in extreme soils can be envisioned to occur in two possible directions, which are simply described as "soil to extremophiles" and "extremophiles to soil". In the first case, a soil and its resident microbial community become exposed to a defining extreme parameter. In the second case, extremophiles from an extreme nonsoil environment colonize a soil sharing the extreme property. Hence, evolution of soil extremophiles occurs in a two-dimensional vectorial space, organized along the soil component and the extreme component axes. Assuming that an organism is already optimally adapted to soil, then evolution will occur strictly towards extremophily. Reciprocally, an optimally adapted extremophile will solely evolve soil microbial characteristics. In practice, however, departure from optimality is expected to occur in the course of adaptation (Fig. 3.1), as the result of genetic and physiological constraints as discussed below.

3.6 The Extreme Soil Adaptive Landscapes

3.6.1 Description of the Landscape

The relationships between environment and genotype have been described in terms of adaptive landscape (Wright 1988). This landscape can be viewed as a series of peaks occupied by optimally fit genotypes, separated by valleys where genotypes are not well adapted to their environment. Epistasis, or the interaction among genetic loci in their effect on phenotypes, results in linkage disequilibria (Maynard Smith et al. 1993) and influences the migration of populations in the adaptive landscape (Whitlock et al. 1995). Thus, as adaptation proceeds and nonrandom association of alleles and unlinked loci occurs, individuals become increasingly committed along evolutionary paths leading to a particular fitness peak. This can be

Fig. 3.1 Simplified view of patterns of evolution for soil micro-organisms and extremophiles colonizing extreme soils. The horizontal shaded arrow indicates that organisms already adapted to soil conditions evolve along a fitness gradient leading to tolerance of a defining extreme soil parameter. The vertical shaded arrow suggests that, reciprocally, organisms already adapted to extreme conditions acquire soil-specific adaptations. The dashed arrows illustrate that both soil organisms and extremophiles might lose some of their original adaptations in the course of this process, through tradeoffs or other mechanisms

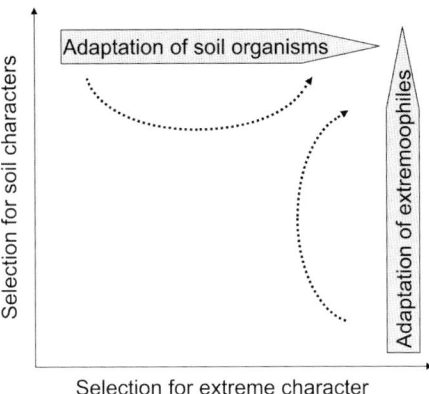

envisioned as a nth-order Markovian process (Usher 1979), whereby a transition from one genotype to another is influenced by an increasing set of previous genotypes, and hence becomes increasingly constrained (Fig. 3.2).

Peaks of adaptive landscapes can be sharp or smooth, depending on whether they are defined by strong or weak selection pressure, and they can be occupied simultaneously by a few or several different genotypes. Rugged fitness landscapes may present an initial nonadapted population with many possible peaks to climb, and under these conditions chance events are important in determining the initial evolutionary course (Korona et al. 1994; Colegrave and Buckling 2005).

The nature of evolutionary events differs in sharp and smooth peaks (Peliti 1997). In the first case, there are strong epistatic interactions, as the strength of directional epistasis is correlated with the average effect of a single mutation (Wilke and Adami 2001). This correlation implies that mutations with strong fitness effects will synergistically or antagonistically affect each other, hence defining the two signs of directional epistasis. On the other hand, in smooth peaks epistatic interactions are not so prevalent initially, but there is strong influence from Muller's ratchet, as slightly deleterious mutations tend to be fixed during evolution in asexual populations (Peliti 1997). The robustness of a system, or its tolerance to deleterious mutations, is correlated with antagonistic epistasis. Furthermore, robustness defines a threshold beyond which synergistic epistasis occurs, where the combined effect of co-existing mutations is larger than expected from the addition of their individual effects (Bershtein et al. 2006). Hence, whether they are expressed immediately or only after a robustness threshold is reached, synergistic epistatic interactions will multiply the effect of deleterious mutations, in both a smooth and a rugged fitness landscape.

Heterogeneity in soils provides for rugged fitness landscapes, with many sharp peaks. Sharpness of the peaks results from a variety of factors determining requisites for soil adaptation. Upon imposition of a defining extreme condition, the soil component of a fitness peak is expected to flatten, as the correspondence between heterogeneous edaphic factors and the intensity of the stress response loosens (see

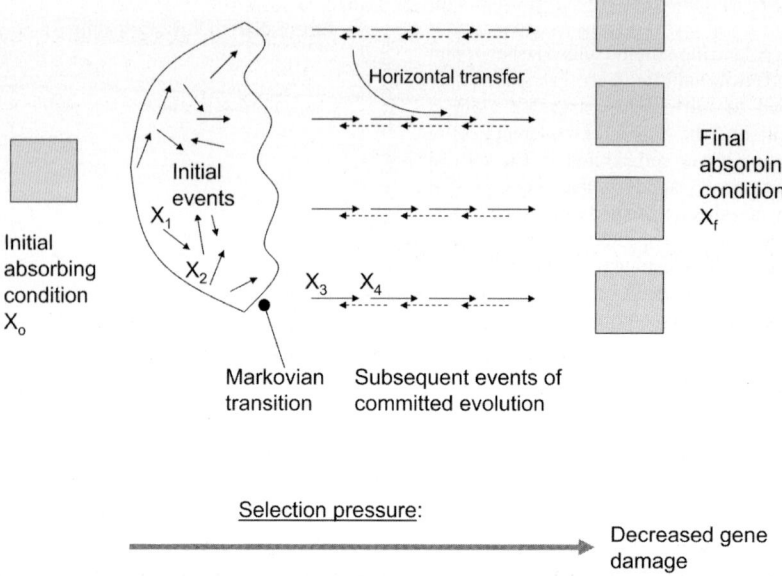

Fig. 3.2 Diversification of an initial population, occurring within a Markovian dependency structure of evolutionary changes, and under selection pressure for decreased gene damage. Evolutionary changes establish a nth-order Markovian chain, in which a given evolutionary condition X_r ($r = 1, 2, 3, \ldots, n$) at a given time depends on an elongating set of previous evolutionary conditions established at sets of previous times, up to the initial absorbing condition X_o. In this way, the stochastic evolutionary process becomes increasingly constrained and committed, through epistatic interactions and the consequent generation of linkage disequilibrium. Upon completion of the adaptation process, each of the final absorbing conditions X_f corresponds to the summit of a particular fitness peak. In the figure, the increased commitment of evolution is indicated as a wavy line of Markovian transition, which in reality would occur gradually. Horizontal transfer between evolving individuals would promote shifting from one Markovian evolutionary path to another

Section 3.5 for a discussion of the combined effects of edaphic and extreme parameters in influencing the stress response). Other factors contributing to the flattening of soil fitness peaks may include an expected decrease in intensity of biological interactions, as fewer soil organisms would tolerate the newly established defining extreme condition. More generally, there will be an equilibration of genotype fitness and adaptation capacity, as initially well-adapted genotypes might have become less capable of further adaptation to the newly changed extreme environment. Indeed, it has been observed that niche specialization may come with a cost of reduced potential to diversify (Buckling et al. 2003). Antagonistic pleiotropy, arising from tradeoffs (Cooper and Lenski 2000), might also contribute to make some of the better soil colonists less capable of adapting to the defining extreme condition.

The importance of biological interactions in defining sharpness of fitness peaks in the soil adaptive landscape is evidenced by the difference in capacity for survival of human intestinal pathogens in autoclaved and nonautoclaved soils (Jamieson et al. 2002; Jiang et al. 2002). It can be expected that the diminished microbial populations and partial or complete removal of plant hosts or other partners would contribute to flatten the fitness landscape by abolishing opportunities for competition, predation, and mutualism.

3.6.2 Lessons from Toxified Soils

The results of studies on herbicide application to soil illustrate the effect of stressors on soil microbial communities. In one such study, herbicide application resulted in a shift in microbial community composition, as some microbial species became less prevalent whereas others were stimulated (Engelen et al. 1998). This may be interpreted as a decrease in the impact of the soil component on community structure, as some genotypes that normally would be away from fitness peaks as defined by the soil factors had an opportunity to become dominant. These particular genotypes were less subjected to soil constraints and responded to a novel selection pressure imposed by the herbicide. Similarly, treatment of soil with Zn resulted in an initial death of microbes due to metal toxicity, followed by regrowth of metal-tolerant bacteria (Díaz-Raviña and Bååth 1996). There are numerous other instances where some groups of soil micro-organisms increased in numbers or proportion following chemical disturbance with heavy metals, pollutants, or herbicides (Kent and Triplett 2002).

If indeed the enhanced growth of some microbial groups in toxified soils reflects a decrease of the relative importance of the soil component of selection and a flattening of the corresponding fitness peaks, then there should be instances where a particular toxic treatment would result in overall decrease of microbial biomass, but without affecting microbial diversity. This was in fact observed in some instances. For example, treatment of soil with the fungicide triadimefon caused a decline in organic carbon and soil microbial biomass but no decline in microbial DNA diversity as measured with RAPD random primer amplification (Yang et al. 2000). In this case, the fungicide treatment may be viewed as having decreased the fitness of dominant bacteria residing at the summit of fitness peaks, thus causing an overall decline in microbial biomass, while allowing a greater number of previously maladapted genotypes to co-occupy the summit of smoother fitness peaks. Similarly, soil microbial biomass was more sensitive to pollution with heavy metals than was microbial diversity (Kandeler et al. 2000), again suggesting a relief from edaphic determinisms upon imposition of a defining extreme parameter. However, the above-cited studies remain specific in scope, and it is agreed with Kent and Triplett (2002) that more should be known on the relationship between amount of biomass and diversity of soil microbial communities.

Thus, introduction of a defining extreme condition in soil results in a remodeling of the adaptive landscape. This landscape now comprises composite fitness peaks, with a smooth soil component and a sharp component defined by the extreme parameter (Fig. 3.3). Initially, a variety of different soil bacteria will reside on the smooth soil component peak. Being away from the ultimate summit defined by the extreme condition, soil bacteria will be subjected to stress and enhanced mutagenesis. In this sense, they will go through a dual selection process, as they are presented with both the edaphic and the extreme components of fitness peaks. In both cases, selection for decreased stress and enhanced genetic stability will be exerted, whereas other organismic processes, such as growth and biotic interactions, will not be as strictly monitored as they are in nonextreme environments.

The extreme component exerts a homogenizing effect on the soil environment (see Section 3.6.1). In this sense, fewer fitness peaks will subsist in extreme soils (Fig. 3.4), bringing various genotypes into coexistence and providing for novel interactions. Numerous examples of interactions occurring on the fitness peaks defined by the extreme parameter are to be found in the present book. In particular, the reader is referred to descriptions of food webs created in extreme soils (see Chapter 4), of lipolithic and endolithic communities from arid soils (Wierzchos et al. 2006; see Chapter 6), of geochemical cycling in cold soils (see Chapter 7) and peatlands (see Chapter 9), and of cometabolic relationships established in hydrocarbon-contaminated desert soils (see Chapter 13).

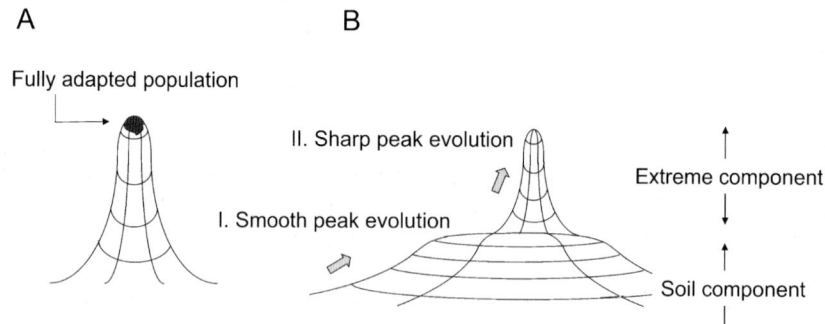

Fig. 3.3 Shape of an individual fitness peak from the adaptive landscape of a nonextreme soil (A) and an extreme soil (B). In nonextreme soils, fitness is determined by a complex series of edaphic and environmental, including climatic, parameters. These interact to mold a fitness landscape composed of simple peaks such as the one shown in (A). According to this simplified fitness topology, a microbial population optimally adapted to one of these peaks would have to lose some of this adaptation if it were to move to another peak and adapt to a different niche. In extreme soils, decreased intensity of interactions and other factors reduce the sharpness of the soil component of composite fitness peaks. Thus, different populations can coexist in a particular soil niche, but also have to adapt to a defining extreme parameter. In extreme soils, adapting populations must follow a path indicated by the arrows, to the summit of composite fitness peaks comprising a flat soil component and a sharp extreme component

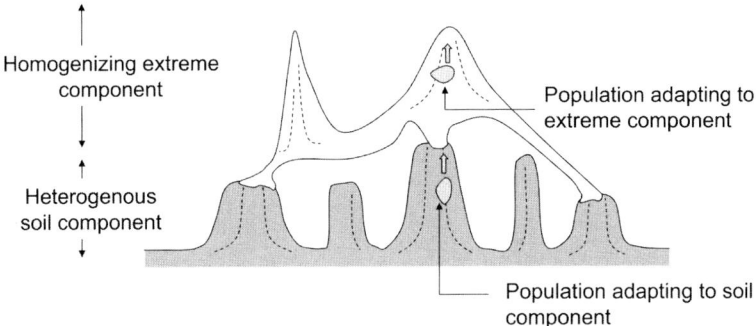

Fig. 3.4 Composite topography of adaptive landscape of an extreme soil. The extreme component of composite fitness peaks acts as a unifying factor, through the creation of new and restricted environments where resident populations are brought together and compete. On the other hand, the extreme component also acts as a discriminating factor, whereby some particular soil adaptations may be incompatible with further adaptation to the defining extreme parameter. This creates evolutionary dead-ends, in the form of soil fitness peaks which are not connected to the superimposing extreme fitness peak

3.7 Conclusions

The evolutionary effect of mutations varies according to the position of a genotype in the fitness landscape. To bacteria finding themselves at the bottom of a peak, mutation offers the promise of adaptation, and environmental challenges create pressure for greater evolvability of individuals. As the genotypes climb the fitness peak, selection for genetic stability occurs. Thus, it appears that evolution acts on evolution itself, in a mirroring effect. In extreme soils, where the selection pressure is in part defined by a clearly identifiable parameter, the mirror of evolution might provide illuminating images. There is truth to be found in such soils, in the absence of those distractions that gentler environments provide.

Acknowledgements The author is thankful to Ziv Arbeli for critical reading of the manuscript.

References

Aertsen A, Michiels CW (2006) Upstream of the SOS response: Figure out the trigger. *Trends Microbiol* 14:421–423

Al Mamun AAM, Gautam S, Humayun MZ (2006) Hypermutagenesis in *mutA* cells is mediated by mistranslational corruption of polymerase, and is accompanied by replication fork collapse. *Molec Microbiol* 62:1752–1763

Allen EE, Tyson GW, Whitaker RJ, Detter JC, Richardson PM, Banfield JF (2007) Genome dynamics in a natural archaeal population. *Proc Natl Acad Sci USA* 104:1883–1888

Andersson DI, Slechta ES, Roth JR (1998) Evidence that gene amplification underlies adaptive mutability of the bacterial *lac* operon. *Science* 282:1133–1135

Baker-Austin C, Dopson M (2007) Life in acid: pH homeostasis in acidophiles. *Trends Microbiol* 15:165–171

Banas JA et al. (2007) Evidence that accumulation of mutants in a biofilm reflects natural selection rather than stress-induced adaptive mutation. *Appl Environ Microbiol* 73:357–361

Bardgett RD (2002) Causes and consequences of biological diversity in soil. *Zoology* 105:367–375

Battista JR, Earl AM, Park M-J (1999) Why is *Deinococcus radiodurans* so resistant to ionizing radiation? *Trends Microbiol* 7:362–365

Baumstark-Khan C, Facius R (2001) Life under conditions of ionizing radiation. In: Horneck G, Baumstark-Khan C (eds) *Astrobiology – The Quest for the Conditions of Life*. Springer-Verlag, Berlin Heidelberg, pp 260–283

Beaber JW, Hochhut B, Waldor MK (2004) SOS response promotes horizontal dissemination of antibiotic resistance genes. *Nature* 427:72–74

Beare MH, Coleman DC, Crossley Jr DA, Hendrix PF, Odum EP (1995) A hierarchical approach to evaluating the significance of soil biodiversity to biogeochemical cycling. *Plant Soil* 170:5–22

Belotte D, Curien J-B, Maclean RC, Bell G (2003) An experimental test of local adaptation in soil bacteria. *Evolution* 57:27–36

Bendich AJ, Drlica K (2000) Prokaryotic and eukaryotic chromosomes: What's the difference? *BioEssays* 22:481–486

Berg OG, Kurland CG (2002) Evolution of microbial genomes: Sequence acquisition and loss. *Mol Biol Evol* 19:2265–2276

Bershtein S, Segal M, Bekerman R, Tokuriki N, Tawfik DS (2006) Robustness-epistasis link shapes the fitness landscape of a randomly drifting protein. *Nature* 444:929–932

Bjedov I et al. (2003) Stress-induced mutagenesis in bacteria. *Science* 300:1404–1409

Boles BR, Thoendel M, Singh PK (2004) Self-generated diversity produces "insurance effects" in biofilm communities. *Proc Natl Acad Sci USA* 101:16630–16635

Brock TD (1985) Life at high temperatures. *Science* 230:132–138

Buckling A, Wills MA, Colegrave N (2003) Adaptation limits diversification of experimental bacterial populations. *Science* 302:2107–2109

Cairns J, Overbaugh J, Miller S (1988) The origin of mutants. *Nature* 335:142–145

Castán P, Casares L, Barbé J, Berenguer J (2003) Temperature-dependent hypermutational phenotype in *recA* mutants of *Thermus thermophilus* HB27. *J Bacteriol* 185:4901–4907

Cho J-C, Tiedje JM (2000) Biogeography and degree of endemicity of fluorescent *Pseudomonas* strains in soil. *Appl Environ Microbiol* 66:5448–5456

Cleaves HJ, Chalmers JH (2004) Extremophiles may be irrelevant to the origin of life. *Astrobiology* 4:1–9

Colegrave N, Buckling A (2005) Microbial experiments on adaptive landscapes. *BioEssays* 27:1167–1173

Cooper VS, Lenski RE (2000) The population genetics of ecological specialization in evolving *Escherichia coli* populations. *Nature* 407:736–739

Cooper VS, Schneider D, Blot M, Lenski RE (2001) Mechanisms causing rapid and parallel losses of ribose catabolism in evolving populations of *Escherichia coli* B. *J Bacteriol* 183:2834–2841

D'Amico S et al. (2002) Molecular basis of cold adaptation. *Phil Trans R Soc B* 357:917–925

Denamur E, Matic I (2006) Evolution of mutation rates in bacteria. *Molec Microbiol* 60:820–827

Des Marais DJ (1998) Earth's early biosphere and its environment. In: Woodward CE, Shull JM, Thronson Jr HA (eds) *Origins (ASP Conference Series 148)*. Astronomical Society of the Pacific, San Francisco, pp 415–434

Di Giulio M (2003) The ancestor of the Bacteria domain was a hyperthermophile. *J Theor Biol* 224:277–283

Díaz-Raviña M, Bååth E (1996) Development of metal tolerance in soil bacterial communities exposed to experimentally increased metal levels. *Appl Environ Microbiol* 62:2970–2977

Dion P (2008) Reconstructing soil biology. In: Nautiyal CS, Dion P (eds) *Molecular Mechanisms of Plant-Microbe Coexistence*. Springer-Verlag, Berlin Heidelberg, in press

Drake JW, Bebenek A, Kissling GE, Peddada S (2005) Clusters of mutations from transient hypermutability. *Proc Natl Acad Sci USA* 102:12849–12854

Drake JW, Charlesworth B, Charlesworth D, Crow JF (1998) Rates of spontaneous mutation. *Genetics* 148:1667–1686

Earl DJ, Deem MW (2004) Evolvability is a selectable trait. *Proc Natl Acad Sci USA* 101:11531–11536

Elena SF, Cooper VS, Lenski RE (1996) Punctuated evolution caused by selection of rare beneficial mutations. *Science* 272:1802–1804

Engelen B, Meinken K, von Wintzingerode F, Heuer H, Malkomes H-P, Backhaus H (1998) Monitoring impact of a pesticide treatment on bacterial soil communities by metabolic and genetic fingerprinting in addition to conventional testing procedures. *Appl Environ Microbiol* 64:2814–2821

Erill I, Campoy S, Mazon G, Barbé J (2006) Dispersal and regulation of an adaptive mutagenesis cassette in the bacteria domain. *Nucl Acids Res* 34:66–77

Erwin DH (2000) Macroevolution is more than repeated rounds of microevolution. *Evol Dev* 2:78–84

Fierer N, Jackson RB (2006) The diversity and biogeography of soil bacterial communities. *Proc Natl Acad Sci USA* 103:626–631

Finkel SE, Kolter R (1999) Evolution of microbial diversity during prolonged starvation. *Proc Natl Acad Sci USA* 96:4023–4027

Fitz-Gibbon ST, Ladner H, Kim U-J, Stetter KO, Simon MI, Miller JH (2002) Genome sequence of the hyperthermophilic crenarchaeon *Pyrobaculum aerophilum*. *Proc Natl Acad Sci USA* 99:984–989

Gelsomino A, Keijzer-Wolters AC, Cacco G, van Elsas JD (1999) Assessment of bacterial community structure in soil by polymerase chain reaction and denaturing gradient gel electrophoresis. *J Microbiol Meth* 38:1–15

Giri B, Giang PH, Kumari R, Prasad R, Varma A (2005) Microbial diversity in soils. In: Buscot F, Varma A. (ed) *Microorganisms in Soils: Roles in Genesis and Functions*. Springer-Verlag, Berlin Heidelberg, pp 19–55

Goldenfeld N, Woese C (2007) Biology's next revolution. *Nature* 445:369

Griffin DW, Kellogg CA, Shinn EA (2001) Dust in the wind: Long range transport of dust in the atmosphere and its implications for global public and ecosystem health. *Global Change Hum Health* 2:20–33

Grogan DW, Carver GT, Drake JW (2001) Genetic fidelity under harsh conditions: Analysis of spontaneous mutation in the thermoacidophilic archaeon *Sulfolobus acidocaldarius*. *Proc Natl Acad Sci USA* 98:7928–7933

Grundmann GL (2004) Spatial scales of soil bacterial diversity - The size of a clone. *FEMS Microbiol Ecol* 48:119–127

Grundmann GL, Debouzie D (2000) Geostatistical analysis of the distribution of NH_4^+ and NO_2^--oxidizing bacteria and serotypes at the millimeter scale along a soil transect. *FEMS Microbiol Ecol* 34:57–62

Grundmann GL, Dechesne A, Bartoli F, Flandrois JP, Chassé JL, Kizungu R (2001) Spatial modeling of nitrifier microhabitats in soil. *Soil Sci Soc Am J* 65:1709–1716

Grundmann GL, Normand P (2000) Microscale diversity of the genus *Nitrobacter* in soil on the basis of analysis of genes encoding rRNA. *Appl Environ Microbiol* 66:4543–4546

Hengge-Aronis R (1999) Interplay of global regulators and cell physiology in the general stress response of *Escherichia coli*. *Curr Opin Microbiol* 2:148–152

Hengge-Aronis R (2002) Signal transduction and regulatory mechanisms involved in control of the σ^S (RpoS) subunit of RNA polymerase. *Microbiol Mol Biol Rev* 66:373–395

Hoitink HAJ, Boehm MJ (1999) Biocontrol within the context of soil microbial communities: A substrate-dependent phenomenon. *Annu Rev Phytopathol* 37:427–446

Ilves H, Hõrak R, Kivisaar M (2001) Involvement of σ^S in starvation-induced transposition of *Pseudomonas putida* transposon Tn*4652*. *J Bacteriol* 183:5445–5448

Ishihama A (1997) Adaptation of gene expression in stationary phase bacteria. *Curr Opin Genet Develop* 7:582–588
Islas S, Velasco AM, Becerra A, Delaye L, Lazcano A (2003) Hyperthermophily and the origin and earliest evolution of life. *Int Microbiol* 6:87–94
Jain R, Rivera MC, Moore JE, Lake JA (2003) Horizontal gene transfer accelerates genome innovation and evolution. *Mol Biol Evol* 20:1598–1602
Jamieson RC, Gordon RJ, Sharples KE, Stratton W, Madani A (2002) Movement and persistence of fecal bacteria in agricultural soils and subsurface drainage water: A review. *Can Biosyst Eng* 44:1.1–1.9
Janion C (2001) Some aspects of the SOS response system — A critical survey. *Acta Biochim Pol* 48:599–610
Jenkins DE, Schultz JE, Matin A (1988) Starvation-induced cross protection against heat or H_2O_2 challenge in *Escherichia coli*. *J Bacteriol* 170:3910–3914
Jiang X, Morgan J, Doyle MP (2002) Fate of *Escherichia coli* O157:H7 in manure-amended soil. *Appl Environ Microbiol* 68:2605–2609
Kandeler E et al. (2000) Structure and function of the soil microbial community in microhabitats of a heavy metal polluted soil. *Biol Fertil Soils* 32:390–400
Kang JM, Iovine NM, Blaser MJ (2006) A paradigm for direct stress-induced mutation in prokaryotes. *FASEB J* 20:2476–2485
Kassen R (2002) The experimental evolution of specialists, generalists, and the maintenance of diversity. *J Evol Biol* 15:173–190
Kellogg CA, Griffin DW (2006) Aerobiology and the global transport of desert dust. *Trends Ecol Evol* 21:638–644
Kent AD, Triplett EW (2002) Microbial communities and their interactions in soil and rhizosphere ecosystems. *Annu Rev Microbiol* 56:211–236
Kharecha P, Kasting J, Siefert J (2005) A coupled atmosphere-ecosystem model of the early Archean Earth. *Geobiology* 3:53–76
King T, Seeto S, Ferenci T (2006) Genotype-by-environment interactions influencing the emergence of *rpoS* mutations in *Escherichia coli* populations. *Genetics* 172:2071–2079
Kirisits MJ, Prost L, Starkey M, Parsek MR (2005) Characterization of colony morphology variants isolated from *Pseudomonas aeruginosa* biofilms. *Appl Environ Microbiol* 71:4809–4821
Kirschvink JL et al. (2000) Paleoproterozoic snowball Earth: Extreme climatic and geochemical global change and its biological consequences. *Proc Nat Acad Sci USA* 97:1400–1405
Kivisaar M (2003) Stationary phase mutagenesis: Mechanisms that accelerate adaptation of microbial populations under environmental stress. *Environ Microbiol* 5:814–827
Knauth LP (2005) Temperature and salinity history of the Precambrian ocean: Implications for the course of microbial evolution. *Palaeogeog Palaeoclimatol Palaeoecol* 219:53–69
Knauth LP, Lowe DR (2003) High Archean climatic temperature inferred from oxygen isotope geochemistry of cherts in the 3.5 Ga Swaziland Supergroup, South Africa. *Geol Soc Am Bull* 115:566–580
Korona R, Nakatsu C, Forney L, Lenski R (1994) Evidence for multiple adaptive peaks from populations of bacteria evolving in a structured habitat. *Proc Nat Acad Sci USA* 91:9037–9041
Kresse AU, Dinesh SD, Larbig K, Römling U (2003) Impact of large chromosomal inversions on the adaptation and evolution of *Pseudomonas aeruginosa* chronically colonizing cystic fibrosis lungs. *Molec Microbiol* 47:145–158
Kültz D (2005) Molecular and evolutionary basis of the cellular stress response. *Annu Rev Physiol* 67:225–257
Kunte HJ, Trüper HG, Stan-Lotter H (2001) Halophilic microorganisms. In: Horneck G, Baumstark-Khan C (eds) *Astrobiology – The Quest for the Conditions of Life*. Springer-Verlag, Berlin Heidelberg, pp 185–199
Kurland CG, Canback B, Berg OG (2003) Horizontal gene transfer: A critical view. *Proc Natl Acad Sci USA* 100:9658–9662

Kussell E, Kishony R, Balaban NQ, Leibler S (2005) Bacterial persistence: A model of survival in changing environments. *Genetics* 169:1807–1814

Kussell E, Leibler S (2005) Phenotypic diversity, population growth, and information in fluctuating environments. *Science* 309:2075–2078

Lazazzera BA (2000) Quorum sensing and starvation: Signals for entry into stationary phase. *Curr Opin Microbiol* 3:177–182

Lewis K (2000) Programmed death in bacteria. *Microbiol Mol Biol Rev* 64:503–514

Lockwood JL (1988) Evolution of concepts associated with soilborne plant pathogens. *Annu Rev Phytopathol* 26:93–121

Loewe L (2004) Response to comment on "High deleterious genomic mutation rate in stationary phase of *Escherichia coli*". *Science* 304:518d

Macario AJL, Lange M, Ahring BK, De Macario EC (1999) Stress genes and proteins in the Archaea. *Microbiol Mol Biol Rev* 63:923–967

Martiny JBH et al. (2006) Microbial biogeography: Putting microorganisms on the map. *Nat Rev Microbiol* 4:102–112

Maynard Smith J, Smith N, O'Rourke M, Spratt B (1993) How clonal are bacteria? *Proc Natl Acad Sci USA* 90:4384–4388

Mazón G, Campoy S, Fernández de Henestrosa AR, Barbé J (2006) Insights into the LexA regulon of *Thermotogales*. *Antonie van Leeuwenhoek* 90:123–137

McArthur JV, Kovacic DA, Smith MH (1988) Genetic diversity in natural populations of a soil bacterium across a landscape gradient. *Proc Natl Acad Sci USA* 85:9621–9624

Michel B (2005) After 30 years of study, the bacterial SOS response still surprises us. *PLoS Biol* 3:e255

Miller C, Thomsen LE, Gaggero C, Mosseri R, Ingmer H, Cohen SN (2004) SOS response induction by β-Lactams and bacterial defense against antibiotic lethality. *Science* 305:1629–1631

Mokkapati SK, Fernández de Henestrosa AR, Bhagwat AS (2001) *Escherichia coli* DNA glycosylase Mug: A growth-regulated enzyme required for mutation avoidance in stationary-phase cells. *Molec Microbiol* 41:1101–1111

Moore FB-G, Woods R (2006) Tempo and constraint of adaptive evolution in *Escherichia coli* (Enterobacteriaceae, Enterobacteriales). *Biol J Linnean Soc* 88:403–411

Moxon ER, Rainey PB, Nowak MA, Lenski RE (1994) Adaptive evolution of highly mutable loci in pathogenic bacteria. *Curr Biol* 4:24–33

Napolitano R, Janel-Bintz R, Wagner J, Fuchs RPP (2000) All three SOS-inducible DNA polymerases (Pol II, Pol IV and Pol V) are involved in induced mutagenesis. *EMBO J* 19:6259–6265

Nicol GW, Glover LA, Prosser JI (2003) Spatial analysis of archaeal community structure in grassland soil. *Appl Environ Microbiol* 69:7420–7429

Noguez AM, Arita HT, Escalante AE, Forney LJ, García-Oliva F, Souza V (2005) Microbial macroecology: Highly structured prokaryotic soil assemblages in a tropical deciduous forest. *Global Ecol Biogeog* 14:241–248

Nohmi T (2006) Environmental stress and lesion-bypass DNA polymerases. *Annu Rev Microbiol* 60:231–253

Norris TB, Wraith JM, Castenholz RW, McDermott TR (2002) Soil microbial community structure across a thermal gradient following a geothermal heating event. *Appl Environ Microbiol* 68:6300–6309

Notley-McRobb L, Pinto R, Seeto S, Ferenci T (2002) Regulation of *mutY* and nature of mutator mutations in *Escherichia coli* populations under nutrient limitation. *J Bacteriol* 184:739–745

Nunan N, Wu K, Young IM, Crawford JW, Ritz K (2002) In situ spatial patterns of soil bacterial populations, mapped at multiple scales, in an arable soil. *Microb Ecol* 44:296–305

Nunan N, Wu K, Young IM, Crawford JW, Ritz K (2003) Spatial distribution of bacterial communities and their relationships with the micro-architecture of soil. *FEMS Microbiol Ecol* 44:203–215

Oren A (2006) Life at high salt concentrations. In: Dworkin M, Falkow S, Rosenberg E, Schleifer K-H, Stackebrandt E (eds) *The Prokaryotes, a Handbook on the Biology of Bacteria*, 3rd edn, vol 2. Springer-Verlag, Berlin Heidelberg, pp 263–282

Peliti L (1997) Introduction to the statistical theory of Darwinian evolution. arXiv:cond-mat/9712027 (http://lanl.arxiv.org/abs/cond-mat/9712027)

Poole AM, Phillips MJ, Penny D (2003) Prokaryote and eukaryote evolvability. *Biosystems* 69:163–185

Price CW, Fawcett P, Cérémonie H, Su N, Murphy CK, Youngman P (2001) Genome-wide analysis of the general stress response in *Bacillus subtilis*. *Molec Microbiol* 41:757–774

Price PB (2007) Microbial life in glacial ice and implications for a cold origin of life. *FEMS Microbiol Ecol* 59:217–231

Ramette A, Tiedje JM (2007) Multiscale responses of microbial life to spatial distance and environmental heterogeneity in a patchy ecosystem. *Proc Natl Acad Sci USA* 104:2761–2766

Ranjard L et al. (2000) Heterogeneous cell density and genetic structure of bacterial pools associated with various soil microenvironments as determined by enumeration and DNA fingerprinting approach (RISA). *Microb Ecol* 39:263–272

Robert F, Chaussidon M (2006) A palaeotemperature curve for the Precambrian oceans based on silicon isotopes in cherts. *Nature* 443:969–972

Rocha EP (2004) Order and disorder in bacterial genomes. *Curr Opin Microbiol* 7:519–527

Ross C, Pybus C, Pedraza-Reyes M, Sung H-M, Yasbin RE, Robleto E (2006) Novel role of *mfd*: effects on stationary-phase mutagenesis in *Bacillus subtilis*. *J Bacteriol* 188:7512–7520

Roth JR, Kugelberg E, Reams AB, Kofoid E, Andersson DI (2006) Origin of mutations under selection: the adaptive mutation controversy. *Annu Rev Microbiol* 60:477–501

Rothschild LJ, Mancinelli RL (2001) Life in extreme environments. *Nature* 409:1092–1101

Saint-Ruf C, Matic I (2006) Environmental tuning of mutation rates. *Environ Microbiol* 8:193–199

Scherer S, Neuhaus K (2006) Life at low temperatures. In: Dworkin M, Falkow S, Rosenberg E, Schleifer K-H, Stackebrandt E (eds) *The Prokaryotes, a Handbook on the Biology of Bacteria*, 3rd edn, vol 2. Springer-Verlag, Berlin Heidelberg, pp 210–262

Schloter M, Lebuhn M, Heulin T, Hartmann A (2000) Ecology and evolution of bacterial microdiversity. *FEMS Microbiol Rev* 24:647–660

Schwartzman D, Lineweaver CH (2005) Temperature, biogenesis and biospheric self-organization. In: Kleidon A, Lorenz R (eds) *Non-Equilibrium Thermodynamics and the Production of Entropy: Life, Earth, and Beyond*. Springer-Verlag, Berlin Heidelberg, pp 207–222

Schwartzman DW, Lineweaver CH (2004) The hyperthermophilic origin of life revisited. *Biochem Soc Trans* 32:168–171

Sessitsch A, Weilharter A, Gerzabek MH, Kirchmann H, Kandeler E (2001) Microbial population structures in soil particle size fractions of a long-term fertilizer field experiment. *Appl Environ Microbiol* 67:4215–4224

Shabala L, Ross T, McMeekin T, Shabala S (2006) Non-invasive microelectrode ion flux measurements to study adaptive responses of microorganisms to the environment. *FEMS Microbiol Rev* 30:472–486

Shapiro JA (2005) Thinking about evolution in terms of cellular computing. *Nat Comput* 4:297–324

Slechta ES, Bunny KL, Kugelberg E, Kofoid E, Andersson DI, Roth JR (2003) Adaptive mutation: General mutagenesis is not a programmed response to stress but results from rare coamplification of *dinB* with *lac*. *Proc Natl Acad Sci USA* 100:12847–12852

Sniegowski PD, Gerrish PJ, Johnson T, Shaver A (2000) The evolution of mutation rates: Separating causes from consequences. *BioEssays* 22:1057–1066

Solomon JM, Grossman AD (1996) Who's competent and when: Regulation of natural genetic competence in bacteria. *Trends Genet* 12:150–155

Staley JT (1997) Biodiversity: Are microbial species threatened? *Curr Opin Biotechnol* 8:340–345

Staley JT, Gosink JJ (1999) Poles apart: Biodiversity and biogeography of sea ice bacteria. *Annu Rev Microbiol* 53:189–215

Stetter KO (2001) Hyperthermophilic microorganisms. In: Horneck G, Baumstark-Khan C (eds) *Astrobiology – The Quest for the Conditions of Life*. Springer-Verlag, Berlin Heidelberg, pp 169–184

Stewart EJ, Madden R, Paul G, Taddei F (2005) Aging and death in an organism that reproduces by morphologically symmetric division. *PLoS Biology* 3:e45

Taddei F, Matic I, Radman M (1995) cAMP-dependent SOS induction and mutagenesis in resting bacterial populations. *Proc Natl Acad Sci USA* 92:11736–11740

Tark M et al. (2005) A DNA polymerase V homologue encoded by TOL Plasmid pWW0 confers evolutionary fitness on *Pseudomonas putida* under conditions of environmental stress. *J Bacteriol* 187:5203–5213

Thattai M, van Oudenaarden A (2004) Stochastic gene expression in fluctuating environments. *Genetics* 167:523–530

Torkelson J, Harris RS, Lombardo MJ, Nagendran J, Thulin C, Rosenberg SM (1997) Genome-wide hypermutation in a subpopulation of stationary-phase cells underlies recombination-dependent adaptive mutation. *EMBO J* 16:3303–3311

Torsvik V, Øvreås L (2002) Microbial diversity and function in soil: From genes to ecosystems. *Curr Opin Microbiol* 5:240–245

Torsvik V, Øvreås L, Thingstad TF (2002) Prokaryotic diversity – Magnitude, dynamics, and controlling factors. *Science* 296:1064–1066

Tringe SG et al. (2005) Comparative metagenomics of microbial communities. *Science* 308:554–557

Usher MB (1979) Markovian approaches to ecological succession. *J Anim Ecol* 48:413–426

Van Belkum A, Scherer S, van Alphen L, Verbrugh H (1998) Short-sequence DNA repeats in prokaryotic genomes. *Microbiol Molec Biol Rev* 62:275–293

Van Gestel M, Merckx R, Vlassak K (1996) Spatial distribution of microbial biomass in microaggregates of a silty-loam soil and the relation with the resistance of microorganisms to soil drying. *Soil Biol Biochem* 28:503–510

Van Veen J, van Overbeek L, van Elsas J (1997) Fate and activity of microorganisms introduced into soil. *Microbiol Mol Biol Rev* 61:121–135

Vogel J, Normand P, Thioulouse J, Nesme X, Grundmann GL (2003) Relationship between spatial and genetic distance in *Agrobacterium* spp. in 1 cubic centimeter of soil. *Appl Environ Microbiol* 69:1482–1487

Vos M, Velicer GJ (2006) Genetic population structure of the soil bacterium *Myxococcus xanthus* at the centimeter scale. *Appl Environ Microbiol* 72:3615–3625

Wardle DA, Bardgett RD, Klironomos JN, Setälä H, van der Putten WH, Wall DH (2004) Ecological linkages between aboveground and belowground biota. *Science* 304:1629–1633

Webster G, Embley TM, Prosser JI (2002) Grassland management regimens reduce small-scale heterogeneity and species diversity of β-proteobacterial ammonia oxidizer populations. *Appl Environ Microbiol* 68:20–30

Whitaker RJ, Banfield JF (2005) Population dynamics through the lens of extreme environments. *Rev Mineral Geochem* 59:259–277

Whitlock MC, Phillips PC, Moore FB-G, Tonsor SJ (1995) Multiple fitness peaks and epistasis. *Annu Rev Ecol Syst* 26:601–629

Wierzchos J, Ascaso C, McKay CP (2006) Endolithic cyanobacteria in halite rocks from the hyperarid core of the Atacama desert. *Astrobiology* 6:415–422

Wilke CO, Adami C (2001) Interaction between directional epistasis and average mutational effects. *Proc R Soc B* 268:1469–1474

Willenbrock H, Friis C, Juncker AS, Ussery DW (2006) An environmental signature for 323 microbial genomes based on codon adaptation indices. *Genome Biol* 7:R114

Wipat A, Harwood CR (1999) The *Bacillus subtilis* genome sequence: The molecular blueprint of a soil bacterium. *FEMS Microbiol Ecol* 28:1–9

Woese CR (1987) Bacterial evolution. *Microbiol Rev* 51:221–271

Woese CR (2002) On the evolution of cells. *Proc Natl Acad Sci USA* 99:8742–8747

Wright S (1988) Surfaces of selective value revisited. *Am Nat* 131:115–123

Yamamoto H (2000) Viable but nonculturable state as a general phenomenon of non-spore-forming bacteria, and its modeling. *J Infect Chemother* 6:112–114

Yang Y-H, Yao J, Hu S, Qi Y (2000) Effects of agricultural chemicals on DNA sequence diversity of soil microbial community: A study with RAPD marker. *Microb Ecol* 39:72–79

Yeiser B, Pepper ED, Goodman MF, Finkel SE (2002) SOS-induced DNA polymerases enhance long-term survival and evolutionary fitness. *Proc Natl Acad Sci USA* 99:8737–8741

Young IM, Crawford JW (2004) Interactions and self-organization in the soil-microbe complex. *Science* 304:1634–1637

Zahradka K et al. (2006) Reassembly of shattered chromosomes in *Deinococcus radiodurans*. *Nature* 443:569–573

Zinser ER, Kolter R (2004) *Escherichia coli* evolution during stationary phase. *Res Microbiol* 155:328–336

Chapter 4
Biodiversity: Extracting Lessons from Extreme Soils

Diana H. Wall

4.1 Introduction

The organisms that live in extreme environments have justifiably captured the imagination of people fascinated with the detection of life and exploration. Reasons for this captivation vary. Some see exploration of these organisms and their environment as a scientific area to provide insight about life on earth, whereas others see economic potential. Whether the extreme environment is human-caused, such as a polluted soil, or a more natural environment (aquatic hot springs, ice, ocean depths, atmosphere, or land), unravelling and understanding the resident organisms, their mechanisms of survival, and the intricate relationship between the habitat and other species, can help us understand life on this planet and elsewhere. Because of global changes, many aspects of extreme environments, such as the identity and types of organisms and communities, the biological traits that allow evolutionary success in a harsh environment, the patterns of distribution of these organisms, the factors controlling their distribution, and their influence on and feedback from ecosystem processes, have increasing relevance to all terrestrial ecosystems. This chapter examines how extreme soils as a habitat for biota can inform our general knowledge of terrestrial biodiversity in many other ecosystems. A brief background on soil biodiversity from other terrestrial systems is presented to set the stage for lessons derived from studies of extreme soils.

Biodiversity is defined by the United Nations Convention on Biodiversity (CBD) as the "variability among living organisms from all sources ... and the biological complexes of which they are a part: this includes diversity within species, between species and of ecosystems" (Convention on Biodiversity 2004). This expansive definition is extremely useful for describing life on earth, determining the biotic composition of an ecosystem, and addressing the rapid changes occurring at temporal and spatial scales to the ecosystem, such as the increasing rate of

---

Diana H. Wall
Department of Biology and Natural Resource Ecology Laboratory, Colorado State University, Fort Collins, CO 80523-1499
e-mail: Diana@nrel.colostate.edu

extinctions of species. Scientists have emphasized that entire populations, as well as single individuals of a species, are being lost at an accelerating rate. The CBD definition is based on both classical morphological and/or genetic taxonomic knowledge of the biological distribution of species, whether endemic or widespread.

The numbers of species of plants and large animals across landscapes and their global distribution are better known than for smaller invertebrates or microbes, and likewise, more is known about biodiversity above- than belowground (Wardle 2002). Whether there are accelerating rates of extinction for less visible organisms such as bacteria or fungi, and particularly for belowground biota, has yet to be determined. Instead, in many cases, protection of a land area for aboveground species assumes belowground species are also conserved. Whether the spatial scale of the protected area is adequate for conservation of both above- and belowground species and food webs is less studied, but land conserved for a plant in a small habitat might be inadequate to conserve significant levels of diversity belowground.

The world's soils have a large abundance and wealth of biotic diversity with species numbers estimated to be greater than aboveground diversity (Wardle 2002). Taxa include microbes (bacteria among which are cyanobacteria and actinomycetes, fungi, Archaea), protozoa, microscopic invertebrates (microarthropods, nematodes, rotifers, tardigrades), large invertebrates (snails, millipedes, centipedes, termites and earthworms), vertebrates (moles, gophers, lizards), vascular plant roots and lichens, cryptogamic crusts, algae, and mosses. Many of these smaller groups can be found in a handful of soil (Wall and Virginia 2000). Because of the abundance and diversity of the multiple taxonomic groups, identifying all species and their interactions in a single soil sample has been problematic. Instead, our understanding of soil biodiversity is largely based on trophic or functional classifications (e.g., herbivore, predator, microbial feeder, detritus feeder) derived from scientific literature on the feeding habits and morphology of a few species. This approach can then be extrapolated to assemble other unnamed species into functional groups in complex food webs. The contribution of soil biota components in ecosystem processes has thus been postulated based on functional grouping of similar species into food webs. This has proven useful to quantify the role of soil biota in processes such as regulating the rate of soil organic matter decomposition, nitrification, primary production, and nutrient cycling (Hunt and Wall 2002). However, research is needed, as assumptions regarding functionality of large groups of soil organisms may not provide realistic measures of these processes.

A further and critical recognition of the dependence of humans on the benefits provided by soil biodiversity is the concept of ecosystem services (Millennium Ecosystem Assessment 2005; Wall 2004). These include carbon sequestration, generation and renewal of soil structure and soil fertility, flood and erosion control, bioremediation of wastes and pollutants, modification of the hydrologic cycle, regulation of atmospheric trace gases, and biocontrol of human, animal, and plant pathogens and parasites. Alterations and loss of the world's terrestrial soils are occurring rapidly, raising concerns that some of these services may be largely interrupted, as they are currently rendered by unsustainable soils (Millennium Ecosystem Assessment 2005).

Given this background on soil biodiversity, can soil biodiversity of extreme environments inform us about biodiversity and ecosystems elsewhere? My intent here is to augment lessons from microbes living in extreme soils with examples of their consumers, primarily invertebrates, in order to better extract and extend lessons to biodiversity inhabiting global soils. It is hoped that these lessons will be further expanded and clarified by the many scientists who are developing exciting new approaches for detecting these amazing organisms and learning how they live in extreme soils and are integral to the working of ecosystems.

4.2 Lesson One: Biodiversity in Soils Is Hidden

Although this statement appears obvious to those working on extreme soils, scientists often bring their biases of larger, visible, and more easily detected organisms to the study of soils. However, most life in soils is microscopic. In extreme soils, an emphasis is to detect and study microbes and microscopic life, particularly at the species or molecular level, whereas in other ecosystems the attention to larger, mostly visible life sometimes dismisses the variety of life below the surface.

In extreme soils of lower plant and animal species diversity, multiple techniques are used to detect life. Without familiarity and testing of the correct extraction technique, microscopic animals and microbes may be considered absent and soils, 'sterile'. Scientists working on extreme soils recognize that techniques used to isolate microscopic invertebrates are varied and require a basic understanding of the limitations of the method as well as the general biology of each particular group. Techniques for extraction, identification, and enumeration (based on classical morphology) are specialized and may differ for each group of taxa occurring in soils. For example, to extract microarthropods and nematode roundworms, two groups of mesofauna that occur in soils worldwide, a single technique should not be used.

Microarthropods (mites and Collembola) live in air-filled pores of soil whereas nematodes are aquatic animals living in water films around soil particles. Nematologists may extract nematodes from soil with methods depending on soil type and organic matter content and whether they want to recover the whole community or just a targeted species. Such methods are based on movement in water by gravity, sieving-centrifugation, or flotation techniques. Microarthropods have a different physiology and behavior and are removed from air pores in soil by methods based on active avoidance (e.g., avoidance of heat using Berlese–Tullgren funnels), aspiration, and flotation (Coleman et al. 1999; Ducarme et al. 1998). Within these two major groups of soil fauna, species differ in body size, movement, life histories, temperature requirements, feeding habits, and physiologies. In addition, many species are rare, and may not be detected without prior evaluation and use of several techniques. For example, the drier, saltier, low carbon soils of the Antarctic Dry Valleys (see Chapters 2 and 12) located away from meltstreams were considered almost sterile until the early 1990s, but different extraction techniques for nematodes and molecular analyses of microbes have shown greater diversity and distribution than previously thought (Aislabie et al. 2006; Barrett et al. 2006; Freckman and Virginia 1997; Wall and Virginia 1999). Rapid faunal analysis from soil using bulk animal DNA for identification is emerging as an addition to classical morphological techniques. However, as with microbial molecular tools, faunal DNA analysis will need to be related to activity of viable populations. As with extreme soils, using numerous techniques in a coordinated manner will increase detection of organisms in all terrestrial soil systems, no matter the location or depth of the soil. This information will increase our knowledge of Earth's biodiversity.

4.3 Lesson Two: Soil Species Have More Than One Survival Strategy

Many survival strategies exist among the organisms in extreme soils that may contribute to evolutionary success. Distantly related organisms may share a strategy, and additionally may have developed multiple adaptations for maintaining populations. Evolution has selected for biota that express ecological traits such as long versus short life cycles, sexual versus other reproductive modes, numerous versus few eggs, multiple dispersal mechanisms, alterations in morphology, and active migration to avoid stress. Types of cryptobiosis, an ametabolic, reversible response to environmental stress known in many taxonomically distinct organisms such as most tardigrades, rotifers, and nematodes, are a response to desiccation (anhydrobiosis), freezing (cryobiosis), and salinity (osmobiosis; Block 1982; Pugh and Dartnall 1994; Sinclair and Sjursen 2001; Treonis and Wall 2005). In the Antarctic, soil nematodes have a variety of strategies including anhydrobiosis, cryobiosis, cold-hardiness (Pickup 1990), intracellular freezing (Wharton 2003), dispersal by wind (Nkem et al. 2006), and life histories. Microarthropods can supercool (Convey et al. 2003), be heat tolerant, or cold-hardy (Sinclair and Sjursen 2001), and can desiccate (Montiel et al. 1998; Worland and Lukesova 2000). Algae and mosses in extreme hot and cold deserts desiccate without water and in the polar deserts become freeze-dried through the long winters until temperature and moisture combine to trigger activity (McKnight et al. 1999). Examples of resistance mechanisms for microbes living in extreme soils are discussed throughout this volume, and add to the synthesis of the underlying evolutionary adaptations of all soil biota.

Survival mechanisms extend to more biodiverse soils in other ecosystems. Even within a diverse phylum such as nematodes, anhydrobiosis is widely distributed. Many nematodes in temperate and tropical soils undergo anhydrobiosis when soils dry, including phylogenetically different species such as the fungal-feeding nematode *Aphelenchus avenae* (Browne et al. 2004; Crowe and Madin 1975; Freckman et al. 1980), bacterial-feeding species, *Panagrolaimus* and *Acrobeloides*, the obligate plant parasites *Rotylenchulus reniformis* and *Scuttellonema brachyurum*, and many others (Demeure et al. 1979; Goyal et al. 2003). However, the degree to which anhydrobiosis, like other forms of cryptobiosis, protects different species can vary (Rothschild and Mancinelli 2001; Wharton 2003).

Thus, a combination of ecological and physiological traits has allowed species to successfully evolve and maintain active populations in extreme soil habitats. These few examples from extreme soils suggest multiple strategies that might also be expressed in nonextreme soils to enable responses to environmental change.

4.4 Lesson Three: Extreme Soils Are Ecosystems

Although there may be fewer species in extreme soils, these particular soils harbor all the characteristics of an ecosystem, for example, species variability, food webs, nutrient cycling, production, decomposition, and interaction with the environment. Food webs in extreme soils are simpler than in other ecosystems and usually have fewer trophic levels because of lower energy or primary production input. In extreme soils, controls on trophic levels in a food web are dependent more on abiotic controls than on top predators of lower trophic levels. Whether the food web is primary producer-based or detritus-based, most food webs will involve only two functional groups, microbes and their consumers (Moore and de Ruiter 2000). Detrital-based food webs could have an additional trophic level if they depend on two types of contemporary primary production: autochthonous (algae growing in soils) or allochthonous (detritus blown in from a nearby source); but if based on ancient legacy carbon alone, there will be only two trophic levels. Protozoa and larger-sized organisms (of size range from 500 µm to 2 mm), such as microfauna (rotifers, tardigrades) and mesofauna (microarthropods, nematodes) consume producers (cyanobacteria or algae), or consume decomposers (bacteria or fungi), and thus regulate the turnover of microbes and nutrients. As these organisms die, organic carbon and nutrients are recycled back to the soil. In some extreme soils such as in the Atacama Desert with their hypolithic communities of phototrophs, the organisms appear to interact solely as primary producers, but more data on heterotrophic microbes are needed (Warren-Rhodes et al. 2006; see also Chapter 6).

The Antarctic Dry Valleys provide examples of soil communities with few species in trophic groups. These have both primary producer-based and detritus-based food webs. These include algal feeders – a single species of nematode, *Eudorylaimus antarcticus* (Wall 2007), bacterial feeders – two nematode species, and more rarely, fungal feeders – a mite and a collembolan species. Tardigrades and rotifers that feed on bacteria or algae occur in about 14% of the wet, organic matter-rich soils across the Dry Valley landscape (Freckman and Virginia 1997). In contrast to those that colonize plant-dominated soils, Dry Valley taxa rarely coexist as a community or more complex food web, and competitive interactions can be limited (Hogg et al. 2006). Extreme soils can also be characterized by an absence of consumer populations and their predators. About 60% of the soils in the Dry Valleys lack nematodes and about 50% of soils in Ellsworth Land, Antarctica, and areas in the Atacama Desert lack soil mesofauna (Convey and McInnes 2005; Freckman and Virginia 1997; Warren-Rhodes et al. 2006). Whether these unsuitable soil habitats are due to soil geochemical and/or food source limitations, or else to other factors, is being studied (Poage et al. in press; Warren-Rhodes et al. 2006).

It is somewhat easier to clarify the food sources within an extreme soil food web, and thus the role of a species in the ecosystem, than it is in highly diverse soils. All the consumers, micro- and/or mesofauna are usually known at the species level in an extreme soil ecosystem. For example, using stable isotopes, Bokhorst et al. (2007) showed that a polar collembolan species feeds preferentially on lichens and algae, rather than moss. Less is known about faunal species feeding on a selective bacterial species, particularly for extreme soils where microbial diversity is appearing to be higher than previously reported (Aislabie et al. 2006; Barrett et al. 2006; Cowan and Tow 2004). Nevertheless, compared to the study of more diverse food webs, analysis of the extreme soil food webs is particularly useful to reveal food web architecture, the role of the species in the ecosystem, and the degree of overlap in geographic species range for soil fauna.

Food webs in nonextreme soils have high energy input from plants and algae, more trophic levels, and potentially hundreds of species in a functional group. Larger macrofauna prey on smaller mesofauna, and so on through the food web. Food webs are thus extremely complex: for example, the diversity of fungal feeding mite species in soils might range in the hundreds of species whereas in an extreme soil, there may be only a single species, if any. Resolving food sources for each species in a functional group for most soils is thus extremely difficult due to the high number of species. Instead, transfer of nutrients, for example, carbon, nitrogen, and phosphorus, through the soil food web can only be estimated based on abundance and biomass of invertebrates within the various functional groups.

Because most functional groups have many species performing the same task or role in highly diverse soil ecosystems, it has been argued that there is considerable redundancy (Loreau and Thebault 2005). If a species were lost, another species would take its place and there would be little change in the ecosystem function. More recently, experiments (Heemsbergen et al. 2004; Roscher et al. 2004) suggest that functional diversity is more important to an ecosystem function than the number of species (see also Hunt and Wall 2002). This is not the case in low-diversity systems where both numbers of functional groups and species are low (e.g., a functional group is represented by one species) and, frequently, one species is key to a process (Wall 2007). Loss of one species could decrease an ecosystem process in an extreme soil. For example, a single nematode species in the Dry Valley soils, the bacterial feeder *Scottnema lindsayae*, is responsible for a disproportionate amount of soil carbon turnover, about 5–7% (Barrett et al. unpublished), such level of activity being unachievable in temperate ecosystems with their highly diverse and greater biomass (Schröter et al. 2003). As the Dry Valleys have cooled, populations of *S. lindsayae* have declined with as yet unknown implications for carbon cycling (Doran et al. 2002).

Knowledge at an ecosystem level gained from studying simple food webs and individual species in extreme soils can be transferred to other terrestrial soils. Simply stated, microscopic species placed in a functional group may not be equal in their roles in an ecosystem process. Their roles may differ on temporal, spatial, physiological, nutritional, and other measurable scales, but may be masked by sheer numbers of species. Combined field and laboratory experimentation to clarify food web interactions will enhance our ability to detect potential ecosystem effects involving loss of species or shift in composition of species (or functional groups). Synthesizing this information will enable us to better monitor how soil biodiversity is altered by global changes, to compare impacts across soil ecosystems, and to better formulate actions to assure long-term soil sustainability.

4.5 Lesson Four: Soils Are Major Drivers of Biodiversity

The geochemical component of extreme soils structures the diversity of life to a greater extent than the corresponding component in nonextreme soils, where biotic influences on soil organic matter and soil structure have masked effects of parent material. Many undisturbed extreme soils today reflect the past geologic history and parent material, and contribute to soil habitats that are highly heterogeneous at small and large spatial scales across the landscape. In ecosystems where plants are absent, for example, the hot hyperarid Atacama desert and the cold polar desert soils of the Dry Valleys, Antarctica, soils are relatively unchanged by centuries of biological (including human) activity and thus, the legacy of previous soil geochemistry patterns still remains. These deserts have extremely low water (<25 mm mean annual rainfall for the Atacama, and <10 cm rainfall equivalent for the Dry Valleys), low soil carbon, low organic matter, high pH, and high salinity (Barrett et al. 2004; Warren-Rhodes et al. 2006) compared to other ecosystems. As with other arid ecosystems, however, there is high spatial variability because soil chemical (e.g., C, N, P, organic matter, pH, salinity) and physical factors (structure, texture, soil type, pore space, bulk density) combine in varying proportions to form numerous habitats for organisms, which can range from suitable to poor (Barrett et al. 2004; Courtright et al. 2001; Wall and Virginia 1999). The soil geochemical heterogeneity affects the abundance of suitable habitats for life and contributes to patchily distributed fauna. Organisms, whether microbes, plants, or invertebrates, are limited by availability of soil resources at centimeter to kilometer scales (Ettema and Wardle 2002; Freckman and Virginia 1989; Poage et al. in press; Schlesinger et al. 1996; Wall and Virginia 1999; Warren-Rhodes et al. 2006).

In this way, spatial segregation of species occurs in extreme soils without the influence of plant roots. For example, in the hyperarid hot Atacama Desert, absence of water determined the spatial scale distribution and presence of photosynthetic and heterotrophic bacteria (Warren-Rhodes et al. 2006). In cold desert Dry Valley soils, where vascular plants are lacking and average mean annual surface soil temperatures are −26°C, four nematode species are distributed across the landscape according to food sources and soil habitat geochemical characteristics (Barrett et al. 2007; Porazinska et al. 2002; Treonis et al. 1999; Wall and Virginia 1999). *S. lindsayae*, the bacterial feeder that is widely distributed and has a greater abundance than the other nematode species, occurs in soils that are drier, saltier, and less organically rich. Another bacteria feeder, *Plectus* spp., is associated with soils that are moist, less saline, and with higher organic carbon; thus, this species rarely overlaps geographically for food with *S. lindsayae*. *Eudorylaimus antarcticus*, the algal feeder (Wall 2007), is found in soil habitats that are moist and highly organic, but its highest abundance occurs in lake sediments and streams (Ayres et al. 2007; Treonis et al. 1999). *Eudorylaimus* and *Plectus* frequently co-occur, but infrequently are found with *S. lindsayae*, or with a rarely found fourth species, *Geomonhystera* sp. This soil food web in the Dry Valley soil ecosystem has no predators and is likely limited by physical constraints rather than species' competition (Wall 2007). Other examples of simple food webs, broad niches, and spatial segregation have been seen in other extreme soils (Convey and McInnes 2005; Richard et al. 1994). These examples illustrate how the heterogeneity in soil habitats alone can be a major driver of local biogeographical patterns.

Globally, geographical patterns of soil biodiversity are driven primarily by climate and vegetation with soil heterogeneity having a variable role in determining biodiversity across spatial scales (Ettema and Wardle 2002). In younger soil systems, the soil biota, including plant roots, have contributed to the organic matter, total carbon, and soil structure by the formation of soil pores and channels. Thus, younger soils are subject to variation in biological, physical, and chemical alterations across shorter temporal and spatial scales than polar deserts, which tend to be older (Young and Crawford 2004). There are, however, examples of plant-dominated ecosystems where the soil substratum may be a stronger driver of belowground biogeographical patterns (Ettema and Wardle 2002). Fierer and Jackson (2006) show the influence of one soil factor, soil pH, as a predictor of soil microbial diversity across ecosystem types in North and South America. However, pH did not explain distribution of hypolithic soil bacteria in the Atacama Desert (Warren-Rhodes et al. 2006). In Arctic soils, where plants occur, variation in soil moisture was a major determinant of CO_2 respiration, which represents an overall measure of soil biotic metabolism (Sjögersten et al. 2006).

Globally, the aboveground distribution patterns of animal and plant diversity generally follow the latitudinal gradient hypotheses (Gaston 1996) of increasing species diversity from the poles to the tropics (Willig et al. 2003). The question of whether microbes are everywhere globally (Fenchel and Finlay 2004; Finlay 2002), or instead have spatial biogeography such as a latitudinal gradient, has resurged as a scientific debate and spawned research to examine constraints to dispersal and colonization for microbes less than 500 µm in addition to organisms of larger size (Fierer and Jackson 2006; Hughes Martiny et al. 2006; Lawley et al. 2004). The discussion on microbes and biogeography has extended to include bacteria, Archea, and some Eukarya (e.g., unicellular algae, Protozoa). In extreme soils, recent studies support unique organisms (Smith et al. 2006; Warren-Rhodes et al. 2006), but it is difficult to prove that these microbes are indeed unique (missing from other ecosystems) because of limited studies using similar molecular detection techniques or descriptions of soil habitat data.

Termites are one of the few groups of soil invertebrates that appear to follow the latitudinal gradient pattern (Eggleton et al. 1996, 1995). Global biogeography for the majority of soil fauna, particularly the micro- and mesofauna is less well known (Bardgett et al. 2005; Hughes Martiny et al. 2006; Maraun et al. 2007) partially because a greater proportion of soils have been sampled in temperate ecosystems (Bardgett 2005). One diverse group of soil microarthropods, oribatid mites, increases in diversity from boreal to temperate ecosystems but this trend does not extend to the tropics (Maraun et al. 2007), which may indicate a sampling problem. This problem has also been noted for global distribution patterns of soil nematodes (Bardgett 2005; Boag and Yeates 1998).

Given the variation in global ecosystems, it is challenging to establish ecological hypotheses explaining patterns of biogeography for soil biodiversity (Willig et al. 2003) or to determine if, at local regional or global scales, species-rich soils correspond to more productive ecosystems (Ettema and Wardle 2002; Young and Crawford 2004). Research in extreme ecosystems has already shown clearly that variation in soil geochemistry alone creates numerous soil habitats that are distinctly different and suitable for some species, but not others. This information, combined with information on species dispersal and colonization and with further knowledge on other drivers (vegetation, climate) of soil biodiversity, should contribute to better predictions of global soil biogeography.

4.6 Lesson Five: Global Changes Are Rapidly Changing Soils

Global changes (climate change, atmospheric change, land use change, species extinctions, invasive species) are having an impact on soils at an increasingly rapid rate (Millennium Ecosystem Assessment 2005). Effects of soil degradation include loss of soil organic matter, erosion, salinisation, compaction, contamination, and sealing (Wall 2004). Several international agreements address the irreversible loss of productive soils and the impact on biodiversity. As an example, the UN Convention to Combat Desertification was signed in 1997 by 178 nations to mitigate the effects of drought by implementation of action plans (UN Convention to Combat Desertification 1997). As recent as September 2006, the European Commission adopted a strategy specifically for soil protection by the EU (European Commission on Soil Protection 2006). These and other agreements have as a basis the knowledge of the benefits to humans (called ecosystem services, see Section 4.1) provided by soils. The understanding that soil life is critically important for provision of ecosystem services is less well accepted as a basis for policy decisions than is the notion of the role of physical degradation (Wall 2004). Additionally, less attention has been drawn to extreme soils, with the exception perhaps of those organisms living in chemically impacted soils (see Chapters 15 and 16), partially because of the magnitude of the ecosystem change occurring to many species and habitats worldwide and because extreme soils do not produce crops (Alley et al. 2007). Can knowledge of biodiversity in extreme soil ecosystems apply elsewhere, when globally there is accumulating evidence that soil functional composition and some soil species are being altered by global changes (Swift et al. 1998; Wardle et al. 2004; Wolters et al. 2000)?

In extreme soils, global changes may homogenize habitat ranges of species, with differing effects on these systems. Rapid environmental climate change could alter species distribution as influenced by habitat requirements, physiological tolerances, and life histories. For example, warming in extremely cold ecosystems might increase soil moisture levels across large spatial scales and blend soil habitat chemistry by affecting decomposition rates and the amount of carbon in the soil, primary production, and salinity levels (Wall 2007). In ecosystems with extreme drought events, soils might have reduced heterogeneity in chemical and physical properties due to high wind erosion. Other global changes in extreme soil systems, such as land use change (resulting from increased human activity) and increased incidence of invasive species, could also alter the food web interactions and relative stability of extreme soils. Thus, the present habitats that specify species range could be altered significantly with consequent changes in species composition and geographic distributions and cascading effects on ecosystem processes across the landscape (Wall 2007).

Little is known about the effect of global changes at the individual species level for most soil systems (Convey and McInnes 2005; Doran et al. 2002). Evidence from the study of extreme soil biodiversity suggests that global change effects will differ with species and ecosystem, and that even species with broad niches can be vulnerable (Barrett et al. unpublished). Thus, in extreme but also in more diverse soils, it will be important to quantify which, if any, species are key to an ecosystem process or ecosystem service and whether they are vulnerable to the many global changes.

4.7 Conclusions

Extreme soils may initially appear to be vastly different from each other and from the rest of the world's soils. There are several unique features of extreme soils: their lack of easily detectable life, reduced number of mesofaunal species within a functional group, fewer trophic levels, less complex food webs, lack of small-scale geographic overlap of species within functional groups, marked periodicity of activity, and food selectivity by invertebrate species (Convey and McInnes 2005; Porazinska et al. 2002; Treonis et al. 1999; Wall 2007). Collectively, the study of extreme soil habitats has revealed information on their biodiversity and on species interactions that are difficult to examine in nonextreme soils. Although much is yet to be learned, there is now sufficient evidence that species diversity in extreme soils has similarities to soil biodiversity elsewhere. Researchers interested in both extreme and nonextreme soils have used some species' traits, revealed at the genetic, population, community, or ecosystem levels, to obtain quantitative measurements of biodiversity. All of these criteria for biodiversity estimation are compatible with the Convention on Biological Diversity definition of biodiversity mentioned earlier, which encompasses variability from species to landscapes levels (Convention on Biological Diversity 2004).

Extreme soil ecosystems are more than model systems or microcosms. The biodiversity found in extreme soil environments is an integral part of the diversity of terrestrial surfaces. The extreme soil ecosystems are not anomalies; they are local to relatively large terrestrial ecosystems with a range of life forms, albeit belonging to relatively few species, having several types of life histories and extraordinary physiological adaptations. For this reason alone, extreme soil habitats are worthy of conservation and protection. Kareiva and Marvier (2003) argued that ecosystems with fewer species, rather than those with high diversity, could be considered higher priorities for conservation. Extreme soils are also critically valuable as indicators of global changes likely to affect other soil systems. Our challenge is to define these habitats in more detail, to quantify their contribution to ecosystem processes and services, and to establish the relevance of biodiversity in extreme soils for sustaining life in all terrestrial soils.

References

Aislabie J, Chhour K-L, Saul D, Miyauchi S, Ayton J, Paetzold R, Balks M (2006) Dominant bacteria in soils of Marble Point and Wright Valley, Victoria Land, Antarctica. *Soil Biol Biochem* 38:3041–3056

Alley R, et al. (2007) Climate Change 2007: The Physical Science Basis. (http://www.usssp-iodp.org/PDFs/SPM2feb07.pdf)

Ayres E, Wall DH, Adams BJ, Barrett JE, Virginia RA (2007) Unique similarity of faunal communities across aquatic-terrestrial interfaces in a polar desert ecosystem. Ecosystems DOI: 10.1007/s10021-007-9035-x

Bardgett RD (2005) *The Biology of Soil*. Oxford University Press, New York

Bardgett RD, Yeates G, Anderson J (2005) Patterns and determinants of soil biological diversity. In: Bardgett R, Usher M, Hopkins D (eds) *Biological Diversity and Function in Soils*. Cambridge University Press, New York, pp 100–118

Barrett JE, Virginia RA, Lyons WB, McKnight DM, Priscu JC, Doran PT, Fountain AG, Wall DH, Moorhead DL (2007) Biogeochemical stoichiometry of Antarctic Dry Valley ecosystems. *J Geophys Res* 112:G01010

Barrett JE, Virginia RA, Wall DH, Cary SC, Adams BJ, Hacker AL, Aislabie JM (2006) Co-variation in soil biodiversity and biogeochemistry in Northern and Southern Victoria Land, Antarctica. *Antarct Sci* 18:535–548

Barrett JE, Virginia RA, Wall DH, Adams BJ (unpublished) Decline of a dominant invertebrate species contributes to altered carbon cycling in low diversity soil ecosystem

Barrett JE, Virginia RA, Wall DH, Parsons AH, Powers LE, Burkins MB (2004) Variation in biogeochemistry and soil biodiversity across spatial scales in a polar desert ecosystem. *Ecology* 85:3105–3118

Block W (1982) Supercooling points of insects and mites on the Antarctic Peninsula. *Ecol Entomol* 7:1–8

Boag B, Yeates G (1998) Soil nematode biodiversity in terrestrial ecosystems. *Biodiv Conserv* 7:617–630

Bokhorst S, Ronfort C, Huiskes A, Convey P, Aerts R (2007) Food choice of Antarctic soil arthropods clarified by stable isotope signatures. *Pol Biol*, Published online DOI 10.1007/s00300-007-256-4

Browne J, Dolan K, Tyson T, Goyal K, Tunnacliffe A, Burnell A (2004) Dehydration-specific induction of hydrophilic protein genes in the anhydrobiotic nematode *Aphelenchus avenae*. *Eukaryot Cell* 3:966–975

Coleman DC, Elliot ET, Blair JM, Freckman DW (1999) Soil invertebrates. In: Robertson G, Coleman D, Bledsoe C, Phillips S (eds) *Standard Soil Methods for Long-Term Ecological Research*. Oxford University Press, New York, pp 349–377

Convention on Biological Diversity (2004) Decisions adopted by the conference of the parties to the convention of biological diversity at its seventh meeting. (http://www.cbd.int/doc/decisions/COP-07-dec-en.pdf)

Convey P, Block W, Peat H (2003) Soil arthropods as indicators of water stress in Antarctic terrestrial habitats? *Global Change Biol* 9:1718–1730

Convey P, McInnes S (2005) Exceptional tardigrade-dominated ecosystems in Ellsworth Land, *Antarct. Ecol* 86:519–527

Courtright EM, Wall DH, Virginia RA (2001) Determining habitat suitability for soil invertebrates in an extreme environment: The McMurdo Dry Valleys, Antarctica. *Antarct Sci* 13:9–17

Cowan D, Tow L (2004) Endangered Antarctic environments. *Annu Rev Microbiol* 58:649–690

Crowe J, Madin K (1975) Anhydrobiosis in nematodes: Evaporative water loss and survival. *J Exp Zool* 193:323–333

Demeure Y, Freckman D, Van Gundy S (1979) *In vitro* response of four species of nematodes to desiccation and discussion of this and related phenomena. *Rev Nématol* 2:203–210

Doran P, Priscu J, Lyons W, Walsh J, Fountain A, McKnight D, Moorhead D, Virginia R, Wall D, Clow G, Fritsen C, McKay C, Parsons A (2002) Antarctic climate cooling and terrestrial ecosystem response. *Nature* 415:517–520

Ducarme X, André HM, Lebrun P (1998) Extracting endogenous microarthropods: A new flotation method using 1,2-dibromoethane. *Eur J Soil Biol* 34:143–150

Eggleton P, Bignell D, Sands W, Mawdsley N, Lawton C, Wood T, Bignell N (1996) The diversity, abundance, and biomass of termites under differing levels of disturbance in the Mbalmayo Forest Reserve, southern Cameroon. *Phil Trans R Soc London* 351:51–68

Eggleton P, Bignell D, Sands W, Waite B, Wood T, Lawton J (1995) The species richness of termites (Isoptera) under differing levels of forest disturbance in the Mbalmayo Forest Reserve, southern Cameroon. *J Trop Ecol* 11:85–98

Ettema CH, Wardle DA (2002) Spatial soil ecology. *Trends Ecol Evol* 17:177–183

European Commission on Soil Protection (2006) A strategy to keep Europe's soils robust and healthy. (http://ec.europa.eu/environments/soil/index.htm#publications)

Fenchel T, Finlay B (2004) The ubiquity of small species: Patterns of local and global diversity. *BioScience* 54:777–784

Fierer N, Jackson R (2006) The diversity and biogeography of soil bacterial communities. *Proc Nat Acad Sci USA* 103:626–631

Finlay B (2002) Global dispersal of free-living microbial eukaryote species. *Science* 296:1061–1063

Freckman DW, Demeure Y, Munnecke D, Van Gundy S (1980) Resistance of the anhydrobiotic *Aphelenchus avenae* to methyl bromide fumigation. *J Nematol* 12:19–22

Freckman DW, Virginia R (1989) Plant-feeding nematodes in deep-rooting desert ecosystems. *Ecology* 70:1665–1678

Freckman DW, Virginia R (1997) Low diversity Antarctic soil nematode communities: distribution and response to disturbance. *Ecology* 78:363–369

Gaston K (1996) Species-range-size distributions: Patterns, mechanisms, and implications. *Trends Ecol Evol* 11:197–201

Goyal K, Tisi L, Basran A, Browne J, Burnell A, Zurdo J, Tunnacliffe A (2003) Transition from natively unfolded to folded state induced by desiccation in an anhydrobiotic nematode protein. *J Biol Chem* 278:12977–12984

Heemsbergen D, Berg M, Loreau M, van Haj J, Faber J, Verhoef H (2004) Biodiversity effects on soil processes explained by interspecific functional dissimilarity. *Science* 306:1019–1020

Hogg I, Cary S, Convey P, Newsham K, O'Donnell A, Adams B, Aislabie J, Frati F, Stevens M, Wall D (2006) Biotic interactions in Antarctic terrestrial ecosystems: Are they a factor? *Soil Biol Biochem* 38:3035–3040

Hughes Martiny J, et al. (2006) Microbial biogeography: Putting microorganisms on the map. *Nature Rev* 4:102–112

Hunt H, Wall D (2002) Modelling the effects of loss of soil biodiversity on ecosystem function. *Global Change Biol* 8:33–50

Kareiva P, Marvier M (2003) Conserving biodiversity coldspots. *Am Sci* 91:344–351

Lawley B, Ripley S, Bridge P, Convey P (2004) Molecular analysis of geographic patterns of Eukaryotic diversity in Antarctic soils. *Appl Environ Microbiol* 70:5963–5972

Loreau M, Thebault E (2005) Food webs and the relationship between biodiversity and ecosystem function. In: de Ruiter P, Wolters V, Moore J (eds) *Dynamic Food Webs: Multispecies Assemblages, Ecosystem Development and Environmental Change*. Academic Press, Amsterdam, pp 270–294

Maraun M, Schatz H, Scheu S (2007) Awesome or ordinary? Global diversity patterns of oribatid mites. *Ecography* 30:209–216

McKnight D, Niyogi D, Alger A, Bomblies A, Conovitz P, Tate M (1999) Dry Valley streams in Antarctica: Ecosystems waiting for water. *BioScience* 49:985–995

Millennium Ecosystem Assessment (2005) *Our Human Planet: Summary for Decision Makers*. World Resources Institute, Washington, DC

Montiel P, Grubor-Lajsic G, Worland M (1998) Partial desiccation induced by sub-zero temperatures as a component of the survival strategy of the Arctic collembolan *Onychiurus arcticus J Insect Physiol* 44:211–219

Moore J, de Ruiter P (2000) Invertebrates in detrital food webs along gradients of productivity. In: Coleman D, Hendrix P (eds) *Invertebrates as Webmasters in Ecosystems*. CABI, New York, pp 161–184

Nkem JM, Wall DH, Virginia RA, Barrett JE, Broos E, Porazinska D, Adams BJ (2006) Wind dispersal of soil invertebrates in the McMurdo Dry Valleys, Antarctica. *Pol Biol* 29:346–352

Pickup J (1990) Strategies of cold-hardiness in three species of Antarctic dorylaimid nematodes. *J Comp Physiol B* 160:167–173

Poage M, Barrett JE, Virginia RA, Wall DH The influence of soil geochemistry on nematode distribution, McMurdo Dry Valleys, Antarctica. *Arct Antarct Alp Res* (in press)

Porazinska DL, Wall DH, Virginia RA (2002) Population age structure of nematodes in the Antarctic Dry Valleys: Perspectives on time, space, and habitat suitability. *Arct Antarct Alp Res* 34:159–168

Pugh P, Dartnall H (1994) The Acari of fresh- and brackish water habitats in the Antarctic and sub-Antarctic regions. *Pol Biol* 14:401–404

Richard K, Convey P, Block W (1994) The terrestrial arthropod fauna of the Byers-Peninsula, Livingston-Island, South-Shetland-Islands. *Pol Biol* 14:371–379

Roscher C, Schumacher J, Baade J, Wilcke W, Gleixner G, Weisser W, Scmid B, Schulze E (2004) The role of biodiversity for element cycling and trophic interactions: an experimental approach in a grassland community. *Basic Appl Ecol* 5:107–121

Rothschild LJ, Mancinelli RL (2001) Life in extreme environments. *Nature* 409:1092–1101

Schlesinger W, Raikes J, Hartley A, Cross A (1996) On the spatial pattern of soil nutrients in desert ecosystems. *Ecology* 77:364–374

Schröter D, Wolters V, de Ruiter P (2003) Carbon and nitrogen mineralisation in the decomposer food webs of a European forest transect. *Oikos* 102:294–308

Sinclair B, Sjursen H (2001) Cold tolerance of the Antarctic springtail *Gomphicephalus hodgsoni* (Collembola, Hypogastruridae). *Antarct Sci* 13:271–279

Sjögersten S, van der Wal R, Woodin S (2006) Small-scale hydrological variation determines landscape carbon dioxide fluxes in the high Arctic. *Biogeochemistry* 80:205–216

Smith J, Tow L, Stafford W, Cary C, Cowan D (2006) Bacterial diversity in three different Antarctic cold desert mineral soils. *Microb Ecol* 51:413–421

Swift MJ, Andren O, Brussaard L, Briones M, Couteaux MM, Ekschmitt K, Kjoller A, Loiseau P, Smith P (1998) Global change, soil biodiversity, and nitrogen cycling in terrestrial ecosystems: three case studies. *Global Change Biol* 4:729–743

Treonis A, Wall D (2005) Soil nematodes and desiccation survival in the extreme arid environment of the Antarctic Dry Valleys. *Integr Comp Biol* 45:741–750

Treonis A, Wall DH, Virginia R (1999) Invertebrate biodiversity in Antarctic Dry Valley soils and sediments. *Ecosystems* 2:482–492

UN Convention to Combat Desertification (1997) Text of the United Nations Convention to Combat Desertification. (http://www.uccd.int/convention/text/convention.php)

Wall DH (ed) (2004) *Sustaining Biodiversity and Ecosystem Services in Soils and Sediments.* Island Press, Washington, D.C.

Wall DH (2007) Global change tipping points: Above- and below-ground biotic interactions in a low diversity ecosystem. *Phil Trans R Soc B* DOI: 10.1098/rstb.2006.1950

Wall DH, Virginia RA (1999) Controls on soil biodiversity: Insights from extreme environments. *Appl Soil Ecol* 13:137–150

Wall DH, Virginia RA (2000) The world beneath our feet: Soil biodiversity and ecosystem functioning. In: Raven PR, Williams T (eds) *Nature and Human Society: The Quest for a Sustainable World.* National Academy of Sciences and National Research Council, Washington, DC, pp 225–241

Wardle D (2002) *Communities and Ecosystems: Linking the Aboveground and Belowground Components.* Princeton University Press, Princeton, NJ

Wardle DA, Brown VK, Behan-Pelletier V, St. John M, Wojtowicz T, Brussaard L, Hunt HW, Paul EA, Wall DH (2004) Vulnerability to global change of ecosystem goods and services driven by soil biota. In: Wall DH (ed) *Sustaining Biodiversity and Ecosystem Services in Soil and Sediments.* Island Press, Washington DC, pp 101–136

Warren-Rhodes KA, Rhodes KL, Pointing SB, Ewing SA, Lacap DC, Gomez-Silva B, Amundson R, Friedmann EI, McKay CP (2006) Hypolithic cyanobacteria, dry limit of photosynthesis, and microbial ecology in the hyperarid Atacama Desert. *Microb Ecol* 52:389–398

Wharton D (2003) The environmental physiology of Antarctic terrestrial nematodes: A review. *J Comp Physiol B* 173:621–628

Willig M, Kaufman D, Stevens R (2003) Latitudinal gradients of biodiversity: Pattern, process, scale, and synthesis. *Annu Rev Ecol Evol Syst* 34:273–313

Wolters V, et al. (2000) Effects of global changes on above- and belowground biodiversity in terrestrial ecosystems: Implications for ecosystem functioning. *BioScience* 50:1089–1098

Worland M, Lukesova A (2000) The effect of feeding on specific algae on the cold-hardiness of two Antarctic micro-arthropods (*Alaskozetes antarcticus* and *Crytopygus antarcticus*). *Pol Biol* 23:766–774

Young I, Crawford J (2004) Interactions and self-organization in the soil-microbe complex. *Science* 304:1634–1637

Part II
Natural Extreme Soils

Chapter 5
Halophilic and Halotolerant Micro-Organisms from Soils

Antonio Ventosa(✉), Encarnacion Mellado, Cristina Sanchez-Porro, and M. Carmen Marquez

5.1 Introduction

Hypersaline environments are extreme habitats in which several other factors, in addition to high salt content, may limit the growth of organisms. These additional factors include temperature, pH, pressure, oxygen, nutrient availability, and solar radiations (Rodriguez-Valera 1988). Hypersaline environments comprise hypersaline waters and soils. Hypersaline waters are defined as those environments that have higher concentrations of salts than seawater (Rodriguez-Valera 1988). However, depending on their origin, the salt composition may differ from that of seawater and on that basis hypersaline water habitats are categorized as thalassohaline, when the relative amounts of the different inorganic salts are approximately equal to those present in seawater, or as athalassohaline, if the proportions of the different salts are markedly different from those of seawater. The later environments are more heterogeneous and may have very different origins. Examples of thalassohaline water habitats, which are typically chroride types, are the Great Salt Lake or the solar salterns used for the industrial production of marine salt by evaporation of seawater; among the athalassohaline waters are the Dead Sea, the Wadi Natrun, Lake Magadi, and several other soda lakes. In contrast to the hypersaline waters, the hypersaline soils are not well defined and in fact there is no clear definition of a saline or hypersaline soil. They are widely represented in our planet. Because most soils contain small amounts of soluble salts, a soil would be considered as hypersaline when its salt concentration is above a certain threshold (Rodriguez-Valera 1988). According to Kaurichev (1980), soils containing more than 0.2% (w/v) soluble salt should be considered as saline soils.

Micro-organisms show quite different responses to salt. According to the particular salt concentration required for their optimal growth, several physiological groups of micro-organisms are considered: (i) nonhalophiles require less than 1%

Antonio Ventosa
Department of Microbiology and Parasitology, Faculty of Pharmacy, University of Sevilla, 41012 Sevilla, Spain
e-mail: ventosa@us.es

NaCl; (ii) halotolerant are nonhalophilic micro-organisms that can tolerate high salt concentrations, in some cases up to 25% NaCl; (iii) slight halophiles grow best in media with 1 to 3% NaCl; (iv) moderate halophiles grow optimally in media with 3 to 15 % NaCl; and finally, (v) extreme halophiles grow best in media containing 15 to 25% NaCl and are able to grow even at saturated salt concentrations (Kushner and Kamekura 1988). The organisms most commonly found in environments with high salt concentrations are the moderately halophilic bacteria and the extremely halophilic bacteria and archaea; however, halotolerant micro-organisms may be also present, although it is considered that they play a minor role in these environments (Rodriguez-Valera 1988).

In this chapter, we review the extremely and moderately halophilic micro-organisms that have been isolated from soils, as well as their ecological distribution. In addition, we present some current or envisioned biotechnological or industrial applications of these interesting micro-organisms.

5.2 Microbial Diversity

Compared to hypersaline aquatic habitats, very little information exists thus far regarding the diversity of halophilic and halotolerant micro-organisms isolated from saline soils, and much work needs to be done on this subject. Pioneering studies on the microbial diversity in saline soils were carried out in the 1980s using conventional procedures and plate count methods. In 1982, Quesada and co-workers studied a hypersaline soil located in an abandoned multipond saltern, near the Mediterranean coast in Alicante, Spain. This kind of saltern provides a range of salt ponds with different salinities, ranging from salt content equivalent to that of seawater up to sodium chloride saturation, and evaporation of the water in the ponds produces soils with a wide range of salt concentrations. The soil studied had NaCl concentrations which ranged from 5.0 to 10.7%; the majority of the organisms isolated from this soil were halophilic bacteria with optimal growth at salt concentration between 5 and 15% and interestingly most of them were able to grow at 0.9% salt. This fact contrasts with the halo-dependence shown by most halophilic bacteria isolated from hypersaline waters or salted food, which in general have higher minimal salt requirements.

One possible explanation for this phenomenon is the heterogeneity of the soil habitats, where the salinity can change markedly in space and time. As a consequence of this heterogeneity, excessively specialized organisms may periodically be eliminated and the euryhaline types more consistently favoured. Most isolates were Gram-negative rods (46.6%) that were identified as members of the genera *Pseudomonas* (22%), *Alcaligenes* (11%), *Vibrio* (currently *Salinivibrio*) (3%), *Flavobacterium* (3%), and *Acinetobacter* (1%). Most of the isolates assigned in this early study to the genera *Pseudomonas*, *Alcaligenes*, and *Flavobacterium* can probably be considered members of the current genus *Halomonas*. Gram-positive rods and Gram-positive cocci represented 35.9% and 17.5% of the isolates, respectively, and were assigned to the genera

Bacillus (19%), *Micrococcus* (currently *Nesterenkonia*) (8%), *Arthrobacter* (6%), *Planococcus* (currently *Marinococcus*) (5%), *Staphylococcus* (3%), *Corynebacterium* (2%), *Brevibacterium* (1%), *Nocardia* (1%), and *Actinomyces* (1%).

A very low proportion of extremely halophilic archaea (1%) was isolated from this soil, probably due to an insufficient incubation period. They were assigned to the genus *Halobacterium* and their presence in this soil suggested the existence of local microsites with sufficiently high salt concentrations to allow the growth of halophilic *Archaea*. This study showed that the taxonomic groups found are the ones that also predominate in nonsaline soils; however, the microbial community structure in the saline soils studied is very different from that of communities in hypersaline waters, where the genera *Salinivibrio* and *Halomonas* are more frequently isolated (Ventosa et al. 1982, 1998; Márquez et al. 1987).

Later studies showed that Gram-positive micro-organisms are extensively represented in saline soils. Ventosa et al. (1983) isolated a group of Gram-positive halophilic cocci from the saline soil near Alicante mentioned before, all of them being moderate halophiles that grew between 2–5% and 25–30% total salts and optimally in media with 10–15% total salts. They were assigned to the species *Planococcus* (*Marinococcus*) *halophilus* and *Sporosarcina* (*Halobacillus*) *halophila*. In this study, no correlation was found between the isolation habitat or the salt range in which growth occurred and the taxonomic affiliation of the isolated strains. Garabito et al. (1998) isolated and studied 71 halotolerant Gram-positive endospore-forming rods from saline soils and sediments of salterns located in different areas of Spain (Huelva, Cádiz, Sevilla, and Mallorca). These isolates were tentatively assigned to the genus *Bacillus,* and the majority of them were classified as extremely halotolerant micro-organisms, being able to grow in most cases in up to 20 or 25% salt. It was not possible to determine if these isolates were normal inhabitants of the environments studied, although, because they showed a very wide salt growth range, their presence as inhabitants of hypersaline environments is quite probable.

Members of the genus *Bacillus* have been isolated from almost all natural habitats and from many other sources, owing to the ability of their spores to be transported and their remarkable capacity for resistance and dormancy, which allows them to survive in unfavorable habitats for long periods (Claus and Berkeley 1986). Extremely halotolerant cocci with similar characteristics to those described for the species *Micrococcus halobius* (currently *Nesterenkonia halobia*) were isolated from saline soils in different regions of the Antarctic continent (Nicolaus et al. 1992). Saline soils are a feature of Antarctic regions and halotolerant cocci had been isolated from this continent before (Miller et al. 1983; Miller and Leschine 1984); however, this is the first study where extremely halotolerant micrococci were reported.

During the last decade, advances in techniques for studying environmental microbiology have changed our views of the microbial communities in soils. Application of molecular biological methods has revealed an astonishing level of microbial diversity in the biosphere, only a small proportion of which has been assessed through cultivation (DeLong and Pace 2001; Rappé and Giovannoni 2003). Soils and sediments are among the most diverse microbial ecosystems and are estimated to contain an order of magnitude more of different prokaryotic "species" than do aquatic environments (Curtis et al. 2002; Torsvik et al. 2002); perhaps this is due to the higher spatial heterogeneity present in soil environments.

Using a cultivation-independent approach based on 16S rRNA gene analysis, Walsh et al. (2005) examined the archaeal diversity along a soil salinity gradient prone to frequent disturbance in the form of salinity fluctuations at Salt Spring, in British Columbia, Canada. Soil samples were collected from three sites along a salinity gradient ranging from 7 to 18% NaCl. The archaeal richness across the different sites studied was similar; however, a significant shift in archaeal community composition was found along the salinity gradient. The haloarchaea (70% of the clones) were the most commonly sampled group in this study, followed by other Euryarchaeota (26%), Crenarchaeota (3%), and halophilic methanogens (1%). It was observed that an increase in the salt concentration of the soils was accompanied by an increase in haloarchaeal diversity and a corresponding decrease in the other archaeal groups. Over 130 unique haloarchaeal sequences, representing 38 ribogroups, were recovered; these data suggest that the haloarchaeal community at Salt Spring is exceptionally diverse in comparison with other previously characterized haloarchaeal communities from solar salterns (Benlloch et al. 2002). The haloarchaeal 16S rRNA gene sequences recovered from Salt Spring soil were comprised of many lineages distributed throughout the haloarchaea, and were related to those of the genera *Halorubrum*, *Natrinema*, *Natronorubrum*, *Haloterrigena*, and *Natronococcus*.

The recovery of such a diversity of haloarchaea was unexpected as the majority of known haloarchaea undergo lysis at NaCl concentrations below 9% NaCl (Grant et al. 2001) and the salinity of Salt Spring appears to be below this threshold level for a considerable period of the year. However, Salt Spring haloarchaea were not closely related to known low salt-adapted or tolerant species (Rodríguez-Valera et al. 1979; Munson et al. 1997; Purdy et al. 2004; Elshahed et al. 2004), suggesting they may be frequently faced with local mortality as a result of frequent declines in soil salinity.

Very recently, Jiang et al. (2006) employed culture-independent (16S rRNA gene analysis) and culture-dependent microbiological techniques to assess microbial diversity and abundance in Lake Chaka, a hypersaline lake in northwestern China. Microbial abundance in the sediments of this lake ranged between 10^8 cells/g at the water–sediment interface and 10^7 cells/g at a sediment depth of 42 cm. The isolates from the sediment showed a halotolerant nature with an optimum salinity of 5% NaCl or lower, consistent with the salinity in the sediment. With respect to the bacterial diversity present in the sediment samples, the majority of the sequences obtained were affiliated with the Firmicutes (low-GC Gram-positive bacteria). This was in contrast with populations from the water samples, dominated by halophilic bacteria with clone sequences related to the *Bacteroides* group. Similar differences were also observed in the archaeal community, with all archaeal clone sequences in the lake water belonging to the order Halobacteriales, whereas most sequences in the sediment libraries were related to Euryarchaeota group III sequences, previously found in a diverse range of environments including methanogenic soils and sediments. A small percentage of sequences from the sediment libraries was related to the Crenarchaeota group. All sequences in Euryarchaeota group III were preliminarily assigned to Thermoplasmales; however, definitive identification was not possible because pure isolates were not available. It is interesting to point out that sequences belonging to Thermoplasmales have been reported to be present in saline soils (Walsh et al. 2005).

5.3 Halophilic Micro-Organisms from Soils

Most halophilic and halotolerant micro-organisms isolated and characterized in detail from saline or hypersaline soils are heterotrophic bacteria, but some haloarchaea have also been described. Here, we review the features of the halophilic archaea and bacteria isolated from soils.

5.3.1 Extremely Halophilic Archaea

With the exception of some halophilic methanogens, the extremely halophilic archaea belong to the haloarchaea, members of the family Halobacteriaceae within the Euryarchaeota. They are aerobic micro-organisms producing red to pink carotenoid pigments and have characteristic archaeal features, such as membranes containing ether-linked isoprenoid chains; their most typical feature is an absolute NaCl requirement, inasmuch as most species require at least 9% NaCl for growth and grow optimally in media with 20–25% NaCl (Grant et al. 2001). However, recent studies on coastal salt-marsh sediments have shown that some isolates are able to grow optimally at 10% NaCl and even grow slowly at 2.5% NaCl (Purdy et al. 2004). Currently, the haloarchaea are represented by more than 60 species grouped in 22 genera (Ventosa 2006). They are common inhabitants of hypersaline environments, and most species have been isolated from aquatic habitats such as salt lakes, soda lakes, or salterns, as well as from salted foods and subterranean salt deposits (Grant et al. 2001).

Only a few haloarchaeal species have been isolated from saline soils: they are *Haloarcula argentinensis* and *Haloarcula mukohatae*, two species isolated in Argentina (Ihara et al. 1997); *Haloarcula japonica*, isolated from saltern soil in Japan (Takashina et al. 1990); *Halorubrum distributum* (formerly *Halobacterium distributus*), isolated from alkaline soils (Zvyagintseva and Tarasov 1987; Oren and Ventosa 1996); and *Haloterrigena turkmenica* (formerly described as *Halococcus turkmenicus*), isolated from a sulfate saline soil in Turkmenistan (Zvyagintseva and Tarasov 1987; Ventosa et al. 1999). In addition to these well-characterized species, several other haloarchaea have been isolated from saline soils, such as *Haloferax* sp. strain D1227, isolated from a soil contaminated with saline oil brine near Grand Rapids, Michigan, USA. This extremely halophilic organism is able to utilize a variety of aromatic compounds; in particular, the degradation of 3-phenylpropionic acid by this haloarchaeon has been studied (Fu and Oriel 1998, 1999).

5.3.2 Moderately Halophilic Bacteria

Most moderately halophilic bacteria isolated and described from soils are heterotrophic bacteria. In contrast to hypersaline water habitats, from which predominantly Gram-negative species were obtained, saline or hypersaline soils have yielded many Gram-positive species, and these have been characterized taxonomically. The microbiota of hypersaline soils are more similar to those of nonsaline soils than to the microbiota from hypersaline waters. This suggests that general features of the environments are more important in determining the microbiota in a particular habitat than are individual factors such as high salinity (Quesada et al. 1983). Lists of validly described halophilic species names and their isolation site are presented here for Gram-positive (Table 5.1) and Gram-negative (Table 5.2) species.

Table 5.1 Selected moderately halophilic Gram-positive species isolated from saline or hypersaline soils

Species	Isolation Source	NaCl Range and Optimum (%)	Reference
Actinopolyspora iraqiensis	Soil sample in Iraq	5–20 (10–15)	Ruan et al. (1994)
Actinopolyspora mortivallis	Soil sample obtained from Death Valley, CA, USA	5–30 (10–15)	Yoshida et al. (1991)
Alkalibacillus haloalkaliphilus	Alkaline, highly saline mud from Wadi Natrun, Egypt	5–20 (10)	Jeon et al. (2005b)
Alkalibacillus salilacus	Soil sediment from a salt lake in Xinjiang Province, China	5–20% (10–12%)	Jeon et al. (2005b)
Bacillus krulwichiae	Soil from Tsukuba, Ibaraki, Japan	0–14	Yumoto et al. (2003)
Bacillus oshimensis	Soil from Oshymanbe, Oshima, Hokkaido, Japan	0–20 (7)	Yumoto et al. (2005)
Bacillus patagoniensis	Rhizosphere of the perennial shrub *Atriplex lampa* in north-eastern Patagonia, Argentina	0–15	Olivera et al. (2005)
Desulfobacter halotolerans	Sediment of Great Salt Lake, Utah, USA	0.5–13 (1.2)	Brandt and Ingvorsen (1997)
Filobacillus milosensis	Beach sediment from Palaeochori Bay, Milos, Greece	2–23 (8–14)	Schlesner et al. (2001)
Halobacillus halophilus	Salt marsh and saline soils	2–15 (3–5)	Spring et al. (1996); Ventosa et al. (1983)
Halobacillus karajensis	Saline soil of the Karaj region, Iran	1–24 (10)	Amoozegar et al. (2003b)
Lentibacillus salarius	Saline sediment of Xinjiang Province, China	1–20(12–14)	Jeon et al. (2005a)

(continued)

Table 5.1 (continued)

Species	Isolation Source	NaCl Range and Optimum (%)	Reference
Lentibacillus salicampi	Salt field in Korea	2–23 (4–8)	Yoon et al. (2002)
Marinococcus halophilus	Saline soil from Alicante and Cadiz, Spain	1–20 (15)	Hao et al. (1984); Marquez et al. (1992)
Marinococcus halotolerans	Saline soil in Qinghai, north-west China	0–25 (10)	Li et al. (2005c)
Microbacterium halotolerans	Soil sediment of Qinghai Province, China	0–15 (5)	Li et al. (2005a)
Nesterenkonia halotolerans	Hypersaline soil from Xinjiang Province, China	0–25	Li et al. (2004b)
Nesterenkonia xinjiangensis	Hypersaline soil from Xinjiang Province, China	0–25	Li et al. (2004b)
Nocardiopsis baichengensis	Saline sediment from Xinjiang Province, China	0–18 (5–8)	Li et al. (2006)
Nocardiopsis chromatogenes	Saline sediment from Xinjiang Province, China	0–18 (5–8)	Li et al. (2006)
Nocardiopsis gilva	Saline sediment from Xinjiang Province, China	0–18 (5–8)	Li et al. (2006)
Nocardiopsis halophila	Saline soil from Iraq	3–20 (5–15)	Al-Tai and Ruam, (1994)
Nocardiopsis halotolerans	Salt marsh soil from Kuwait	0–15 (10)	Al-Zarban et al. (2002a)
Nocardiopsis rhodophaea	Saline sediment from Xinjiang Province, China	0–18 (5–8)	Li et al. (2006)
Nocardiopsis rosea	Saline soil from Xinjiang Province, China	0–18 (5–8)	Li et al. (2006)
Nocardiopsis salina	Saline soil from Xinjiang Province, China	3–20 (10)	Li et al. (2004a)
Nocardiopsis xinjiangensis	Saline soil from Xinjiang Province, China	10	Li et al. (2003a)
Prauserella alba	Saline soil from Xinjiang Province, China	0–25 (10–15)	Li et al. (2003c)
Prauserella halophila	Saline soil from Xinjiang Province, China	5–25 (10–15)	Li et al. (2003c)
Saccharomonospora halophila	Marsh soil in Kuwait	10–30 (10)	Al-Zarbam et al. (2002b)
Saccharomonospora paurometabolica	Soil from Xinjiang Province, China	5–20 (10)	Li et al. (2003b)
Salinicocccus hispanicus	Saline soil from Alicante and Cadiz, Spain	0.5–25 (10)	Marquez et al. (1990)
Sporohalobacter lortetii	Dead Sea sediment	6–12 (8.7)	Oren (1983)
Streptomonospora alba	Soil from Xinjiang Province, China	5–25 (10–15)	Li et al. (2003d)
Streptomonospora salina	Soil from Xinjiang Province, China	15	Cui et al. (2001)

(continued)

Table 5.1 (continued)

Species	Isolation Source	NaCl Range and Optimum (%)	Reference
Tenuibacillus multivorans	Soil from Xinjiang Province, China	1–20 (5–8)	Ren and Zhou (2005)
Thalassobacillus devorans	Saline soil in South Spain	0.5–20 (7.5–10)	García et al. (2005a)
Virgibacillus koreensis	Salt field near Taean-Gun on the Yellow Sea in Korea	0.5–20 (5–10)	Lee et al. (2006)
Virgibacillus salexigens	Soil from Huelva, Cadiz, Sevilla, and Mallorca, Spain	7–20 (8–10)	Garabito et al. (1997)

Table 5.2 Selected moderately halophilic Gram-negative species isolated from saline or hypersaline soils

Species	Isolation Source	NaCl Range and Optimum (%)	Reference
Halanaerobacter salinarius	Salt ponds in Camargue, France	5–30 (14–15)	Moune et al. (1999)
Halanaerobium alcaliphilum	Sediment from Great Salt Lake, Utah, USA	2.5–25 (10)	Tsai et al. (1995)
Halanaerobium lacusrosei	Sediment of Retba Lake, Senegal	7.5–34 (18–20)	Cayol et al. (1995)
Halanaerobium praevalens	Sediment of Great Salt Lake. Utah, USA	2–30 (20)	Zeikus et al. (1983)
Halanaerobium saccharolytica subsp. senegalensis	Sediments of Retba Lake, Senegal	5–25 (7.5–12.5)	Cayol et al. (1994b)
Halomonas anticariensis	Soil from Fuente de Piedra. Málaga, Spain	0.5–15 (7.5)	Martínez-Cánovas et al. (2004a)
Halomonas boliviensis	Soil around the lake Laguna Colorada, Bolivia	0–25 (5)	Quillaguaman et al. (2004)
Halomonas campisalis	Soil sample collected from a dry salt flat south of Alkali Lake, USA	1–25 (8.7)	Mormile et al. (1999)
Halomonas eurihalina	Hypersaline soil in Alicante, Spain	0.5–30 (7.5)	Quesada et al (1990); Mellado et al. (1995)
Halomonas halophila	Hypersaline soil located near Alicante, Spain	2–30 (7.5)	Quesada et al. (1984)
Halomonas maura	Soil from a solar saltern at Asilah, Morocco	1–15 (7.5–10)	Bouchotroch et al. (2001)
Halomonas organivorans	Saline soil from Isla Cristina, Huelva, Spain	1.5–30 (7.5–10)	Garcia et al. (2004)
Halomonas salina	Saline soils located near Alicante, Spain	2.5–20 (5)	Valderrama et al. (1991)
Halomonas ventosae	Saline soil in Jaen, Spain	3–15 (6–9)	Martinez-Canovas et al. (2004b)

(continued)

Table 5.2 (continued)

Species	Isolation Source	NaCl Range and Optimum (%)	Reference
Halothermotrix orenii	Sediment of a Tunisian salt lake	4–20 (5–10)	Cayol et al. (1994)
Marinobacter excellens	Sediment collected from Chazhman Bay, Sea of Japan	1–15	Gorshkova et al. (2003)
Marinobacter koreensis	Sea sand in Pohang, Korea	0.5–20	Kim et al. (2006)
Marinobacter lipolyiticus	Saline soil from Cadiz, Spain	1–15 (7.5)	Martin et al. (2003)
Marinobacter sediminum	Marine coastal sediment from Peter the Great Bay, Sea of Japan	0.5–18	Romanenko et al. (2005)
Natroniella acetigena	Mud from the soda Lake Magadi, Kenya	10–26 (12–15)	Zhilina et al. (1999)
Orenia salinaria	Salt ponds in salterns in Camargue, France	2–25 (5–10)	Moune et al. (2000)
Palleronia marisminoris	Hypersaline soil bordering a solar saltern in Murcia, Spain	0.5–15 (5)	Martinez-Checa et al. (2005)
Salipiger mucosus	Hypersaline soil from a solar saltern in Calblanche, Murcia, Spain	0.5–20 (3–6)	Martinez-Canovas et al. (2004c)

5.3.2.1 Gram-Positive Bacteria

In nature, salinity often goes together with alkalinity. Several alkaliphilic *Bacillus* species have been identified to date (Vedder 1934; Spanka and Fritze 1993; Nielsen et al. 1995; Agnew et al. 1995; Fritze, 1996; Switzer et al. 1998; Yumoto et al. 2003; Olivera et al. 2005). Some of these have been isolated from soil samples and show halophilic characteristics. *Bacillus krulwichiae* (Yumoto et al. 2003), isolated in Tsukuba, Japan, is a facultatively anaerobic, straight rod with peritrichous flagella that produces ellipsoidal spores and utilizes benzoate or *m*-hydroxybenzoate as the sole carbon source. *Bacillus patagoniensis* (Olivera et al. 2005) was isolated from the rhizosphere of the perennial shrub *Atriplex lampa* in north-eastern Patagonia. Finally, *Bacillus oshimensis* (Yumoto et al. 2005) is a halophilic nonmotile, facultatively alkaliphilic species. It exhibits a NaCl requirement for growth, and salt may be required for pH homeostasis, adaptation to an alkali environment, or energy production through the respiratory chain (Tokuda and Unemoto 1981, 1984) or ATPase function (Ueno et al. 2000).

Some moderately halophilic, spore-forming, Gram-positive rods were originally assigned to the genus *Bacillus*, but have been reclassified within new genera by the application of molecular methods and improved phenotypic approaches (Heyndrickx et al. 1998; Yoon et al. 2001, 2004). Indeed, 16S rRNA sequence and chemotaxonomic analyses revealed the existence of several phylogenetically distinct lineages within the genus *Bacillus* (Ash et al. 1991; Nielsen et al. 1994). One example of such independent lineage is a group comprising the species *Alkalibacillus haloalkaliphilus* (formerly *Bacillus haloalkaliphilus*; Fritze, 1996; Jeon et al. 2005b) isolated from alkaline, high saline mud from Wadi Natrun, Egypt, and *Alkalibacillus salilacus*, isolated from soil sediment of a salt lake in China (Jeon et al. 2005b).

Another example is the genus *Virgibacillus*. This genus was first proposed by Heyndrickx et al. (1998) based on polyphasic data and its description was later emended by Heyrman et al. (2003). Members of the genus *Virgibacillus* produce oval to ellipsoidal endospores, are Gram-positive motile rods, and have DNA G+C content ranging from 36 to 43 mol% (Heyrman et al. 2003). Currently, this genus comprises eight species, two of which are moderately halophilic and have been isolated from soil samples: *Virgibacillus salexigens* (Garabito et al. 1997; Heyrman et al 2003) and the recently described *Virgibacillus koreensis* (Lee et al. 2006).

Several other aerobic or facultatively anaerobic, moderately halophilic, endospore-forming, Gram-positive bacteria have been classified within genera related to *Bacillus*. Genera that include halophilic species isolated from soil samples are *Halobacillus*, *Filobacillus*, *Tenuibacillus*, *Lentibacillus*, and *Thalassobacillus*. The genus *Halobacillus* is clearly differentiated from other related genera on the basis of its cell-wall peptidoglycan type; members of this genus have peptidoglycan based on L-Orn-D-Asp (Spring et al. 1996). The genus *Filobacillus* has peptidoglycan based on L-Orn-D-Glu (Schlesner et al. 2001). Within these genera, the halophilic species isolated from soils are: *Halobacillus halophilus* (Spring et al. 1996), *Halobacillus karajensis* (Amoozegar et al. 2003b), and *Filobacillus milolensis* (Schlesner et al. 2001). With respect to the genus *Lentibacillus*, two halophilic soil species are recognized, *Lentibacillus salicampi* isolated from a salt field in Korea (Yoon et al. 2002), and *Lentibacillus salarius* from a saline sediment in China (Jeon et al. 2005a). Finally, the genera *Tenuibacillus* and *Thalassobacillus* are comprised of only one species, *Tenuibacillus multivorans* isolated from a saline soil in XinJiang, China (Ren and Zhou 2005) and *Thalassobacillus devorans*, isolated from a saline soil in South Spain (García et al. 2005a). All these genera belong to the family Bacillaceae, included in the phylogenetic group of the low GC Gram-positive bacteria, and are closely related.

The genus *Marinococcus* was proposed in 1984 to accommodate two moderately halophilic species, *Marinococcus halophilus* and *Marinococcus albus* (Hao et al. 1984). A third species, *Marinococcus halotolerans*, has been described recently (Li et al. 2005c). They are motile cocci that grow over a wide range of salt concentrations and up to 20% NaCl, have *meso*-diaminopimelic acid in their cell wall, a DNA G+C content ranging between 43.9 and 46.6 mol%, and MK-7 as the menaquinone system. Of these species, *M. halophilus* is the most commonly isolated and it occurs in most hypersaline environments (Ventosa et al. 1983; Márquez et al. 1992). *M. halophilus* and *M. halotolerans* were originally isolated from soil samples located in Spain and China, respectively. *Salinicoccus hispanicus* (formerly *Marinococcus hispanicus*; Marquez et al. 1990; Ventosa et al. 1992) was isolated from a hypersaline soil located near Alicante (Spain). This species comprises Gram-positive, nonmotile, and non-spore-forming cocci.

The genus *Microbacterium* was established by Orla-Jensen (1919) and emended by Collins et al. (1983). It comprises a diverse group of Gram-positive, non-spore-forming rods. More recently, this genus was emended again to unify the genera *Microbacterium* and *Aureobacterium* (Takeuchi and Hatano 1998). Members of the genus *Microbacterium* can be isolated from a wide range of habitats, including soil, water, plants, steep liquor, milk products, and humans. *Microbacterium halotolerans*, an aerobic, Gram-positive, nonmotile, non-spore-forming short rod, is the only halophilic species of this genus isolated from a soil sample, specifically from a saline soil in the west of China (Li et al. 2005a).

The genus *Micrococcus* was first described more than 100 years ago, and since then its description has been revised several times. A phylogenetic and chemotaxonomic reanalysis of the genus *Micrococcus* resulted in the proposal of the genus *Nesterenkonia* (Stackebrandt et al. 1995) constituted by coccoid or short Gram-positive rods, non-spore-forming, chemo-organotrophic with strictly respiratory metabolism. Species of this genus are aerobic, catalase-positive, and moderately halophilic or halotolerant (Stackebrand et al. 1995; Collins et al. 2002; Li et al. 2005b). Two species isolated from soil habitats are *Nesterenkonia halotolerans* and *Nesterenkonia xinjiangensis* (Li et al. 2004a).

The presence and activity of dissimilatory sulfate-reducing bacteria in hypersaline environments have been reported by several investigators (Trüper 1969; Caumette 1993; Oren 1988; Tardy-Jacquenod et al. 1996). Some of the highest sulfate reduction rates recorded have been measured in hypersaline microbial mats with salinities up to 18–21% (Caumette et al. 1994). Active sulfate reduction has also been documented in hypersaline sediments (Zeikus 1983; Nissenbaum 1975; Skyring 1987). In 1983, Oren isolated a sulfate-reducing, anaerobic, halophilic, rod-shaped, endospore-forming bacterium from Dead Sea sediments. This bacterium was designated as *Clostridium lortetii* and later it was reclassified as *Sporohalobacter lortetii* (Oren 1983; Oren et al. 1987). The cells of this species are characterized by the production of gas vesicles, generally near the developing terminal endospore, and these vesicles remain attached to the mature endospore after degeneration of the vegetative cell.

The genus *Desulfobacter* comprises nutritionally specialized sulfate reducers with acetate as their characteristic substrate, which is oxidized via a modified citric acid cycle (Hansen 1994; Möller et al. 1987). *Desulfobacter halotolerans* is the first acetate-oxidizing, sulfate-reducing, and halophilic bacterium to have been recognized. It was isolated from sediments of Great Salt Lake, Utah (Brandt and Ingvorsen 1997). This bacterium reduces sulfate, thiosulfate, or sulfite. Acetate, ethanol, and pyruvate are used as electron donors and carbon sources.

The occurrence of actinomycetes in highly saline environments and the tolerance of these organisms to high salt concentrations was first described by Tresner et al. (1968) and Gottlieb (1973). The family Nocardiopsaceae contains three genera, namely *Nocardiopsis* (Meyer 1976), *Thermobifida* (Zhang et al. 1998), and *Streptomonospora* (Cui et al. 2001). At present, the genus *Nocardiopsis* comprises 19 validly published species names (Li et al. 2003a, 2004b; Al-Zarban et al. 2002a). These species comprise aerobic, Gram-positive, non-acid-fast, and nonmotile organisms. Originally, members of this genus had been isolated from mildewed

So far, the order Halanaerobiales contains 24 species, grouped into 11 genera and 2 families, *Halanaerobiaceae* and *Halobacteroidaceae* (Oren et al. 1984; Rainey et al. 1995; Oren 2000). Studies on solar salterns of the French Mediterranean coast (Salin-de-Giraud, Carmargue, Rhône Delta), permitted the isolation of several fermentative and halophilic bacteria from the sediments of hypersaline lagoons or from the surface of certain hypersaline ponds characterized by gypsum deposits. The salinities ranged from 13 to 34% NaCl (Caumette 1993; Caumette at al. 1994).

Most fermentative bacterial isolates were assigned to the family *Halobacteroidaceae*. One of these isolates was described as a new species, *Halanaerobacter salinarius* (Moune et al. 1999), isolated from the black anoxic sediment of these ponds, where it co-exists with the sulfate reducers. Among the other strains isolated in this sampling, strain SG 3902 was phylogenetically related to the genus *Orenia* according to 16S rRNA gene sequence comparison. Whereas this genus was initially represented by a single species, *Orenia marismortui*, isolated from the Dead Sea (Rainey et al. 1995), isolate SG 3902 showed sufficient physiological and genetic differences from the species *O. marismortui* to be considered as a new member of the genus *Orenia*, designated *Orenia salinaria* (Moune et al. 2000). Another halophilic species of the family *Halobacteroidaceae* is *Natroniella acetica*, an extremely haloalkaliphilic, chemo-organotrophic, homoacetogenic bacterium isolated from the bottom mud of the soda-depositing Lake Magadi in Kenya (Zhilina et al. 1996).

The other family of the order Halanaerobiales is Halanaerobiaceae, that comprises one genus, *Halanaerobium*. This genus was described by Zeikus and coworkers in 1983 and comprises nine species, some of which are halophilic species isolated from soil samples. The first species included in this genus was *Haloanaerobium praevalens*, a species with an extremely halophilic response (Zeikus et al. 1983), isolated during a study carried out in 1979 on the microbial ecology of anaerobic decomposition processes in Great Salt Lake (Utah). The objective of the study was to understand how organic matter is degraded in a productive hypersaline environment and what bacterial species actively participate in this process (Zeikus, 1983). A subsequent screening study of the Great Salt Lake led to the description of *Halanaerobium alcaliphilum*, an alkalitolerant species able to grow at a range of pH values from pH 5.8 to 10, with an optimum between pH 6.7 to 7.0 (Tsai et al. 1995).

Two other species, *Halanaerobium saccharolytica* subsp. *senegalensis* (formerly *Haloincola saccharolytica* subsp. *senegalensis*; Cayol et al. 1994b) and *Halanaerobium lacusrosei* (Cayol et al. 1995), were isolated from 1.5-m deep sediment of hypersaline Lake Retba, near Dakar (Senegal); this lake is located 100 m from the Atlantic Ocean. Finally, within this family, the genus *Halothermothrix* comprises only one halophilic species, *H. orenii*, to have been isolated from a soil habitat. This was the first described moderately halophilic (with an optimal growth between 5 and 17.5% NaCl) thermophile that showed a capacity to ferment carbohydrates to acetate, ethanol, H_2, and CO_2. This species, which was isolated from sediments of a hypersaline Tunisian lake (Chott El Guettar), extended the temperature limit of halophilic anaerobic bacteria to 68°C. The optimum temperature for growth of this species is 60°C (Cayol et al. 1994a).

5.4 Biotechnological Applications

A great metabolic diversity has been found in halophilic and halotolerant micro-organisms isolated from soils, and in fact many of them have biotechnological potential. Thus, some of them produce halophilic proteins useful in specific transformations, and others excrete products such as compatible solutes and biopolymers of great interest in different industries. In addition, these micro-organisms could be used in environmental bioremediation processes. The biotechnological applications of halophilic micro-organisms have been reviewed in detail (Ventosa et al. 1998; Margesin and Schinner 2001; Mellado and Ventosa 2003). In this chapter, we focus on interesting aspects of halophilic and halotolerant micro-organisms from saline soils.

5.4.1 Extracellular Enzymes

Industrial biocatalysis has found in the halophilic micro-organisms a source of enzymes with novel properties of high interest. Over the years, different enzymes of halotolerant and halophilic micro-organisms isolated from saline soils have been described and a number of new possibilities for industrial processes have emerged due to their overall inherent stability at high salt concentrations. These enzymes could be used in harsh industrial processes such as food processing, biosynthetic processes, and washing (Ventosa et al. 2005).

Halophilic enzymes are active and stable at high salt concentrations, showing specific molecular properties that allow them to cope with osmotic stress. In general, these enzymes present an excess of acidic residues over basic residues and a low content of hydrophobic residues at their surface (Mevarech et al. 2000). The hydrolases able to break down various polymers constitute the group of highest biotechnological interest, catalyzing conversions under conditions compatible with industrial applications. So far, most of the characterized halophilic enzymes have been obtained from micro-organisms isolated from hypersaline or saline water systems, such as lakes or salterns (Ventosa et al. 2005). We focus only on the few halophilic enzymes produced by micro-organisms isolated from saline soils.

Amylases produced by several halophilic bacteria isolated from saline soils have been reported. *Nesterenkonia halobia* (formerly *Micrococcus halobius*) produces an amylase showing high dependency on salt for activity; thus, in the absence of high concentrations of NaCl or KCl the enzyme loses its activity (Onishi and Kamekura 1972; Onishi 1972). Onishi and Hidaka (1978) reported the purification of two amylases from the culture filtrate of *Acinetobacter* sp., a moderately halophilic bacterium isolated from sea-sands. The enzymes showed maximal activity in saline media containing 0.2–0.6 M NaCl or KCl, at pH 7.0 and 50–55°C.

Halothermothrix orenii, a thermophilic, halophilic anaerobic bacterium isolated from the sediment of a Tunisian salted lake, produces two amylases, designated as AmyA and AmyB. The gene encoding AmyA has been cloned, characterized, and expressed in *Escherichia coli* (Mijts and Patel 2002). The recombinant enzyme presented optimal activity at 65°C in 5% NaCl and pH 7.5, although a significant activity was also measured in the presence of up to 25% NaCl. Li et al. (2002) have performed a crystallographic study of AmyA. The crystal structure of the enzyme reveals the lack of the acidic surface which is characteristic of halophilic proteins. AmyB is a 599-residue protein active and stable at 10% NaCl, the structure of which has also been determined by X-ray crystallography (Tan et al. 2003).

Another extracellular amylase produced by the moderately halophilic, Gram-positive, spore-forming bacterium *Halobacillus karajensis* (Amoozegar et al. 2003b) has been studied. This bacterium was isolated from surface saline soil of the Karaj region, Iran. The maximum protease activity was achieved in the presence of 5% NaCl at pH 7.5–8.5 and 50°C (Amoozegar et al. 2003a).

A halophilic member of the genus *Bacillus*, *Bacillus* sp. no. 21-1, produces a protease showing maximal activity at 5 M NaCl and 0.75 M KCl (Kamekura and Onishi 1974). Another extracellular protease has been characterized from a haloalkaliphilic bacterium isolated from salt-enriched soil samples collected from saline habitats from the Veraval costal region of Gujarat, India. This bacterium has been identified as *Bacillus pseudofirmus* and the maximal enzyme production (410 U/ml) is achieved during the early stationary phase of growth (Patel et al. 2005).

Several moderately halophilic bacteria able to produce extracellular enzymes of industrial interest have been isolated in the course of a screening program performed in saline waters and soils in South Spain (Sánchez-Porro et al. 2003). A total of 892 strains able to produce different hydrolases were isolated: amylase (269 strains), lipase (207 strains), protease (201 strains), DNAse (118 strains), and pullulanase (97 strains). Most of the strains showing hydrolytic activities were isolated from saline aquatic habitats, although surprisingly a higher number of isolates from saline soils produced lipolytic activity.

The lipolytic strain SM19 isolated from saline soil was selected for further characterization. This bacterium has been proposed as a new species and named *Marinobacter lipolyticus* (Martín et al. 2003). The gene encoding the enzyme was cloned by screening an expression library in *E. coli*. This lipase gene, designated *lipM*, encodes a protein of 271 amino acids, with an estimated molecular mass of 30.5 kDa. The deduced amino acid sequence of the *lipM* gene exhibited significant amino acid sequence identity with lipases belonging to the α/β-hydrolase super-family (Martín et al. unpublished results).

Several extracellular enzymes from haloarchaea have been characterized (Ventosa et al. 2005), however, these have been isolated from micro-organisms inhabiting hypersaline aquatic systems. In this sense, the future characterization of novel enzymes from halophilic bacteria or archaea isolated from saline soils constitutes an interesting research topic due to their potential biotechnological applications.

5.4.2 Production of Compatible Solutes

Compatible solutes are low-molecular weight organic compounds such as polyols, amino acids, sugars, and betaines that the halophilic and halotolerant bacteria accumulate intracellularly to achieve osmotic balance (Brown 1976). These compounds have industrial applications as stabilizers of enzymes, nucleic acids, membranes, and whole cells against stresses such as high temperature, desiccation, and freezing (Louis et al. 1994; Nieto and Vargas 2002). Only a few studies deal with the biosynthesis of compatible solutes by moderate halophiles isolated from soils.

Ectoines constitute the predominant class of osmolytes accumulated by bacteria grown at high salt concentrations (Galinski and Tindall 1992). The accumulation of ectoine has been probed in the Gram-positive moderate halophile *Marinococcus halophilus* and the genes encoding the biosynthetic pathway of the compatible solute have been characterized (Louis and Galinski, 1997). *Halomonas elongata* strain KS3 is a moderately halophilic bacterium isolated from a salty soil in Thailand. This strain accumulates ectoine and hydroxyectoine as compatible solutes. The accumulation of these compounds is stimulated by hyperosmotic stress induced by salt. Ectoine production reaches approximately $120\,\mu g\,mg^{-1}$ of dry cells at a concentration of 2.56 M NaCl and hydroxyectoine accumulates to $45\,\mu g\,mg^{-1}$ of dry cells at a concentration of 2.56 M NaCl (Ono et al. 1998). The enzymes involved in the biosynthesis of ectoine in *H. elongata* KS3 have been characterized (Ono et al. 1999).

A similar ectoine biosynthetic pathway has been found in *Chromohalobacter salexigens*, a moderately halophilic bacterium that has been used as a model micro-organism for the study of osmoadaptation mechanisms in moderate halophiles (Cánovas et al. 1997, 1998). The use of ectoine as a stabilizer of different enzymes (amylases, lipases, cellulases, proteases) has been patented (Toyoda et al. 1997). Moreover, ectoine and ectoine derivatives have been patented as moisturizers in cosmetics (Motitschke et al. 2001). On the other hand, Nakayama et al. (2000) expressed the ectoine biosynthetic genes in cultured tobacco, obtaining an increase in the tolerance of the transgenic plant to hyperosmotic stress. This is an interesting result, especially considering that in arid and semi-arid lands environmental stress limits plant growth and productivity.

Halobacillus halophilus (formerly *Sporosarcina halophila*) is a moderately halophilic bacterium isolated from salt marsh soils of the North Sea coast of Germany (Claus et al. 1983). This bacterium produces glutamate and glutamine as its main compatible solutes and thus it could be used for the industrial production of these compounds. Recently, Saum et al. (2006) described the routes for the biosynthesis of these solutes and characterized the genes involved in their synthesis.

5.4.3 *Production of Exopolysaccharides*

Several moderately halophilic bacteria isolated from saline soils are able to produce exopolysaccharides which can be secreted in the extracellular medium in the form of amorphous slime. The exopolysaccharides have biotechnological interest due to both viscosifying and emulsifying properties, which allow them to be used in medicine, pharmacy, foodstuffs, cosmetics, and the petroleum industry (Quesada et al. 2004).

The genus *Halomonas* groups several species able to produce exopolysaccharides. *Halomonas eurihalina* (formerly *Volcaniella eurihalina*) has been described as a producer of a exopolysaccharide with the ability to increase the viscosity of solutions at low pH values (Martínez-Checa et al. 2002). *Halomonas maura*, isolated from soil samples collected from a saltern at Asilah (Morocco), produces an exopolysaccharide designated Mauran that has been studied in detail. This polymer is highly viscous and shows thixotropic and pseudoplastic behaviour presenting interesting properties for biotechnology (Arias et al. 2003). The genes involved in the biosynthesis of this exopolysaccharide have been cloned and the maximum induction is reached during the stationary phase in the presence of 5% marine salts (Arco et al. 2005). *Halomonas ventosae*, isolated from saline soils in Jaen (Spain), and *Halomonas anticariensis*, isolated from soil samples taken from Fuente de Piedra, a saline wetland in Malaga, South Spain, produce exopolysaccharides showing emulsifying activity on several hydrophobic substrates (Mata et al. 2006). In addition to the genus *Halomonas*, other moderately halophilic bacteria isolated from hypersaline soils, such as *Salipiger mucescens* (Martínez-Canovas et al. 2004c) and *Palleronia marisminoris* (Martínez-Checa et al. 2005), have also been described as exopolysaccharide producers; however, the features of these exopolysaccharides have not been investigated in detail.

5.4.4 Production of Carotenoids

Most haloarchaea inhabiting hypersaline habitats synthesize carotenoids that contribute to the red color presented by many saltern crystallizer ponds and hypersaline lakes (Kushwaha et al. 1974; Oren 2002). Industrial applications of carotenoids include their use as nutrient supplements, as food colorants, in the pharmaceutical industry, or in animal feeds. In this sense, Asker and Ohta (1999) used soil samples from a salt farm as a source for the isolation of carotenoid-producing micro-organisms; they isolated 31 red haloarchaea that might represent valuable sources of carotenoids.

5.4.5 Biodegradation of Toxic Compounds

Industrial processes such as the production of pesticides, herbicides, and pharmaceutical products, in addition to paper mills and petrochemical industries generate wastewaters containing toxic compounds and with different levels of salinities. According to environmental regulations, these residual waters need to be treated prior to their release in the environment. Some of the treatments conventionally used combine physicochemical and biological methods. In the biological treatment, the micro-organisms conventionally used show only poor degradative efficacy due to the highly saline conditions. The degradation of aromatic compounds in saline conditions using the potential of halophilic bacteria is of great interest and offers the promise of innovative research in saline soil microbiology and of new treatment strategies for saline effluents generated by different industrial activities.

The metabolism of aromatic compounds in halophilic archaea was demonstrated for the first time in *Haloferax* sp. D1227, isolated from soil contaminated with highly saline oil brine (Fu and Oriel 1998, 1999). Several studies have dealt with the isolation, from saline contaminated soils, of moderate halophiles with the ability to degrade different organic compounds. The moderately halophilic bacterium *Arhodomonas aquaeoli*, isolated from subterranean brines, has the ability to aerobically degrade oil (Adkins et al. 1993). Recently, a study focused on the characterization of an active and acclimated bacterial population able to degrade aromatic compounds in saline habitats of South Spain has been carried out (García et al. 2005b). Two novel moderately halophilic bacteria isolated from saline soils, designated as *Halomonas organivorans* and *Thalassobacillus devorans* have been described. *H. organivorans* is a Gram-negative bacterium with the ability to use a broad variety of organic compounds (benzoic acid, *p*-hydroxybenzoic acid, cinnamic acid, salicylic acid, phenylacetic acid, phenylpropionic acid, phenol, *p*-coumaric acid, ferulic acid, and *p*-aminosalicylic acid; García et al. 2004). *Thalassobacillus devorans* is a Gram-positive bacterium isolated by phenol enrichment of samples collected in contaminated hypersaline soils (García et al. 2005a). Both novel species have the potential to be used for the biodegradation of contaminated saline habitats.

Organophosphonates constitute toxic contaminants due mainly to the stable covalent carbon to phosphorus (C–P) bond. The contamination of environments with these recalcitrant compounds constitutes a serious environmental problem. Hayes et al. (2000) isolated a halophilic strain from soil beneath a road gritting salt pile with the ability to utilize phosphonoacetate, 2-aminoethyl-, 3-aminopropyl-, 4-aminobutyl-, methyl- and ethyl-phosphonates as phosphorus sources for growth. The novel isolate was shown to be probably *Chromohalobacter marismortui* or *Chromohalobacter* (*Pseudomonas*) *beijerinckii* on the basis of 16S rRNA analysis.

Recently, Kleinsteuber et al. (2006) have performed microbiological examination of an exploited oil field with high soil salinity near Comodoro Rivadaria in Patagonia (Argentina). They characterized a community of halophilic bacteria with a capacity for diesel fuel degradation and identified the most active species by high-resolution cell sorting and analysis of the 16S rRNA gene.

On the other hand, halophilic bacteria tolerant to heavy metals could be used as bioassay indicator organisms in saline-polluted environments. Several halotolerant and halophilic bacteria isolated from hypersaline soils tolerate high concentrations of different metals, such as Co, Ni, Cd, or Cr (Nieto et al. 1989; Ríos et al. 1998).

5.5 Conclusions

Saline and hypersaline soils are increasingly abundant as a consequence of the irrigation and desertization processes. The study of the microbiological activities in these habitats yields results of great interest. Halophilic and halotolerant micro-organisms are the most important microbial population of these soils. The microbial diversity of saline soils is more similar to that of nonsaline soils than to the microbiota found in hypersaline water systems. Many studies have been focused on the isolation and characterization of halophilic bacteria and archaea from saline or hypersaline soils. Biotechnological applications of these micro-organisms have been or are under investigation. However, more extensive studies on the ecology, biochemistry, and physiology of organisms belonging to the different microbial groups present in these soils are urgently required.

Acknowledgements The authors' research was supported by grants from the Quality of Life and Management of Living Resources Programme of the European Commission (QLK3-CT-2002-01972), Spanish Ministerio de Educación y Ciencia (projects BIO2006-06927 and CTM2006-03310) and Junta de Andalucía.

References

Adkins JP, Madigan MT, Mandelco L, Woese CR, Tanner RS (1993) *Arhodomonas aquaeolei* gen. nov., sp. nov., an aerobic, halophilic bacterium isolated from a subterranean brine. *Int J Syst Bacteriol* 43:514–520

Agnew MD, Koval SF, Jarrell KF (1995) Isolation and characterization of novel alkaliphiles from bauxite-processing waste and description of *Bacillus vedderi* sp. nov., a new obligate alkaliphile. *Syst Appl Microbiol* 18:221–230

Al-Tai AM, Ruan JS (1994) *Nocardiopsis halophila* sp. nov., a new halophilic actinomycete isolated from soil. *Int J Syst Bacteriol* 44:474–478

Al-Zarban SS, Abbas I, Al-Musallam AA, Steiner U, Stackebrandt E, Kroppenstedt RM (2002a) *Nocardiopsis halotolerans* sp. nov., isolated from salt marsh soil in Kuwait. *Int J Syst Evol Microbiol* 52:525–529

Al-Zarban SS, Al-Musallam AA, Abbas I, Stackebrandt E, Kroppenstedt RM (2002b) *Saccharomonospora halophila* sp. nov., a novel halophilic actinomycete isolated from marsh soil in Kuwait. *Int J Syst Evol Microbiol* 52:555–558

Amoozegar MA, Malekzadeh F, Malik KA (2003a) Production of amylase by newly isolated moderate halophile, *Halobacillus* sp. strain MA-2. *J Microbiol Meth* 52:353–359

Amoozegar MA, Malekzadeh F, Malik KA, Schumann P, Spröer C (2003b) *Halobacillus karajensis* sp. nov., a novel moderate halophile. *Int J Syst Evol Microbiol* 53:059–1063

Arahal DR, Castillo AM, Ludwig W, Schleifer KH, Ventosa A (2002a) Proposal of *Cobetia marina* gen. nov., comb. nov., within the family *Halomonadaceae*, to include the species *Halomonas marina*. *Syst Appl Microbiol* 25:207–211

Arahal DR, Ludwig W, Schleifer KH, Ventosa A (2002b) Phylogeny of the family *Halomonadaceae* based on 23S and 16S rDNA sequence analyses. *Int J Syst Evol Microbiol* 52:241–249

Arahal DR, Ventosa A (2005) The family *Halomonadaceae*. In: Dworkin M, Falkow S, Rosenberg E, Schleifer KH, Stackebrandt E (eds) *The Prokaryotes: An Evolving Electronic Resource for the Microbial Community*, Release 3.20. Springer, New York, (http://141.150.157.117:8080/prokPUB/index.htm)

Arco Y, Llamas I, Martínez-Checa F, Argandona M, Quesada E, del Moral A (2005) *epsABCJ* genes are involved in the biosynthesis of the exopolysaccharide mauran produced by *Halomonas maura*. *Microbiology* 151: 2841–2851

Arias S, del Moral A, Ferrer MR, Tallon R, Quesada E, Béjar V (2003) Mauran, an exopolysaccharide produced by the halophilic bacterium *Halomonas maura*, with a novel composition and interesting properties for biotechnology. *Extremophiles* 7: 319–326

Ash C, Farrow JAE, Wallbanks S, Collins MD (1991) Phylogenetic heterogeneity of genus *Bacillus* as revealed by comparative analysis of small-subunit-ribosomal RNA sequence. *Lett Appl Microbiol* 13: 202–206

Asker D, Ohta Y (1999) Production of canthaxanthin by extremely halophilic bacteria. *J Biosc Bioeng* 88:617–621

Benlloch S, López-López A, Casamayor EO, Øvreås L, Goddard V, Daae FL, et al. (2002) Prokaryotic genetic diversity througout the salinity gradient of a coastal solar saltern. *Environ Microbiol* 4: 349–360

Bouchotroch S, Quesada E, del Moral A, Llamas I, Béjar V (2001) *Halomonas maura* sp. nov., a novel moderately halophilic, exopolysaccharide-producing bacterium. *Int J Syst Evol Microbiol* 51:1625–1632

Brandt KK, Ingvorsen K (1997) *Desulfobacter halotolerans* sp. nov., a halotolerant acetate-oxidizing sulfate-reducing bacterium isolated from sediments of Great Salt Lake, Utah. *Syst Appl Microbiol* 20:366–373

Brocq-Rousseau D (1904) Sur un Streptothrix. *Ref Gen Botanique* 16:20–26

Brown AD (1976) Microbial water stress. *Bacteriol Rev* 40:803–846

Cánovas D, Vargas C, Calderón MI, Ventosa A, Nieto JJ (1998) Characterization of the genes for the biosynthesis of the compatible solute ectoine in the moderately halophilic bacterium *Halomonas elongata* DSM 3043. *Syst Appl Microbiol* 21:487–497

Cánovas D, Vargas C, Iglesias-Guerra F, Csonka LN, Rhodes D, Ventosa A, Nieto JJ (1997) Isolation and characterization of salt-sensitive mutants of the moderate halophile Halomonas elongata and cloning of the ectoine synthesis genes. *J Biol Chem* 272:25794–25801

Caumette P (1993) Ecology and physiology of phototrophic bacteria and sulphate-reducing bacteria in marine salterns. *Experientia* 49:473–481

Caumette P, Matheron R, Raymond N, Relexeans J-C (1994) Microbial mats in the hypersaline ponds of Mediterranean salterns (Salin-de-Giraud, France). *FEMS Microbiol Ecol* 13:273–286

Cayol JL, Ollivier B, Patel BK, Ageron E, Grimont PA, Prensier G, García JL (1995) *Haloanaerobium lacusroseus* sp. nov., an extremely halophilic fermentative bacterium from the sediments of a hypersaline lake. *Int J Syst Bacteriol* 45:790–797

Cayol JL, Ollivier B, Patel BK, Prensier G, Guezennec J, García JL (1994a) Isolation and characterization of *Halothermothrix* orenii gen. nov., sp. nov., a halophilic, hermophilic, fermentative, strictly anaerobic bacterium. *Int J Syst Bacteriol* 44:534–540

Cayol JL, Ollivier B, Soh ALA, Fardeau ML, Ageron E, Grimont PAD, Prensier G, Guezennec J, Magot M, García, JL (1994b) *Haloincola saccharolytica* subsp. *senegalensis* subsp. nov., isolated from the sediments of a hypersaline lake, and emended description of *Haloincola saccharolytica*. *Int J Syst Bacteriol* 44:805–811

Chun J, Bae KS, Moon EY, Jung SO, Lee HK, Kim SJ (2000) *Nocardiopsis kunsanensis* sp. nov., a moderately halophilic actinomycete isolated from a saltern. *Int J Syst Evol Microbiol* 50:1909–1913

Claus D, Berkeley RCW (1986) Genus *Bacillus* Cohn 1872. In: Sneath PHA (ed) *Bergey'S Manual of Systematic Bacteriology*, vol. 2. Williams & Wilkins, Baltimore, pp 1105–1139

Claus D, Fahmy F, Rolf HJ, Tosunoglu N (1983) *Sporosarcina halophila* sp. nov., an obligate, slightly halophilic bacterium from salt marsh soils. *Syst Appl Microbiol* 4:496–506

Collins MD, Jones D, Kroppenstedt RM (1983) Reclassification of *Brevibacterium imperiale* (Steinhaus) and 'Corynebacterium laevaniformans' (Dias and Bhat) in a redefined genus *Microbacterium* (Orla-Jensen), as *Microbacterium imperiale* comb. nov. and *Microbacterium laevaniformans* nom. rev.; comb. nov. *Syst Appl Microbiol* 4:65–78

Collins MD, Lawson PA, Labrenz M, Tindall BJ, Weiss N, Hirsch P (2002) *Nesterenkonia lacusekhoensis* sp. nov., isolated from hypersaline Ekho Lake, east Antarctica, and emended description of the genus *Nesterenkonia*. *Int J Syst Evol Microbiol* 52:1145–1150

Cui XL, Mao PH, Zeng M, Li WJ, Zhang LP, Xu LH, Jiang CL (2001) *Streptimonospora salina* gen. nov., sp. nov., a new member of the family *Nocardiopsaceae*. *Int J Syst Evol Microbiol* 51:357–363

Curtis TP, Sloan WT, Scannell JW (2002) Estimating prokaryotic diversity and its limits. *Proc Natl Acad Sci USA* 99: 10494–10499

DeLong EF, Pace NR (2001) Environmental diversity of bacteria and archaea. *Syst Biol* 50: 470–478

Dobson SJ, Franzmann PD (1996) Unification of the genera *Deleya* (Baumann et al. 1983), *Halomonas* (Vreeland et al. 1980) and *Halovibrio* (Fendrich, 1988) and the species *Paracoccus halodenitrificans* (Robinson and Gibbons, 1952) into a single genus, *Halomonas* and placement of the genus *Zymobacter* in the family *Halomonadaceae*. *Int J Syst Bacteriol* 46:550–558

Elshahed MS, Najar FZ, Roe BA, Oren A, Dewers TA, Krumholz LR (2004) Survey of archaeal diversity reveals an abundance of halophilic Archaea in a low-salt, sulphide- and sulphur-rich spring. *Appl Environ Microbiol* 70: 2230–2239

Fritze D (1996) *Bacillus haloalkaliphilus* sp. nov. *Int J Syst Bacteriol* 46:98–101

Fu W, Oriel P (1998) Gentisate 1,2-dioxygenase from *Haloferax* sp. D1227. *Extremophiles* 2:439–446

Fu W, Oriel P (1999) Degradation of 3-phenylpropionic acid by *Haloferax* sp. D1227. *Extremophiles* 3:45–53

Galinski EA, Tindall BJ (1992) Biotechnological prospects for halophiles and halotolerant microorganisms. In: Herbert RD, Sharp RJ (eds) *Molecular Biology and Biotechnology of Extremophiles*. Blackie, London, pp 76–114

Garabito MJ, Arahal DR, Mellado E, Márquez MC, Ventosa A (1997) *Bacillus salexigens* sp. nov., a new moderately halophilic *Bacillus species*. *Int J Syst Bacteriol* 47: 735–741

Garabito MJ, Márquez MC, Ventosa A (1998) Halotolerant *Bacillus* diversity in hypersaline environments. *Can J Microbiol* 44: 95–102

García MT, Gallego V, Ventosa A, Mellado E (2005a) *Thalassobacillus devorans* gen. nov., sp. nov., a moderately halophilic, phenol-degrading, Gram-positive bacterium. *Int J Syst Evol Microbiol* 55:1789–1795

García MT, Mellado E, Ostos JC, Ventosa A (2004) *Halomonas organivorans* sp. nov., a moderate halophile able to degrade aromatic compounds. *Int J Syst Evol Microbiol* 54:1723–1728

García MT, Ventosa A, Mellado E (2005b) Catabolic versatility of aromatic compound-degrading halophilic bacteria. *FEMS Microbiol Ecol* 54:97–109

Garriga M, Ehrmann MA, Arnau J, Hugas M, Vogel RF (1998) *Carnimonas nigrificans* gen. nov., sp. nov., a bacterial causative agent for black spot formation on cured meat products. *Int J Syst Bacteriol* 48:677–686

Gauthier MJ, Lafay B, Christen R, Fernández L, Acquaviva M, Bonin P, Bertrand JC (1992) *Marinobacter hydrocarbonoclasticus* gen. nov., sp. nov., a new, extremely halotolerant, hydrocarbon-degrading marine bacterium. *Int J Syst Bacteriol* 42:568–576

Gorshkova NM, Ivanova EP, Sergeev AF, Zhukova NV, Alexeeva Y, Wrighy JP, Nicolau DV, Mikhailov VV, Christen R (2003) *Marinobacter excellens* sp. nov., isolated from sediments of the Sea of Japan. *Int J Syst Evol Microbiol* 53:2073–2078

Gottlieb D (1973) General considerations and implications of the actinomycetes. In: Sakes G, Skinner FA (eds) *Actinomycetales: Characteristics and Practical Importance*. Academic Press, New York, pp 1–10

Grant WD, Kamekura M, McGenity TJ, Ventosa A (2001) Class III. Halobacteria *class. nov*. In: Boone DR, Castenholz RW (eds) *Bergey's Manual of Systematic Bacteriology*. Springer, New York, pp 294–334

Hansen TA (1994) Metabolism of sulfate-reducing prokariotes. *Antonie van Leeuwenhoek* 66: 165–185

Hao MV, Kocur M, Komagata K (1984) *Marinococcus* gen. nov., a new genus for motile cocci with meso-diaminopimelic acid in the cell wall; and *Marinococcus albus* sp. nov. and *Marinococcus halophilus* (Novitsky and Kushner) comb. nov. *J Gen Appl Microbiol* 30:449–459

Hayes VE, Ternan NG, McMullan G (2000) Organophosphonate metabolism by a moderately halophilic bacterial isolate. *FEMS Microbiol Lett* 186:171–175

Heyndrickx M, Lebbe L, Kersters K, De Vos P, Forsyth C, Logan NA (1998) *Virgibacillus*: A new genus to accommodate *Bacillus pantothenticus* (Proom and Knight 1950). Emended description of *Virgibacillus pantothenticus Int J Syst Bacteriol* 48: 99–106

Heyrman J, Logan NA, Busse HJ, Balcaen A, Lebbe L, Rodríguez-Diaz M, Swings J, De Vos P (2003) *Virgibacillus carmonensis* sp. nov., *Virgibacillus necropolis* sp. nov. and *Virgibacillus picturae* sp. nov., three novel species isolated from deteriorated mural paintings, transfer of the species of the genus *Salibacillus* to *Virgibacillus*, as *Virgibacillus marismortui* comb. nov. and *Virgibacillus salexigens* comb.nov., and emended description of the genus *Virgibacillus*. *Int J Syst Evol Microbiol* 53: 501–511

Ihara K, Watanabe S, Tamura T (1997) *Haloarcula argentinensis* sp. nov. and *Haloarcula mukohataei* sp. nov., two new extremely halophilic archaea collected in Argentina. *Int J Syst Bacteriol* 47: 73–77

Jeon CO, Lim J-M, Lee J-C, Lee GS, Lee J-M, Xu L-H, Jiang C-L, Kim C-J (2005a) *Lentibacillus salarius* sp. nov., isolated from saline sediment in China, and emended description of the genus *Lentibacillus*. *Int J Syst Evol Microbiol* 55: 1339–1343

Jeon CO, Lim J-M, Lee J-M, Xu L-H, Jiang C-L, Kim C-J (2005b) Reclassification of *Bacillus haloalkaliphilus* Fritze 1996 as *Alkalibacillus haloalkaliphilus* gen. nov., comb. nov. and the description of *Alkalibacillus salilacus* sp. nov., a novel halophilic bacterium isolated from a salt lake in China. *Int J Syst Evol Microbiol* 55:1891–1896

Jiang H, Dong H, Zhang G, Yu B, Chapman LR, Fields MW (2006) Microbial diversity in water and sediment of Lake Chaka, an athalassohaline lake in northwestern China. *Appl Environ Microbiol* 72: 3832–3845

Kamekura M, Onishi H (1974) Halophilic nuclease from a moderately halophilic *Micrococcus varians*. *J Bacteriol* 119:339–344

Kaurichev IS (1980) *Practicas de Edafologia*. Mir, Moscow

Kim BY, Weon HY, Yoo SH, Kim JS, Kwon SW, Stackebrandt E, Go SJ (2006) *Marinobacter koreensis* sp. nov., isolated from sea sand in Korea. *Int J Syst Evol Microbiol* 56: 2653–2656

Kleinsteuber S, Riis V, Fetzer I, Harms H, Muller S (2006) Population dynamics within a microbial consortium during growth on diesel fuel in saline environments. *Appl Environ Microbiol* 72:3531–3542

Kushner DJ, Kamekura M (1988) Physiology of halophilic eubacteria. In: Rodríguez-Valera F (ed) *Halophilic Bacteria*, vol. 1. CRC Press, Boca Raton, FL, pp 87–103

Kushwaha SC, Gochnauer MB, Kushner DJ, Kates M (1974) Pigments and isoprenoid compounds in extremely and moderately halophilic bacteria. *Can J Microbiol* 20:241–245

Lee J-S, Lim J-M, Lee KC, Lee J-C, Park Y-H, Kim C-J (2006) *Virgibacillus koreensis* sp. nov., a novel bacterium from salt field, and transfer of *Virgibacillus picturae* to the genus *Oceanobacillus* as *Oceanobacillus picturae* comb. nov. with emended descriptions. *Int J Syst Evol Microbiol* 56:251–257

Li M-G, Li W-J, Xu P, Cui X-L, Xu L-H, Jiang C-L (2003a) *Nocardiopsis xinjiangensis* sp. nov., a halophilic actinomycete isolated from a saline soil sample in China. *Int J Syst Evol Microbiol* 53:317–321

Li N, Patel BK, Mijts BN, Swaminathan K (2002) Crystallization of an alpha-amylase, AmyA, from the thermophilic halophile *Halothermothrix orenii*. *Acta Crystallogr D Biol Crystallogr* 58:2125–2126

Li W-J, Chen H-H, Kim C-J, Park D-J, Tang S-K, Lee J-C, Xu L-H, Jiang C-L (2005a) *Microbacterium halotolerans* sp. nov., isolated from a saline soil in the west of China. *Int J Syst Evol Microbiol* 55:67–70

Li W-J, Chen H-H, Kim C-J, Zhang Y-Q, Park D-J, Lee J-C, Xu L-H, Jiang C-L (2005b) *Nesterenkonia sandarakina* sp. nov. and *Nesterenkonia lutea* sp. nov., novel actinobacteria, and emended description of the genus *Nesterenkonia*. *Int J Syst Evol Microbiol* 55:463–466

Li W-J, Chen H-H, Zhang Y-Q, Schumann P, Stackebrandt E, Xu L-H, Jiang C-L (2004a) *Nesterenkonia halotolerans* sp. nov. and *Nesterenkonia xinjiangensis* sp. nov., actinobacteria from saline soils in the west of China. *Int J Syst Evol Microbiol* 54:837–841

Li W-J, Kroppenstedt RM, Wang D, Tang S-K, Lee J-C, Park D-J, Kim C-J, Xu L-H, Jiang C-L (2006) Five novel species of the genus *Nocardiopsis* isolated from hypersaline soils and emended description of *Nocardiopsis salina* Li et al. 2004. *Int J Syst Evol Microbiol* 56:1089–1096

Li W-J, Park D-J, Tang S-K, Wang D, Li J-C, Lee J-C, Xu L-H, Kim C-J, Jiang C-L (2004b) *Nocardiopsis salina* sp. nov., a novel halophilic actinomycete isolated from saline soil in China. *Int J Syst Evol Microbiol* 54:1805–1809

Li W-J, Schumann P, Zhang Y-Q, Chen G-Z, Tian X-P, Xu L-H, Stackebrandt E, Jiang C-L (2005c) *Marinococcus halotolerans* sp. nov., isolated from Qinghai, north-west China. *Int J Syst Evol Microbiol* 55:1801–1804

Li W-J, Tang S-K, Stackebrandt E, Kroppenstedt RM, Schumann P, Xu L-H, Jiang, C-L (2003b) *Saccharomonospora paurometabolica* sp. nov., a moderately halophilic actinomycete isolated from soil in China. *Int. J Syst Evol Microbiol* 53:1591–1594

Li W-J, Xu P, Tang S-K, Xu L-Hu, Kroppenstedt RM, Stackebrandt E, Jiang C-L (2003c) *Prauserella halophila* sp. nov. and *Prauserella alba* sp. nov., moderately halophilic actinomycetes from saline soil. *Int J Syst Evol Microbiol* 53:1545–1549

Li W-J, Xu P, Zhang L-P, Tang S-K, Cui X-L, Mao P-H, Xu L-H, Schumann P, Stackebrandt E, Jiang, C-L (2003d) *Streptomonospora alba* sp. nov., a novel halophilic actinomycete, and emended description of the genus *Streptomonospora* Cui et al. 2001. *Int J Syst Evol. Microbiol* 53:1421–1425

Louis P, Galinski EA (1997) Characterization of genes for the biosynthesis of the compatible solute ectoine from *Marinococcus halophilus* and osmoregulated expression in *Escherichia coli*. *Microbiology* 143:1141–119

Louis P, Trüper HG, Galinski EA (1994) Survival of *Escherichia coli* during drying and storage in the presence of compatible solutes. *Appl Microbiol Biotechnol* 41:648–688

Margesin R, Schinner F (2001) Potential of halotolerant and halophilic microorganisms for biotechnology. *Extremophiles* 5:73–83

Márquez MC, Ventosa A, Ruiz-Berraquero F (1987) A taxonomic study of heterotrophic halophilic and non-halophilic bacteria from a solar saltern. *J Gen Microbiol* 133:45–56

Márquez MC, Ventosa, A, Ruiz-Berraquero F (1990) *Marinococcus hispanicus*, a new species of moderately halophilic gram-positive cocci. *Int J Syst Bacteriol* 40:165–169

Márquez MC, Ventosa A, Ruiz-Berraquero F (1992) Phenoypic and chemotaxonomic characterization of *Marinococcus halophilus*. *Syst Appl Microbiol* 15:63–69

Martín S, Márquez MC, Sánchez-Porro C, Mellado E, Arahal DR, Ventosa A (2003) *Marinobacter lipolyticus* sp. nov., a novel moderate halophile with lipolytic activity. *Int J Syst Evol Microbiol* 53:1383–1387

Martínez-Cánovas MJ, Béjar V, Martínez-Checa F, Quesada E (2004a) *Halomonas anticariensis* sp. nov., from Fuente de Piedra, a saline-wetland wildfowl reserve in Malaga, southern Spain. *Int J Syst Evol Microbiol* 54:1329–1332

Martínez-Cánovas MJ, Quesada E, Llamas I, Béjar V (2004b) *Halomonas ventosae* sp. nov., a moderately halophilic, denitrifying, exopolysaccharide-producing bacterium. *Int J Syst Evol Microbiol* 54:733–737

Martínez-Cánovas MJ, Quesada E, Martinez-Checa F, del Moral A, Béjar V (2004c) *Salipiger mucescens* gen. nov., sp. nov., a moderately halophilic, exopolysaccharide-producing bacterium isolated from hypersaline soil, belonging to the alpha-Proteobacteria. *Int J Syst Evol Microbiol* 54:1735–1740

Martínez-Checa F, Quesada E, Martínez-Cánovas MJ, Llamas I, Béjar V (2005) *Palleronia marisminoris* gen. nov., sp. nov., a moderately halophilic, exopolysaccharide-producing bacterium belonging to the 'Alphaproteobacteria', isolated from a saline soil. *Int J Syst Evol Microbiol* 55:2525–2530

Martínez-Checa F, Toledo FL, Vilchez R, Quesada E, Calvo C (2002) Yield production, chemical composition, and functional properties of emulsifier H28 synthesized by *Halomonas eurihalina* strain H-28 in media containing various hydrocarbons. *Appl Microbiol Biotechnol* 58:358–363

Mata JA, Béjar V, Llamas I, Arias S, Bressollier P, Tallon R, Urdaci MC, Quesada E (2006) Exopolysaccharides produced by the recently described halophilic bacteria *Halomonas ventosae* and *Halomonas anticariensis*. *Res Microbiol* 157:827–835

Mellado E, Ventosa A (2003) Biotechnological potential of moderately and extremely halophilic microorganisms. In: Barredo JL (ed) *Microorganisms for Health Care, Food and Enzyme Production*. Research Signpost, Kerala, pp 233–256

Mevarech M, Frolow F, Gloss LM (2000) Halophilic enzymes: Proteins with a grain of salt. *Biophys Chem* 86:155–164

Meyer J (1976) *Nocardiopsis dassonvillei*, a new genus of the order *Actinomycetales*. *Int J Syst Bacteriol* 26:487–493

Mijts BN, Patel BK (2002) Cloning, sequencing and expression of an alpha-amylase gene, amyA, from the thermophilic halophile *Halothermothrix orenii* and purification and biochemical characterization of the recombinant enzyme. *Microbiology* 148:2343–2349

Miller KJ, Leschine S (1984) A halotolerant *Planococcus* from Antarctic Dry Valleys. *Curr Microbiol* 11:205–210

Miller KJ, Leschine S, Huguenin RL (1983) Characterization of a halotolerant-psychroloterant bacterium from Dry Valley Antarctic soil. *Adv Space Res* 3:43–47

Möller DR, Schauder R, Fuchs G, Thauer RK (1987) Acetate oxidation to CO_2 via a citric acid cycle involving an ATP-citrate lyase: A mechanism for the synthesis of ATP via substrate level phosphorylation in *Desulfobacter postgatei* growing on acetate and sulfate. *Arch Microbiol* 148:202–207

Mormile MR, Romine MF, García MT, Ventosa A, Bailey TJ, Peyton BN (1999) *Halomonas campisalis* sp. nov., a denitrifying, moderately haloalkaliphilic bacterium. *Syst Appl Microbiol* 22:551–558

Motitschke L, Driller H, Galinski E (2001) Patent US6267973

Moune S, Eatock C, Matheron R, Willison JC, Hirschler A, Herbert R, Caumette P (2000) *Orenia salinaria* sp. nov., a fermentative bacterium isolated from anaerobic sediments of Mediterranean salterns. *Int J Syst Evol Microbiol* 50:721–729

Moune S, Manac'h N, Hirschler A, Caumette P, Willison JC, Matheron R (1999) *Haloanaerobacter salinarius* sp. nov., a novel halophilic fermentative bacterium that reduces glycine-betaine to trimethylamine with hydrogen or serine as electron donors; emendation of the genus *Haloanaerobacter*. *Int J Syst Bateriol* 49:103–112

Munson MA, Nedwell DB, Embley TM (1997) Phylogenetic diversity of Archaea in sediment samples from a coastal salt marsh. *Appl Environ Microbiol* 63:4729–4733

Nakayama H, Yoshida K, Ono H, Murooka Y, Shinmyo A (2000) Ectoine, the compatible solute of *Halomonas elongata*, confers hyperosmotic tolerance in cultured tobacco cells. *Plant Physiol* 122:1239–1247

Nicolaus B, Marsiglia F, Esposito E, Trincone A, Lama L, di Prisco G, Gambacorta A, Valderrama MJ, Grant WD (1992) Isolation of extremely halotolerant cocci from Antarctica. *FEMS Microbiol Lett* 99:145–150

Nielsen P, Fritze D, Priest FG (1995) Phenetic diversity of alkaliphilic *Bacillus* strains: Proposal for nine species. *Microbiology* 141:1745–1761

Nielsen P, Rainey FA, Outtrup H, Priest FG, Fritze D (1994) Comparative 16S rDNA sequence analysis of some alkaliphilic bacilli and the establishment of a sixth rRNA group within the genus *Bacillus*. *FEMS Microbiol Lett* 117:61–66

Nieto JJ, Fernández-Castillo R, Márquez MC, Ventosa A, Quesada E, Ruíz-Berraquero F (1989) Survey of metal tolerance in moderately halophilic eubacteria. *Appl Environ Microbiol* 55:2385–2390

Nieto JJ, Vargas C (2002) Synthesis of osmoprotectants by moderately halophilic bacteria: Genetic and applied aspects. In Pandalai SG (ed) *Recent Res Devel Microbiol*. Research Signpost, Kerala, pp 403–418

Nissenbaum A (1975) The microbiology and biogeochemistry of the Dead Sea. *Microbial Ecol* 2:139–161

Nonomura H, Ohara Y (1971) Distribution of actinomycetes in soil. X. New genus and species of monosporic actinomycetes. *J Ferment Technol* 49:895–903

Okamoto T, Taguchi H, Nakamura K, Ikenaga H, Kuraishi H, Yamasato K (1993) *Zymobacter palmae* gen. nov., sp. nov., a new ethanol-fermenting peritrichous bacterium isolated from palm sap. *Arch Microbiol* 160:333–337

Olivera N, Sineriz F, Breccia JD (2005) *Bacillus patagoniensis* sp. nov., a novel alkalitolerant bacterium from the rhizosphere of *Atriplex lampa* in Patagonia, Argentina. *Int J Syst Evol Microbiol* 55:443–447

Onishi H (1972) Halophilic amylase from a moderately halophilic *Micrococcus. J Bacteriol* 109:570–574

Onishi H, Hidaka O (1978) Purification and properties of amylase produced by a moderate halophilic *Acinetobacter* sp. *Can J Microbiol* 24:1017–1023

Onishi H, Kamekura M (1972) *Micrococcus halobius* sp. n. *Int J Syst Bacteriol* 22:233–236

Ono H, Okuda M, Tongpim S, Imai K, Shinmyo A, Sakuda S, Kaneko Y, Murooka Y, Takano M (1998) Accumulation of compatible solutes, ectoine and hydroxyectoine, in a moderate halophile, *Halomonas elongata* KS3 isolated from dry salty land in Thailand. *J Ferm Bioeng* 85:362– 368

Ono H, Sawada K, Khunajakr N, Tao T, Yamamoto M, Hiramoto M, Shinmyo A, Tacano M, Murooka Y (1999) Characterization of biosynthetic enzymes for ectoine as a compatible solute in a moderately halophilic eubacterium, *Halomonas elongata. J Bacteriol* 181:91–99

Oren A (1983) *Clostridium lortetii* sp. nov., a halophilic obligatory anaerobic bacterium producing endospores with attached gas vacuoles. *Arch Microbiol* 136:42–48

Oren A (1988) Anaerobic degradation of organic compounds at high salt concentrations. *Antonie van Leeuwenhoek* 54:267–277

Oren A (2000) Change of the names *Haloanaerobiales, Haloanaerobiaceae* and *Haloanaerobium* to *Halanaerobiales, Halanaerobiaceae* and *Halanaerobium,* respectively, and further nomenclatural changes within the order *Halanaerobiales. Int J Syst Evol Microbiol* 50: 2229–2230

Oren A (2002) Diversity of halophilic microoganisms: Environments, phylogeny, physiology, and applications. *J Ind Microbiol Biotechnol* 28:56–63

Oren A, Paster BJ, Woese CR (1984) *Haloanaerobiaceae*: A new family of moderately halophilic, obligately anaerobic bacteria *Syst Appl Microbiol* 5:71–80

Oren A, Pohla H, Stackebrandt E (1987) Transfer of *Clostridium lortetii* to a new genus *Sporohalobacter* gen. nov. as *Sporohalobacter lortetii* comb. nov. and description of *Sporohalobacter marismortui* sp. nov. *Syst Appl Microbiol* 9:239–246

Oren A, Ventosa A (1996) A proposal for the transfer of *Halorubrobacterium distributum* and *Halorubrobacterium coriense* to the genus *Halorubrum* as *Halorubrum distributum* comb. nov. and *Halorubrum coriense* comb. nov., respectively. *Int J Syst Bacteriol* 46:1180

Orla-Jensen S (1919) *The Lactic Acid Bacteria.* Host & Sons, Copenhagen

Patel R, Dodia M, Singh SP (2005) Extracellular alkaline protease from a newly isolated haloalkaliphilic *Bacillus* sp.: Production and optimization. *Proc Biochem* 40:3569–3575

Purdy KJ, Cresswell-Maynard TD, Nedwell DB, McGenity TJ, Grant WD, Timmis KN, Embley TM (2004) Isolation of haloarchaea that grow at low salinities. *Environ Microbiol* 6:591–595

Quesada E, Béjar V, Calvo, C (1993) Exopolysaccaharide production by *Volcaniella eurihalina. Experientia* 49:1037–1041

Quesada E, Béjar V, Ferrer MR, Calvo C, Llamas I, Martínez-Checa F, Arias S, Ruíz-García C, Paez R, Martínez-Cánovas MJ, del Moral A (2004) Moderately halophilic, exopolysaccharide-producing bacteria. In: Ventosa A (ed) *Halophilic Microorganisms.* Springer-Verlag, Berlin, pp 285–295

Quesada E, Valderrama MJ, Béjar V, Ventosa A, Gutiérrez MC, Ruíz-Berraquero F, Ramos-Cormenzana A (1990) *Volcaniella eurihalina* gen. nov., sp. nov., a moderately halophilic non-motile Gram-negative rod. *Int J Syst Bacteriol* 40:261–267

Quesada E, Ventosa A, Rodríguez-Valera F, Megías L, Ramos-Cormenzana A (1983) Numerical taxonomy of moderate halophiles from hypersaline soils. *J Gen Microbiol* 129:2649–2657

Quesada E, Ventosa A, Rodríguez-Valera F, Ramos-Cormenzana A (1982) Types and propierties of some bacteria isolated from hypersaline soils. *J Appl Bacteriol* 53:155–161

Quesada E, Ventosa A, Ruíz-Berraquero F, Ramos-Cormenzana A (1984) *Deleya halophila*, a new species of moderately halophilic bacteria. *Int J Syst Bacteriol* 34:287–292

Quillaguaman J, Hatti-Kaul R, Mattiasson B, Alvarez MT, Delgado O (2004) *Halomonas boliviensis* sp. nov., an alkalitolerant, moderate halophile isolated from soil around a Bolivian hypersaline lake. *Int J Syst Evol Microbiol* 54:721–725

Rainey FA, Zhilina TN, Boulygina ES, Stackebrandt E, Tourova TP, Zavarzin GA (1995) The taxonomic status of the fermentative halophilic anaerobic bacteria: Description of *Haloanaerobiales* ord. nov., *Halobacteroidaceae* fam. nov., *Orenia* gen. nov. and further taxonomic rearrangements at the genus and species level. *Anaerobe* 1:185–199

Rappé MS, Giovannoni SJ (2003) The uncultured microbial majority. *Annu Rev Microbiol* 57: 369–394

Ren P-G, Zhou P-J (2005) *Tenuibacillus multivorans* gen. nov., sp. nov., a moderately halophilic bacterium isolated from saline soil in Xin-Jiang, China. *Int J Syst Evol Microbiol* 55:95–99

Ríos M, Nieto JJ, Ventosa A (1998) Numerical taxonomy of heavy metal-tolerant nonhalophilic bacteria isolated from hypersaline environments. *Int Microbiol* 1:45–51

Rodríguez-Valera F (1988) Characteristics and microbial ecology of hypersaline environments. In: Rodríguez-Valera F (ed) *Halophilic Bacteria*, vol. 1. CRC Press, Boca Raton, Fla, pp 3–30

Rodríguez-Valera F, Ruíz-Berraquero F, Ramos-Cormenzana, A (1979) Isolation of extreme halophiles from seawater. *Appl Environ Microbiol* 38:164–165

Romanenko LA, Schumann P, Rohde M, Zhukova NV, Mikhailov VV, Stackebrandt E (2005) *Marinobacter bryozoorum* sp. nov. and *Marinobacter sediminum* sp. nov., novel bacteria from the marine environment. *Int J Syst Evol Microbiol* 55:143–148

Ruan JS, Al-Tai AM, Zhou ZH, Qu LH (1994) *Actinopolyspora iraqiensis* sp. nov., a new halophilic actinomycete isolated from soil. *Int J Syst Bacteriol* 44:759–763

Sánchez-Porro C, Martin S, Mellado E, Ventosa A (2003) Diversity of moderately halophilic bacteria producing extracellular hydrolytic enzymes. *J Appl Microbiol* 94:295–300

Saum SH, Sydow JF, Palm P, Pfeiffer F, Oesterhelt D, Muller V (2006) Biochemical and molecular characterization of the biosynthesis of glutamine and glutamate, two major compatible solutes in the moderately halophilic bacterium *Halobacillus halophilus*. *J Bacteriol* 188:6808–6815

Schlesner H, Lawson PA, Collins MD, Weiss N, Wehmeyer U, Volker H, Thomm M (2001) *Filobacillus milensis* gen. nov., sp. nov., a new halophilic spore-forming bacterium with Orn-D-Glu-type peptidoglycan. *Int J Syst Evol Microbiol* 51:425–431

Skyring GW (1987) Sulfate reduction in coastal ecosystem. *Geomicrobiol J* 5:295–374

Spanka R, Fritze D (1993) *Bacillus cohnii* sp. nov., a new, obligately alkaliphilic, oval-spore-forming *Bacillus* species with ornithine and aspartic acid instead of diaminopimelic acid in the cell wall. *Int J Syst Bacteriol* 43:150–156

Spring S, Ludwig W, Márquez MC, Ventosa A, Schleifer KH (1996) *Halobacillus* gen. nov., with descriptions of *Halobacillus litoralis* sp. nov., and *Halobacillus trueperi* sp. nov., and transfer of *Sporosarcina halophila* to *Halobacillus halophilus* comb. nov. *Int J Syst Bacteriol* 46:492–496

Stackebrandt E, Koch C, Gvozdiak O, Schumann P (1995) Taxonomic dissection of the genus *Micrococcus*: *Kocuria* gen. nov., *Nesterenkonia* gen. nov., *Kytococcus* gen. nov., *Dermacoccus* gen. nov., and *Micrococcus* Cohn 1872 gen. emend. *Int J Syst Bacteriol* 45:682–692

Switzer BJ, Burns BA, Buzzelli J, Stolz JF, Oremland RS (1998) *Bacillus arsenicoselenatis*, sp. nov., and *Bacillus selenitireducens*, sp. nov.: Two haloalkaliphiles from Mono Lake, California that respire oxyanions of selenium and arsenic. *Arch. Microbiol* 171:19–30

Takashina T, Hamamoto T, Otozai K, Grant WD, Horikoshi K (1990) *Haloarcula japonica* sp. nov., a new triangular halophilic archaebacterium. *Syst Appl Microbiol* 13:177–181

Takeuchi M, Hatano K (1998) Union of the genera *Microbacterium* Orla-Jensen and *Aureobacterium* Collins et al. in a redefined genus *Microbacterium*. *Int J Syst Bacteriol* 48:739–747

Tan TC, Yien YY, Patel BK, Mijts BN, Swaminathan K (2003) Crystallization of a novel alpha-amylase, AmyB, from the thermophilic halophile *Halothermothrix orenii*. *Acta Crystallogr D Biol Crystallogr* 59:2257–2258

Tang S-K, Li W-J, Wang D, Zhang Y-G, Xu L-H, Jiang C-L (2003) Studies of the biological characteristic of some halophilic and halotolerant actinomycetes isolated from saline and alkaline soils. *Actinomycetologica* 17:6–10

Tardy-Jacquenod C, Caumette P, Matheron R, Lanau C, Arnauld O, Magot M (1996) Characterization of sulphate-reducing bacteria isolated from oil-field water. *Can J Microbiol* 42: 259–266

Tokuda H, Unemoto T (1981) A respiration-dependent primary sodium extrusion system functioning at alkaline pH in the marine bacterium *Vibrio alginolyticus*. *Biochem Biophys Res Commun* 102:265–271

Tokuda H, Unemoto T (1984) Na+ is translocated at NADH: quinone oxidoreductase segment in the respiratory chain of *Vibrio alginolyticus*. *J Biol Chem* 259:7785–7790

Torsvik V, Øvreås L, Thingstad TF (2002) Prokaryotic diversity – Magnitude, dynamics, and controlling factors. *Science* 296:1064–1066

Toyoda Y, Oowaya K, Tacano M, Shibata S (1997) Patent JP9143167

Tresner HD, Hayes JS, Backus EJ (1968) Differential tolerance of streptomycetes to sodium chloride as a taxonomic tool. *Appl Microbiol* 16:1134–1136

Trüper HG (1969) Bacterial sulphate reduction in the Red Sea hot brines. In: Degens ET, Ross DA (eds) *Hot Brines and Heavy Metal Deposits in the Red Sea*. Springer-Verlag, New York, pp 263–271

Tsai CR, García JL, Patel BK, Cayol JL, Baresi L, Mah RA (1995) *Haloanaerobium alcaliphilum* sp. nov., an anaerobic moderate halophile from the sediments of Great Salt Lake, Utah. *Int J Syst Bacteriol* 45:301–307

Ueno S, Kaieda N, Koyama N (2000) Characterization of a P-type Na$^+$-ATPase of a facultatively anaerobic alkaliphile, *Exiguobacterium aurantiacum*. *J Biol Chem* 275:14537–14540

Valderrama MJ, Quesada E, Béjar V, Ventosa A, Gutiérrez MC, Ruíz-Berraquero F, Ramos-Cormenzana A (1991) *Deleya salina* sp. nov., a moderately halophilic Gram-negative bacterium. *Int J Syst Bacteriol* 41:377–384

Vedder A (1934) *Bacillus alcalophilus* n. sp.; benevens enkele ervaringen met sterk alcalische voedingsbodems. Antonie van Leeuwenhoek *J Microbiol Serol* 1:141–147

Ventosa A (2006) Unusual micro-organisms from unusual habitats: Hypersaline environments. In: Logan NA, Lappin-Scott HM, Oyston PCF (eds) *Prokaryotic Diversity: Mechanisms and Significance*. Cambridge University Press, Cambridge, pp 223–253

Ventosa A, Gutierrez MC, García MT, Ruiz-Berraquero F (1989) Classification of *Chromobacterium marismortui* in a new genus, *Chromohalobacter* gen. nov., as *Chromohalobacter marismortui* comb. nov., nom. rev. *Int J Syst Bacteriol* 39:382–386

Ventosa A, Gutierrez MC, Kamekura M, Dyall-Smith ML (1999) Proposal to transfer *Halococcus turkmenicus, Halobacterium trapanicum* JCM 9743 and strain GSL-11 to *Haloterrigena turkmenica* gen. nov., comb. nov. *Int J Syst Bacteriol* 49:131–136

Ventosa A, Márquez MC, Weiss N, Tindall BJ (1992) Transfer of *Marinococcus hispanicus* to the genus *Salinicoccus* as *Salinicoccus hispanicus* comb. nov. *Syst Appl Microbiol* 15:530–534

Ventosa A, Nieto JJ, Oren A (1998) Biology of moderately halophilic aerobic bacteria. *Microbiol Mol Biol Rev* 62: 504–544

Ventosa A, Quesada E, Rodríguez-Valera F, Ruíz-Berraquero F, Ramos-Cormenzana A (1982) Numerical taxonomy of moderately halophilic Gram-negative rods. *J Gen Microbiol* 128: 1959–1968

Ventosa A, Ramos-Cormenzana A, Kocur M (1983) Moderately halophilic gram-positive cocci from hypersaline environments. *Syst Appl Microbiol* 4: 564–570

Ventosa A, Sánchez-Porro C, Martín S and Mellado E (2005) Halophilic archaea and bacteria as a source of extracellular hydrolytic enzymes. In: Gunde-Cimerman A, Oren A, Plemenitas A (eds) *Adaptation of Life at High Salt Concentrations in Archaea, Bacteria and Eukarya*. Springer-Verlag, Heidelberg, pp 337–354

Walsh DA, Papke RT, Doolittle WF (2005) Archaeal diversity along a soil salinity gradient prone to disturbance. *Environ Microbiol* 7:1655–1666

Yakimov MM, Golyshin PN, Lang S, Moore ER, Abraham WR, Lunsdorf H, Timmis KN (1998) *Alcanivorax borkumensis* gen. nov., sp. nov., a new, hydrocarbon-degrading and surfactant-producing marine bacterium. *Int J Syst Bacteriol* 48:339–348

Yoon JH, Kang KH, Park YH (2002) *Lentibacillus salicampi* gen. nov., sp. nov., a moderately halophilic bacterium isolated from a salt field in Korea. *Int J Syst Evol Microbiol* 52:2043–2048

Yoon J-H, Oh T-K, Park Y-H (2004) Transfer of *Bacillus halodenitrificans* Denariaz et al. 1989 to the genus *Virgibacillus* as *Virgibacillus halodenitrificans* comb. nov. *Int J Syst Evol. Microbiol* 54:2163–2167

Yoon JH, Weiss N, Lee KC, Lee IS, Kang KH, Park YH (2001) *Jeotgalibacillus alimentarius* gen. nov., sp. nov., a novel bacterium isolated from jeotgal with L-lysine in the cell wall, and reclassification of *Bacillus marinus* Rueger 1983 as *Marinibacillus marinus* gen. nov., comb. nov. *Int J Syst Evol Microbiol* 51:2087–2093

Yoshida M, Matsubara K, Kudo T, Horikoshi K (1991) *Actinopolyspora mortivallis* sp. nov., a moderately halophilic actinomycete. *Int J Syst Bacteriol* 41:15–20

Yumoto I, Hirota K, Goto T, Nodasaka Y, Nakajima K (2005) *Bacillus oshimensis* sp. nov., a moderately halophilic, non-motile alkaliphile. *Int J Syst Evol Microbiol* 55:907–911

Yumoto I, Yamaga S, Sogabe Y, Nodasaka Y, Matsuyama H, Nakajima K, Suemori A (2003) *Bacillus krulwichiae* sp. nov., a halotolerant obligate alkaliphile that utilizes benzoate and m-hydroxybenzoate. *Int J Syst Evol Microbiol* 53:1531–1536

Zeikus G, Hegge PW, Thompson TE, Phelps TJ, Langworthy TA (1983) Isolation and description of *Haloanaerobium praevalens* gen. nov. and sp. nov., an obligately anaerobic halophile common to Great Salt Lake sediments. *Curr Microbiol* 9:225–234

Zeikus JG (1983) Metabolic communication between biodegradative populations in nature. In: Slater JH, Whittenbury R, Wimpenny JWT (eds) *Microbes in Their Natural Environments*. University Press, Cambridge, pp. 423–462

Zhang Z, Wang Y, Ruan J (1998) Reclassification of *Thermomonospora* and *Microtetraspora*. *Int J Syst Bacteriol* 48:411–422

Zhilina TN, Zarvazin GA, Detkova EN, Rainey FA (1996) *Natroniella acetigena* gen. nov. sp. nov., an extremely haloalkaliphilic, homoacetic bacterium: A new member of *Haloanaerobiales* *Curr Microbiol* 32:320–326

Zvyagintseva IS, Tarasov AL (1987) Extreme halophilic bacteria from saline soils. *Microbiologiya* 56:839–844

Chapter 6
Atacama Desert Soil Microbiology

Benito Gómez-Silva(✉), Fred A. Rainey, Kimberley A. Warren-Rhodes, Christopher P. McKay, and Rafael Navarro-González

6.1 Introduction

The Atacama Desert is an ancient temperate desert (mean annual temperature of 14–16°C) that extends across 1,000 km from 30°S to 20°S along the Pacific coast of South America (McKay et al. 2003; Fig. 6.1). As discussed by Rundel et al. (1991) and Miller (1976) the desert owes its extreme aridity to the climatic regime dominated by a constant temperature inversion due to the cool north-flowing Humboldt Current and the presence of the strong Pacific anticyclone. The position of the Pacific anticyclone is generally stable with a small shift of a few degrees south in the summer (Trewartha 1961). Geological and soil mineralogical evidence suggests that extreme arid conditions have persisted in the Southern Atacama for 10–15 million years (Myrs; Ericksen 1983; Houston and Hartley 2003; Clarke 2006) making it one of the oldest deserts on Earth.

The driest parts of the Atacama Desert are located between approximately 22°S to 26°S in the broad valley formed by the coastal range and the medial range (the Cordillera de Domeyko) (Fig. 6.2). One of the most striking and unusual features of the Atacama Desert is the presence of large nitrate deposits. Early in the last century, nitrate mining operations were conducted in this area (Woodcock and Hill 1901), but, currently, there are no permanent human settlements in this hyperarid region. Although nitrate deposits are found in many deserts, significant accumulations are found only in the Atacama. Current understanding suggests that the nitrate is of atmospheric (lightning) origin based on stable isotope evidence (as suggested by Böhlke et al. (1997)) and that lightning-related nitrate production is not unusually intense over the Atacama. What is unusual is that there are no removal mechanisms due to the lack of water activity and resultant lack of microbial denitrification. Over the history of the Atacama (10–15 Myr or more) the accumulation has resulted in large concentrated deposits.

Within the driest part of the Atacama there exists a region with "Mars-like" soils (Navarro-González et al. 2003). There are three characteristics that make these soils

---

Benito Gómez-Silva
Instituto del Desierto y Unidad de Bioquímica, Facultad Ciencias de la Salud, Universidad de Antofagasta, Casilla 170, Antofagasta, Chile
e-mail: bgomez@uantof.cl

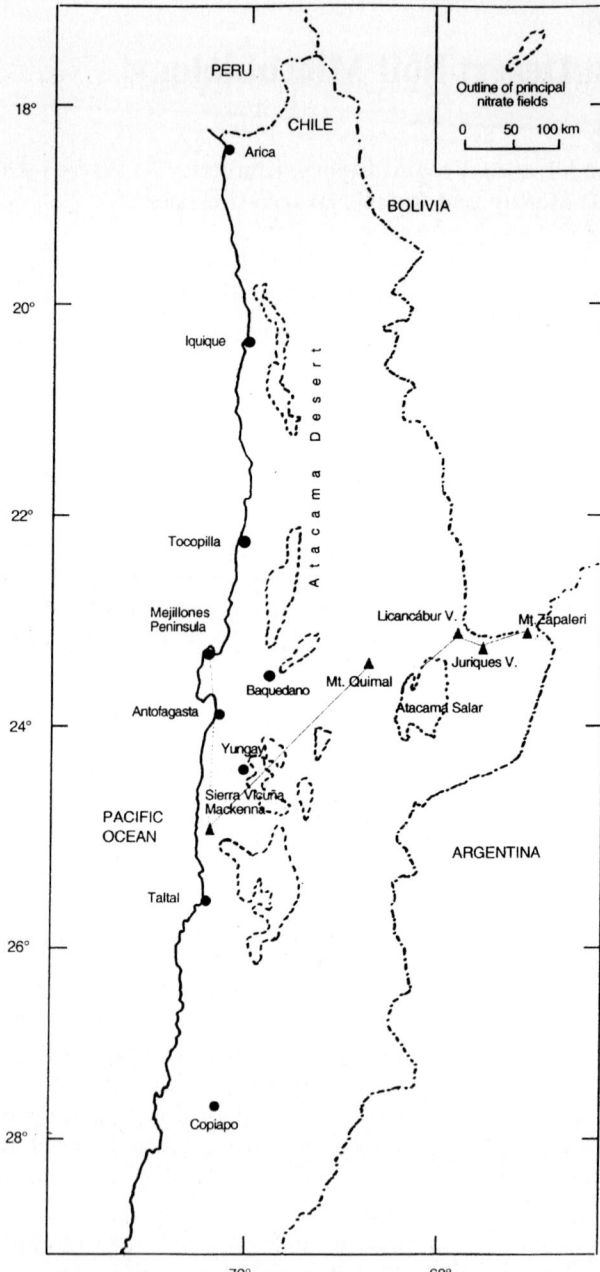

Fig. 6.1 Map of the Atacama Desert in Northern Chile showing the location of the hyperarid core near Yungay (24°S). (Adapted from McKay et al. 2003)

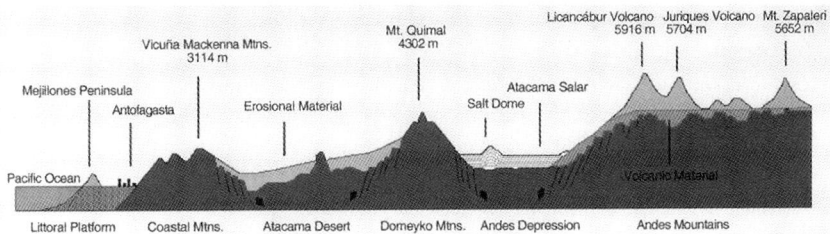

Fig. 6.2 Cross-section of the Atacama Desert region near Antofagasta, Chile. (Adapted from Guías Turistel, Ediciones Turiscom)

Mars-like: first, there are very low levels of organic material and the organics that are present are refractory. They do not decompose at the temperatures reached by the Viking GCMS (500°C). Second, there are very low levels of soil bacteria, and, in some locations, these are undetectable either by culture or DNA amplification (Navarro-González et al 2003), or by Limulus Amebocyte Lysate, a sensitive and specific assay to screen for the presence of endotoxin (lipopolysaccharide) from Gram-negative bacteria (M. Turnbull, personal communication). Third, the soil contains an oxidizing agent with the ability to oxidize at equal rates L- and D- amino acids, as well as L- and D- sugars.

Soils to the south of the arid core region do not show these characteristics. We think these Mars-like characteristics result from the extreme aridity of the core region of the Atacama. The entire Atacama is arid and receives very little rain. However, many locations in the desert receive marine fog providing sufficient moisture for hypolithic algae, lichens, and even cacti (Rundel et al. 1991; Warren-Rhodes et al. 2006). However, in the region south of Antofagasta the coastal range blocks the marine fog. The crest-line of the coastal range averages 3,000 m for about 100 km south of Antofagasta. The region that is in the "fog shadow" of this high coastal crest-line is the region that contains the Mars-like soils. Rech et al. (2003) have used sulfur isotopes to trace out the areas affected by marine input. Their results are consistent with the coastal range blocking the marine fog in the region of Mars-like soils. Thus the Atacama Desert is in the rain shadow of the Andes and the cold Humboldt Current. The Mars-like soils are found in those regions that are even drier yet because they are also in the fog shadow of the coastal range.

In this chapter we discuss the microbiology of the extreme arid core of the Atacama Desert and compare it to less arid regions of that desert. We then discuss how the arid core region provides a test bed for the future search for life on Mars.

6.2 Atacama Soil Organics

Soil organics are a mixture of recognizable biological material whose components have been altered to the degree that it no longer retains its original structural or chemical organization (Oades 1989). At any given time the amount of soil organics reflects the long-term balance between input and loss rates (Olson 1963). The input of soil organics increases with mean annual precipitation and temperature, whereas their residence time decreases with mean annual precipitation and mean annual temperature (Lieth 1973; Amundson 2001). Soil organics are ultimately degraded to carbon dioxide through microbial respiration and abiotic processes (Bunt and Rovira 1954; Parsons et al. 2004).

Organics in surface samples of Atacama Desert soils have been studied in detail using dry analytical methods, namely pyrolysis-gas chromatography-mass spectrometry (pyr-GC-MS) at atmospheric pressure in an inert atmosphere by Navarro-González et al. (2003). Samples exposed to flash heating at 500°C in a He atmosphere revealed that the most arid zone of the Atacama, the Yungay area (~24°S) is depleted of organic molecules. At a higher temperature pyrolysis (750°C), only formic acid and benzene are detectable. In contrast, samples from less arid sites (28°S) release a complex mixture of organic compounds upon pyrolysis at 750°C (formic acid, propenenitrile, 1,2-butadiene, 2-butene, 1,3-pentadiene, 2-methylfuran, benzene, methylbenzene, benzenenitrile, ethylbenzene, 1,2-dimethylbenzene, and styrene). A comparative analysis of this thermal treatment with proteins, peptides, and free amino acids yielded a series of carboxylic acids, saturated nitriles, and saturated, unsaturated, and aromatic hydrocarbons; carbohydrates were degraded to a series of aliphatic aldehydes, ketones, carboxylic acids, aromatic compounds, and furan derivatives; fatty acids were pyrolyzed to alkanes, alkenes, aromatic compounds, and short chain carboxylic acids; porphyrins were degraded to pyrroles; and nucleic acid bases released unsaturated nitriles and substituted furans. Pyrolysis of Atacama Desert bacterial isolates (e.g., strain AT01-3, isolated from soil samples at S 24° 4′ 9.6″, W 69° 51′ 58.8″, Navarro-González et al. 2003) released a mixture of all the above classes of organic compounds.

From these results, it was concluded that Atacama sites at 28°S contain bacteria and/or all of the major classes of biomolecules at levels within the detection limits of the pyr-GC-MS protocol applied (Navarro-González et al. 2003). The two characteristic compounds released by pyrolysis at 750°C in a He atmosphere by all samples of the Atacama Desert along a precipitation gradient from 24°S to 28°S are formic acid, a highly oxidized organic compound, and benzene, a thermally stable aromatic compound. These two compounds are typically released by the thermal treatment of monocarboxylic acids, polycarboxylic acids, carbohydrates, polysaccharides, amino acids, and proteins.

Formic acid is present at concentrations of ~50 µg/g in the Yungay area (~24°S), then decreases by an order of magnitude at 26°S, and increases again in the less arid zone. In contrast, benzene is present at trace levels (~1–2 µg/g) at ~24°S with its concentration increasing to ~80 µg/g in the less arid zone (~28°S). The ratio between formic acid and benzene reaches its highest value (≥15 units) in the Yungay area, and then sharply drops to ≤0.3 from 25°S to 28°S. A high formic acid/benzene ratio indicates that the organic matter present in the region is oxidized, and possibly composed of refractory organics such as aliphatic and aromatic mono- and polycarboxylic acids (Navarro-González et al. 2003).

The organics present in the Yungay soils are dominated by carboxylic acids and polycyclic aromatic hydrocarbons as determined by nuclear magnetic resonance

(NMR) and infrared spectroscopy (IR) analysis. The $\delta^{13}C$ values of soil organics varied from −26.1‰ at 24°S to −28.9‰ at 28°S (Navarro-González et al. 2006), a typical range for organic matter produced by C_3 photosynthesis (O'Leary 1998). Similarly the C/N ratio for these samples varied from 8.2 at 24°S to 16.7 at 28°S (Navarro-González et al. 2006), a typical range for soil organic matter (Batjes 1996). Surprisingly, the levels of total organics determined by wet analytical methods (titration by oxidation with permanganate) in the Atacama Desert soil are much higher than the levels derived by pyr-GC-MS (Navarro-González et al. 2006). The concentration of organics, as determined by permanganate oxidation, is ~0.04 mg of carbon per gram of soil in the Yungay area, and increases with precipitation to ~0.70 mg of carbon per gram of soil at ~28°S. Considering that 1 µg of benzene is equivalent to 0.92 µg of carbon, it appears that the organic matter content is underestimated by pyr-GC-MS by a factor varying from 20 to 40 in the Yungay region to about 10 at ~28°S. These results show an important limitation of the pyr-GC-MS technique for detection of organic material. When organics are present as low-level refractory substances, the temperatures reached by the pyrolysis method (500–750°C) are inadequate to release all the organics for detection.

Free carboxylic and amino acids present in the Atacama soil at a sample site situated at approximately ~27°S have been analyzed by solvent extraction (1:1 mixture isopropanol and water assisted by ultrasonic treatment), followed by chemical derivatization and gas chromatography-mass spectrometry (Buch et al. 2006). The following compounds were detected at µg/g levels: benzoic acid, 0.33; nonanoic acid, 0.32; 3-methyl benzoic acid, 0.01; 2-butanoic acid, 0.01; hexanoic acid, 0.52; glycine, 0.02; alanine, 0.09; sarcosine, 0.02; valine, 0.10; and norvaline, <0.01 (Buch et al. 2006). Total amino acids (free and chemically bound) in the Atacama soil precipitation gradient have been analyzed by acid hydrolysis followed by capillary electrophoresis and/or HPLC. The level of glycine varies from 0.03 µg/g, at 24.1°S to 0.02 µg/g at 25.8°S, and 0.32 µg/g at 28.2°S (Skelley et al. 2005).

Although amino acids are the single most abundant compounds in bacteria (e.g., *E. coli*) cells (Neidhardt et al.1990), it is known that most (>98%) of the amino acids originally present in bacterial cells do not undergo sublimation and are destroyed during pyrolysis under vacuum (Glavin et al. 2001). Even though nucleic acid bases are less abundant than amino acids, accounting for only 24% of the dry *E. coli* cell weight (Neidhardt et al. 1990), purines and pyrimidines are much more resistant to thermal degradation than protein-bound amino acids and readily sublimate directly from DNA and RNA when cells are heated (Glavin et al. 2002). Trace levels of nucleobases (5 to 76 ng/g) were identified in the sublimed extract of the Atacama Desert in the Flat Top Hill site (approximately 160 km south of the Yungay region). Based on the amount of adenine sublimed (5.4 ng/g), Glavin at al. (2004) estimated a total bacterial density of 4.4×10^6 cells per gram.

Because of the lack of water availability in the soil, there is virtually no primary productivity in this zone. Hypolithic cyanobacteria colonize only about 28% of translucent stones in the less arid zone of the Atacama Desert, around ~27°S, but these bacteria are extremely rare in the core region of the Atacama, colonizing less than 0.1% of the translucent rocks in the Yungay region (~24°S; Warren-Rhodes et al. 2006). Culturable heterotrophic bacteria are present in the less arid region of the Atacama Desert at levels of 10^7 colony-forming units per gram of soil. However, in the core region the levels of culturable heterotrophic bacteria are extremely low, corresponding to between $\leq 10^2$ to 10^4 colony-forming units per gram of soil from 24° to 25°S (Navarro-González et al. 2003). Therefore, microbial degradation of soil organic matter (SOM) in this region is extremely restricted, implying that the degradation of SOM must be controlled mainly by abiotic processes.

Photodegradation is thought to be the dominant abiotic process controlling the decomposition of SOM (Austin and Vivanco 2006). The ultraviolet flux along the transect from 24° to 25°S is not significantly different because the region is only 1 km above sea level, and the zone has clear blue skies all year around (McKay et al. 2003). Therefore, the hyperarid conditions in the Yungay area must inhibit microbial survival and the net result is that photochemical processes dominate. This is the first known region of the planet in which the degradation of SOM is mainly controlled by abiotic processes. In another extreme hyperarid area, the Dry Valleys in the Antarctica, it has been shown that the abiotic processes may be as important as the biological processes in the degradation of SOM to carbon dioxide (Parsons et al. 2004). These soils have significantly higher levels of organics, from 0.24 to 0.34 mg of carbon per gram of soil (Parsons et al. 2004), as compared to ~0.04 mg of carbon per gram of soil in the Yungay area (see above, this section).

6.3 Isolation and Detection of Heterotrophic Bacteria

There is currently a limited amount of published information on the levels of heterotrophic bacteria and their identity from the Atacama Desert. Studies to date have estimated that desert soil populations ranged from 0 to 10^7 colony-forming units per gram (CFU/g). This wide range of values is not surprising considering the extent of the desert, the various levels of precipitation, as well as the chemical compositions and elevations of various sample sites. Apparent differences in population size might also correspond to differences in efficiency of the recovery methods used. Population size estimates were obtained using both culture-dependent (Cameron 1969a,b; Navarro-González et al. 2003; Maier et al. 2004; Lester et al. 2007) and culture-independent approaches (Glavin et al. 2004; Drees et al. 2006). The culture-dependent approaches which involve the growth of viable cells on artificial culture media are for the most part in agreement and conclude that the bacterial numbers contained within the soils from the most arid region of the Atacama are extremely low and in most cases close to or below the detection limit of the methods applied. The population density values reported from culture-dependent studies of Atacama soil samples are also very low when compared to results from similar studies of other arid or nonarid soils (Navarro-González et al. 2003).

In the recent literature, the first estimation of bacterial population density comes from the study of Navarro-González et al. (2003) in which samples along a precipitation gradient from north to south were studied. The CFU/g values from that study ranged from < 10^3 (in the hyperarid core region at 24°S) to > 10^5 (in the most southern sample site at 28°S). The bacteria isolated in that study were mainly species of the Actinobacteria and Firmicutes (or Gram-positive bacteria with a low G+C content), with only a small number of Proteobacteria being recovered. This study showed a decrease in the number of cultivable heterotrophic bacteria in surface soils along the south to north precipitation gradient of the Atacama Desert. Together with this decrease in numbers, a corresponding decrease in diversity based on taxonomic assignment was observed.

In the study of Lester et al. (2007), a total of 20 isolates was identified by 16S rRNA gene sequence comparisons and found to be Alphaproteobacteria (12 strains), Betaproteobacteria (1 strain), and *Bacillus* species (7 strains). Of the seven *Bacillus* strains, five were identified as *B. pumilus* strains, and the Alphaproteobacteria were related to the genera *Sphingomonas, Asticcacaulis, Mesorhizobium, Bradyrhizobium,* and *Afipia*. These isolates came from three samples at a specific site and were isolated on Trypticase Soy Agar and R2A agar (Difco Catalog). In the same study (Lester et al. 2007), culture-independent methods yielded a small number of 16S rRNA gene sequences from 12 denaturing gradient gel electrophoresis (DGGE) bands. These were shown to belong to the Gemmatimonadetes, Actinobacteria, Planctomycetes, Thermomicrobia, and Proteobacteria.

Although a number of these studies have been carried out in the most arid core of the desert, these have been conducted at sites distant from one another and bacteria were recovered using a variety of different culture media, making direct comparisons difficult. In addition, the number of samples studied at each site has been limited.

To obtain results that would be more amenable to comparative analyses, we performed extensive sampling within the hyperarid core region of the desert close to Yungay. Within the hyperarid core region of the desert, the number of culturable heterotrophic bacteria in surface soil samples is extremely low. No heterotrophic bacteria could be cultured from two-thirds of the samples from this area that were used to inoculate at least one culture medium. The CFU/g values determined within this core region range from 0 to 10^5 per gram of soil. The sample containing 10^5 CFU/g is an outlier from a region in which the majority of the samples have CFU/g in the range 0 to < 10^3. The organisms cultured from these surface samples collected within the core region are limited in diversity and the majority belongs to the class Actinobacteria and more specifically to a particular group within this class, the family *Geodermatophilaceae*. The remaining organisms that were recovered belonged to species of the genera *Sphingomonas, Bacillus, Arthrobacter, Brevibacillus, Kocuria, Cellulomonas,* and *Hymenobacter*. A considerable degree of spatial heterogeneity in distribution of the CFU/g values is observed among the surface samples, which cannot be easily explained and is not highly correlated to soil chemistry.

The same situation is found in the subsurface samples (Navarro-González et al. 2004). Indeed, soil pits constructed within the hyperarid core region of the Atacama Desert show a pattern of spatial heterogeneity similar to the surface samples. The CFU/g values range from 0 to 10^3 in all layers of the soil pits sampled. No pattern is discernible in the spatial distribution of the bacteria isolated from the soil pits, with the majority of layers containing no culturable bacteria. This is in stark contrast to a soil pit constructed and studied at a more southern point along the precipitation gradient, where CFU/g values in the range 10^3 to 10^6 are detected in all layers. The diversity of the bacteria isolated from the subsurface samples was similar to that recovered from the surface samples, except that no Proteobacteria were recovered. The examination of samples from the hyperarid core region consistently demonstrates that the soils of this region have extremely low numbers of culturable heterotrophic bacteria and that the diversity of these organisms is limited when compared to microbial populations from other desert soils.

Many unanswered questions remain regarding the reasons for these low numbers and the limited diversity within the Atacama Desert soil microbial populations. Lester et al. (2007) conclude that the bacteria isolated from the single site that they examined were recently transported into the desert. The subsurface data would point against this conclusion because there are viable bacterial strains at one meter depth that have not been introduced to the environment recently. Future studies of organisms recovered from the hyperarid core region should concentrate on their desiccation resistance and survival under low nutrient conditions.

6.4 Cyanobacteria-Dominated Microbial Consortia

Liquid water is a paramount condition for life on our planet. Long-term mean annual rainfall in the Atacama Desert is only a few millimeters in its driest core (24°–25°S, 69–70°W) and increases with latitude (McKay et al. 2003). At this biotic extreme environment, desert pavement (surface soils mantled by gravels) is colonized by hypolithic or endolithic bacterial consortia forming biofilms on or within rock substrates such as quartz (Warren-Rhodes et al. 2006), halite (Wierzchos et al. 2006), and gypsum (J. Wierzchos and B. Gómez-Silva, unpublished results), as has been reported in other hot and cold deserts (Friedmann 1982; Schlesinger et al. 2003). The cyanobacterium *Chroococcidiopsis* sp. is the ubiquitous primary producer of these communities inhabiting porous and translucent stones which retain sufficient moisture and filter excessive solar radiation.

Abundance of colonized stones correlates positively with the increase in the latitudinal rainfall gradient, as does microbial community diversity estimated on the basis of the recovery of 16S rRNA gene-defined genotypes (Warren-Rhodes et al. 2006). At wetter sites, microbial consortia contain genotypes belonging to cyanobacteria (*Chroococcidiopsis* sp. and *Nostoc* sp.), Alpha- and Gammaproteobacteria, Acidobacteriales, and *Phormidium*.

Based on radiocarbon data from hypolithic biofilms and soils, microbial activity shows a sharp decline from wetter sites (27°S, 70°W) to the driest site at Yungay (24°S, 70°W), with steady-state residence times of one year and over 3,000 years, respectively (Warren-Rhodes et al. 2006; Fig. 6.3). At Yungay, a site with Mars-like soils (Navarro-González et al. 2003), cyanobacterial colonization of stones may be at least 12,000 years old, comparable to Antarctic cryptoendolithic consortia (Bonani et al. 1988). Across the latitudinal precipitation gradient, total organic carbon of hypolithic soils is five to fifteenfold greater than organic carbon of non-hypolithic surface soils (Warren-Rhodes et al. 2006). Radiocarbon values also indicate that organic carbon cycling is four times greater in the hypolithic community of the Atacama hyperarid core than in surrounding soils, whereas at a wetter site (27°S, 70°W) mean turnover times were indistinguishable. These facts indicate that colonized stones in the Atacama are islands of fertility (Schlesinger at el. 1990) whose existence and spatial location is a consequence of nonuniform liquid water distribution.

The spatial distribution of hypolithic communities in the Atacama is generally nonrandom. Transect data and spatial studies at several sites revealed that cyanobacterial spatial pattern is clumped, or aggregated and distinct from the background quartz stone pattern (Warren-Rhodes et al. 2007a). These patterns are similar to 'island-patch' distributions shown elsewhere for both desert vegetation and cyanobacterial soil crusts (Ludwig et al. 2005; Belnap et al. 2005). A detailed examination of one such patch in the hyperarid core (25°S, 69°W) revealed that hypolithic communities tend to preferentially colonize large quartz stones (typically larger than the mean quartz stone size), and this may be an ecological response to extremely low water availability, with large stones collecting and retaining more scarce water and thus contributing

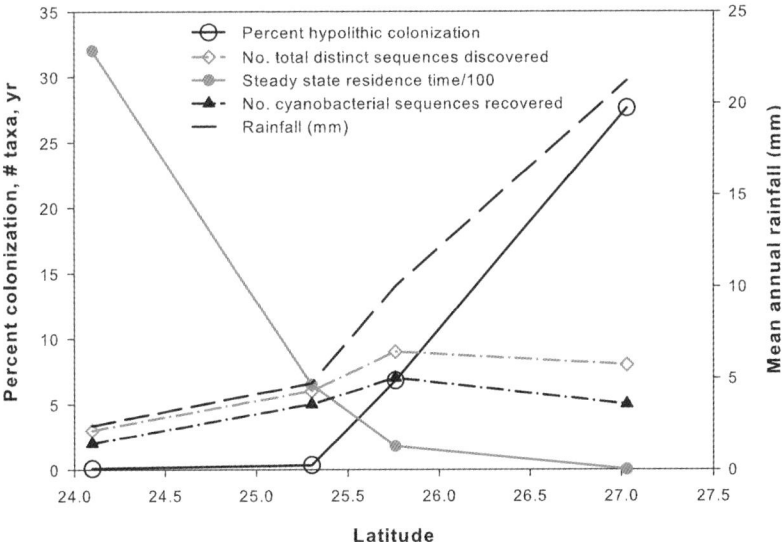

Fig. 6.3 The hypolithic cyanobacterial community across the moisture gradient in the Atacama Desert. The hyperarid core of the Atacama Desert is at 24°S. As shown in the dashed thick line, rainfall increases markedly south of 25°S. Observed along this gradient were: a decrease in steady-state residence times of organic carbon (solid gray line, filled grey circles); an increase in fraction of suitable stones that are colonized by hypolithic cyanobacteria (solid black line, open circles); an increase in the total number of distinct sequences recovered from hypolithic communities (dashed thin grey line, open diamonds); and an increase in total number of cyanobacterial sequences recovered (dashed thin black line, filled black triangles). (Adapted from Warren-Rhodes et al. 2006)

to the survival of the hypolithic or endolithic community (Warren-Rhodes et al. 2007a). These organisms also tend to colonize particular stone aspects and certain locations within the environment that favor maximum moisture availability, with the heterogeneity in hypolithic community spatial patterns linked to nonrandom water distribution patterns and the factors that determine such patterns. As an example, many hypolithic patches in the core are located either within the landscape's extreme topographic lows, where moisture from rainfall converges, or in topographic highs, where fog moisture concentration is greatest, again suggesting that fog may play a role in the survival of hypolithic communities in the Atacama.

McKay et al. (2003) have reported high nighttime relative humidity and dew in the central Atacama, a relatively frequent event that depends on the movement of moist air masses or dense fogs called *camanchaca* from the Pacific that reach the Yungay area mostly at night hours. Warren-Rhodes et al. (2006) recorded the presence of liquid water from fog on the sides of colonized quartz stones at several sites along the north–south transect centered at 70°W. At these sites, a major proportion (60–95%) of hypolithic communities are located on stone sides, in contrast to other deserts such as the Mojave and Negev where hypoliths under stones are the predominant colonization form. These ecological variations may indicate exploitation of more frequent fog versus rainfall moisture sources by hypolithic communities in the Atacama, or may be a function of other possible factors, such as rock thickness and orientation in relation to light availability.

Although fog and dew are a more reliable and frequent source of moisture than rainfall in the Atacama's hyperarid core, these phenomena did not contribute as significantly to measurable moisture on subsurface soils under stones as rare rain events did. Taken together, liquid water from all sources in this region is still exceedingly low. At Yungay, fog and rain provide liquid water to hypolithic cyanobacterial communities during an average of 400h per year but water is only available for photosynthesis during 75h per year (McKay et al. 2003, Warren-Rhodes et al. 2006). This results in an extreme environment with the lowest percentage of quartz stones colonized by cyanobacterial communities measured to date in the world's deserts (Warren-Rhodes et al. 2007b).

Another ecological niche exploited by cyanobacterial communities may afford greater water availability in the Atacama's extremely dry core. Recently, Wierzchos et al. (2006) have reported endolithic colonization of halite rocks at the Yungay area of the Atacama. These halite nonmarine evaporites are bottom-grown crusts shaped by aeolian action and long-term dissolution by rain and fog events. Using various electron microscopy techniques and fluorescent reagents, aggregates of *Chroococcidiopsis* and associated rod-shaped heterotrophic bacteria were found to be adhered to the crystal surfaces within the porous halite rocks. Endoevaporitic colonization of this crystalline and porous salt crust microhabitat may rely on the relative abundance of fog and dew events as a source of moisture. Deliquescence and capillarity are two processes probably involved in water absorption and retention. Our preliminary evidence indicates that endoevaporitic colonization is more broadly present in the Atacama Desert than previously envisioned. Long-term meteorological monitoring, as well as culturing and molecular approaches for the study of microbial communities are presently being carried out to obtain a more complete picture of this new Atacama niche.

The aridity gradient in the Atacama Desert parallels the decline in the number of hypolithic communities, diversity, and activity, leading from soils with a relative abundance of microbial life (27°S) to Mars-like soils (24°S) where life is at a critical biological threshold. Compared with the Antarctic desert where life is temperature-limited (Friedmann 1982), liquid water input in the Atacama is the key environmental factor that controls microbial ecology. In this warm desert environment, temperature, habitat availability, soil pH, soil toxicity, and carbon inflow have a minor or null impact on photoautotrophic life.

Contrasting with the scarcity of liquid water found along the aridity gradient and particularly in the hyperarid Atacama core (24°–25°S), coastal Atacama hillslopes are moisture-rich ecological environments owing to their function as topographic barriers to cloudbanks and marine air moving eastwards from the Pacific Ocean. Maximum daily air relative humidity in these ecosystems is greater than 80% during most of the year, a condition that is maintained from late afternoon to early morning hours (Cáceres et al. 2004). Water condensation from fog and heavy dew provides regular moisture to subsurface soils in these environments, and unique coastal fog oases, or lomas, have developed, although their existence is generally limited to particularly fog-prone areas (Rundel et al. 1991; Muñoz-Schick et al. 2001; Thompson et al. 2003).

Apart from coastal lomas, the remaining and larger portion of the Atacama's arid coastal desert, where mean annual rainfall is often <5 mm/yr (18–25°S), seems devoid of life. However, as mentioned above (this section), hypolithic communities are also present here. Recent study of these coastal hypolithic communities shows that cultivable members of these bacterial consortia include *Bacillus* sp., *Streptomyces* sp. and the ubiquitous cyanobacterial primary producer *Chroococcidiopsis* sp. (Gómez-Silva et al. unpublished results). We have confirmed that accompanying *Bacillus* species biosynthesize exopolysaccharides, the composition, structure, and rheological properties of which are currently being assessed in order to understand their role in the desiccation tolerance of these micro-organisms.

Atacama cyanobacteria-supported microbial consortia spend long periods of time under high solar insolation within extremely dry habitats which render cells metabolically inactive and at high risk of undergoing macromolecular damage (Dose et al. 2001). For example, exposure of dry *B. subtilis* spores to Atacama Desert's full sunlight at solar fluences equivalent to 16 and 300 kJ m^{-2} (at 280–320 nm) caused UV- B cell inactivation and 63% and 100% loss in cell viability respectively, due to DNA double-strand breaks (Dose et al. 2001). During rewetting periods, endolithic and hypolithic communities must promptly resume metabolism to sustain damage repair, growth, and cell division. Friedmann et al. (1993) have suggested that, in rapidly fluctuating environments such as the Antarctic cryptoendolithic ecosystem, net photosynthesis gain is mostly used to compensate the metabolic cost of survival.

Scytonemin is a stable and passive sunscreen produced by cyanobacteria that protects cells even during long periods of metabolic inactivity (Dillon and Castenholz 1999). Its synthesis in *Chroococcidiopsis* sp. has been shown to be modulated by environmental factors such as UV-A irradiance, temperature, photooxidative conditions, and osmotic stress (Dillon et al. 2002). These capabilities, coupled with the extreme desiccation resistance of photosynthetic micro-organisms such as *Chroococcidiopsis*, suggest possible models for past life on Mars (Friedmann and Ocampo-Friedmann 1995). Given the extreme duration and degree of hyperarid conditions in the Atacama Desert, the above and other survival and adaptation strategies of lithobiont communities from the Atacama should be further evaluated.

6.5 Patchiness of Transition from High- to Low-Density Populations

As discussed above, the distribution of soil bacteria in the extreme arid core region of the Atacama has considerable variability. This is in contrast with the more uniform, and higher, density of soil bacteria in the wetter regions of the Atacama. The transition between bacteria-rich soils and Mars-like soils is not yet understood. We have considered several possibilities: (1) the transition could be gradual with the number of organisms and the concentration of organic material dropping off monotonically with the decrease in water availability; (2) the transition could be sharp, analogous to a tree-line on a mountain slope; (3) the transition could be patchy with 'islands' of bacteria-rich soil in otherwise Mars-like terrain. Preliminary data suggest that the transition is patchy.

In the extreme arid core, the soil is dominated by Mars-like conditions with isolated islands of bacteria-rich soil. Moving toward wetter regions, these isolated island presumably become more numerous and eventually merge to form the continuously habitable soils we observe in the regions of the Atacama that receive >25mm/yr of rain. In the arid core area the patchiness could follow geological patterns in the soils or could follow subtle environmental patterns of water availability. Alternatively, the patchiness could be an intrinsic response of the microbial ecosystem. Rietkerk et al. (2004) have pointed out that self-organized patchiness and the resource concentration mechanisms involved have been reported from various ecosystems, but are most prominent in arid ecosystems with water as the controlling resource. Typically, this observation has been applied to grasses, bushes, and trees in arid regions but may also apply to bacteria facing the low water availability of the Atacama.

6.6 Relevance to Exobiology

The Atacama Desert, one of the oldest and driest deserts on Earth, provides an analogue for life in dry conditions on early or present Mars. We have used this analogue to explore the limits of life under Mars-like conditions for heterotrophic bacteria (Navarro-González et al. 2003) and for phototrophic cyanobacteria (Warren-Rhodes et al. 2006). The core of the Atacama is the end member in mineralogical comparisons between Mars and Earth soils and represents soils that differ qualitatively from soils in wetter desert environments (Ewing et al. 2006).

Our biological and chemical results suggest that, if the Viking lander had landed in the arid core region of the Atacama, it would have been unable to detect any evidence of life. Indeed, the lander instruments would have produced results similar to what they produced on Mars. The pyrolysis GCMS would have been unable to detect organic material (Navarro-González et al. 2006), and the Labeled Release experiment would have shown CO_2 release indicating decomposition of added organics (Navarro-González et al. 2003). Yet there is evidence of life remaining in the arid Atacama soils in the form of refractory organic material. These soils can be used as an analogue for the development and testing of instruments for organic and biological analysis on Mars. In addition, these soils can be used to develop and test methods for characterizing the soil oxidant on Mars. Although the presence of a soil oxidant has been known in the Atacama now for several years, its nature and mechanism of formation is still uncertain. Clearly, we should be able to characterize soil oxidants here on Earth before we venture into an examination of Mars chemistry.

6.7 Conclusions: NASA-Supported Research in the Atacama Desert

The relevance of the Atacama Desert to exobiology was appreciated in the 1960s when R. Cameron from the Jet Propulsion Laboratory collected samples from several sites in this hostile environment for use in the Viking testing protocols. The microbiological work was done using culturing methods. However, at that time very little was known about the heterogeneity of the Atacama due to variations in the penetration of the coastal fog as discussed above. The next phase of research began in 1994, when NASA and the University of Antofagasta established a year-round environmental monitoring station at the University of Antofagasta Research site near Yungay. In addition to recording the air temperature and humidity, rain, dew, light, and wind speed and direction, the unit also recorded the soil moisture with two different sensors and at three different depths. The results from this station for the years 1994–1998 have been published (McKay et al. 2003). The data showed how profoundly dry the Atacama was, compared to other deserts that were studied by identical instruments (e.g., Mojave, Negev, Gobi, etc.). During these same early expeditions (ca. 1994, 1996, and 1998), soil samples were collected for microbial analysis using culture-dependent and culture-independent techniques from a variety of sites around the Atacama. An unexpected lack of bacteria was demonstrated in some of the samples. Combined with the results from the environmental monitoring station, indicating that the core region of the Atacama near Yungay was 50 times drier than the Mojave, these observations led us to focus on this area for further investigation.

References

Admundson R (2001) The carbon budget in soils. *Ann Rev Earth Planet Sci* 29:535–562
Austin AT, Vivanco L (2006) Plant litter decomposition in a semi-arid ecosystem controlled by photodegradation. *Nature* 442:555–558
Batjes NH (1996) Total carbon and nitrogen in the soils of the world. *Eur J Soil Sci* 47:151–163
Belnap J, Welter J, Grimm N, Barger N, Ludwig J (2005) Linkages between microbial and hydrologic processes in arid and semiarid watersheds. *Ecology* 86:298–307
Bölhke JK, Ericksen GE, Revesz K (1997) Stable isotopic evidence for an atmospheric origin of desert nitrate deposits in northern Chile and southern California, USA. *Chem Geol* 136:135–152
Bonani G, Friedmann EI, Ocampo-Friedmann R, McKay CP, Woelfli W (1988) Preliminary report on radiocarbon dating of cryptoendolithic microorganisms. *Polarforschung* 58:199–200
Buch A, Glavin DP, Sternberg R, Szopa C, Rodier C, Navarro-González R, Raulin F, Cabane M, Mahaffy PR (2006) A new extraction technique for in situ analyses of amino acids and carboxylic acids on Mars by gas chromatography mass spectrometry. *Planetary Space Sci* 54:1592–1599
Bunt JS, Rovira AD (1954) Oxygen uptake and carbon dioxide evolution of heat-sterilized soil. *Nature* 173:1242
Cáceres L, Delatorre J, Gómez-Silva B, Rodriguez V, McKay CP (2004) Atmospheric moisture collection from a continuous air flow through a refrigerated tube. *Atmospheric Res* 71:127–137
Cameron R (1969a) Cold desert characteristics and problems relevant to the other arid lands. In: McGinnies W, Goldman B (eds) *Arid Lands in Perspective*. The University of Arizona Press, Tucson, pp 167–205
Cameron R (1969b) Abundance of microflora in soils of desert regions. NASA Technical Report 32-1378, pp 1–16
Clarke JDA (2006) Antiquity of aridity in the Chilean Atacama Desert. *Geomorphology* 73:101–114
DIFCO Catalog (http://www.bd.com/ds/technicalCenter/inserts/R2Aagar.pdf)
Dillon JG, Castenholz RW (1999) Scytonemin, a cyanobacterial sheath pigment, protects against UVC radiation: Implications for early photosynthetic life. *J Phycol* 35:673–681
Dillon JG, Tatsumi CM, Tandingan PG, Castenholz RW (2002) Effect of environmental factors on the synthesis of scytonemin, a UV-screening pigment, in a cyanobacterium (*Chroococcidiopsis* sp.). *Arch Microbiol* 177:322–331
Dose K, Bieger-Dose A, Fiester U, Gómez-Silva B, Klein A, Risi S, Stridde C (2001) Survival of microorganisms under the extreme conditions of the Atacama Desert. *Origins Life Evol Biosphere* 31:287–303
Drees KP, Neilson JW, Betancourt JL, Quade J, Henderson DA, Pryor BM, Maier RM (2006) Bacterial community structure in the hyperarid core of the Atacama Desert, Chile. *Appl Environ Microbiol* 72:7902–7908
Ericksen GE (1983) The Chilean nitrate deposits. *Am Sci* 71:366–374
Ewing SA, Sutter B, Owen J, Nishiizumi K, Sharp W, Cli SS, Perry K, Dietrich W, McKay CP, Amundson R (2006) A threshold in soil formation at Earth's arid-hyperarid transition. *Geochim Cosmochem Acta* 70:5293–5322
Friedmann EI (1982) Endolithic microorganisms in the Antarctic cold desert. *Science* 215:1045–1053
Friedmann EI, Kappen L, Meyer MA, Nienow JA (1993) Long-term productivity in the cryptoendolithic microbial community of the Ross Desert, Antarctica. *Microb Ecol* 25:51–69
Friedmann EI, Ocampo-Friedmann R (1995) A primitive cyanobacterium as pioneer microorganism for terraforming Mars. *Adv Space Res* 15:243–246
Glavin DP, Cleaves HJ, Schubert M, Aubrey A, Bada JL (2004) New methods for estimating bacterial cells abundances in natural samples by use of sublimation. *Appl Environ Microbiol* 70:5923–5928
Glavin DP, Schubert M, Bada JL (2002) Direct isolation of purines and pyrimidines from nucleic acids using sublimation. *Anal Chem* 74:6408–6412

Glavin DP, Schubert M, Botta O, Kminek G, Bada JL (2001) Detecting pyrolysis products from bacteria on Mars. *Earth Planet Sci Lett* 185:1–5

Houston J, Hartley AJ (2003) The central Andean west-slope rainshadow and its potential contribution to the origin of hyperaridity in the Atacama Desert. *Int J Climatol* 23:1453–1464

Lester ED, Satomi M, Ponce A (2007) Microflora of extreme arid Atacama Desert soils. *Soil Biol Biochem* 39:704–708

Lieth J (1973) Primary production: Terrestrial ecosystems. *Hum Ecol* 1:303–332

Ludwig J, Wilcox B, Breshears D, Tongway D, Imeson A (2005) Vegetation patches and runoff-erosion as interacting eco-hydrological processes in semi-arid landscapes. *Ecology* 86:288–297

Maier LM, Drees KP, Neilson JW, Handerson DA, Quade J, Betancourt JL (2004) Microbial life in the Atacama Desert. *Science* 306:1289

McKay CP, Friedmann EI, Gómez-Silva B, Cáceres L, Andersen DT, Landheim R (2003) Temperature and moisture conditions for life in the extreme arid region of the Atacama Desert: Four years of observations including the El Niño of 1997-98. *Astrobiology* 3:393–406

Miller A (1976) The climate of Chile. In: Schwerdfeger W (ed) *World Survey of Climatology*, vol 12 *Climate of Central and South America*. Elsevier. Amsterdam, pp 113–145

Munoz-Schick M, Pinto R, Mesa A, Moreira-Muñoz A (2001) Fog oases during El Niño Southern Oscillation 1997–1998 in the coastal hills south of Iquique, Tarapacá Region, Chile. *Rev Chi Hist Nat* 74:389–405

Navarro-González R, Navarro KF, De la Rosa J, Molina P, Iñiquez E, Miranda LD, Morales P, Cienfuegos E, Coll P, Raulin F, Amils R, McKay CP (2006). The limitations on organic detection in Mars-like soils by termal volatilization-gas chromatography-MS and their implications for the Viking results. *Proc Nat Acad Sci USA* 103:16089–16094

Navarro-González R, Rainey FA, McKay CO (2004) Microbial life in the Atacama Desert. *Science* 396:1289

Navarro-González R, Rainey FA, Molina P, Bagaley DR, Hollen BJ, De la Rosa J, Small AM, Quinn RC, Grunthaner FJ, Cáceres L, Gómez-Silva B, McKay CP (2003) Mars-like soils in the Atacama Desert, Chile, and the dry limit of microbial life. *Science* 302:1018–1021

Neidhardt FC, Ingraham JL, Schaechter M (1990) *Physiology of Bacterial Cell: A Molecular Approach*. Sinauer, Sunderland, MA

Oades JM (1989) An introduction to organic matter in mineral soils. In: Cixon JB, Weed SB (eds) *Minerals in Soil Environments*, 2nd edn, Soil Sci Soc Am. Madison, WI, pp 1244

O'Leary MH (1998) Carbon isotopes in photosynthesis. *BioScience* 38:328–336

Olson JS (1963) Energy storage and the balance of producers and decomposers in ecological systems. *Ecology* 44:322–331

Parsons AN, Barrett JE, Wall DH, Virginia RA (2004) Soil carbon dioxide flux in Antarctic Dry Valley ecosystems. *Ecosystems* 7:286–295

Rech J, Quade J, Hart W (2003) Isotopic evidence for the source of Ca and S in soil gypsum, anhydrite and calcite in the Atacama Desert, Chile. *Geochim Cosmochim Acta* 67:575–586

Rietkerk M, Dekker S, Ruiter P, van de Koppel J (2004) Self-organized patchiness and catastrophic shifts in ecosystems. *Science* 305:1926–1929

Rundel PW, Dillon MO, Palma B, Mooney HA, Gulmon SL, Ehleringer JR (1991) The phytogeography and ecology of the coastal Atacama and Peruvian deserts. *Aliso* 13:1–49

Schlesinger WH, Pippin J, Wallenstein M, Hofmockel K, Klepeis D, Hahall B (2003) Community composition and photosynthesis by photoautotrophs under quartz pebbles, southern Mojave Desert. *Ecology* 84:3222–3231

Schlesinger WH, Reynolds J, Cunningham G, Huenneke L, Jarrel W, Virginia R, Whitford W (1990) Biological feedbacks in global desertification. *Science* 247:1043–1048

Skelley A, Scherer JR, Aubrey AD, Grover WH, Ivester RHC, Ehrenfreund P, Gruthener FJ, Bada JL, Mathies RA (2005) Development and evaluation of a microdevice for amino acid biomarker detection and analysis on Mars. *Proc Nat Acad Sci USA* 102:1041–1046

Thompson WV, Palma B, Knowles JT, Holbrook NM (2003) Multi-annual climate in Parque Nacional Pan de Azúcar, Atacama Desert, Chile. *Rev Chi Hist Nat* 76:235–254

Trewartha GT (1961) *The Earth's Problem Climates*. University of Wisconsin Press, Madison

Warren-Rhodes KA, Dungan JL, Piatek J, Stubbs K, Gómez-Silva B, Chen Y, McKay CP (2007a) Ecology and spatial pattern of cyanobacterial island patches in the Atacama Desert, Chile. *JGR Biogeosciences* 112, G04S15, doi:10.1029/2006JG000305

Warren-Rhodes KA, Rhodes KL, Liu SJ, Zhou PJ, McKay CP (2007b) Nanoclimate monitoring of cyanobacterial communities in China's hot and cold deserts. *JGR Biogeosciences* 112, G01016, doi:10.1029/2006JG000260

Warren-Rhodes KA, Rhodes KL, Pointing SB, Ewing S, Lacap DC, Gómez-Silva B, Amundson R, Friedmannn EI, Mc Kay CP (2006) Hypolithic cyanobacteria, dry limit of photosynthesis and microbial ecology in the hyperarid Atacama Desert. *Microb Ecol* 52:389–398

Wierzchos J, Ascaso C, McKay CP (2006) Endolithic cyanobacteria in halite rocks from the hyperarid core of the Atacama Desert. *Astrobiology* 4:415–422

Woodcock WH, Hill A (1901) Improvements in the method of and apparatus for extracting nitrate soda from the raw material "caliche" and from the refuse thereof known as "ripio". UK Patent GB 190003918

Chapter 7
Microbial Communities and Processes in Arctic Permafrost Environments

Dirk Wagner

7.1 Introduction

The Arctic plays a key role in Earth's climate system, as global warming is predicted to be most pronounced at high latitudes and because one third of the global carbon pool is stored in ecosystems of the northern latitudes. Global warming will have important implications for the functional diversity of microbial communities in these systems. It is likely that temperature increases at high latitudes will stimulate microbial activity and carbon decomposition in Arctic environments and accelerate climate change by increasing trace gas release (Melillo et al. 2002, Zimov et al. 2006).

In polar regions, huge layers of frozen ground – termed permafrost – cover more than 25% of the land surface (Zhang et al. 1999) and significant parts of the coastal sea shelves (Romanovskii et al. 2005; Fig. 7.1). Permafrost can extend hundreds to more than 1,000 m into the subsurface (Williams and Smith 1989). This environment is controlled by extreme climate and terrain conditions. Particularly, seasonal freezing and thawing lead to distinct gradients in temperature and geochemistry in the upper active layer of permafrost. As it was thought that these conditions were hostile to life, permafrost was considered as uninhabitable even for micro-organisms. However, we now know that microbial communities in permafrost environments exist and are composed by members of all three domains of life (Archaea, Bacteria, and Eukarya), with a total biomass comparable to that of communities of temperate soil ecosystems (Wagner et al. 2005).

The permafrost microbial communities have to overcome the combined action of extremely cold temperature, freeze–thaw cycles, desiccation, and starvation (Gilichinsky and Wagener 1994; Morozova and Wagner 2007). Recent studies indicated that micro-organisms not only survive under permafrost conditions, but also may sustain active metabolism (Rivkina et al. 2004; Wagner et al. 2007). Although methods of modern molecular ecology are still rarely used to study diversity and

---

Dirk Wagner
Alfred Wegener Institute for Polar and Marine Research, Research Unit Potsdam, Telegrafenberg A45, 14473 Potsdam, Germany
e-mail: Dirk.Wagner@awi.de

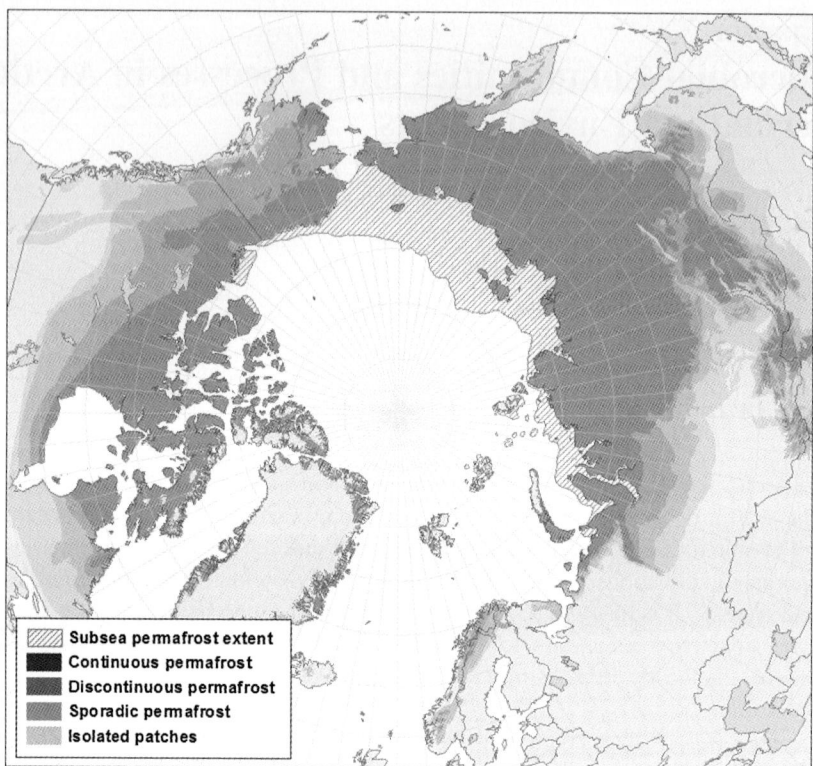

Fig. 7.1 Terrestrial and submarine permafrost distribution in the northern hemisphere. (International Permafrost Association Standing Committee on Data Information and Communication 2003)

community structure in permafrost environments (for recent examples of such use, see Colwell et al. 1999; Rivkina et al. 2000; Wartiainen et al. 2003; Vishnivetskaya et al. 2006; Ganzert et al. 2007; Steven et al. 2007), a diverse range of micro-organisms have been discovered in the different ecosystems (Shi et al. 1997; Kobabe et al. 2004; Wagner et al. 2005). Whereas microbial metabolism has been rather well studied in temperate environments, little corresponding information is available for the Arctic environments. In particular, the importance of microbial diversity for the functioning and stability of the Arctic ecosystem, the carbon dynamics controlled by micro-organisms, and the reaction of these micro-organisms to changing environmental conditions at high latitudes remain poorly understood.

In addition to its global relevance as a large carbon reservoir, the permafrost extreme environment is of particular interest to astrobiological research as an analogue for extraterrestrial permafrost habitats, which are common in our solar system. Since the current ESA mission *Mars Express* detected methane in the Martian atmosphere for the first time (Formisano 2004), recent studies have focused on methanogenic archaea from permafrost environments as potential candidates for life on Mars (Wagner et al. 2001; Morozova et al. 2007; see Chapter 11).

This review first describes the environmental conditions in permafrost. It then deals with the microbial communities established in these environments, examining their function (as far as it is known), and their role and significance in the biogeochemical cycles.

7.2 The Permafrost Environment

Permafrost is defined as ground, comprised of soil, sediment, or rock, and includes ice and organic material, that remains at or below 0°C for at least two consecutive years (van Everdingen 2005). Arctic permafrost regions are characterized by low mean annual air temperatures, low mean annual precipitation (Table 7.1), and poor to missing vegetation. During the relatively short period of arctic summer, only the surface zone (a few dm thick) of permafrost sediments thaws: this is called the *active layer*. Active layer depths range from a few cm in the high Arctic to more than 2 m in subarctic regions. Permafrost can be cemented by ice, which is typical for Arctic regions, or, in the case of insufficient interstitial water, may be dry as occurs in the Antarctic polar deserts (see Chapter 12) or rocky areas.

The permafrost environment can be divided into three temperature-depth layers, characterized by different living conditions. These are: the active layer, with an extreme temperature regime from about +15 to −35°C depending on air temperature fluctuations; the upper, perennially frozen permafrost sediments (of 10–20 m thickness), with smaller seasonal temperature variations of about 0 to −15°C; and the deeper permafrost sediments, which are characterized by a stable temperature

Table 7.1 Climate data for selected localities in circumarctic permafrost environments

Locality	Coordinates	Mean Annual Temperature [°C]	Minimum/ Maximum Temperature [°C]	Total Precipitation [mm]	Reference
Green Harbour, Spitzbergen	78°N, 15°E	−8	−19/+6	370	French 1996
Severnay Zemlya, Krasnoyarsk	79°N, 91°E	−14	−45/+6	97	Orvig 1970
Lena Delta, Yakutsk	73°N, 126°E	−15	−48/+18	320	ROSHYDROMET 2004
Lake El'gygytgyn, Chukotka	67°N, 172°E	−10	−40/+26	178	Nolan and Brigham-Grette 2007
Dawson City, Canada	64°N, 139°W	−5	−31/+14	343	French 1996
Sachs Harbour, Canada	71°N, 125°W	−14	−29/+5	93	French 1996

regime of about −5 to −10°C (French 1996). The boundary between the active layer and the perennially frozen ground is called the *permafrost table*, which acts as a physical and chemical barrier. Intensive physicochemical processes under extreme conditions take place in the active layer and upper permafrost sediments (Ostroumov 2004). In the deeper permafrost layers, conditions have been stable for long periods of time and microbial processes are limited (French 1996; Wagner et al. 2007).

Other components of the specific stratigraphy of permafrost are patterned ground formation and various cryogenic structures such as ice wedges, taliks, and cryopegs (Fig. 7.2). Thermal conditions determine the presence of such formations and structures. For instance, the large differences between summer and winter temperature in permafrost leads to the formation of typical patterned grounds (e.g., sorted circle and high- and low-centered polygons) with a prominent microrelief (Fig. 7.3a–c). The development of these structures is often related to the processes of ground ice formation. The term "ground ice" describes all types of ice in permafrost deposits, ranging from poor ice crystals to massive horizontal layers of ice with a thickness of several decameters.

Ice wedges occur typically in tundra environments with polygonal patterned grounds. In the cold winter season, thermal contraction cracks form polygonal nets. These cracks are then filled with snow melt water at the beginning of spring. Repeated cracking, filling with water and freezing can produce low-centered

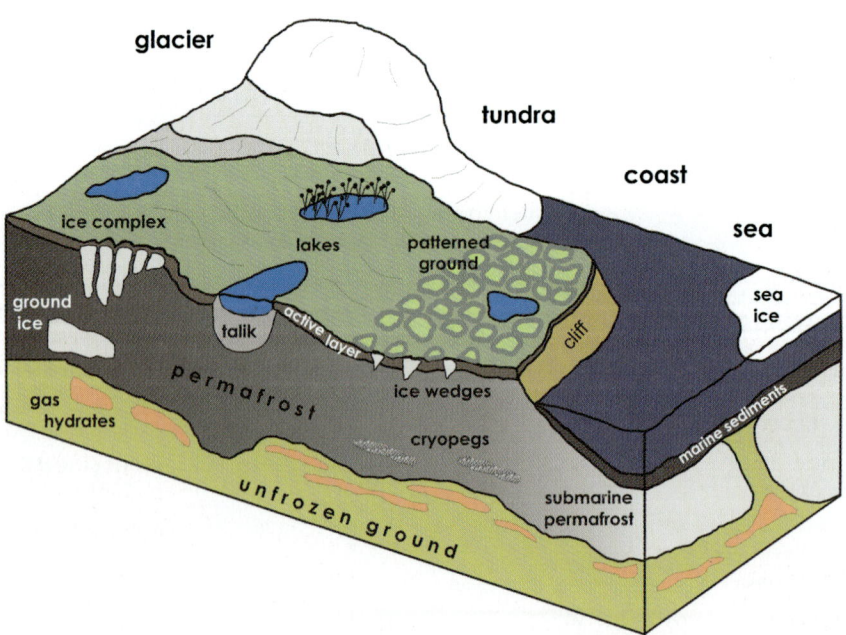

Fig. 7.2 Block diagram of an Arctic permafrost environment showing the different landscape units (glacier, tundra, coast, and sea) with the potential cryogenic features (ice complexes and wedges, massive ground ice, taliks, cryopegs), differentiated by their thermal regime

Fig. 7.3 Patterned grounds, cryogenic structures, and permafrost soils of Arctic polar regions: (**a**) sorted nets, Dawson City, Canada (photo E.-M. Pfeiffer, University of Hamburg); (**b**) sorted circle, Spitsbergen (photo J. Boike, AWI); (**c**) low-centered polygons, Lena Delta, Siberia (photo D. Wagner, AWI); (**d**) permafrost soil (*Glacic Aquiturbel*) of the polygon rim, Lena Delta, Siberia (photo L. Kutzbach, University of Greifswald); (**e**) ice-wedge, Lena Delta, Siberia (photo D. Wagner, AWI); (**f**), ice complex, Lena Delta, Siberia (photo V. Rachold, IASC); (**g**) permafrost soils (*Ruptic-Histic Aquiturbel*) of an ice complex area, Lena Delta, Siberia (photo D. Wagner, AWI)

polygonal microrelief with ice wedges of several meters in width and two to three decameters in depth over geological times of tens of thousands of years (Fig. 7.3e; Washburn 1978). Pleistocene ice-rich erosional remains of such a polygonal landscape form an *ice complex*, or Yedoma (Fig. 7.3f). An unfrozen sediment layer or body in the perennially frozen ground, located mostly below water bodies, is called a *talik*: this occurs because of local anomalies in thermal, hydrological, hydrogeological, or hydrochemical conditions (van Everdingen 2005). *Cryopegs* (overcooled water brine lenses) are defined as a layer of unfrozen ground that is perennially cryotic, forming part of the permafrost (van Everdingen 2005). Freezing of cryopegs is prevented by freezing-point depression due to the high salt content (140–300 g l^{-1}) of the pore water (Gilichinsky et al. 2005).

It has been known for some time that the shallow shelves of the Arctic coastal seas are underlined by submarine permafrost (see Figs. 7.1 and 7.2), which was formed during the Holocene sea level rise by flooding of the formerly terrestrial permafrost (Romanovskii et al. 2005). The flooding of the cold terrestrial permafrost (the temperature of which was −5 to −15°C) with relatively warm (−0.5 to −2°C) saline sea water changed the system profoundly and resulted in a warming of the permafrost (Overduin 2007).

Permafrost soils (*cryosols*) have been developed in the upper zone of the cryolithosphere (active layer and upper permafrost sediments) where the temperatures range from −50 to +30°C (Yershov 1998). Therefore, permafrost soils are mainly formed by cryopedogenesis, which involves freezing and thawing, frost stirring, mounding, fissuring, and solifluction. The repeating cycles of freezing and thawing lead to cryoturbation features (frost churning) that includes irregular, broken, or involuted horizons (Fig. 7.3d) and an enrichment in organic matter and inorganic compounds, especially on top of the permafrost table (Van Vliet-Lanoë 1991, Bockheim et al. 1999). As a result of cryopedogenesis, many permafrost soils are influenced by a strong microrelief, which causes small-scale variations in soil types (Fig. 7.3d,g) and vegetation characteristics, as well as in the microclimatic conditions. This affects the abundance, processes and diversity of microbial communities in permafrost environments. Table 7.2 summarizes the physiochemical properties of a typical permafrost soil, established on the dry rim part of a low-centered polygon from the Lena Delta, Siberia.

The seasonal variation of soil temperature also influences the availability of pore water. The presence of unfrozen water is an essential biophysical requirement for the survival of micro-organisms in permafrost. As temperatures drop below zero, free water is increasingly lost. At the same time, freezing of water leads to an increase of salt content in the remaining pore solution. In clayey permafrost soils, liquid water was found at temperatures down to −60°C (Ananyan 1970). The most important biological feature of this water is its possible role in transfer of ions and nutrients (Ostroumov and Siegert 1996).

Permafrost ecosystems are therefore extremely heterogeneous in nature. They are influenced by the regional climatic conditions, which provide harsh and strongly fluctuating conditions to their inhabitants. In these habitats, the extraordinarily high content of solid components randomly intermixed with gaseous and liquid components

Table 7.2 Selected physiochemical properties of a permafrost soil (*Glacic Aquiturbel*) of the Lena Delta, northeast Siberia[a]

Horizon[b]	Depth [cm]	T [°C]	pH	TOC [%]	TN [%]	DOC [mg l^{-1}]	CH$_4$ µmol [g^{-1}]	Sand [%]	Silt [%]	Clay [%]
Ajj	0–5	6.4	n.d.	2.1	0.12	7.3	0.4	85.7	10.4	3.9
Bjjg1	5–12	5.0	n.d.	2.0	0.11	7.1	0.3	74.3	20.6	5.0
Bjjg2	12–20	4.0	n.d.	2.4	0.14	9.0	35.3	68.0	25.8	6.3
	20–27	3.4	7.9	3.0	0.09	7.3	65.8	63.7	30.3	6.0
	27–35	2.4	6.7	2.4	0.07	4.0	153.5	56.5	34.5	9.1
Bjjg3	35–42	1.7	6.8	2.7	0.15	8.7	224.7	59.3	34.0	6.7
	42–49	1.0	n.d.	3.3	0.18	17.3	478.7	43.7	43.8	12.5

[a] Modified from Wagner et al. (2005).
[b] Horizon nomenclature according to Soil Survey Staff (1998); T, temperature; TOC, total organic carbon; TN, total nitrogen; DOC, dissolved organic carbon.

hampers the movement of micro-organisms, the mixing of substrates, and physical interaction with other organisms. This stimulates the formation of spatially separated microcolonies, which are subject to location-based adaptation and microevolutionary processes.

7.3 The Permafrost Microbiota

7.3.1 General Presentation

The first report on viable micro-organisms in permafrost was given in 1911 by Omelyansky. This pioneering investigation was followed by a number of studies revealing the presence of significant cell counts and various types of micro-organisms, including bacteria, yeasts, fungi, and protozoa, within the active layer and the perennially frozen ground of permafrost soils (Kris 1940; James and Sutherland 1942; Boyd 1958; Boyd and Boyd 1964). Since that time, a number of investigations on microbial abundance and physiology within different circumarctic environments have been carried out (e.g., Zvyagintsev et al. 1985; Khlebnikova et al. 1990; Rivkina et al. 2000; Kobabe et al. 2004; Gilichinsky et al. 2005; Zak and Kling 2006; Liebner and Wagner 2007).

Using classical isolation strategies, the most important physiological groups of micro-organisms could be recognized, including aerobic and anaerobic heterotrophs, methane oxidizers, nitrifying and nitrogen-fixing bacteria, sulfate and iron reducers, acetogens, and methanogens. The dominant microbial genera are *Acetobacterium, Acinetobacter, Arthrobacter, Bacillus, Cellulomonas, Flavobacterium, Methanosarcina, Methylobacter, Micrococcus, Nitrobacter, Nitrosomonas, Pseudomonas, Rhodococcus,* and *Streptomyces* (e.g., Gilichinsky et al. 1995; Kotsyurbenko et al. 1995; Omelchenko et al.

1996; Shi et al. 1997; Simankova et al. 2000; Suzuki et al. 2001; Wartiainen et al. 2006a). Total microbial counts obtained for permafrost soils gave high numbers of micro-organisms in the range from 10^8 to 10^9 cells g^{-1} soil (Kobabe et al. 2004) and for the perennially frozen ground between 10^3 and 10^8 cells g^{-1} sediment (Rivkina et al. 1998). Table 7.3 presents some examples of micro-organisms isolated from permafrost environments.

It is notoriously difficult to obtain a wide diversity of micro-organisms from environmental samples in culture, especially from low-temperature habitats, and the biogeochemical roles of Bacteria, Archaea, and Fungi have consequently been studied using black-box techniques such as epifluorescence direct counts, DNA and protein synthesis rates, enzyme activity, and a host of other methods that are inherently blind to variations in community composition (e.g., Vorobyova et al. 1997; Spirina and Fedorov-Davydov 1998; Bakermans et al. 2003; Šantrůčková et al. 2003; Colwell et al. 1999; Liebner and Wagner 2007; Panikov and Sizova 2007).

Much of what is now known of environmental microbial diversity is based on distinguishing between different organisms, as represented by their nucleic acids

Table 7.3 Examples of micro-organisms isolated from permafrost environments

Micro-organism	Description	Environment	Reference
Acetobacterium tundrae	Psychrophilic acetogenic bacterium	Permafrost soil (tundra wetland), Polar Ural	Simankova et al. 2000
Acinetobacter strain no. 6	Psychrotrophic, lipolytic bacterium	Permafrost soil (tundra), Siberia	Suzuki et al. 2001
Candidatus Nitrotoga arctica	Cold-adapted, nitrite-oxidizing bacterium	Permafrost soil (polygonal tundra), Lena Delta, Siberia	Alawi et al. 2007
Carnobacterium pleistocenium	Psychrotolerant, facultative anaerobe	Permafrost (32,000 years old), Fox tunnel, Alaska	Pikuta et al. 2005
Clostridium algoriphilum	Psychrophilic, anaerobic, spore-forming bacterium	Cryopeg (100–120,000 years old), Yakutskoe Lake, East Siberian sea coast	Shcherbakova et al. 2005
Exiguobacterium strain 255-15	Psychrophilic, osmo-tolerant, facultative anaerobe	Permafrost (2–3 million years old), Kolyma-Indigirka, Siberia	Vishnivetskaya et al. 2000
Methanosarcina strain SMA-21	Cold-adapted, stress-tolerant methanogen	Permafrost soil (polygonal tundra), Lena Delta, Siberia	Morozova and Wagner 2007
Methylobacter psychrophilus	Psychrophilic, methane-oxidizing bacterium	Permafrost soil (dwarf birch tundra), Northern Russia	Omelchenko et al. 1996
Methylobacter tundripaludum	Cold-adapted, methane-oxidizing bacterium	Permafrost soil, Ny-Ålesund, Svalbard	Wartiainen et al. 2006b
Methylocystis rosea	Cold-adapted, methane-oxidizing bacterium	Permafrost soil, Ny-Ålesund, Svalbard	Wartiainen et al. 2006a
Psychrobacter strain 273-4	Cold-adapted Proteobacterium	Permafrost (20–40,000 years old), Kolyma-Indigirka, Siberia	Vishnivetskaya et al. 2000

or their lipid composition, without actually culturing these organisms or having any direct knowledge of their morphology, physiology, or ecology. However, modern molecular-ecological studies of diversity and community structure in permafrost environments are still rare (for some examples, see Zhou et al. 1997; Høj et al. 2005; Neufeld and Mohn 2005; Ganzert et al. 2007; Steven et al. 2007).

Both with fluorescence in situ hybridization (FISH) and with DNA-based investigations, all relevant groups of micro-organisms (alpha, beta, gamma, and delta subclasses of the Proteobacteria, Bacteroidetes division, Gram-positive bacteria with low and high GC content and Archaea) could be detected at high cell numbers in the active layer and in the frozen ground of permafrost (Shi et al. 1997; Zhou et al. 1997; Kobabe et al. 2004). Despite differences in the requirements of the specific groups, which influence their abundances in the soils, the total diversity and quantity of active cells was strongly related to the content and quality of organic matter (Kobabe et al. 2004; Wagner et al., 2005).

In spite of the harsh environmental conditions prevailing in the deeper horizons of the active layer close to the permafrost table, there is evidence for a high amount of cells (4×10^7 cells g^{-1} soil) in this zone, that maintain at least minimal activity (Kobabe et al. 2004). Detailed bacterial 16S rDNA clone library analyses of a polygonal tundra soil from the Lena Delta (northern Siberia) revealed a great variability of colonization by representatives of the main phyla (Actinobacteria, Bacteroidetes, Chloroflexi, Firmicutes, Gemmatimonadetes, Planctomycetes, Proteobacteria, and Verrucomicrobia) within the soil of the polygon rim. The community composition in the center soil was more homogeneous, although remaining influenced by small-scale variations in environmental conditions (S. Liebner, personal communication). These communities were dominated by Bacteroidetes, Actinobacteria, Proteobacteria, and Firmicutes (in order of decreasing prevalence), with a distinct shift along the vertical temperature gradient profile.

Another study carried out in Northeast Siberia showed that the alpha and delta subclasses of the Proteobacteria dominated the microbial community, representing a proportion of about 50% of the detected organisms (Zhou et al. 1997). Composition of a microbial community from a frozen ground on Ellesmere Island, Canada was similar to that of the active layer, but in this case the dominating phyla possessed Actinobacteria- and Proteobacteria-related sequences (Steven et al. 2007). The archaeal community in this study was composed of 61% Euryarchaeota and 39% Crenarchaeota, suggesting the presence of a diverse archaeal population. In ancient permafrost sediments from Northeast Siberia, the following major groups were found: Actinomycetales (*Arthrobacter* and Microbacteriaceae), other Actinobacteria, Bacteroidetes (*Flavobacterium*), Firmicutes (*Exiguobacterium* and *Planomicrobium*), Alphaproteobacteria (*Sphingomonas*), and Gammaproteobacteria (*Psychrobacter* and Xanthomonadaceae; Vishnivetskaya et al. 2006). In these various studies, a sizable part of the microbial community belonged to thus far unclassified micro-organisms, which indicates the existence of large unknown communities in permafrost environments. Thus, the physiology and function of these presumably dominant micro-organisms remain unknown.

7.3.2 Methane-Cycling Micro-Organisms

Methanogenic archaea and methane-oxidizing bacteria were the object of particular attention in permafrost studies, because of their key role in the Arctic methane cycle and consequently of their significance for the global methane budget.

Microbial methane production (or methanogenesis) is a prominent process during the anaerobic decomposition of organic matter. Methanogenesis is solely driven by a small group of strictly anaerobic organisms called methanogenic archaea, which belong to the kingdom Euryarchaeota (Garcia et al. 2000; see Chapter 9).

The highest cell counts of methanogenic archaea were detected in the active layer of permafrost, with numbers of up to 3×10^8 cells g^{-1} soil (Kobabe et al. 2004). These represented between 0.5% and 22.4% of the total cell counts. Phylogenetic analyses revealed a great diversity of methanogens in the active layer, with species belonging to the families Methanobacteriaceae, Methanomicrobiaceae, Methanosarcinaceae, and Methanosaetaceae (Høj et al. 2005; Ganzert et al. 2007; Metje and Frenzel 2007). Other sequences detected were affiliated with the euryarchaeotal Rice Cluster II and V (Hales et al. 1996; Grosskopf et al. 1998; Ramakrishnan et al. 2001) as well as with Group I.3b of the uncultured Crenarchaeota (non-methanogenic archaea; Ochsenreiter et al. 2003).

The detected families were not restricted to specific depths of the soil profiles. Environmental sequences from the Laptev Sea coast form four specific permafrost clusters (Ganzert et al. 2007). Permafrost Cluster I was recovered mainly from cold horizons (with temperatures of less than 4°C) of the active layer and related to Methanosarcinacea. Permafrost Clusters II and III related to Methanomicrobiales and Permafrost Cluster IV related to Rice Cluster II. It was hypothesized by the authors that these clusters comprise methanogenic archaea with a specific physiological potential to survive under harsh environmental conditions. The phylogenetic affiliation of recovered sequences indicated a potential for both hydrogenotrophic and acetoclastic methanogenesis in permafrost soils.

Methanosarcina sp. SMA-21, which is closely related to *Methanosarcina mazei*, was recently isolated from a Siberian permafrost soil in the Lena Delta. The organism grows well at 28°C and slowly at low temperatures (4°C and 10°C) with H_2/CO_2 (80:20, v/v, pressurised at 150 kPa) as a substrate. The cells grow as cocci, with a diameter of 1–2 μm. Cell aggregates were regularly observed (Fig. 7.4a). *Methanosarcina* SMA-21 is characterized by an extreme tolerance to very low temperatures (−78.5°C), high salinity, starvation, desiccation, and oxygen exposure (Morozova and Wagner 2007). Furthermore, this archaeon survived for three weeks under simulated thermophysical Martian conditions (Morozova et al. 2007; see Chapter 12 for a discussion on the possible presence of methanogens on Mars).

The biological oxidation of methane by methane-oxidizing (or methanotrophic) bacteria, which represent very specialized Proteobacteria, is the only sink for methane in permafrost habitats (Trotsenko and Khmelenina 2005). Methanotrophic bacteria are common in almost all environments, where they can survive under unfavorable living conditions through the formation of exospores and cysts.

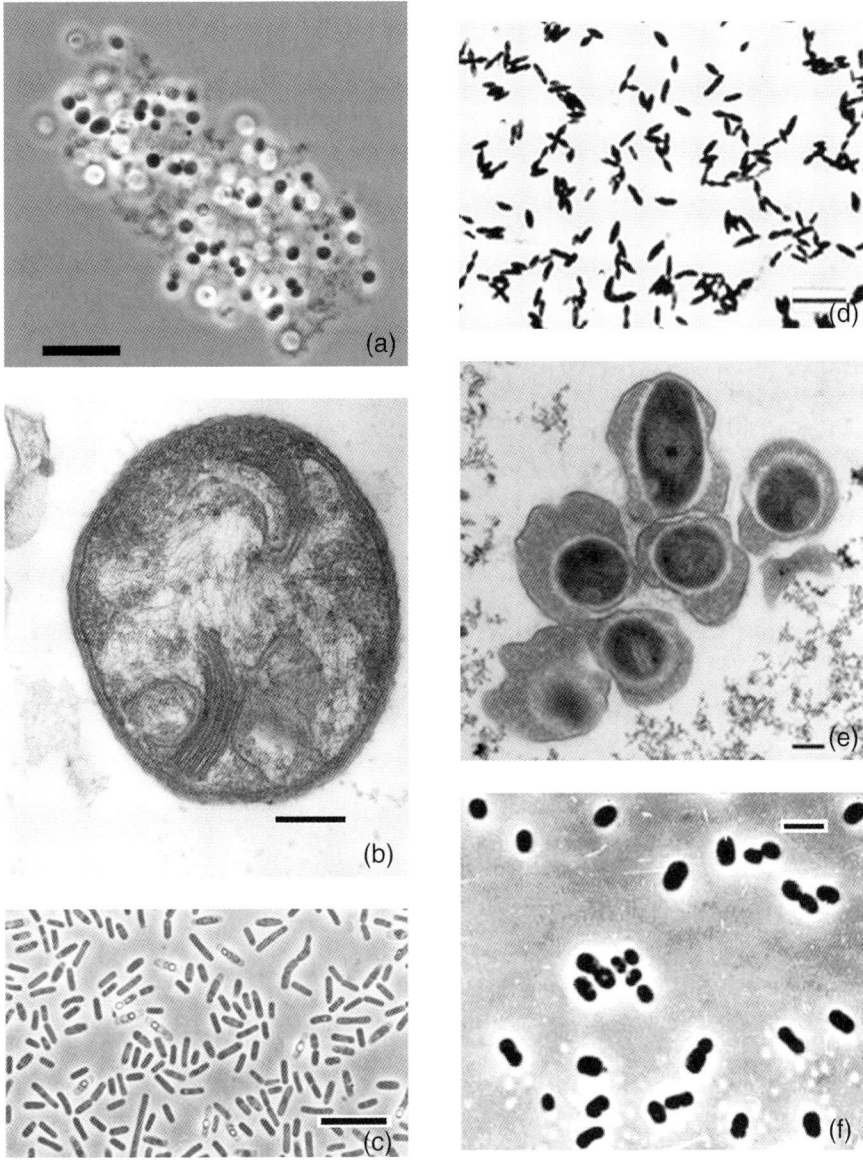

Fig. 7.4 Selected micro-organisms (Bacteria, Archaea) isolated from different permafrost environments: (**a**) *Methanosarcina* sp. SMA-21 (D. Wagner and D. Morozova, AWI; bar: 10 μm); (**b**) *Methylobacter tundripaludum* (Wartiainen et al. 2006b; bar: 200 nm); (**c**) *Clostridium algoriphilum* (Shcherbakova et al. 2005; bar: 1 μm); (**d**) *Acetobacterium tundrae* (Simankova et al. 2000; bar: 10 μm) (**e**) *Candidatus* Nitrotoga arctica (by courtesy of E. Spieck and T. Sanders, University Hamburg; bar: 200 nm); (**f**) *Psychrobacter* sp. 273-4 (Vishnivetskaya et al. 2000; bar: 5 μm)

Up to 2×10^8 cells of methane-oxidizing bacteria g^{-1} soil were detected in the active layer of permafrost soils by fluorescence in situ hybridization (Liebner and Wagner 2007). Most horizons of the soils were dominated by type-I methanotrophic bacteria (see Chapter 9 for a presentation of the various types of methanotrophs). Only in samples close to the permafrost table were type-II more abundant than type-I methanotrophs. In contrast with this, another study using phospholipid fatty acid (PLFA) concentrations and stable isotope probing showed that the community growing at low in situ temperatures was dominated by type-I methanotrophs (C. Knoblauch pers. communication). This was further confirmed by phylogenetic analyses of methanotrophic bacteria in Arctic wetland soils of Svalbard, indicating the dominance of type-I over type-II methanotrophs.

Irrespective of whether type-I or type-II methanotrophs are dominant in any particular cold location, the analyses revealed the two genera *Methylobacter* (type I) and *Methylosinus* (type II) in all studied localities (Wartiainen et al. 2003). Phospholipid fatty acid analyses revealed the PLFA 18:1Δcis10, a signature for the two methanotrophic genera *Methylosinus* and *Methylocystis* of the Alphaproteobacteria, only in the drier sites of polygonal tundra. In contrast, the PLFA 16:1Δcis8, indicative for the genera *Methylomonas, Methylomicrobium, Methylosarcina*, and *Methylosphaera*, were detected in all sites of the polygonal tundra in the Lena Delta (Wagner et al. 2005).

Methylobacter psychrophilus, isolated from a Siberian tundra soil, represents a cold-loving type-I methane-oxidizing bacterium (Omelchenko et al. 1996). Recently, two new species of methanotrophs were isolated from an Arctic wetland soil in Svalbard. *Methylobacter tundripaludum* (Fig. 7.4b) belongs to type I. This Gram-negative, rod-shaped, pale-pink pigmented bacterium grows optimally at 23°C, but with a minimal temperature well down to 10°C (Wartiainen et al. 2006a). *Methylocystis rosea* is a Gram-negative, pink-red pigmented, polymorphic rod belonging to type II. It can grow between 5 and 37°C, with optimal growth occurring at 27°C (Wartiainen et al. 2006b).

7.3.3 Other Observations on Permafrost Biodiversity

Some recent studies dealt with the biodiversity of 100,000–120,000 year-old cryopegs in Siberian permafrost (Gilichinsky et al. 2005). Direct microbial cell counts revealed numbers in the range of 10^7 cells ml^{-1} saline water. A variety of aerobic and anaerobic, sporeless and spore-forming, halophilic and psychrophilic bacteria as well as mycelial fungi and yeasts have been isolated, including genera such as *Arthrobacter, Bacillus, Erwinia, Frigoribacterium, Microbacterium, Psychrobacter, Paenibacillus, Rhodococcus*, and *Subtercola. Clostridium algoriphilum* sp. nov. was isolated, and shown to be adapted to low nutrient concentrations (Fig. 7.4c; Shcherbakova et al. 2005). The metabolic end products of this anaerobic bacterium are lactate and butyrate, which can be used as substrates by heterotrophic *Psychrobacter* isolates, indicating the possibility of a trophic food chain within the microbial communities of cryopegs.

Additional novel micro-organisms were isolated recently from various habitats. For example, *Acetobacterium tundrae* (DSM 9173) was isolated from tundra wetlands of Polar Ural (Simankova et al. 2000). The organism is cold-adapted with a growth temperature optimum of 20°C, and a temperature range of 1 to 30°C. It is a Gram-positive, oval shaped, flagellated rod (Fig. 7.4d), fermenting H_2/CO_2, formate, methanol, and several sugars to acetate as the sole end product. *Carnobacterium pleistocenium*, a novel psychrotolerant, facultative anaerobic bacterium, was isolated from Pleistocene ice from the Fox tunnel in Alaska (Pikuta et al. 2005). A Gram-positive, motile rod, the organism grows best at 24°C, with a range of 0 to 28°C. Metabolic end products are acetate, ethanol, and CO_2. *Exiguobacterium* sp. 255-15 is a nonspore-forming, Gram-positive bacterium isolated from a 2–3 million-year permafrost core (Vishnivetskaya et al. 2000). Its cells are short rods about 1 μm in length with rounded ends. It is a facultative anaerobe but grows most profusely aerobically.

A novel nitrite-oxidizing bacterium was obtained by enrichment culture and provisionally classified as "*Candidatus* Nitrotoga arctica" (Fig. 7.4e). The organism was cultured at 10°C and is characterized by a fatty acid profile which is different from those of known nitrite oxidizers but similar to fatty acid profiles of Betaproteobacteria (Alawi et al. 2007). *Psychrobacter* sp. 273-4 is a small, non-motile coccoid rod (Fig. 7.4f) often found in pairs, isolated from a 20–40 thousand year-old Siberian permafrost core (Vishnivetskaya et al. 2000). The strain is characterized by rapid growth at low temperatures and excellent survival after exposure to long-term freezing.

Viable green algae were isolated from Arctic deep sediments frozen for 5–7 thousand years (Vorobyova et al. 1997). All isolates grew slowly at 20–25°C and were sensitive to high light intensities. The photosynthetic pigments chlorophyll a, chlorophyll b, and pheophytin were found in a wide range of sediments of different genesis and age.

Both in the active layer and in the perennially frozen sediments, a large variety of fungi was detected. In the active layer of Arctic tundra tussock and shrub soils, the fungal community was composed of Ascomycota, Basidiomycota, Zygomycota, Chytridiomycota, and Glomeromycota (Wallenstein et al. 2007). Although the tussock communities had higher proportions of Ascomycota (Dothideomycetes, Pezizomycetes, and Sordariomycetes), the shrub soils were dominated by Zygomycota (Zygomycetes). Another study performed in Alaska reported the dominance of basidiomycetous dimorphic yeasts (*Mrakia* and *Leucosporidium*) and ascomycetous mycelial fungi (*Geomyces*; Panikov and Sizova 2007). In ancient permafrost deposits up to 400,000 years old, only the groups Ascomycota, Basidiomycota, and Zygomycota could be detected (Lydolph et al. 2005).

The absence of a wide spectrum of cultured organisms is a recurrent theme in these studies and suggests that many micro-organisms from permafrost environments are either unculturable or the appropriate methods of enrichment and cultivation have not been used.

7.4 Role and Significance of the Microbiota

7.4.1 Temperature Effects

Certain key processes of global biogeochemical cycles (e.g., C, N, and S cycles) are carried out exclusively by highly specialized micro-organisms (e.g., methanogenic archaea, acetogenic, nitrifying, and sulfate-reducing bacteria), which play a quantitatively dominant role in mineralization processes (Hedderich and Whitman 2006; Drake et al. 2006; Bock and Wagner 2006; Rabus et al. 2006). Although the physiology and ecology of various micro-organisms from temperate environments are well studied, little is known about the activity and function of many of the phyla and species in permafrost habitats described in the previous section (see Section 7.3).

The active layer of permafrost is subjected to annual freezing and thawing cycles, which result in large temperature and geochemistry gradients along the depth profile of the soils. The extreme temperature regime is one of the most important parameters regulating the metabolic activity and survival of micro-organisms. Several recent studies demonstrated activities of micro-organisms from the active layer and the perennially frozen ground at subzero temperatures. Various micro-organisms isolated from Siberian permafrost exhibit metabolic activities down to $-10°C$ (Bakermans et al. 2003; Jakosky et al. 2003). The incorporation of ^{14}C-labeled acetate into bacterial lipids examined in microcosm experiments at temperatures varying between $+5°C$ and $-20°C$ revealed activity of the indigenous micro-organisms over this entire range (Rivkina et al. 2000).

The minimum temperature for growth of micro-organisms was recently reported to be $-35°C$ (Panikov and Sizova 2007). Growth yields of isolated micro-organisms similar to those shown above the freezing point were maintained down to $-17°C$. Between $-18°C$ and $-35°C$, growth was only detectable for three weeks after cooling. Then metabolic activity declined to zero, and micro-organisms entered a state of reversible dormancy. Studies on methanogenic activity and biomass in a Holocene permafrost core from the Lena Delta (Siberia) showed that the methane found at certain depths of the sediments originated from modern methanogenesis by cold-adapted methanogenic archaea (Wagner et al. 2007). These findings are in accordance with the grouping of microbial metabolic rates of cold-adapted micro-organisms that was proposed by Price and Sowers (2004): rates of the first group are sufficient for microbial growth; those of the second group are sufficient for metabolism but too low for growth, whereas rates belonging to the third group allow survival in a dormant state accompanied by macromolecular damage repair. The reviewed results of microbial metabolism at subzero temperatures contradict the idea of the 'community of survivors' in permafrost soils (Gounot 1999; Rothschild and Mancinelli 2001), which are not thought to 'prefer' their environment but are rather said to be more resistant than others that were similarly challenged.

7.4.2 Carbon Cycling

Currently, the question that is most hotly debated with respect to permafrost ecosystems is this one: 'What will happen to the carbon stored in permafrost, in the event of a climate change?' The relevance of Arctic carbon reservoirs is highlighted by current climate models that predict significant changes in temperature and precipitation in the northern hemisphere (Kattenberg et al. 1996; Smith et al. 2002). In particular, the degradation of permafrost and the associated release of climate-relevant trace gases from intensified microbial turnover of organic carbon and from destabilized gas hydrates represent a potential environmental hazard.

Carbon cycling under anoxic conditions within the predominantly wet permafrost soils is mainly performed via methane production, which is the final process in a sequence of hydrolysis and fermentation (Schink and Stams 2006). Thus, methanogenic archaea stand in close relationship with other micro-organisms of the anaerobic food chain, comprising, in particular, acetogenic bacteria, clostridia, and other bacteria (Kotsyurbenko et al. 1993; Stams 1994). In cold environments two main pathways of energy-metabolism by methanogens dominate: (i) the reduction of CO_2 to CH_4 using H_2 as a reductant (hydrogenotrophic methanogenesis) and (ii) the fermentation of acetate to CH_4 and CO_2 (acetoclastic methanogenesis; Conrad 2005).

Methanogenic activity was observed at low in situ temperatures with rates of up to 39 nmol CH_4 h^{-1} g^{-1} soil in the active layer of permafrost (Wagner et al. 2003; Høj et al. 2005; Metje and Frenzel 2007). The highest activities were measured in the coldest zones of the profiles. Furthermore, it could be shown that the methane production is regulated more by the quality of soil organic carbon than by the in situ temperature (Wagner et al. 2005; Ganzert et al. 2007). Another important factor affecting archaeal communities in permafrost soils is the water regime. Along a natural soil moisture gradient, changes in archaeal community composition were observed, which suggest that the differences in these communities were responsible for the large-scale variations in methane emissions observed with changes in soil hydrology (Høj et al. 2006).

Microbial methane oxidation in the oxic zones of the active layer is of great importance to the control of methane releases from permafrost environments. Methane-oxidizing bacteria are using methane as their sole carbon source, with consequent energy production by the oxidation of CH_4 to CO_2 (Hanson and Hanson 1996). Methane oxidation rates in Canadian permafrost soils ranged from 58 to 92% depending on the environmental conditions (Popp et al. 2000). However, the methane oxidation activities showed vertical shifts with respect to optimal temperature and distribution of type I and type II methanotrophs in Siberian permafrost soils (Liebner and Wagner 2007). In the upper active layer, maximum methane oxidation potentials were detected at 21°C. Deep active layer zones that are constantly exposed to temperatures below 2°C show a maximum potential for methane oxidation at 4°C. This indicates a dominance of psychrophilic methanotrophs close to the permafrost table.

A close relationship exists between methane fluxes and microbiological processes and communities in permafrost soils. Micro-organisms do not only survive in their extreme habitat but also are metabolically active under in situ conditions, which shows that the microbial communities are well adapted to low temperatures and extreme geochemical gradients. However, they are also tolerant to temperature increases. This is evidenced by results showing that a slight temperature increase can lead to a substantial increase in methanogenic activity within perennially frozen deposits (Wagner et al. 2007). In case of permafrost degradation by thermokarst or coastal erosion processes, this would lead to an extensive expansion of the methane deposits and fluxes with a subsequent impact on the total atmospheric methane budget.

7.4.3 Nitrogen Cycling

Nitrogen turnover is strongly correlated with the carbon cycle but little is known about nitrogen fluxes in Arctic ecosystems and the organisms involved. Low temperature and poor substrate quality often limit decomposition and nitrogen mineralization in many arctic ecosystems (Jonasson et al. 1993). However, higher rates of nitrogen fixation were observed in climate change simulation experiments on Ellesmere Island, Canada (Deslippe et al. 2005). Nitrifying bacteria were detected in permafrost soils and sediments (Bartosch et al. 2002; Alawi et al. 2007). Even in old deep permafrost sediments, nitrifiers can survive long periods of starvation and dryness (Soina et al. 1991). Nearly nothing is known about the Arctic source strength for the long-life greenhouse gases NO and N_2O. Furthermore, other climate-relevant processes such as microbial methane oxidation are influenced by the activity of ammonia oxidizers. More generally, Arctic carbon fluxes and turnover rates are limited by microbial-mediated nitrogen mineralization.

7.4.4 Sulfur Cycling

Sulfur plays a key role in marine biogeochemical cycles, in particular in anaerobic sediments of the marine shelfs. About 50% of the carbon mineralization in shelf sediments is oxidized via the reduction of sulfate to sulfide by sulfate-reducing bacteria (Jørgensen 1982). The released sulfide can be oxidized chemically or by sulfide-oxidizing bacteria in aerobic sediment layers. However, coastal erosion and sea level rise created the shallow shelves of the Arctic Ocean, for example those of the Laptev Sea, the bottom of which corresponds to formerly terrestrial permafrost (Rachold et al. 2005; Romanovskii et al. 2005). Flooding of the cold (−5 to −15°C) terrestrial permafrost with relatively warm (−0.5 to −2°C) saline, sulfur-rich water from the Laptev Sea changed the system profoundly and resulted in a warming of the permafrost (Rachold et al. 2007). Studies on the microbial diversity and activity in submarine permafrost have been conducted neither by culture-dependent methods nor by culture-independent molecular approaches. Therefore, response of microbial mineralization and other processes to rising temperatures in these carbon-rich permafrost ecosystems, as well as effects on microbial abundance and diversity, are totally unknown.

The permafrost environment forces the adaptation of the microbial communities to low temperature conditions and promotes the growth of species that thus far remain undetected in temperate ecosystems. Therefore, Arctic permafrost environments can be seen as active microbial ecosystems rather than frozen habitats with microbial survivors. The evaluation of microbiological data and their correlation with climatic and geochemical results represent the basis for the understanding of the role of permafrost in the global system. Of particular relevance are feedback mechanisms related to nutrient cycles, biogeochemical processes, and greenhouse gas emissions in the context of a warming Earth.

7.5 Conclusions: Future Directions for Research

Although one fourth of the Earth land surface and distinct areas of the coastal sea shelves are affected by permafrost, the physiology, function, and diversity of microbial communities in these ecosystems is sparsely investigated thus far. This may be partially caused by the relative inaccessibility of the investigation areas and the associated logistic problems. However, the main difficulty lies in the lack of methodologies specific for permafrost sampling and isolation of cold-adapted micro-organisms from Arctic soils and sediments. This is shown by the discrepancy between the small numbers of psychrophilic micro-organisms isolated thus far from permafrost environments in contrast to the observed significant metabolic rates under in situ conditions. Methodological developments should consider the following aspects: enrichment of micro-organisms should be performed directly in the field or in batch or continuous laboratory culture; culture techniques should be developed for the enrichment of 'syntrophically associated' micro-organisms; subzero culturing methods are needed; and state-of-the-art culture-independent molecular techniques for diversity and functional analyses of microbial communities should be applied on permafrost.

The lack of isolates from permafrost also limits possible biotechnological uses. Cold-adapted micro-organisms from permafrost exhibit properties very different from those of other thermal classes. Therefore, the vast genetic resources of micro-organisms from permafrost environments remain nearly unexploited. It is likely that mainly extremophilic microbes could offer technologically and/or economically significant products such as enzymes, polysaccharides, osmoprotectors, and liposomes (Cavicchioli et al. 2002). Therefore, one essential goal of microbial diversity exploration in cold regions will be to recover new isolates, some of which will prove useful for biotechnology processes or medicine.

Apart from the global relevance of permafrost as a large carbon reservoir, this extreme environment is also of particular interest to astrobiological research, as an analogue for extraterrestrial permafrost habitats, which are a common occurrence in our solar system (Gilichinsky 2001; Wagner et al. 2001). Particularly, the observation of methane in the Martian atmosphere by the current mission of the European Space Agency, *Mars Express* (Formisano 2004), has stimulated the debate over possible microbial life on Mars (see also Chapter 11). Recently, it has been shown that methanogenic archaea isolated from Siberian permafrost environments are more tolerant to environmental stress and simulated thermophysical Martian conditions than methanogens from temperate ecosystems (Morozova and Wagner 2007, Morozova et al. 2007). Micro-organisms from terrestrial permafrost are valuable model organisms in our effort to investigate the possibility of microbial life in extraterrestrial permafrost ecosystems.

References

Alawi M, Lipski A, Sanders T, Pfeiffer E-M, Spieck E (2007) Cultivation of a novel cold-adapted nitrite oxidizing Betaproteobacterium from the Siberian Arctic. *ISME J* 1:256–264
Ananyan AA (1970) Unfrozen water content in frozen clay at a temperature from $-0.6°C$ to $-40°C$ – $-60°C$. *Merzlotnye Issledovaniya* 10:267–270 (in Russian)
Bakermans C, Tsapin AI, Souza-Egipsy V, Gilichinsky DA, Nealson KH (2003) Reproduction and metabolism at $-10°C$ of bacteria isolated from Siberian permafrost. *Environ Microbiol* 5: 321–326
Bartosch S, Hartwig C, Spieck E, Bock E (2002) Immunological detection of *Nitrospira*-like bacteria in various soils. *Microbiol Ecol* 43:26–33
Bock E, Wagner M (2006) Oxidation of inorganic nitrogen compounds as an energy source. In: Dworkin M, Falkow S, Rosenberg E, Schleifer K-H, Stackebrandt E (eds.) *Prokaryotes*, vol 2, Springer, New York, pp 457–495
Bockheim JG, Everett LR, Hinkel KM, Nelson FE, Brown J (1999) Soil organic storage and distribution in Arctic Tundra, Barrow, Alaska. *Soil Sci Soc Am J* 63:934–940
Boyd WL (1958) Microbiological studies of arctic soils. *Ecol* 39:332–336
Boyd WL and Boyd JW (1964) The presence of bacteria in permafrost of the Alaskan arctic. *Can J Microbiol* 10:917–919
Cavicchioli R, Siddiqui KS, Andrews D, Sowers KR (2002) Low-temperature extremophiles and their applications. *Curr Opin Biotech* 13:253–261
Colwell FS, Delwiche ME, Blackwelder D, Wilson MS, Lehman RM, Uchida T (1999) Microbial communities from core intervals, JAPEX/JNOC/GSC Mallik 5L-38 gas hydrate research well. In: Scientific Results from Mallik 2002 Gas Hydrate Production Research Well Program, Mackenzie Delta, Northwest Territories, Canada. Dallimore SR, Uchida T, Collett TS (eds), Bulletin 544. Geological Survey of Canada, Microbiology section, pp. 189–195
Conrad R (2005) Quantification of methanogenic pathways using stable carbonisotopic signatures: A review and a proposal. *Organ Geochem* 36:739–752
Deslippe JR, Egger KN, Henry HR (2005) Impact of warming and fertilization on nitrogen-fixing microbial communities in the Canadian High Arctic. *FEMS Microbiol Ecol* 53:41–50
Drake H, Küsel K, Matthies C (2006) Acetogenic prokaryotes. In: Dworkin M, Falkow S, Rosenberg E, Schleifer K-H, Stackebrandt E (eds.) *Prokaryotes*, vol 2, Springer, New York, pp 354–420
Formisano V (2004) Detection of methane in the atmosphere of Mars. *Science* 306:1758–1761
French HM (1996) *The Periglacial Environment*. Longman, London
Ganzert L, Jurgens G, Münster U and Wagner D (2007) Methanogenic communities in permafrost-affected soils of the Laptev Sea coast, Siberian Arctic, characterized by 16S rRNA gene fingerprints. *FEMS Microbiol Ecol* 59:476–488

Garcia JL, Patel BKC and Olliver B (2000) Taxonomic, phylogenetic and ecological diversity of methanogenic archaea. *Anaerobe* 6:205–226

Gilichinsky DA (2001) Permafrost model of extraterrestrial habitat. In: Horneck G, Baumstark-Khan C (eds) *Astrobiology. The Quest for the Conditions of Life*, Springer, Berlin, pp. 271–295

Gilichinsky DA, Rivkina E, Bakermans C, Shcherbakova V, Petrovskaya L, Ozerskaya S, Ivanushkina N, Kochkina G, Laurinavichius K, Pecheritsina S, Fattakhova R, Tiedje JM (2005) Biodiversity of cryopegs in permafrost. *FEMS Microbiol. Ecol* 53:117–128

Gilichinsky DA, Wagener S (1994) Microbial life in permafrost. In: Gilichinsky D (ed) *Viable Microorganisms in Permafrost*, Pushchino Research Center, pp. 7–20

Gilichinsky DA, Wagener S, Vishnivetskaya TA (1995) Permafrost microbiology. *Permafrost Periglac Process* 6:281–291

Gounot AM (1999) Microbial life in permanently cold soils. In: Margesin R, Schinner F (eds) *Cold-Adapted Organisms*, Springer, Berlin, pp. 3–16

Grosskopf R, Stubner S, Liesack W (1998) Novel euryarchaeotal lineages detected on rice roots and in the anoxic bulk soil of flooded rice microcosms. *Appl Environ Microbiol* 64:4983–4989

Hales BA, Edwards C, Ritchie DA, Hall G, Pickup RW, Saunders JR (1996) Isolation and identification of methanogen-specific DNA from blanket bog peat by PCR amplification and sequence analysis. *Appl Environ Microbiol* 62:668–675

Hanson RS, Hanson TE (1996) Methanotrophic bacteria. *Microbiol Rev* 60:439–471

Hedderich R, Whitman W (2006) Physiology and biochemistry of the methane-producing archaea. In: Dworkin M, Falkow S, Rosenberg E, Schleifer K-H, Stackebrandt E (eds.) *Prokaryotes*, vol 2, Springer, New York, pp 1050–1079

Høj L, Olsen RA, Torsvik VL (2005) Archaeal communities in High Arctic wetlands at Spitsbergen, Norway (78°N) as characterised by 16S rRNA gene fingerprinting. *FEMS Microbiol Ecol* 53:89–101

Høj L, Rusten M, Haugen LE, Olsen RA, Torsvik VL (2006) Effects of water regime on archaeal community composition in Arctic soils. *Environ. Microbiol* 8: 984–996

International Permafrost Association Standing Committee on Data Information and Communication (comp.) (2003) *Circumpolar Active-Layer Permafrost System, Version 2.0*. Edited by M. Parsons and T. Zhang. Boulder, CO: National Snow and Ice Data Center/World Data Center for Glaciology. CD-ROM

Jakosky BM, Nealson KN, Bakermans C, Ley RE, Mellon MT (2003) Subfreezing activity of microorganisms and the potential habitability of Mars' polar regions. *Astrobiology* 3:343–350

James N, Sutherland ML (1942) Are there living bacteria in permanently frozen subsoil? *Can J Res Sect Bot Sci* 20:228–235

Jonasson S, Havstrom M, Jensen M, Callaghan TV (1993) In situ mineralization of nitrogen and phosphorus of arctic soils after perturbations simulating climate-change. *Oecologia* 95: 179–186

Jørgensen BB (1982) Mineralization of organic matter in the sea bed – The role of sulphate reduction. *Nature* 296:643–645

Kattenberg A, Giorgi F, Grassel H, Meehl GA, Michell JFB, Stoufer RJ, Tokioka T, Weaver AJ, Wigley TML (1996) Climate models – Projections of future climate. In: Houghton JT (ed) *Climate Change 1995*, University Press, Cambridge, pp 285–357

Khlebnikova GM, Gilichinsky DA, Fedorov-Davydov DC, Vorobyova EA (1990) Quantitative evaluation of microorganisms in permafrost deposits and buried soils. *Microbiology* 59:106–112

Kobabe S, Wagner D, Pfeiffer EM (2004) Characterization of microbial community composition of a Siberian tundra soil by fluorescence in situ hybridization. *FEMS Microbiol Ecol* 50:13–23

Kotsyurbenko OR, Nozhevnikova AN, Zavarzin GA (1993) Methanogenic degradation of organic matter by anaerobic bacteria at low temperature. *Chemosphere* 27:1745–1761

Kotsyurbenko OR, Simankova MV, Nozhevnikova AN, Zhilina TN, Bolotina NP, Lysenko AM, Osipov GA (1995) New species of psychrophilic acetogens: *Acetobacterium bakii* sp. nov., *A. paludosum* sp. nov., *A. fimetarium* sp. nov. *Arch Microbiol* 163:29–34

Kris AE (1940) Microorganisms in permafrost. *Microbiology* 9:879–886 (in Russian)

Liebner S, Wagner D (2007) Abundance, distribution and potential activity of methane oxidizing bacteria in permafrost soils from the Lena Delta, Siberia. *Environ Microbiol* 9:107–117

Lydolph MC, Jacobsen J, Arctander P, Thomas M, Gilbert P, Gilichinsky DA, Hansen AJ, Willerslev E, Lange L (2005) Beringian paleoecology inferred from permafrost-preserved fungal DNA. *Appl Environ Microbiol* 71:1012–1017

Melillo JM, Steudler PA, Aber JD, Newkirk K, Lux H, Bowles FP, Catricala C, Magill A, Ahrens T, Morrisseau S (2002) Soil warming and carbon-cycle feedbacks to the climate system. *Science* 298:2173–2176

Metje M, Frenzel P (2007) Methanogenesis and methanogenic pathways in a peat from subarctic permafrost. *Environ Microbiol* 9:954–964

Morozova D, Möhlmann D, Wagner D (2007) Survival of methanogenic archaea from Siberian permafrost under simulated Martian thermal conditions. *Orig Life Evol Biosph* 37:189–200

Morozova D, Wagner D (2007) Stress response of methanogenic archaea from Siberian permafrost compared to methanogens from non-permafrost habitats. *FEMS Microbiol Ecol* 61:16–25

Neufeld JD, Mohn WW (2005) Unexpectedly high bacterial diversity in Arctic tundra relative to boreal forest soils, revealed by serial analysis of ribosomal sequence tags. *Appl. Environ Microbiol* 71:5710–5718

Nolan M, Brigham-Grette J (2007) Basic hydrology, limnology, and meteorology of modern Lake El'gygytgyn, Siberia. *J Paleolimnol* 37:17–35

Ochsenreiter T, Selezi D, Quaiser A, Bonch-Osmolovskaya L, Schleper C (2003) Diversity and abundance of Crenarchaeota in terrestrial habitats studied by 16S RNA surveys and real time PCR. *Environ Microbiol* 5:787–797

Omelchenko MB, Vasilieva LV, Zavarzin GA, Savel'eva ND, Lysenko AM, Mityushina LL, Khmelenina VN, Trotsenko YA (1996) A novel psychrophilic methanotroph of the genus *Methylobacter*. *Microbiology* 65:339–343

Omelyansky VL (1911) Bacteriological investigation of the Sanga mammoth and surrounding soil. *Arkhiv Biologicheskikh Nauk (Scientific Biological Archive)* 16:335–340 (in Russian)

Orvig S (1970) *Climates of the Polar Regions. World Survey of Climatology*. Elsevier, New York

Ostroumov V (2004) Physico-chemical processes in cryogenic soils. In: Kimble JM (ed) *Cryosols*, Springer, Berlin, pp 347–364

Ostroumov V, Siegert C (1996) Exobiological aspects of mass transfer in microzones of permafrost deposits. *Adv Space Res* 18:79–86

Overduin P (2007) The expedition COAST I. In: Schirrmeister L. (ed.) Expeditions in Siberia in 2005. Russian-German Cooperation System Laptev Sea. Reports on Polar Research 550:1–39

Panikov NS, Sizova MV (2007) Growth kinetics of microorganisms isolated from Alaskan soil and permafrost in solid media frozen down to −35°C. *FEMS Microbiol Ecol* 59:500–512

Pikuta EV, Marsic D, Bej A, Tang J, Krader P, Hoover RB (2005) *Carnobacterium pleistocenium* sp. nov., a novel psychrotolerant, facultative anaerobic bacterium isolated from permafrost of the Fox tunnel in Alaska. *Int J Syst Evol Microbiol* 55:473–478

Popp TJ, Chanton JP, Whiting GJ, Grant N (2000) Evaluation of methane oxidation in the rhizosphere of a *Carex* dominated fen in north central Alberta, Canada. *Biogeochem* 51:259–281

Price PB, Sowers T (2004) Temperature dependence of metabolic rates for microbial growth, maintenance, and survival. *Proc Natl Acad Sci* 101:4631–4636

Rabus R, Hansen T, Widdel F (2006) Dissimilatory sulfate- and sulfur-reducing prokaryotes. In: Dworkin M, Falkow S, Rosenberg E, Schleifer K-H, Stackebrandt E (eds.) *Prokaryotes*, vol 2, Springer, New York, pp 659–768

Rachold V, Are FE, Atkinson DE, Cherkashov G, Solomon SM (2005) Arctic coastal dynamics (ACD): An introduction. *Geo-Marine Lett* 25:63–68

Rachold V, Bolshiyanov DY, Grigoriev MN, Hubberten HW, Junkers R, Kunitsky VV, Merker F, Overduin P, Schneider W (2007) Nearshore Arctic subsea permafrost in transition. *EOS* 88:149–150

Ramakrishnan B, Lueders T, Dunfield PF, Conrad R, Friedrich MW (2001) Archaeal community structures in rice soils from different geographical regions before and after initiation of methane production. *FEMS Microbiol Ecol* 37:175–186

Rivkina EM, Friedmann EI, McKay CP, Gilichinsky DA (2000) Metabolic activity of permafrost bacteria below the freezing point. *Appl Environ Microbiol* 66:3230–3233

Rivkina EM, Gilichinsky D, Wagener S, Tiedje J and McGrath J (1998) Biochemical activity of anaerobic microorganisms from buried permafrost sediments, *Geomicrobiol* 15:187–193

Rivkina E, Laurinavichius K, McGrath J, Tiedje J, Shcherbakova V, Gilichinsky D (2004) Microbial life in permafrost. *Adv Space Res* 33:1215–1221

Romanovskii NN, Hubberten H-W, Gavrilov AV, Eliseeva AA, Tipenko GS (2005) Offshore permafrost and gas hydrate stability zone on the shelf of East Siberian Seas. *Geo-Marine Lett* 25:167–182

ROSHYDROMET (2004) Russian Federal Service for Hydrometeorology and Environmental Monitoring, http://www.worldweather.org/107/c01040.htm

Rothschild LJ, Mancinelli RL (2001) Life in extreme environments. *Nature* 409:1092–1101

Šantrůčková H, Bird MI, Kalaschnikov YN, Grund M, Elhottova D, Šimek M, Grigoryev S, Gleixner G, Arneth A, Schulze E-D, Lloyd J (2003) Microbial characteristics of soils on a latitudinal transect in Siberia. *Global Change Biol* 9:1106–1117

Schink B, Stams AJM (2006) Syntrophism among Prokaryotes. In: Dworkin M, Falkow S, Rosenberg E, Schleifer K-H, Stackebrandt E (eds.) *Prokaryotes*, vol 2, Springer, New York, pp 309–335

Shcherbakova VA, Chuvilskaya NA, Rivkina EM, Pecheritsyna SA, Laurinavichius KS, Suzina NE, Osipov GA, Lysenko AM, Gilichinsky DA, Akimenko VK (2005) Novel psychrophilic anaerobic spore-forming bacterium from the overcooled water brine in permafrost: Description *Clostridium algoriphilum* sp. nov. *Extremophiles* 9:239–246

Shi T, Reevers R, Gilichinsky D, Friedmann EI (1997) Characterization of viable bacteria from Siberian permafrost by 16S rDNA sequencing. *Microbial Ecol* 33:167–179

Simankova MV, Kotsyurbenko OR, Stackebrandt E, Kostrikina NA, Lysenko AM, Osipov GA, Nozhevnikova AN (2000) *Acetobacterium tundrae* sp. nov., a new psychrophilic acetogenic bacterium from tundra soil. *Arch Microbiol* 174:440–447

Smith J, Stone R, Fahrenkamp-Uppenbrink J (2002) Trouble in polar paradise: Polar science. *Science* 297:1489–1492

Soil Survey Staff (1999) Soil Taxonomy - A basic system of soil classification for making and interpreting soil surveys, 2nd ed. US Government printing Office, Washington D.C

Soina VS, Lebedeva EV, Golyshina OV, Fedorov-Davydov DG, Gilichinsky DA (1991) Nitrifying bacteria from permafrost deposits of the Kolyma lowland. *Microbiologia* 60:187–190 (in Russian)

Spirina EV, Fedorov-Davydov DG (1998) Microbiological characterization of cryogenic soils in the Kolymskaya Lowland. *Eurasian Soil Sci* 31:1331–1344

Stams AJM (1994) Metabolic interactions between anaerobic bacteria in methanogenic environments. *Antonie van Leeuwenhoek* 66:271–294

Steven B, Briggs G, McKay CP, Pollard WH, Greer CW, Whyte LG (2007) Characterization of the microbial diversity in a permafrost sample from the Canadian high Arctic using culture-dependent and culture-independent methods. *FEMS Microbiol Ecol* 59:513–523

Suzuki T, Nakayama T, Kurihara T, Nishino T, Esaki N (2001) Cold-active lipolytic activity of psychrotrophic *Acinetobacter* sp. strain no. 6. *J Biosci Bioeng* 92:144–148

Trotsenko YA, Khmelenina VN (2005) Aerobic methanotrophic bacteria of cold ecosystems. *FEMS Microbiol Ecol* 53:15–26

Van Everdingen R (2005) Multi-language glossary of permafrost and related ground-ice terms. National Snow and Ice Data Center/World Data Center for Glaciology. Boulder, Colorado, USA

Van Vliet-Lanoë B (1991) Differential frost heave, load casting and convection: Converging mechanisms; a discussion of the origin of cryoturbations. *Permafrost Periglac Process* 2:123–139

Vishnivetskaya T, Kathariou S, McGrath J, Gilichinsky D, Tiedje J (2000) Low-temperature recovery strategies for the isolation of bacteria from ancient permafrost sediments. *Extremophiles* 4:165–173

Vishnivetskaya TA, Petrova MA, Urbance J, Ponder M, Moyer CL, Gilichinsky DA, Tiedje JM (2006) Bacterial community in ancient Siberian permafrost as characterized by culture and culture-independent methods. *Astrobiology* 6:400–414

Vorobyova E, Soina V, Gorlenko M, Minkovskaya N, Zalinova N, Mamukelashvih A, Gilichinsky D, Rivkina E, Vishnivetskaya T (1997) The deep cold biosphere: Facts and hypothesis. *FEMS Microbiol Rev* 20:277–290

Wagner D, Gattinger A, Embacher A, Pfeiffer EM, Schloter M, Lipski A (2007) Methanogenic activity and biomass in Holocene permafrost deposits of the Lena Delta, Siberian Arctic and its implication for the global methane budget. *Global Change Biol* 13:1089–1099

Wagner D, Kobabe S, Pfeiffer EM, Hubberten HW (2003) Microbial controls on methane fluxes from a polygonal tundra of the Lena Delta, Siberia. *Permafrost Periglac Process* 14:173–185

Wagner D, Lipski A, Embacher A, Gattinger A (2005) Methane fluxes in permafrost habitats of the Lena Delta: Effects of microbial community structure and organic matter quality. *Environ Microbiol* 7:1582–1592

Wagner D, Spieck E, Bock E, Pfeiffer EM (2001) Microbial life in terrestrial permafrost: Methanogenesis and nitrification in Gelisols as potentials for exobiological processes. In: Horneck G, Baumstark-Khan C (eds) *Astrobiology: The Quest for the Conditions of Life*. Springer. Berlin, pp 143–159

Wallenstein MD, McMahon S, Schimel J (2007) Bacterial and fungal community structure in Arctic tundra tussock and shrub soils. *FEMS Microbiol Ecol* 59:428–435

Wartiainen I, Hestnes AG, McDonald IR, Svenning MM (2006a) *Methylocystis rosea* sp. nov., a novel methanotrophic bacterium from Arctic wetland soil, Svalbard, Norway (78° N). *Int J Sys Evol Microbiol* 56:541–547

Wartiainen I, Hestnes AG, McDonald IR, Svenning MM (2006b) *Methylobacter tundripaludum* sp. nov., a methane-oxidizing bacterium from Arctic wetland soil on the Svalbard islands, Norway (78° N). *Int J Sys Evol Microbiol* 56:109–113

Wartiainen I, Hestnes AG, Svenning MM (2003) Methanotrophic diversity in high arctic wetlands on the islands of Svalbard (Norway) – Denaturing gradient gel electrophoresis analysis of soil DNA and enrichment cultures. *Can J Microbiol* 49:602–612

Washburn AL (1978) *Geocryology. A Survey of Periglacial Processes and Environments*. Arnold, London

Williams PJ, Smith MW (1989) *The Frozen Earth: Fundamentals of Geocryology*. Cambridge University Press, Cambridge

Yershov ED (1998) *General Geocryology*. Cambridge University Press, Cambridge

Zak DR, Kling GW (2006) Microbial community composition and function across an arctic tundra landscape. *Ecology* 87:1659–1670

Zhang T, Barry RG, Knowles K, Heginbotton JA, Brown J (1999) Statistics and characteristics of permafrost and ground-ice distribution in the northern hemisphere. *Polar Geography* 23:132–154

Zhou J, Davey ME, Figueras JB, Rivkina E, Gilichinsky D, Tiedje JM (1997) Phylogenetic diversity of a bacterial community determined from Siberian tundra soil DNA. *Microbiology* 143:3913–3919

Zimov SA, Schuur EAG, Chapin III FS (2006) Permafrost and the global carbon budget. *Science* 312:1612–1613

Zvyagintsev DG, Gilichinsky DA, Blagodatskii SA, Vorobyeva EA, Khlenikovam GM, Arkhangelov AA, Kudryavtseva NN (1985) Survival time of microorganisms in permanently frozen sedimentary rock and buried soils. *Microbiology* 54:155–161

Chapter 8
Aerobic, Endospore-Forming Bacteria from Antarctic Geothermal Soils

Niall A. Logan(✉) and Raymond N. Allan

8.1 Introduction: Taxonomy of *Bacillus* Species and Related Genera

The term 'aerobic endospore-forming bacteria' is used to embrace *Bacillus* species and related genera, for which the production of resistant endospores in the presence of oxygen remains the defining feature. They are also expected to possess Gram-positive cell wall structures (but staining reactions, even in young cultures, may be Gram-variable or frankly Gram-negative), and may be aerobic or facultatively anaerobic. These characters have formed part of the definition of the group for many years, but some exceptions have emerged. *Bacillus infernus* and *B. arseniciselenatis* are strictly anaerobic, and spores have not been detected in *B. infernus*, *B. subterraneus*, and *B. thermoamylovorans*.

Molecular taxonomic methods have had a huge impact on the classification of these organisms, and the number of taxa, including thermophiles, has increased greatly. The 1986 edition of *Bergey's Manual of Systematic Bacteriology* (Claus and Berkeley 1986) listed 40 valid *Bacillus* species, of which only four were true thermophiles. These were *Bacillus stearothermophilus*, *B. acidocaldarius*, and *B. schlegelii*, each with strains reported from geothermal soils (see Table 8.1), and *B. thermoglucosidasius*. "*Bacillus caldolyticus*", "*B. caldotenax*", and "*B. caldovelox*" (Heinen and Heinen 1972) were isolated from naturally heated waters and were listed as '*Species incertae sedis*.' These still await validation.

Since that 1986 edition of *Bergey's Manual* and up to late 2006, 229 further species have been newly described or revived among *Bacillus* and the 11 genera derived from it (subsequently two of these new genera were merged, so there are now 10 new genera recognized). Furthermore, 20 new genera containing 59 species have been proposed to accommodate novel aerobic endospore formers not previously assigned to *Bacillus*. Overall, then, there have been proposals for 30 new genera and 288 new

Niall A. Logan
Department of Biological and Biomedical Sciences, Glasgow Caledonian University, Cowcaddens Road, Glasgow G4 0BA, United Kingdom
e-mail: nalo@gcal.ac.uk

Table 8.1 Thermophilic, aerobic endospore-forming bacteria from geothermal sources

Genus and Species of Organism (and Number of Species in Genus)	Original Reference	Original Source[a]	Other Sources[a,b]
Geobacillus (17)			
G. stearothermophilus	Donk 1920	Canned corn and beans	Food, milk, water, soil, hot compost, sugar beet juice, hot spring, **geothermal soil, Southern Urals** (Golovacheva et al. 1965), hydrothermal vents
G. gargensis	Nazina et al. 2004	Hot spring	
G. tepidamans	Schäffer et al. 2004	**Geothermal soil, Yellowstone, USA**, beet sugar factory	
G. thermodenitrificans	Ambroz 1913	Soil	Uncultivated soil, sugar beet juice, hot compost, **geothermal soil, South Sandwich Islands** (Logan et al. 2000), hydrothermal vents
G. thermoleovorans	Zarilla & Perry 1987	Soil, muds, sludge	Uncultivated soil, hydrothermal vents, petroleum reservoirs, hot springs
"G. thermoleovorans subsp. stromboliensis"	Romano et al. 2005	**Geothermal soil, Eolian Islands, Italy**	
G. vulcani	Caccamo et al. 2000	Hydrothermal vent	
Geobacillus spp.			Deep-sea hydrothermal vents, sea mud
Bacillus (134)			
B. aeolius	Gugliandolo et al. 2003	Shallow marine vent	
B. coagulans	Hammer 1915	Evaporated milk	Soil, canned foods, tomato juice, gelatin, milk, medical preparations, silage, **geothermal soil, Southern Urals** (Golovacheva et al. 1965)
B. fumarioli	Logan et al. 2000	**Geothermal soil, Antarctica**	Gelatin
B. infernus	Boone et al. 1995	Deep terrestrial subsurface	

(continued)

Table 8.1 (continued)

Genus and Species of Organism (and Number of Species in Genus)	Original Reference	Original Source[a]	Other Sources[a,b]
B. schlegelii	Schenk & Aragno 1979	Lake sediment	Geothermal water, mud and ash, glacier ice, air, **Antarctic geothermal soil** (Hudson et al. 1988)
B. thermantarcticus	Nicolaus et al. 1996	**Geothermal soil, Antarctica**	
B. tusciae	Bonjour & Aragno 1984	Geothermal pond	
***Alicyclobacillus* (11)**			
A. acidocaldarius	Darland & Brock 1971	Hot acid springs, **acid geothermal soil, Hawaii**	**Antarctic geothermal soil** (Hudson & Daniel 1988), gelatin
A. acidocaldarius subsp. *rittmannii*	Nicolaus et al. 1998	**Geothermal soil, Antarctica**	
A. hesperidum	Albuquerque et al. 2000	**Geothermal soil, Azores**	
A. vulcanalis	Simbahan et al. 2004	Hot spring	
Alicyclobacillus sp.			**Antarctic geothermal soil** (Bargagli et al. 2004)
***Aneurinibacillus* (5)**			
An. terranovensis	Allan et al. 2005	**Geothermal soil, Antarctica**	
***Anoxybacillus* (10)**			
Anox. amylolyticus	Poli et al. 2006	**Geothermal soil, Antarctica**	
Anox. ayderensis	Dulger at al. 2004	Hot spring	
Anox. gonensis	Belduz et al. 2003	Hot springs	
Anox. flavithermus	Heinen et al. 1982	Hot spring	
Anox. kestanbolensis	Dulger at al. 2004	Hot spring	
Anox. voinovskiensis	Yumoto et al. 2004	Hot spring	
***Brevibacillus* (13)**			
Br. levickii	Allan et al. 2005	**Geothermal soil, Antarctica**	
***Caldalkalibacillus* (1)**			
C. thermarum	Xue et al. 2006	Hot spring	
***Sulfobacillus* (4)**			
Sulfobacillus spp.			Geothermal waters, Yellowstone National Park, USA, Montserrat Island
***Vulcanibacillus* (1)**			
Vulc. modesticaldus	L'Haridon et al. 2006	Deep-sea hydrothermal vent	

[a] Sources of organisms corresponding to geothermal soils are indicated in boldtype.
[b] References for isolation from geothermal soil are given.

or revived species or new combinations, and yet only seven proposals for merging species or subspecies were made in that time. Nearly 50 of these new species are thermophiles, and several moderate thermophiles have also been described, but rather few have been isolated from geothermal soils (see Table 8.1).

8.2 Habitats of Thermophiles

Most aerobic endospore formers are saprophytes widely distributed in the natural environment, but some are opportunistic or obligate pathogens of animals, including humans, other mammals, and insects. The main habitats are soils of all kinds, ranging from acid to alkaline, hot to cold, and fertile to desert, and the water columns and bottom deposits of fresh and marine waters. The bacteriology of geothermal soils has received much less attention than that of hot springs and pools, hydrothermal vents, and other heated aqueous environments. There is a long history of studies on thermophilic *Bacillus*, dating from the proposal of the best-known species, *B. stearothermophilus* (now *Geobacillus stearothermophilus*), from canned food in 1920. Of the present 17 valid species in *Geobacillus*, there are reports of eight with isolates from unheated soils, six from aqueous environments associated with oil or gas fields, six from composts, five from hot springs and shallow hydrothermal vents, four from foods, including canned foods, milk, and sugar processing, and only three (*G. stearothermophilus*, *G. tepidamans*, and *G. thermodenitrificans*) from geothermal soils (see Table 8.1).

A further taxon from geothermal soil, "*G. thermoleovorans* subsp. *stromboliensis*", awaits validation. Further, unidentified, thermophilic endospore formers have been reported from hydrothermal vents, geothermal soils, and sea muds (White et al. 1993; Marteinsson et al. 1996; Takami et al. 1997). Although the strains of White et al. (1993) were from several geothermal sites around the world, including Iceland, it is not clear which of their isolates were from heated soils. Other thermophiles are found in the genera *Alicyclobacillus*, *Aneurinibacillus*, *Anoxybacillus*, *Bacillus*, *Brevibacillus*, *Caldalkalibacillus*, *Sulfobacillus*, *Thermobacillus*, *Ureibacillus*, and *Vulcanibacillus*, but of the newer thermophilic taxa of aerobic endospore formers, fewer than 20% include strains that have been isolated from geothermal soils, and it is remarkable that over half of those were first isolated from Antarctic geothermal soils (see Table 8.1).

Although thermophilic aerobic endospore formers and other thermophiles might be expected to be restricted to hot environments, they are also widespread in cold environments and appear to be ubiquitously distributed in soils worldwide. Indeed, Weigel (1986) described how easy it is to isolate such organisms from cold soils and even from Arctic ice. Endospores readily survive distribution from natural environments to a wide variety of other habitats, as the example of *B. fumarioli*, described in detail below, demonstrates.

Strains of *Geobacillus* with growth temperature ranges of 40 to 80°C can be isolated from subsurface layers of soils whose temperatures never exceed 25°C

8 Aerobic, Endospore-Forming Bacteria from Antarctic Geothermal Soils

(Marchant et al. 2002). That spores may survive in such cool environments without any metabolic activity is understandable, but their wide distribution and contribution of up to 10% of the cultivable flora suggest that they do not merely represent contamination from hot environments (Marchant et al. 2002). A study of some Irish temperate soils found aerobic thermophile counts of $1.5\text{--}8.8 \times 10^4$ colony-forming units per gram, and similar results were obtained for other temperate soils from Europe (McMullan et al. 2004), suggesting that these are part of the autochthonous flora (Rahman et al. 2004). It is possible that the direct heating action of the sun on the upper layers of the soil, and local heating from the fermentative and putrefactive activities of mesophiles, might be sufficient to allow the multiplication of thermophiles, but perhaps these organisms are capable of very low levels of activity at normal environmental temperatures (McMullan et al. 2004).

8.3 Antarctic Geothermal Soils

There was constant volcanic activity in Antarctica during the Cenozoic period, and steaming ground is to be found in a number of circumpolar islands and on the continent (Fig. 8.1). Thus, although Antarctica is largely an ice-bound continent that

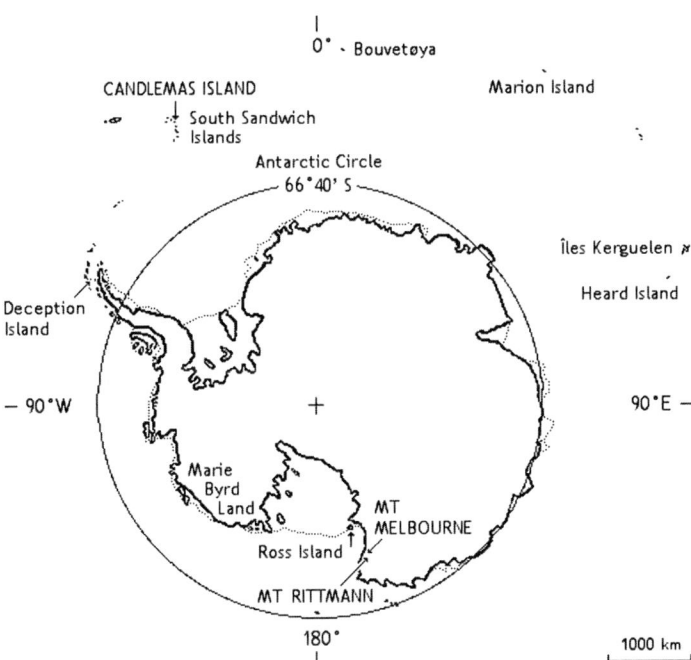

Fig. 8.1 Map of Antarctica and the sub-Antarctic islands, with geothermal sites named. Names in capital letters indicate the sites from which *Bacillus fumarioli*, *Brevibacillus levickii* and *Aneurinibacillus terranovensis* were isolated

relies upon solar heating during the summer to support a sparse growth of terrestrial life, several sites exist where volcanic activity warms the soil and steam emissions from fumaroles condense to maintain relatively steady water supplies that may support the growth of vegetation. All of these places are remote, and are costly and difficult to visit, and therefore no comprehensive study of the microbiologies of their geothermal soils has been made, but we are fortunate to have some information on the aerobic endosporeforming floras of five of these sites.

8.3.1 The Antarctic Continent

Three sites, Mt Erebus, Mt Melbourne, and Mt Rittmann, are the only known high-altitude localities of fumarolic activity and associated vegetation within Antarctica. The unique selective pressures of such sites make the organisms that live there of special biological interest (Broady 1993), and they can harbour unique vegetation communities which appear to have formed following colonisation by propagules from circumpolar continents (Linskens et al. 1993).

Mounts Erebus and Melbourne represent two of the four provinces of the McMurdo Volcanic Group, which is one of the most extensive alkali volcanic provinces in the world. These two volcanoes were named in 1841 by James Clark Ross after his expedition ship *Erebus,* and the British Prime Minister Lord Melbourne. It is worth noting in passing that some of the expedition's specimens were submitted to the eminent botanist Christian Ehrenberg, in Berlin, for microbiological examination. It was Ehrenberg who had proposed *Vibrio subtilis* (now *Bacillus subtilis,* and the type species of the genus) in 1835.

Mt Erebus (3,794 m; 77° 32′S; 167° 8′E) is the most active volcano on the Antarctic continent, and the largest of four cones on Ross Island. Its cone rises 200 to 300 m above the summit plateau and it is bordered by a side crater, which is the rim of a filled-in older crater or caldera, to the SW. The floor of the main crater has an inner crater, whose floor bears a fumarolic ridge. The walls of the main crater also bear fumaroles, and discontinuous lines of fumaroles and ice towers radiate from the crater rim and across the plateau. Patches of warm ground lie on the rim of the main crater, within the side crater, and on the plateau. An area of warm ground to the NW of the main crater, with surface ground temperatures that may reach 75°C, is an Antarctic Specially Protected Area (ASPA) called Tramway Ridge; the ice-free ground here is terraced, and the steep sides of the terraces bear the main crusts of vegetation.

Mt Melbourne (2,733 m; 74° 21′S; 164° 42′E) is situated in the centre of a relatively young (probably $2-3 \times 10^6$ years old) volcanic field that has been formed by a large number of small, individual eruptive centres (Broady et al. 1987). Located on the southern rim of the main summit crater of Mt Melbourne is an ASPA called Cryptogam Ridge (Fig. 8.2). This is a deglaciated site with soil temperatures typically reaching 40–50°C at depths of a few centimetres; it supports a unique community including algal and bryophyte species unknown elsewhere in Antarctica

8 Aerobic, Endospore-Forming Bacteria from Antarctic Geothermal Soils

Fig. 8.2 Mount Melbourne, Northern Victoria Land, Antarctica, looking towards Wood Bay in the Ross Sea. The deglaciated Antarctic Specially Protected Area called Cryptogam Ridge is seen in the foreground. The icy hummock in the centre of the picture is formed by condensate freezing above a fumarole

Fig. 8.3 Mount Rittmann, Northern Victoria Land, Antarctica, showing areas of heated soil with fumaroles, and ice towers formed from frozen fumarole condensate, interspersed with patches of permafrost

(Nicolaus et al. 1991). The flora of the geothermally heated area of the northwest (NW) slope of the mountain, lying at 2,400 to 2,500 m, is less well developed than that of Cryptogam Ridge.

Mt Rittmann (2,600 m; 73° 27′S; 165° 30′E) (Fig. 8.3) was discovered during the fourth Italian Antarctic Expedition (1988–1989). The soil surface temperature ranges from 34.4–41.5°C and there is patchy development of vegetation. The geothermally heated biosystem at Mt Rittmann has been described by Bargagli et al. (1996). It lies at an altitude of about 2,600m, and has small fumaroles whose internal temperatures (at 10cm depth) range between 50°C and 63°C; patches of moss grow on the warm soil which has a relatively high moisture content (by Antarctic standards) and a pH of around 5.4.

Soil samples were collected from the NW slope of Mt Melbourne and from the Mt Rittmann geothermal site by British Antarctic Survey (BAS) members of the international BIOTEX 1 expedition during the 1995–1996 austral summer, and further samples were taken from these sites and from Cryptogam Ridge during the 1998–1999 austral summer by one of the authors (NAL).

Although Mt Melbourne and Mt Rittmann lie in the same volcanic province, and their soils may appear similar in some respects (they have very low concentrations of essential nutrients such as N and P), Bargagli et al. (1996) showed that there were differences in the mineral contents of soil samples collected from Mt Melbourne (higher Cu and Zn) and Mt Rittmann (higher Cd and Pb), and Pepi et al. (2005) reported that mossy and unvegetated soil samples from the NW slope of Mt Melbourne had higher iron contents than did soils from Cryptogam Ridge and Mt Rittmann. For other major elements, soils from the NW slope of Mt Melbourne also had the highest contents of Na and Al; soils from Cryptogam Ridge had the highest content of Mg; and soils from Mt Rittmann had the lowest contents of Ca, Al, and Fe. For trace elements, soils from the NW slope of Mt Melbourne also had the highest contents of Cd, Cr, Pb, and Zn; soils from Cryptogam Ridge had the highest contents of Cu and Hg; and soils from Mt Rittmann had the lowest contents of Cu and Hg. Thus organisms in these soils may be constrained by low concentrations of essential nutrients and, in some cases, relatively high concentrations of toxic minerals.

8.3.2 The South Sandwich Archipelago

The South Sandwich archipelago comprises 11 islands, all of which exhibit recent or continuing volcanic activity. They lie on the Scotia Ridge between latitude 56°18′ and 59°28′S and longitude 26°14′ and 28°11′W; this volcanic arc has a deep-sea trench on its convex (eastern) side descending to over 7,000m. The fauna and flora of Candlemas Island (Fig. 8.4; map in Logan et al. 2000) are more extensive than those of other islands in the arc, and include penguins, petrels, skuas, and seals, grass, bryophytes, and lichens. The southern massif forms the largest part of the island, and is an ice-capped remnant of an extinct volcano. In the younger and actively volcanic northern part lies Lucifer Hill (232m; 57°04′S, 26°42′W), a complex of scoria cones surrounded by a mass of five main lava flows. The oldest of these bear ash mantles, whereas the youngest is very recent. There are patches of moss around active fumaroles high on the hill, around inactive fumaroles at Clinker Gulch, and on lava soils at the base of the volcanic cone, with temperatures

Fig. 8.4 Candlemas Island, South Sandwich Archipelago, viewed from the east. Lucifer Hill, in the unglaciated northern part of the island, is seen on the right-hand side of the picture, with its summit partially obscured by cloud. (Aerial photograph kindly provided by Dr John L. Smellie, British Antarctic Survey. © British Crown Copyright/MOD.)

ranging from 85°C down to 0°C. The northern and southern parts of the island are linked by an area of low flat sand containing two large lagoons. Mossy soil samples were collected from the summit and base of Lucifer Hill by British Antarctic Survey personnel on behalf of one of the authors (NAL) during the 1996–1997 austral summer.

8.3.3 Deception Island

Deception Island (62°57′S, 60°38′W) is one of seven islands that comprise the South Shetland archipelago. It is 17-km across and ring-shaped, being the rim of a caldera that has been flooded by the sea, and it lies on the expansion axis of the Bransfield Rift that separates the archipelago from the Antarctic Peninsula. A narrow gap in the caldera rim leads to the basin of Port Foster, a perfect natural harbour. It has seen human activity since the 1820s, including sealing, whaling, and scientific research; however, there were major eruptions in 1967, 1969, and 1970, and the island is less frequented now. Several ASPAs have been established, and the whole island is an Antarctic Specially Managed Area. It was visited by a Spanish scientific expedition in the austral summer of 1989–1990, and water and sand samples were collected from four geothermal sites: a fumarole at Cerro Caliente, Irizar Lake, Kroner Lake, and Whalers Bay (Llarch et al. 1997).

8.4 Endospore Formers from Antarctic Geothermal Soils

The first report of an aerobic endospore former isolated from Antarctica was by Dr A. L. McLean, who studied the bacteriology of ice and snow during Douglas Mawson's Australasian Antarctic Expedition of 1911–1914. Little other Antarctic bacteriology was published until Darling and Siple (1941) described the isolation of 178 strains from various Antarctic environments, and found 66% of them to belong to nine *Bacillus* species. Ugolini and Starkey (1966) isolated bacteria from fumaroles of Mt Erebus, but according to Hudson and Daniel (1988) these were probably mesophiles.

The first isolations of thermophilic endospore formers were made by Hudson and Daniel (1988) from Tramway Ridge and its locality, and from the side and west craters of Mt Erebus. They found strains of *Bacillus* in most of their samples. In particular, they isolated strains resembling *B. acidocaldarius* (now in *Alicyclobacillus*) in the more acid soils of Tramway Ridge and its neighborhood (Hudson et al. 1989), *Bacillus schlegelii* (capable of utilizing thiosulfate; Hudson et al. 1988) in an acid soil near Tramway Ridge, and anaerobes resembling *Clostridium thermohydrosulfuricum* (now *Thermoanaerobacter thermohydrosulfuricus*) from most of their samples. *Bacillus schlegelii* was first isolated from the sediment of a Swiss lake, and *B. tusciae* was discovered when *Bacillus schlegelii* was first sought in geothermal environments (Bonjour and Aragno, 1984).

Nicolaus et al. (1991) isolated four strains of thermophilic eubacteria from Cryptogam Ridge, Mt Melbourne, and one strain from near the seashore at the foot of the mountain, and the effects of growth temperature on polar lipid patterns and fatty acid compositions of these and other strains were compared (Nicolaus et al. 1995). One of the Cryptogam Ridge isolates was proposed as *Bacillus thermoantarcticus* (Nicolaus et al. 1996). Its exopolysaccharide chemistry has been investigated (Manca et al. 1996) and its digestion of xylan studied (Lama et al. 2004). The proposal was later validated and the name of the isolate corrected to *Bacillus thermantarcticus*. However, it should probably belong in *Geobacillus*.

Nicolaus et al. (1998) also isolated strains of *Alicyclobacillus* from Mt Rittmann, and proposed the new subspecies, *Alicyclobacillus acidocaldarius* subsp. *rittmannii*; strains had ω-cyclohexyl fatty acids, MK-7 quinones, and hopanoids characteristic of *Alicyclobacillus* species, but lacked amylolytic activity. The effects of growth temperature on lipid modulation were studied in detail (Nicolaus et al. 2002). In 30 geothermal soil samples from Mt Rittmann, these authors found only aerobic endospore formers belonging to *Alicyclobacillus*; it will be recalled (see above, this section) that Hudson et al. (1989) found strains similar to *Alicyclobacillus* (then *Bacillus acidocaldarius*) on Mt Erebus. Bargagli et al. (2004) isolated a strain tentatively identified as an *Alicyclobacillus* species from the iron-rich NW slope of Mt Melbourne, and found that it needed iron supplements in its growth media (Pepi et al. 2005).

Llarch et al. (1997) studied six isolates from sand, sediment, and water from geothermal sites on Deception Island. None of these strains was identifiable as a member of an established thermophilic species of *Bacillus*, but two strains from fumarolic water showed some relationship with *B.* (now *Geobacillus*) *stearothermophilus*. More interesting was the identification of two strains as *B. licheniformis* and *B. megaterium*, and of two others as outliers of the species *B. firmus* and *B. lentus*; all four species are known as mesophiles, but the Deception Island isolates all had optimal growth temperatures between 60 and 65°C, which considerably extend the maximum growth temperatures known for these species.

Logan et al. (2000) examined geothermal soil samples from Mt Melbourne (both Cryptogam Ridge and the NW slope), Mt Rittmann, and Candlemas Island. They used a variety of growth conditions in order to isolate aerobic endospore formers but they did not find strains of *Bacillus thermantarcticus* or *Alicyclobacillus acidocaldarius* subsp. *rittmannii*. Instead, they isolated an organism that grew optimally at pH 5.5 and 50°C on a nutritionally weak, solid medium (*Bacillus fumarioli* agar, or BFA: 0.4% yeast extract, 0.3% KH_2PO_4, 0.2% $(NH_4)_2SO_4$, and traces of $MgSO_4$ and $CaCl_2$; with 5 mg/l $MnSO_4$ to enhance sporulation). The organism subsequently grew and sporulated better on a medium of half this nutrient strength. The organism was isolated from Cryptogam Ridge on Mt Melbourne (but not the NW slope of Mt Melbourne), and from Mt Rittmann and Candlemas Island and was proposed as *Bacillus fumarioli* (Fig. 8.5). It was found on Mt Rittmann both as spores and vegetative cells, in soils whose temperatures ranged from 3.4°C to 62.5°C, the proportions of sporulated cells tending to be higher at the temperature extremes (9% at 8.3°C; 29% at 58.5°C), and lower (3% at 42.5°C) at temperatures approaching the growth optimum.

Finding *B. fumarioli* on two volcanoes 110 km apart and on Candlemas Island, which is about 5,500 km distant from Mt Melbourne, was quite striking, and it seemed remarkable that it could not be isolated from the NW slope of Mt Melbourne,

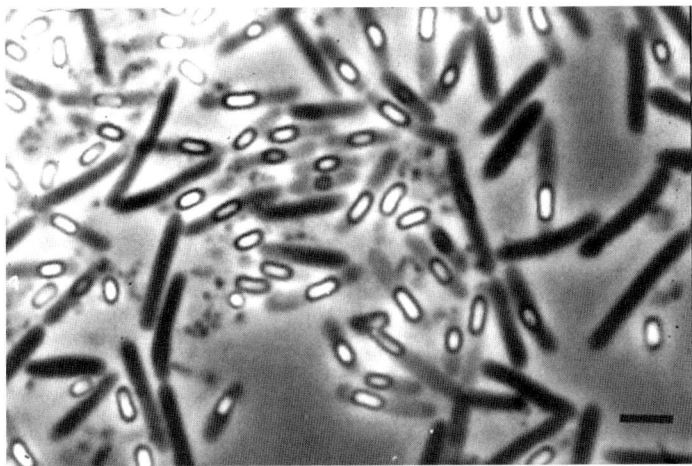

Fig. 8.5 Photomicrograph of sporangia and vegetative cells of *Bacillus fumarioli* viewed by phase contrast microscopy; ellipsoidal and cylindrical spores lie paracentrally and subterminally in unswollen sporangia. Bar represents 2 μm

despite repeated sampling; nor was it isolated from 25 cold soils local to Mt Melbourne. It was even more surprising, therefore, that using BFA De Clerck et al. (2004) were able to isolate *B. fumarioli* from gelatin production plants in Belgium, France, and the USA. Although both ecosystems have similarly low pH and moderately high temperatures, their geographical separations are huge, and the organic load in Antarctic soils is very low compared with that of gelatin. A polyphasic taxonomic comparison showed very close relationships between the Antarctic and gelatin isolates; the latter organisms did, however, produce an abundant protein with high similarity to a stress response protein, and subtractive DNA hybridization revealed genomic differences between the two sets of isolates that might indicate adaptive evolution to a specific environment (De Clerck et al. 2004).

It is of special interest that the NW ridge of Mt Melbourne failed to yield *B. fumarioli* from samples taken on two occasions. Why this particular geothermal site, lying a short distance from Cryptogam Ridge, should not yield the organism is not understood. It was noted that moss was absent from the NW ridge, yet *B. fumarioli* was isolated from both mossy and moss-free areas of Cryptogam Ridge and Mt Rittmann. Broady et al. (1987) remarked on the low diversity of Victoria Land warm ground bryophytes compared with Deception Island and the South Sandwich Islands in the maritime Antarctic, and suggested that as none of the local cold-ground bryophytes of Victoria Land had colonized the local volcanoes, it may be inferred that the soil chemistries of the fumarole environments might be unsuitable; indeed, as noted above, Bargagli et al. (1996) and Pepi et al. (2005) found appreciable differences in the mineral contents of these different soils. Broady et al. (1987) also noted that the geothermal areas of Mt Erebus and Mt Melbourne are, in comparison with maritime Antarctica, much farther from the rich propagule sources of more temperate lands to the north and west, and well south of the circumpolar westerly airstream that might carry and deposit such propagules; however, the discovery of *B. fumarioli* in Europe and America suggests that this organism is widely dispersed. Llarch et al. (1997) did not find *B. fumarioli* in their geothermal soils from Deception Island, but they did not cultivate at pH 5.5 from their samples.

Two other kinds of aerobic endospore formers were isolated from Mt. Melbourne during the expedition of 1998–1999 (Logan et al. 2000). They were scanty and difficult to cultivate, and so were not studied further at that time, but additional strains were isolated from the same soil samples back in Glasgow. Allan et al. (2005) subjected 13 such strains to polyphasic taxonomic study, and proposed seven isolates from the NW slope of Mt Melbourne as the new species *Brevibacillus levickii* (Fig. 8.6), and proposed six isolates from Cryptogam Ridge and the vents and summit of Mt Rittmann as another new species, *Aneurinibacillus terranovensis* (Fig. 8.7). *Brevibacillus* strains were not isolated from the sites at Mt Rittmann or Cryptogam Ridge and *Aneurinibacillus* strains were not isolated from the NW slope of Mt Melbourne. The distribution of *An. terranovensis* thus correlates with that of *B. fumarioli* on Mts Melbourne and Rittmann, whereas *Br. levickii* was only found in soils of the NW slope of Mt Melbourne, from which *An. terranovensis* and *B. fumarioli* could not be isolated; perhaps these observations owe something to the differences in soil chemistries that Bargagli et al. (1996) and Pepi et al. (2005) discovered at these sites.

8 Aerobic, Endospore-Forming Bacteria from Antarctic Geothermal Soils 167

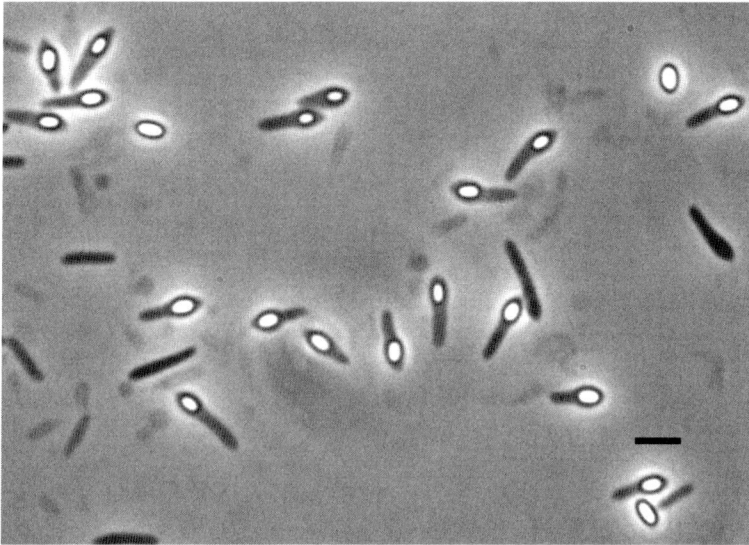

Fig. 8.6 Photomicrograph of sporangia and vegetative cells of *Brevibacillus levickii* viewed by phase-contrast microscopy; ellipsoidal spores lie subterminally and terminally in swollen sporangia. Bar represents 2 μm

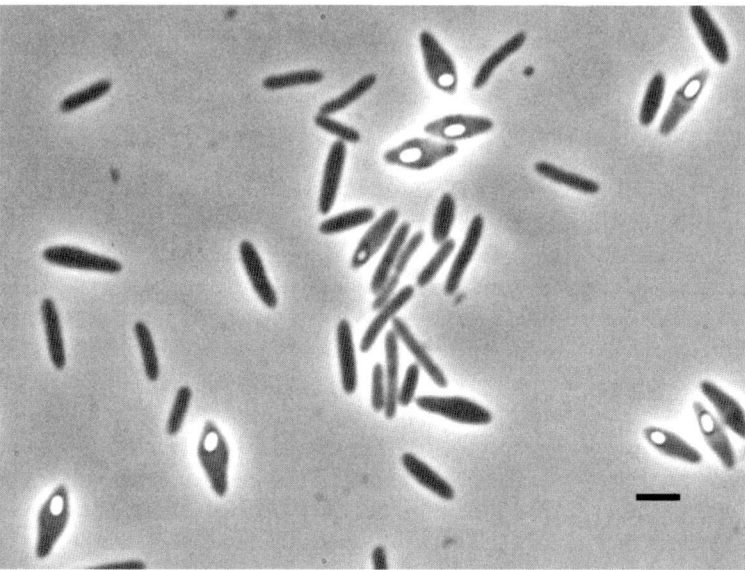

Fig. 8.7 Photomicrograph of sporangia and vegetative cells of *Aneurinibacillus terranovensis* viewed by phase-contrast microscopy; ellipsoidal spores lie centrally, paracentrally and subterminally in swollen sporangia. Bar represents 2 μm

Our emphasis thus far has been on thermophilic or thermotolerant organisms, but the same geothermal soils have also yielded mesophilic species. Bargagli et al. (2004) isolated strains related to *Paenibacillus validus* (a species that we have found repeatedly in unheated Antarctic soils) and a strain they identified as *P. apiarius* on Mt Melbourne, and found that these organisms, from iron-rich soils, often benefited from iron supplements in their growth media (Pepi et al. 2005). The novel species *B. luciferensis*, *B. shackletonii* (Logan et al. 2002, 2004a), *Paenibacillus cineris*, and *P. cookii* were isolated from Candlemas Island; *P. cookii*, like *B. fumarioli*, has also been isolated from a gelatin production plant (Logan et al. 2004b). The presence of *Paenibacillus* species in these soils is of particular interest, as these organisms often fix nitrogen; Rodríguez-Díaz et al. (2005) demonstrated the presence of the *nifH* gene in *Paenibacillus wynnii*, from unheated soil on Alexander Island, and capacity for acetylene reduction in *P. cineris* and *P. cookii*. Logan et al. (2000) also found strains of the mesophiles *B. sphaericus* (also found on Alexander Island by Rodríguez-Díaz et al. (2005)) and *B. cereus*, and the thermophile *B.* (now *Geobacillus*) *thermodenitrificans* on Candlemas Island.

8.5 Adaptations for Growth at High Temperatures

Organisms growing at high temperatures need enzyme adaptations to give molecular stability as well as structural flexibility, heat-stable protein-synthesizing machinery, and adaptations of membrane phospholipid composition. They differ from their mesophilic counterparts in the fatty acid and polar headgroup compositions of their phospholipids. The effect of temperature on the membrane composition of *G. stearothermophilus* has been intensively studied. Phosphatidyl glycerol (PG) and cardiolipin (CL) comprise about 90% of the phospholipids, but as the growth temperature rises the PG content increases at the expense of the CL content. The acyl-chain composition of all the membrane lipids also changes; the longer, saturated-linear and iso fatty acids with relatively high melting points increase in abundance, and anteiso fatty acids and unsaturated components with lower melting points decrease. As a result, the organism is able to maintain nearly constant membrane fluidity across its whole growth temperature range; this has been termed homeoviscous adaptation. An alternative theory, homeophasic adaptation, considers that maintenance of the liquid-crystalline phase is more important than an absolute value of membrane fluidity in bacteria (Tolner et al. 1997).

The major cellular fatty acid components of *Geobacillus* species following incubation at 55°C are (with ranges as percent of total given in parentheses) iso-$C_{15:0}$ (20–40%; mean 29%), iso-$C_{16:0}$ (6–39%; mean 25%) and iso-$C_{17:0}$ (7–37%; mean 19.5%), that account for 60–80% of the total (Nazina et al. 2001). The figures given by Fortina et al. (2001) for *G. caldoxylosilyticus* and Sung et al. (2002) for *G. toebii* generally lie within these ranges, with the exception that strains of the former species showed 45–57% of iso-$C_{15:0}$. Such higher levels of iso-$C_{15:0}$ are also found in *Anoxybacillus* species (Dulger et al. 2004). *Geobacillus thermoleovorans* subsp. *stromboliensis* (Romano et al. 2005), isolated from Italian geothermal soil, showed fatty acid patterns within the ranges seen for other *Geobacillus* species. *Thermobacillus xylanilyticus* Touzel et al. (2000) shows a fatty acid profile dominated by iso $C_{16:0}$, whereas the profile of *Vulcanibacillus modesticaldus* is dominated by iso $C_{15:0}$ (L'Haridon et al. 2006). Direct comparison of profiles between the obligately thermophilic species and mesophilic aerobic endospore formers is not normally possible, as the assays of members of the two groups have not usually been done at the same temperature.

Nicolaus et al. (1995) studied the effects of growth temperature on polar lipid patterns of aerobic endospore formers from geothermal and unheated soils from Antarctica; at 60°C, the strain subsequently proposed as *Bacillus thermantarcticus* (and which presently awaits reassignment to *Geobacillus*) showed a level (27%) of iso-$C_{17:0}$ fatty acid which is similar to that of *Bacillus thermoglucosidasius,* but a high level (36%) of anteiso $C_{17:0}$ fatty acid in comparison with *Bacillus* and *Geobacillus* species. Nicolaus et al. (2002) reviewed their lipid studies on Antarctic isolates. Strains tentatively identified as *Bacillus* showed increased phosphoglycolipid contents with increased growth temperature, at the expense of phosphoaminolipid and phospholipids; higher-melting point acyl chains such as iso-$C_{17:0}$ were favoured at maximum growth temperatures, whereas iso-$C_{15:0}$ was synthesized at minimum growth temperature.

Llarch et al. (1997) compared the fatty acid profiles of aerobic endospore formers isolated from Antarctic geothermal environments; their six isolates had temperature ranges with minima between 17 and 45°C and maxima between 62 and 73°C, with optima of 60 to 70°C. Two strains (temperature ranges 37–70 and 45–73°C) were found to lie nearest to *G. stearothermophilus* in a phenotypic analysis, and two other isolates could be identified as strains of *B. licheniformis* (temperature range 17–68°C) and *B. megaterium* (temperature range 17–63°C) whose maximum growth temperatures were extended beyond those seen in strains from temperate environments. The fatty acid profiles for all of these strains were compared following incubation at 45°C, and the results suggested that any potential distinctions between the rather variable fatty acid profiles of *Geobacillus* species and *Bacillus* species are largely lost when strains of each group are incubated at the same temperature.

Members of *Alicyclobacillus* possess an apparently unique phenotype, as the main membranous lipid components of most species are ω-alicyclic (ω-cyclohexane or ω-cycloheptane) fatty acids; it has been shown that ω-alicyclohexyl fatty acids pack densely, resulting in low diffusion at high temperatures (Kannenberg et al. 1984). Polar lipids based upon hopanoids are also important chemotaxonomic markers for this genus, and hopanoid content is increased in response to elevated temperatures at low pH. Together with lipids containing ω-cyclohexane fatty acids, hopanoids are important for forming a biological membrane stable enough to withstand extreme temperature and pH conditions.

Nicolaus et al. (2002) found that the percentage of cyclohexyl fatty acids increased as the growth temperature was raised for their Antarctic *Alicyclobacillus* isolates. However, an *Alicyclobacillus* isolate from fruit juice, that did not possess ω-alicyclic fatty acids, showed a fatty acid profile similar to that of *Bacillus tusciae* (Goto et al. 2003).

8.6 Nutrition and Growth Conditions of Thermophiles

Most aerobic endospore formers are chemo-organotrophs and, despite the very wide diversity of the genus, will grow well on routine media such as nutrient agar or trypticase soy agar. However, some isolates, particularly those from nutritionally poor environments, may grow poorly if at all on these standard media because of their neutral pH, and/or insufficient salinity, or because they are nutritionally too rich. Most species will use glucose and/or other fermentable carbohydrates as sole sources of carbon and energy. Inorganic and organic sources of nitrogen are used. Many species will utilise an ammonium salt as their sole nitrogen source; amino acids are also widely utilized, and strains of some species can use urea. Most thermophilic species conform to this general nutritional pattern for aerobic endospore formers.

Geobacillus species utilize a wide range of substrates, including carbohydrates, organic acids, peptone, tryptone, and yeast extract; the ability to utilize hydrocarbons as carbon and energy sources is a widely distributed property in the genus (Nazina et al. 2001; Marchant et al. 2006). A strain of *G. thermoleovorans* has been found to have extracellular lipase activity and high growth rates on lipid substrates such as olive oil, soybean oil, mineral oil, tributyrin, triolein, and Tweens 20 and 40 (Lee et al. 1999).

Alicyclobacillus species are obligately acidophilic, and have been found particularly in fruit juices and solfataric environments; isolates from the latter show different patterns of resistance to metal salts (Simbahan et al. 2004).

Bacillus schlegelii grows chemolithoautotrophically, using H_2 as electron donor and CO_2 as carbon source (CO will satisfy both requirements), or chemo-organoheterotrophically; it can also grow autotrophically on thiosulfate (Hudson et al. 1988). Hydrogenase is constitutive and has a temperature optimum between 70 and 75°C. Carbohydrates are not used. Organic acids and a small number of amino acids are utilized as sole carbon sources, whereas ammonium ions, asparagine, and urea can be utilized as sole nitrogen sources. *Bacillus tusciae* also grows chemolithoautotrophically, using H_2 as electron donor and CO_2 as carbon source, or chemoorganoheterotrophically. Carbon and nitrogen sources are similar to *B. schlegelii*, and some alcohols can also be used as carbon sources.

Members of the genus *Sulfobacillus* are facultative autotrophs that can obtain their energy by oxidizing ferrous iron, as well as elemental sulfur or its reduced compounds. None of the three established species from this genus have been reported from natural geothermal environments, but unidentified *Sulfobacillus* strains have been reported from geothermal waters (see Table 8.1).

Amino acid transport in *G. stearothermophilus* is Na^+-dependent, which is unusual for neutrophilic terrestrial organisms, but common among marine bacteria and alkalophiles; however, the possession of primary and secondary Na^+-transport systems may be advantageous to the organism by allowing energy conversion via Na^+-cycling. Although the phospholipid adaptations needed to give optimal membrane fluidity at the organism's growth temperature also result in increased proton permeability, this may be counteracted by increased proton pumping activity using the less permeable sodium ions as coupling ions (de Vrij et al. 1990; Tolner et al. 1997).

Allan (2006) used a [^{14}C] l-glutamic acid tracing technique to study the type strains of *Br. levickii* and *An. terranovensis* with a view to understanding why they could not be isolated from the same habitats on Mt Melbourne and Mt Rittmann (see Section 8.4 above and Fig. 8.1). Results for both species showed distinct differences in the mechanisms used for l-glutamic acid uptake; *Br. levickii* possesses a secondary uptake system specific for l-glutamic acid which is dependent on K$^+$ and possibly H$^+$, and *An. terranovensis* possesses multiple uptake systems which are capable of transporting other amino acids as well as l-glutamic acid, a process which appears to be dependent on multiple factors including Na$^+$, K$^+$, H$^+$, the electrical gradient across the cell membrane, and osmotic conditions. These studies showed that both strains utilize glutamate, which is probably available from cyanobacteria and microalgae in their natural habitats (Siebert and Hirsch 1988). Glutamate is nonessential, however, as the organisms grew on a defined medium from which it had been omitted. For comparative purposes, l-glutamic acid uptake by strains of *B. fumarioli* and *B. cereus* were also investigated. *B. fumarioli* was found to possess an uptake system similar to that of *An. terranovensis* (consistent, perhaps, with their similar habitats), whereas the *B. cereus* strain possessed multiple uptake systems capable of transporting both the d and l isomers of glutamic acid.

8.7 Conclusions

It is clear that our understanding of the bacteriology of Antarctic geothermal soils remains in its infancy. Although several new species of aerobic endospore formers have been discovered in these niches in the last 15 years, almost nothing is known about the relationships of the prokaryotes and eukaryotes within these environments, whereas knowledge of the ecologies of deep-sea hydrothermal vents is ever increasing (Reysenbach et al. 2006).

Acknowledgements We thank P. De Vos and colleagues at the University of Gent, Belgium, N. R. Russell, and BAS (especially the late D. D. Wynn-Williams, and P. Convey), and the Italian (especially R. Bargagli and B. Nicolaus) and Spanish Antarctic research programmes for helping us to make our contribution.

References

Albuquerque L, Rainey FA, Chung AP, Sunna A, Nobre MF, Grote R, Antranikian G, Da Costa MS (2000) *Alicyclobacillus hesperidum* sp. nov., and a related genomic species from solfataric soils of São Miguel in the Azores. *Int J Syst Evol Microbiol* 50:451–457
Allan RN (2006) PhD thesis. Glasgow Caledonian University
Allan RN, Lebbe L, Heyrman J, De Vos P, Buchanan CJ, Logan NA (2005) *Brevibacillus levickii* sp. nov. and *Aneurinibacillus terranovensis* sp. nov., two new thermoacidophiles isolated from geothermal soils of northern Victoria Land, Antarctica. *Int J Syst Evol Microbiol* 55:1039–1050

Ambroz A (1913) *Denitrobacterium thermophilum* spec nova, ein Beitrag zur Biologie der thermophilen Bakterien. *Zentralbl Bakteriol Parasitenkd Infektionskr Hyg Abt.II* 37:3–16

Bargagli R, Broady PA, Walton DWH (1996) Preliminary investigation of the thermal biosystem of Mount Rittmann fumaroles (northern Victoria Land, Antarctica). *Antarctic Science* 8:121–126

Bargagli R, Skotnicki ML, Pepi M, Mackenzie A, Agnorelli C (2004) New record of moss and thermophilic bacteria species and physico-chemical properties of geothermal soils on the northwest slope of Mt Melbourne (Antarctica). *Polar Biol* 27:423–536

Belduz AO, Dulger S, Demirbag Z (2003) *Anoxybacillus gonensis* sp. nov., a moderately thermophilic, xylose-utilizing, endospore-forming bacterium. *Int J Syst Evol Microbiol* 53: 1315–1320

Bonjour F, Aragno M (1984) *Bacillus tusciae,* a new species of thermoacidophilic, facultatively chemolithotrophic, hydrogen oxidizing sporeformer from a geothermal area. *Arch Microbiol* 139:397–401

Boone DR, Liu Y, Zhao Z-J, Balkwill DL, Drake GR Stevens TO, Aldrich HC (1995) *Bacillus infernus* sp. nov., an Fe(III)- and Mn(IV)-reducing anaerobe from the deep terrestrial subsurface. *Int J Syst Bacteriol* 45:441–448

Broady P, Given D, Greenfield L, Thompson K (1987) The biota and environment of fumaroles on Mt Melbourne, northern Victoria Land. *Polar Biol* 7:97–113

Broady PA (1993) Soils heated by volcanism. In: E. I. Friedmann (ed) *Antarctic Microbiology.* Wiley-Liss, New York, pp. 413–432

Caccamo D, Gugliandolo C, Stackebrandt E, Maugeri TL (2000) *Bacillus vulcani* sp. nov., a novel thermophilic species isolated from a shallow marine hydrothermal vent. *Int J Syst Evol Microbiol* 50:2009–2012

Claus D, Berkeley RCW (1986) Genus *Bacillus* Cohn 1872. In: Sneath PHA, Mair NS, Sharpe ME, Holt JG (ed) *Bergey's Manual of Systematic Bacteriology,* vol. 2. Williams & Wilkins, Baltimore, pp.1105–1139

Darland G, Brock TD (1971) *Bacillus acidocaldarius* sp. nov., an acidophilic, thermophilic sporeforming bacterium. *J Gen Microbiol* 67:9–15

Darling CA, Siple PA (1941) Bacteria of Antarctica. *J Bact* 42: 83–98

De Clerck E, Gevers D, Sergeant K, Rodríguez-Díaz M, Herman L, Logan NA, Van Beeumen J, De Vos P (2004) Genotypic and phenotypic comparison of *Bacillus fumarioli* isolates from geothermal Antarctic soil and gelatine. *Res Microbiol* 155: 483–490

de Vrij W, Speelmans G, Heyne RIR, Konings WN (1990) Energy transduction and amino acid transport in thermophilic aerobic and fermentative bacteria. *FEMS Microbiol Rev* 75:183–200

Donk PJ (1920) A highly resistant thermophilic organism. *J Bacteriol* 5:373–374

Dulger S, Demirbag Z, Belduz AO (2004) *Anoxybacillus ayderensis* sp. nov., and *Anoxybacillus kestanbolensis* sp. nov. *Int J Syst Evol Microbiol* 54:1499–1503

Fortina MG, Mora D, Schumann P, Parini C, Manachini PL, Stackebrandt E (2001) Reclassification of *Saccharococcus caldoxylosilyticus* as *Geobacillus caldoxylosilyticus* (Ahmad et al. 2000) comb. nov. *Int J Syst Evol Microbiol* 51:2063–2071

Golovacheva RS, Egorova LA, Loginova LG (1965) Ecology and systematics of aerobic obligate-thermophilic bacteria isolated from thermal localities on Mount Yangan-Tau and Kunashir Isle of the Kuril chain. *Microbiology* (English translation of *Mikrobiologiya*) 34: 693–698

Goto K, Mochida K, Asahara M, Suzuki M, Kasai H, Yokota A (2003) *Alicyclobacillus pomorum* sp. nov., a novel thermo-acidophilic bacterium that does not possess ω-alicyclic fatty acids, and emended description of the genus *Alicyclobacillus. Int J Syst Evol Microbiol* 53:1537–1544

Gugliandolo C, Maugeri TL, Caccamo D, Stackebrandt E (2003) *Bacillus aeolius* sp. nov. a novel thermophilic, halophilic marine *Bacillus* species from Eolian Islands (Italy). *Syst Appl Microbiol* 26:172–176

Hammer BW (1915) Bacteriological studies on the coagulation of evaporated milk. *Iowa Agr Exp Sta Res Bull* 19:119–131

Heinen UJ, Heinen W (1972). Characteristics and properties of a caldoactive bacterium producing extracellular enzymes and two related strains. *Arch Mikrobiol* 82: 1–23

Heinen W, Lauwers AM, Mulders JWM (1982) *Bacillus flavothermus*, a newly isolated facultative thermophile. *Antonie van Leeuwenhoek J Microbiol Serol* 48:265–272

Hudson JA, Daniel RM (1988) Enumeration of thermophilic heterotrophs in geothermal soils from Mount Erebus, Ross Island, Antarctica. *Appl Env Microbiol* 54:622–624

Hudson JA, Daniel RM, Morgan HW (1988) Isolation of a strain of *Bacillus schlegelii* from geothermally heated antarctic soil. *FEMS Microbiol Lett* 51: 57–60

Hudson JA, Daniel RM, Morgan HW (1989) Acidophilic and thermophilic *Bacillus* strains from geothermally heated antarctic soil. *FEMS Microbiol Lett* 60: 279–282

Kannenberg E, Blume A, Poralla K (1984) Properties of ω-cyclohexane fatty acids in membranes. *FEBS Lett* 172: 331–334

L'Haridon S, Miroshnichenko ML, Kostrikina NA, Tindall BJ, Spring S, Schumann P, Stackebrandt E, Bonch-Osmlovskaya EA, Jeanthon C (2006) *Vulcanibacillus modesticaldus* gen. nov., sp. nov., a strictly anaerobic, nitrate-reducing bacterium from deep-sea hydrothermal vents. *Int. J. Syst. Evol. Microbiol.* 56:1047–1053

Lama L, Calandrelli V, Gambacorta A, Nicolaus B (2004) Purification and characterization of thermostable xylanase and β-xylosidase by the thermophilic bacterium *Bacillus thermantarcticus*. *Res Microbiol* 155: 283–289

Lee D-W, Koh Y-S, Kim K-J, Kim B-C, Choi H-J, Kim D-S, Suhatono MT, Pyun Y-R (1999) Isolation and characterization of a thermophilic lipase from *Bacillus thermoleovorans* ID-1. *FEMS Microbiol Lett* 179: 393–400

Linskens HF, Bargagli R, Cresti M, Forcardi S (1993) Entrapment of long-distance transported pollen grains by various moss species in coastal Victoria Land, Antarctica. *Polar Biol* 13: 81–87

Llarch Á, Logan NA, Castelví J, Prieto MJ, Guinea J (1997) Isolation and characterization of thermophilic *Bacillus* spp. from geothermal environments on Deception Island, South Shetland archipelago. *Microb Ecol* 34: 58–65

Logan NA, Lebbe L, Hoste B, Goris J, Forsyth G, Heyndrickx M, Murray BL, Syme N, Wynn-Williams DD, De Vos P (2000) Aerobic endospore-forming bacteria from geothermal environments in northern Victoria Land, Antarctica, Candlemas Island, South Sandwich archipelago, with the proposal of *Bacillus fumarioli* sp. nov. *Int J Syst Evol Microbiol* 50:1741–1753

Logan NA, Lebbe L, Verhelst A, Goris J, Forsyth G, Rodríguez-Díaz M, Heyndrickx M, De Vos P (2002) Proposal of *Bacillus luciferensis* sp. nov., from volcanic soil on Candlemas Island, South Sandwich archipelago. *Int J Syst Evol Microbiol* 52: 1985–1989

Logan NA, Lebbe L, Verhelst A, Goris J, Forsyth G, Rodríguez-Díaz M, Heyndrickx M, De Vos P (2004a) Proposal of *Bacillus shackletonii* sp. nov. from volcanic soils on Candlemas Island, South Sandwich archipelago. *Int J Syst Evol Microbiol* 54: 373–376

Logan NA, De Clerck E, Lebbe L, Verhelst A, Goris J, Forsyth G, Rodríguez-Díaz M, Heyndrickx M, De Vos P (2004b) Proposal of *Paenibacillus cineris* sp. nov. and *Paenibacillus cookii* sp. nov., from Antarctic volcanic soils and a gelatin-processing plant. *Int J Syst Evol Microbiol* 54: 1071–1076

Manca MC, Lama L, Improta R, Esposito E, Gambacorta A, Nicolaus B (1996) Chemical composition of two exopolysaccharides from *Bacillus thermantarcticus*. *Appl Env Microbiol* 62:3265–3269

Marchant R, Banat IM, Rahman TJS, Berzano M 2002. The frequency and characteristics of highly thermophilic bacteria in cool soil environments. *Env Microbiol* 4:595–602

Marchant R, Sharkey FH, Banat IM, Rahman TJ, Perfumo A (2006) The degradation of *n*-hexadecane in soil by thermophilic geobacilli. *FEMS Microbiol Ecol* 56: 44–54

Marteinsson VG, Birrien J-L, Jeanthon C, Prieur D (1996) Numerical taxonomic study of thermophilic *Bacillus* isolated from three geographically separated deep-sea hydrothermal vents. *FEMS Microbiol Ecol* 21: 255–266

McMullan G, Christie JM, Rahman TJ, Banat IM, Ternan NG, Marchant R (2004) Habitat, applications and genomics of the aerobic, thermophilic genus *Geobacillus*. *Biochem Soc Trans* 32: 214–217

Nazina TN, Lebedeva EV, Poltaraus AB, Tourova TP, Grigoryan AA, Sokolova DS, Lysenko AM, Osipov GA (2004) *Geobacillus gargensis* sp. nov., a novel thermophile from a hot spring, and the reclassification of *Bacillus vulcani* as *Geobacillus vulcani* (Caccamo et al. 2000) comb. nov. *Int J Syst Evol Microbiol* 54: 2019–2024

Nazina TN, Tourova TP, Poltaraus AB, Novikova EV, Grigoryan AA, Ivanova AE, Lysenko AM, Petrunyaka VV, Osipov GA, Belyaev SS, Ivanov MV (2001) Taxonomic study of aerobic thermophilic bacilli: Descriptions of *Geobacillus subterraneus* gen nov, sp. nov. and *Geobacillus uzenensis* sp. nov. from petroleum reservoirs and transfer of *Bacillus stearothermophilus, Bacillus thermocatenulatus, Bacillus thermoleovorans Bacillus kaustophilus, Bacillus thermoglucosidasius, Bacillus thermodenitrificans* to *Geobacillus* as *Geobacillus stearothermophilus, Geobacillus thermocatenulatus, Geobacillus thermoleovorans, Geobacillus kaustophilus, Geobacillus thermoglucosidasius, Geobacillus thermodenitrificans*. *Int J Syst Evol Microbiol* 51: 433–446

Nicolaus B, Improta R, Manca MC, Lama L, Esposito E, Gambacorta A (1998) Alicyclobacilli from an unexplored geothermal soil in Antarctica: Mount Rittmann. *Polar Biol* 19: 133–141

Nicolaus B, Lama L, Esposito E, Manca MC, di Prisco G, Gambacorta A (1996) "*Bacillus thermoantarcticus*" sp. nov. from Mount Melbourne, Antarctica: A novel thermophilic species. *Polar Biol* 16: 101–104

Nicolaus B, Lama L, Gambacorta A (2002) Thermophilic *Bacillus* isolates from Antarctic environments. In: Berkeley RCW, Heyndrickx M, Logan NA, De Vos P (eds) *Applications and Systematics of Bacillus and Relatives*. Blackwell Science, Oxford, UK, pp. 123–140

Nicolaus B, Manca MC, Lama L, Esposito E, Gambacorta A (1995) Effects of growth temperature on the polar lipid pattern and fatty acid composition of seven thermophilic isolates from the Antarctic continent. *Syst Appl Microbiol* 18: 32–36

Nicolaus B, Marsiglia F, Esposito E, Trincone A, Lama L, Sharp R, di Prisco G, Gambacorta A (1991) Isolation of five strains of thermophilic bacteria in Antarctica. *Polar Biol* 11: 425–429

Pepi M, Agnorelli C, Bargagli R (2005) Iron demand by thermophilic and mesophilic bacteria isolated from an antarctic geothermal soil, *Biometals* 18: 529–536.

Poli A, Esposito E, Lama L, Orlando P, Nicolaus G, de Appolonia F, Gambacorta A, Nicolaus B (2006) *Anoxybacillus amylolyticus* sp. nov., a thermophilic amylase producing bacterium isolated from Mount Rittmann (Antarctica). *Syst Appl Microbiol* 29:300–307

Rahman TJ, Marchant R, Banat IM (2004) Distribution and molecular investigation of highly thermophilic bacteria associated with cool soil environments. *Biochem Soc Trans.* 32: 209–213

Reysenbach A-L, Liu Y, Banta AB, Beveridge TJ, Kirshtein JD, Schouten S, Tivey M K, Von Damm K, Voytek MA (2006) Isolation of a ubiquitous obligate thermoacidophilic archaeon from deep-sea hydrothermal vents. *Nature* 442: 444–447

Rodríguez-Díaz M, Lebbe L, Rodelas-González B, Heyrman J, De Vos P, Logan NA (2005) *Paenibacillus wynnii* sp. nov., a new species harbouring the *nifH* gene, isolated from Alexander Island, Antarctica. *Int J Syst Evol Microbiol* 55: 2093–2099

Romano I, Poli A, Lama L, Gambacorta A, Nicolaus B (2005) *Geobacillus thermoleovorans* subsp. *stromboliensis* subsp. nov., isolated from the geothermal volcanic environment. *J Gen Appl Microbiol* 51:183–189

Schäffer C, Franck WL, Scheberl A, Kosma P, McDermott TR, Messner P (2004) Classification of isolates from Austria and Yellowstone National Park as *Geobacillus tepidamans* sp. nov. *Int J Syst Evol Microbiol* 54:2361–2368

Schenk A, Aragno M (1979). *Bacillus schlegelii*, a new species of thermophilic, facultatively chemolithoautotrophic bacterium oxidizing molecular hydrogen. *J Gen Microbiol* 115:333–341

Siebert J, Hirsch P (1988) Characterization of 15 selected coccal bacteria isolated from Antarctic rock and soil samples from the McMurdo-Dry Valleys (South-Victoria Land), *Polar Biol* 9: 37–44

Simbahan J, Drijber R, Blum P (2004) *Alicyclobacillus vulcanalis* sp. nov., a thermophilic, acidophilic bacterium isolated from Coso Hot Springs, California, USA. *Int J Syst Evol Microbiol* 54:1703–1707

Sung M-H, Kim H, Bae J-W, Rhee S.K, Jeon CO, Kim K, Kim J-J, Hong S-P, Lee S-G, Yoon J-H, Park Y-H, Baek D-H (2002) *Geobacillus toebii* sp. nov., a novel thermophilic bacterium isolated from hay compost. *Int J Syst Evol Microbiol* 52: 2251–2255

Takami H, Inoue A, Fuji F, Horikoshi K (1997) Microbial flora in the deepest sea mud of the Mariana Trench. *FEMS Microbiol Lett* 152: 279–285

Tolner B, Poolman B, Konings WN (1997) Adaptation of microorganisms and their transport systems to high temperatures. *Comp Biochem Physiol* 118A: 423–428

Touzel JP, O'Donohue M, Debeire P, Samain E, Breton C (2000) *Thermobacillus xylanilyticus* gen. nov., sp. nov., a new aerobic thermophilic xylan-degrading bacterium isolated from soil. *Int J Syst Evol Microbiol* 50:315–320

Ugolini FC, Starkey RL (1966) Soils and microorganisms from Mount Erebus, Antarctica. *Nature* 211:440–441

Weigel J (1986) Methods for isolation and study of thermophiles. In: Brock (ed), *Thermophiles: General, Molecular and Applied Microbiology*. Wiley, New York, pp. 17–37

White D, Sharp RJ, Priest FG (1993) A polyphasic taxonomic study of thermophilic bacilli from a wide geographical area. *Antonie van Leeuwenhoek J Microbiol Serol* 64: 357–386

Xue Y, Zhang X, Zhou C, Zhao Y, Cowan DA, Heaphy S, Grant WD, Jones BE, Ventosa A, Ma Y (2006) *Caldalkalibacillus thermarum* gen. nov., sp. nov., a novel alkalithermophilic bacterium from a hot spring in China. *Int J Syst Evol Microbiol* 56: 1217–1221

Yumoto I, Hirota K, Kawahara T, Nadasaka Y, Okuyama H, Matsuyama H, Yokota Y, Nakajima K, Hoshino T (2004) *Anoxybacillus voinovskiensis* sp. nov., a moderately thermophilic bacterium from a hot spring in Kamchatka. *Int J Syst Evol Microbiol* 54:1239–1242

Zarilla K, Perry JJ (1987). *Bacillus thermoleovorans,* sp. nov., a species of obligately thermophilic hydrocarbon utilizing endospore-forming bacteria. *System Appl Microbiol* 9: 258–264

Chapter 9
Peatland Microbiology

Shwet Kamal and Ajit Varma(✉)

9.1 Introduction

Peat contains a high proportion of dead organic matter, mainly plants, accumulated over thousands of years. Peatlands cover about 5–8% of the world's surface. Peatlands are important wetland areas maintaining supplies of clean water to rivers and acting as a carbon sink. Peatlands may contain 3–3.5 times the amount of carbon stored in tropical rain forests and act as an ancient habitat containing many rare and threatened species. As such, they are an unparalleled record of our past.

9.2 General Description of Peatlands

9.2.1 What Is Peat?

Peat is a brownish-black material that is formed in acidic, anaerobic wetland (peatland) conditions. It consists mainly of partially decomposed, loosely compacted organic matter with more than 50% carbon. It is made up of *Sphagnum* mosses, stems and roots of sedges and reeds, animal remains, plant remains, fruits, and pollens. Unlike most other ecosystems, the dead plants/animals in peatlands remain without decomposing for hundreds or thousands of years. This is because of waterlogged conditions, where the lack of oxygen prevents micro-organisms from rapidly decomposing the dead plants/animals. As a result, the organic matter in peatland is easily identifiable. The formation of peat is a very slow process, and it takes approximately 10 years for 1 cm of peat to form.

Ajit Varma
Amity Institute of Microbial Sciences, Amity University Uttar Pradesh, Noida-201303 (India)
e-mail: ajitvarma@aihmr.amity.edu

9.2.2 Description of Peatlands

A peatland is an ecosystem where the production rate of biomass exceeds its decomposition rate. The result is the accumulation of organic matter coming from plant debris, animal remains, and microbes. This more or less decomposed biomass forms the peat. Peatlands can also be defined as anaerobic and acidic wetlands containing peat.

In a natural peatland ecosystem, the water table level is near the soil surface, which creates anaerobic conditions lowering the microbial activity. Peatlands are distinguished from marshes and swamps by the lower decomposition/production ratio.

9.2.3 Types of Peatlands

9.2.3.1 Bogs (Ombrotrophic Peatlands)

Ombrotrophic peatlands are fed only by rainwater or snow falling on their surface. The groundwater has no contact with the wetland and no nutrients come through water supplies from adjacent ecosystems. It is characterized by mineral deficiency and acidic conditions (pH 4.2) due to a dominant vegetation of *Sphagnum*. Because of their high ion exchange capacity and release of organic acids, *Sphagnum* species can lower the pH of their environment. The process of bog formation is called as ombrotrophication.

Bogs have a diplotelmic soil structure, characterized by the presence of an acrotelm and a catotelm (Ingram, 1978). The catotelm is the bottom layer of peat that is permanently below the water table resulting in anaerobic conditions, low microbial activity, and peat decomposition. The acrotelm (30–50 cm), overlying the catotelm, is characterized by periodically alternating anaerobic and aerobic conditions, as determined by water table fluctuations. This alternation accelerates microbial activity. The acrotelm has a loose structure comprising living parts of mosses, as well as dead and poorly decomposed plant material. It can contain and release large quantities of water that limit variations of the water table in peat bogs.

9.2.3.2 Fens (Minerotrophic Peatlands)

Fens are peatlands fed by precipitation (rainwater) as well as surface run-off water and are rich in basic cations due to contact with run-off water from mineral soil. For this reason fens are also called minerotrophic peatlands. The trophic status of fens varies between oligotrophic (poor) to eutrophic (rich), depending upon the concentration of basic cations (such as Ca, Mg, and Na). The pH of water and substrate of oligotrophic fen varies from 3.8 to 6.5, whereas for eutrophic fens it varies from pH 5.8 to 8.4 (Sjors 1950). The vegetation of poor fens consists of *Sphagnum* mosses, whereas on rich fens herbaceous species of Cyperaceae family and brown mosses from Amblystegiaceae family are dominant. Dwarf shrubs and trees are also present on both poor and rich fens.

9.2.4 Geographical Distribution of Peatlands

Peatlands cover almost 5–8% (500 million hectares) of the world's surface. The occurrence, extent, and types of peatland depend on the prevailing climate. Most of the temperate and boreal regions of the Northern Hemisphere offer favorable conditions for peatland development.

The largest peatland areas of Asia are found in Indonesia, China, and Malaysia, covering a total area of 24.4 million ha, most of which occurs in the tropical zone of Indonesia. Tropical peatlands are most diverse, comprising fresh water swamp, peat swamps, as well as eutrophic, mesotrophic, and oligotrophic types.

European peatlands are most extensive in Finland, Sweden, and Norway. Much of the European peat resources are threatened due to development. Natural peatlands in the Netherlands are lost, Switzerland and Germany each have only 500 ha remaining, whereas the United Kingdom has seen a loss of >90% of bog peatland. Ireland has only 18% of its original peatland area left.

North America has almost 40% of the total world's peatlands with 173 million ha. The main peatland areas of Canada are found in Manitoba, Ontario, Alberta, and Saskatchewan and are the least threatened peatlands. Over half of the United States peatlands occur in nearly undisturbed settings in Alaska. In South America peatland occurs in Brazil, the Falklands, Mexico, Chile, Guyana, and Cuba.

Peatlands in Australia and New Zealand occur in temperate zones, covering an area of 2.4 million ha. The bogs are found in Patagonia, Tasmania, New Zealand, and the Eastern Australia highlands. The most characteristic bog type in these regions is that of the cushion-plant bog, which does not occur elsewhere in the world.

African peatlands are classified into five types: blanket bogs, raised bogs, valley peatlands, fens, and reed swamps, covering an area of 5.8 million ha. The major peatland areas are in the highlands of Burundi, Rwanda, South Africa, southwestern Uganda, the highland area of Zaire, and the Aberdane Range and the Cherangani Hills in western Kenya.

The peatlands of Siberia cover a vast area, with 103 million ha existing in western Siberia alone.

9.2.5 Peatland Development

Peatlands formed originally either by filling in of shallow water bodies by peat and organic matter, or they form in areas where river flood plains, forests, grasslands, or rocky areas previously existed. These changes occur because of tectonic processes, local climatic conditions, and geology resulting in the formation of peatlands.

Peatlands can develop by two processes:

(a) Terrestrialization or infilling of shallow lakes: Terrestrialization occurs when a peatland is initiated from a lake or a water body
(b) Paludification of poorly drained land: Paludification is a process where *Sphagnum* mosses invade and begin to accumulate peat on uplands.

In the tropics, paludification is a common process of peatland formation (Gorham et al. 2003), whereas in temperate regions, terrestrialization serves as the primary mechanism of peatlands initiation followed by paludification (Anderson et al. 2003). Indeed, the water level influences peatland formation and development, but allogenic (external influences such as atmospheric humidity and temperature) and autogenic factors (internal influences such as succession of vegetation, topography, substrate, and hydrology) also play an important role. A positive hydric regime (water input higher than water output), low temperature, and high precipitation favor paludification, whereas for terrestrialization allogenic factors can be in a wider range because the water table is already at the soil surface (Payette 2001) and even in drought periods the initiation of peatlands by terrestrialization is possible.

9.2.6 Chemical Composition of Tropical and Temperate Peatland

9.2.6.1 Organic Composition

The chemical composition of peat materials is predominantly influenced by the parent vegetation, the degree of decomposition, and the original chemical environment. Peat materials can be grouped into five fractions:

(a) Water-soluble compounds
(b) Ether and alcohol-soluble materials
(c) Cellulose and hemicellulose
(d) Lignin and lignin-derived substances
(e) Nitrogenous materials or crude proteins

The content of water-soluble compounds, mainly polysaccharides, mono-sugars, and some tannins, generally varies depending on the stage of decomposition. The ether and alcohol extracts contain fatty acids, waxlike components, resins, and nitrogenous fats and some waxes, tannins, various pigments, alkaloids, and soluble carbohydrates. The amounts are strongly related to the original vegetation. For example, *Sphagnum* peats may contain as much as 15% of soluble carbohydrates, reeds and sedge peats less than 5%. The cellulose and hemicellulose fraction decomposes easily and the content in the original vegetation is therefore usually greater than that in the derived peat. The lignin and lignin-derived materials commonly constitute the largest portion of the peat because of poor decomposition (range 20–50%). Finally, the nitrogenous constituents are small with respect to other fractions and mostly proteinaceous in nature. Total nitrogen may vary from 0.3–4.0% (of dry weight).

9.2.6.2 Elemental Composition

There is generally a wide variation in mineral composition between different peats, but the principal constituents other than carbon, hydrogen, nitrogen, and oxygen are either silicon or calcium. The silicon usually comes from wind-blown minerals or washed-in sediments and is therefore of low abundance. The calcium content can be high in eutrophic environments. Together with magnesium, this element in its ionic form is strongly adsorbed onto the colloidal organic particles. Contents of iron, aluminium, sodium, and sulfur reach high levels in some peats due to environmental conditions. It is not clear whether tropical peats have elemental compositions essentially different from temperate peats. It is suggested, however, that because tropical soils commonly have larger sesquioxide contents than temperate regions, the tropical peat-forming environments and the peats themselves might generally reflect such a difference by higher iron and aluminium contents.

9.2.7 Significance of Peatlands

Functions and values of peatlands make them valuable ecosystems. Their role as a carbon sink has gained visibility recently because of its impact on the greenhouse effect and climatic changes. Natural peatland emits greenhouse gases such as methane (CH_4), but they also stock a large amount of carbon present in plant debris and peat.

Peat bogs also play a role in regulating water flow by stocking water in cases of abundant precipitation and releasing it during dry periods. Peat and peatland vegetation both are extremely good water filters, making peatlands highly effective in removing sediments, pollutants, and pathogens.

The minerals present in the peatlands contribute to the removal of the pollutants through chemical reactions such as denitrification and sulfate reduction. Some of the reactions require energy inputs, which are drawn from the large amount of carbon stored in the peatlands. Presence of peatlands downstream of industries and mines ensures filtering out and temporary storage of pollutants, including waste matter such as uranium from gold mining operations, thus protecting river systems.

Due to slow decomposition rate and prevailing anoxic conditions, plant parts, especially seeds and pollen, are preserved in peat for thousands of years and thus these ecosystems act as paleoarchives. With modern dating techniques, it is possible to reconstruct past environment and climate through the identification of seeds and pollen present within the superposed peat layers.

Biodiversity is another criterion on which base peatlands deserve special status. Because they are unique acidic ecosystems, peat bogs support specific plant communities. A number of plant and bird species are found only in peatlands.

9.2.8 Peatland Vegetation

As with any other ecosystem, peatland vegetation depends upon the availability of nutrients in the soil, groundwater source, and base rock type. Plants that formed the earlier layers of peat hundreds of years ago may not be identical to present vegetation of the peatland. The vegetation includes sedges, reeds, bulrushes, swamp forest trees, herbs, orchids, shrubs, and ferns.

Peatland ecosystems represent a harsh environment for plants because of acidic and nutrient-poor conditions, a high water table, and exposition to desiccation due to the absence of protection against wind and sun, limiting the growth and survival of many plant species. A few plant communities such as *Sphagnum* mosses are the dominant vegetation, maintaining a water-logged condition due to their ability to retain water. They cause acidification by releasing humic acids and are efficient at absorbing and keeping nutrients.

The shrubs include small-leaved cranberry, rhodora, Labrador tea, and leather leaf. These plants can tolerate very wet to very dry conditions, wind, and ice, and produce a toxin that prevents browsing by mammals.

Three types of carnivorous plants are found in peatlands: pitcher plants, sundews, and bladderworts. Although these plants still photosynthesize, their carnivorous habits allow them to grow more vigorously in a competitive environment. The diet of carnivorous plants, which includes not only insects, but also mites, spiders, and even small frogs, has evolved to supply nitrogen which is lacking in the soil. The plant's prey, attracted by the red-veined leaves, is trapped by downward pointing bristles which not only prevent escape, but force it further down into the rain and dew held by the "pitcher." It is then digested by acid and proteolytic enzymes secreted by digestive glands of the plants and absorbed into the plant as food. In some cases certain bacteria are also helpful in digestion of the insect inside the plant.

Trees often show unusual growth forms in peatlands, particularly in open areas of bogs where peat layers are often thickest. Orchids, with their elegant stalks and showy flowers, contrast with the thick-leaved shrubs and spiky sedges.

9.3 Cycles in Peatlands

9.3.1 Carbon Cycle in Peatlands

Peatlands exchange carbon dioxide with the atmosphere. Carbon dioxide is sequestered by vegetation by way of photosynthesis. Carbon is then accumulated as a peat deposit in the anaerobic catotelm. Carbon dioxide is released in the atmosphere through plant and microbial respiration; the two components together are referred to as "total respiration". Plants release CO_2 through mitochondrial respiration and photorespiration. Soil also emits CO_2 following the aerobic decomposition of organic matter (Clymo et al. 1998) and methane (CH_4) oxidation (methanotrophy) by bacteria (Sundh et al. 1995), two processes that occur in the acrotelm. However, the conditions in peatland limit decomposition, so most of the carbon is retained and stored in the peat. Figure 9.1 presents an overview of the carbon cycle in peatlands.

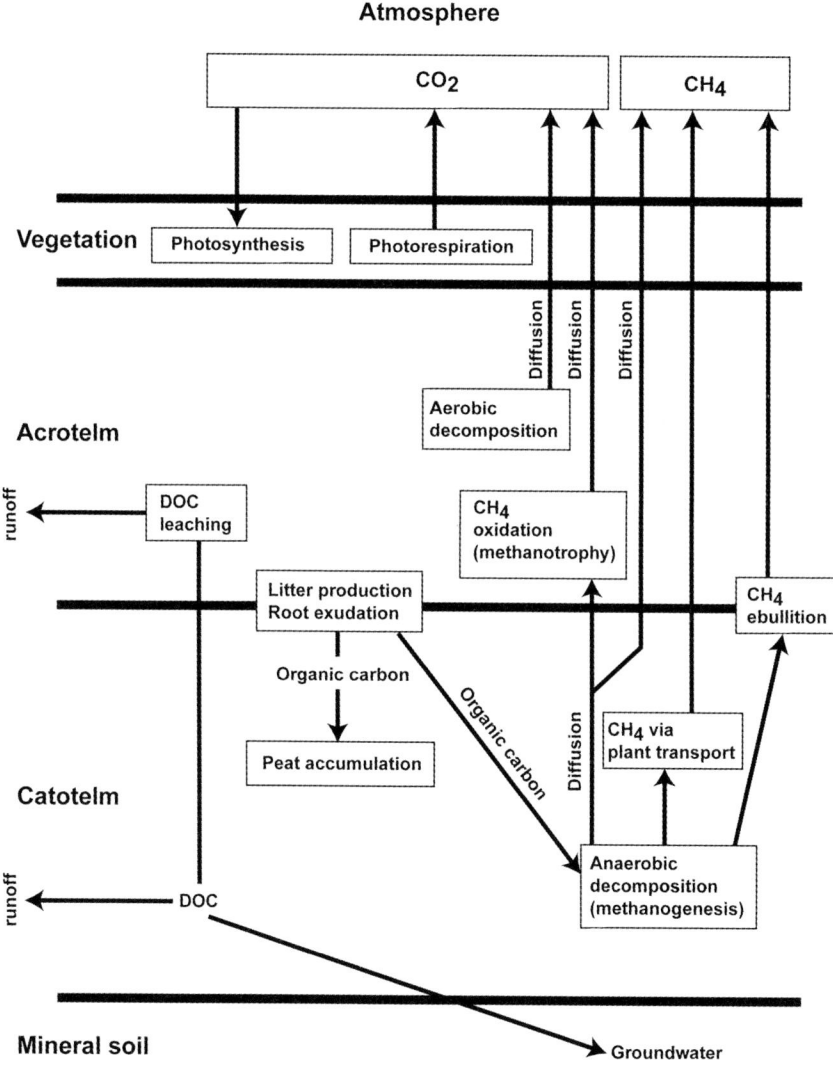

Fig. 9.1 The carbon cycle in a natural peatland. Methane production from peatlands, exchange of carbon dioxide with the atmosphere, and dissolved organic carbon (doc) processes are presented. (Used with permission, Faubert (2004).)

9.3.2 Cycle of Dissolved Organic Carbon in Peatlands

Organic material is also released from the peatland as dissolved organic carbon (DOC; see Fig. 9.1), which gives a characteristic brown color to the peatland water. DOC is released by leaching and enters into downstream aquatic ecosystems (Dalva and Moore 1991). DOC exports from temperate and boreal peatlands range between 4 and 20 g m^{-2} yr^{-1} (Moore 2001). Billett et al. (2006) investigated the spatial and temporal connectivity between concentrations of organic carbon in stream water and the soil carbon pool in a small (130 ha) upland catchment in northeastern Scotland by comparing downstream changes in dissolved organic carbon with spatial changes in the soil carbon pool and concluded that the linkage between stream water DOC concentrations and the soil carbon pool in the upper 1.5 km of the stream are likely to be driven by temperature-related DOC production in near-surface peatlands.

9.3.3 Methane Cycle

Methane is produced by anaerobic decomposition of the organic matter (Cao et al. 1996) in the catotelm by the process called methanogenesis. Plants stimulate methanogenesis by releasing "young" organic material as root exudates in the catotelm. This fresh carbon is easily decomposable by methanogenic bacteria forming methane, which is released in the atmosphere through three processes: diffusion, ebullition, and plant transport. Diffusion of CH_4 occurs throughout the peat column when this substrate is not oxidized by methanotrophic bacteria. Ebullition of CH_4 happens via bubbles that are released to the atmosphere from water-saturated peat (Rosenberry et al. 2003) whereas vascular plants transport CH_4 to the atmosphere via the aerenchymatic tissues from the roots that are in the catotelm (Thomas et al. 1996). These components of the CH_4 cycle (presented in Fig. 9.1) are controlled by temperature (Dunfield et al. 1993), water table position (Moore and Dalva 1993), peat chemistry (Svensson and Sundh 1992), and plant community factors such as bryophyte distribution (Bubier et al. 1995). Peatlands contribute almost 3 to 7% of the global annual emission of greenhouse gas methane.

9.4 Diversity of Organisms Colonizing Peatlands

The microflora in peatland soil consists of prokaryotes (bacteria and Archaea) and fungi (saprophytes and mycorrhizae). A soil fauna comprising nematodes, mites, colembola, and other insects is also present. All of these organisms play important roles in carbon cycling and interact with plants via exchange of organic and inorganic compounds. They also alter the physicochemical environment of the interacting organisms.

Direct counts of bacteria using epifluorescence microscopy and radioisotopic measurements of microbial degradative processes indicated bacterial densities of about 10^8 cells ml^{-1} of peatland water, irrespective of depth. Radioisotopic most-probable-number (MPN) counts of heterotrophs able to mineralize ^{14}C-labeled substrates to $^{14}CO_2$ showed significant populations of glucose degraders (10^4–10^6 cells ml^{-1}) as well as degraders of benzoate (10^2–10^3 cells ml^{-1}), 2,4-dichlorophenoxy acetate (10^2–10^5 cells ml^{-1}) and *Sphagnum* material (10^3–10^7 cells ml^{-1}) in the various peatlands examined. The MPN counts of NO_3^- reducers varied from 10^3–10^6 cells ml^{-1}, SO_4^- reducers from 10^2–10^3 cells ml^{-1}, methanogenic bacteria from 10^3–10^6 cells ml^{-1}, and methane oxidizers (methanotrophs) from 10^3–10^4 cells ml^{-1}, depending on sampling site and depth. Many pure cultures of aerobic bacteria and fungi have been isolated from various peatlands. Most of them can grow on organic compounds (carbohydrates, aromatic molecules, hydrocarbons, etc.) as sources of carbon. Some *Bacillus* bacteria are able to fix atmospheric N_2.

9.4.1 Insects

Peatlands contain distinctive insects in addition to widely distributed generalists, including species restricted to bogs (tyrphobionts) and species characteristic of bogs but not confined to them (tyrphophiles). Bogs raised above the water table form characteristic habitat islands in southern boreal and temperate forest zones. The historical development and nature of individual bogs are reflected by differences among their insects, which are of great biogeographical and ecological interest. The environmental sensitivity of bogs also makes insects valuable as bioindicators. Some of the important peatland insects are the water beetles (whirling beetles and great diving beetles), water bugs (pond skaters), water boatmans, and water scorpions. They usually feed upon small flying insects and small animals.

9.4.2 Fungi

9.4.2.1 Mushrooms and Ectomycorrhizae

Mushrooms typically seen in peatlands are dominated by saprotrophs and species of *Galerina* are particularly common. Another common species is the Yellowleg Bonnet, *Mycena epipterygia*. It is, however, very easy to identify as it has a thin jellylike layer on the cap which can be peeled off. Waxcaps are sometimes called the orchids of the fungal world and some small bright red species can be found in bogs. These are likely to be the Vermilion Waxcap, *Hygrocybe miniata*, and the Goblet Waxcap, *Hygrocybe cantharellus* or *Hygrocybe helobia*. Other fungi that can be found are species in the groups *Stropharia* (often on dung) and *Hypholoma*. There are the oddities such as the Bog Beacon, *Mitrula paludosa* and *Sarcoleotia turficola,* which demonstrate the variety of forms of fungi. Any lone birch tree or clumps of birch around the bog edge will host a different suite of fungi with leaf litter decayers, species such as the Birch Polypore, *Piptoporus betulinus,* or the Birch Woodwart, *Hypoxylon multiforme* on the trees, or various ectomycorrhizal fungi around their base. The latter group help the tree to take up nutrients from the soil and protect it against soil pathogens in exchange for carbon that the tree produces by photosynthesis. Typical ectomycorrhizal fungi include the Tawny Grisette, *Amanita fulva,* the Birch Milkcap, *Lactarius tabidus,* the Ghost Bolete, *Leccinum holopus,* the Birch Brittlegill, *Russula betularum,* and the Yellow Swamp Brittlegill, *Russula claroflava.*

Some mycorrhizal type associations have evolved to help the woody plants to survive in low nutrient conditions, where shrubs and conifers co-exist in an organic matrix dominated by *Sphagnum* mosses. Ectomycorrhizal fungi produce an elaborate hyphal network, thus vastly expanding the root exploration zone (Fig. 9.2).

Fig. 9.2 Extensive network of mycorrhizal hyphae radiating from roots of a larch (*larix*) seedling grown in peat. (Used with permission http://www.biology.ed.ac.uk/research/groups/jdeacon/mrhizas/ecbmycor.htm#top)

9.4.2.2 Orchid Mycorrhizae

Orchid mycorrhizae are a symbiotic relationship between the roots of plants of the family Orchidaceae and a variety of fungi. All orchids are mycoheterotrophic at some point in their life cycle. Orchid mycorrhiza activity is critically important during orchid germination, as an orchid seed has virtually no energy reserve and obtains its carbon from the fungal symbiont. Many adult orchids retain their fungal symbionts, although the respective benefits to the adult photosynthetic orchid and the fungus remain largely unexplored (Fig. 9.3). The fungi that form orchid mycorrhiza are typically basidiomycetes. These fungi come from a range of taxa including *Ceratobasidium* (*Rhizoctonia*), *Sebacina, Tulasnella*, and *Russula*. Some orchids associate with saprotrophic or pathogenic fungi, whereas other orchids associate with ectomycorrhizal fungal species. These latter associations are often called tripartite associations as they involve the orchid, the ectomycorrhizal fungus, and the ectomycorrhizal host plant.

9.4.2.3 Ericoid Mycorrhizae

Ericoid mycorrhizae are a symbiotic relationship between fungi and the roots of plants from the order Ericales. Ericoid mycorrhiza is considered crucial for the success of the family Ericaceae in a variety of edaphically stressful environments worldwide. Ericaceous plants commonly co-occur in soils with leguminous or

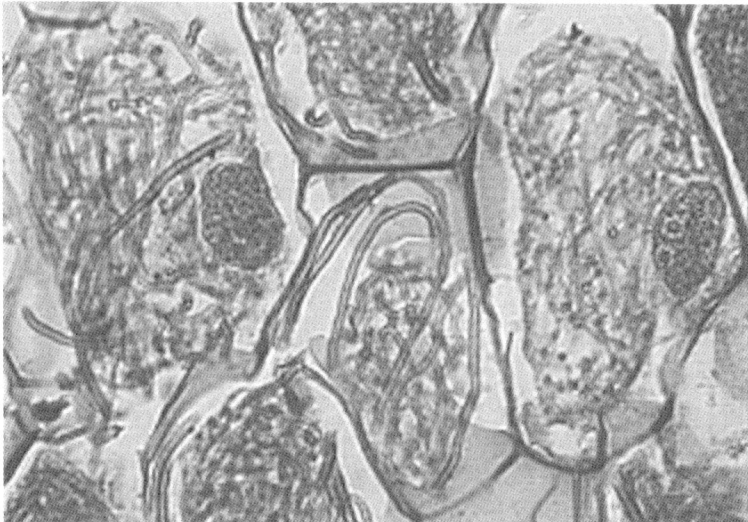

Fig. 9.3 Orchid mycorrhiza, showing coils of fungal hyphae in cells of the protocorm. (Used with permission: http://www.biology.ed.ac.uk/research/groups/jdeacon/mrhizas/ecbmycor.htm#top)

carnivorous plants, further highlighting the low nutrient status of these soils. Ericoid mycorrhizal fungi enable their host plants to obtain nutrients in these poor soils. Ericoid mycorrhizas are formed between ascomycetous, and more rarely hyphomycetous fungi, and species of the Ericaceae and Epacridaceae. Members of the Ericaceae exhibit three types of mycorrhiza: ericoid, arbutoid, and monotropoid. The ericoid type is the most important and is found in some genera as *Calluna, Erica, Rhododendron*, and *Vaccinium*.

Ericoid mycorrhizas have evolved in association with plants that often dominate some of the most climatically and edaphically stressed environments. These plants characteristically become dominant when levels of acidity become extreme. Under these very low pH conditions, mineralization of nutrients is inhibited and metallic elements show maximum solubility. There is evidence that species of *Calluna* and *Vaccinium* have a very high constitutional tolerance to some stresses. Usually, the formation of ericoid mycorrhizae increases this tolerance and there is good reason to believe that their presence is essential for survival of the plant partner (Bhatnagar and Varma 2006).

Generally, fungi forming ericoid mycorrhizae are from order Helotiales of the ascomycetes. These fungi include *Rhizoscyphus ericae, Hymenoscyphus ericae, Oidiodendron maius, Myxotrichum arcticum, Phialocephala fortinii*, and *Leptodontidium orchidicola*.

9.4.3 Methanogens

Methanogens are Archaea (Woese 1981) that occupy narrow ecological niches in peatland and other environments (see Chapter 7 for a discussion of methanogens in permafrost). They mediate the formation of methane from simple substrates (e.g., H_2–CO_2, formate, methanol, and acetate) in highly reducing and anaerobic environments (Garcia et al. 2000). They have the most stringent requirements of all anaerobes for the absence of oxygen (<2 ppm) and for growth they require a redox potential of less than −330 mV.

Peatlands are wetland ecosystems where productivity exceeds biodegradation and except for the surface water layer, peats are anaerobic environments. As such, they represent suitable habitats for methanogenic bacteria. Williams and Crawford (1984) found a high population of 10^6 methanogen cells ml^{-1} of peatland interstitial water. The methanogenic food chain is a microbial system that mediates the biodegradation of organic compounds in many anaerobic environments. The carbon flow through this chain avoids a buildup of inhibitory metabolic end products. Because methanogenesis is the terminal step in this anaerobic food chain, any perturbation of the chain should be reflected by altered methane production. Consequently, methanogenesis is a key process to study, reflecting the combined activities of many different microbial groups. Over 65 species of methanogens belonging to 20 different genera are known today (Sowers, 1995).

In peatlands, H_2 and CO_2-dependent methanogenesis is thought to be the main pathway for CH_4 production, but in some minerotrophic peatland (fens) acetoclastic methanogenesis is often predominant in upper peat layers (Popp and Chanton, 1999). The diversity of methanogenic communities of fen and bogs (Basiliko et al. 2003) has recently been described, but data on methanogenic pathways and methanogen populations are scarce. Galand et al. (2005) studied the methanogenic pathways in three peatland systems (i.e., mesotrophic fens, oligotrophic fens, and ombrotrophic bogs) and found that mesotrophic fens harbour the lowest production rates of CH_4 from H_2 and CO_2. Because H_2–CO_2 and acetate are the main precursors of CH_4, the fraction of CH_4 not produced by H_2–CO_2 is predominantly the result of acetoclastic methanogenesis.

Methanogenesis occurs in three steps. Decomposition of organic matter in anaerobic environments first occurs by hydrolysis, performed by bacteria. This is followed by fermentation, performed by bacteria and some Archaea, and methanogenesis, performed exclusively by members of the domain Archaea. Methanogens produce methane by the reduction of carbon dioxide using hydrogen as an electron donor, or through the cleaving of acetate into methane and carbon dioxide (Luton et al. 2002). Sauer and Thauer (2000) showed that methyl-coenzyme M (2-methylthioethane sulfonate) is the key intermediate of methane formation in a methanogenic Archaea (*Methanosarcina barkeri*). It is generated from coenzyme M (*B*-2-mercaptoethane sulfonate) in methyl transfer reactions catalyzed by proteins containing zinc.

The methanogens of peatland ecosystems mainly belong to five orders. Members of the Methanomicrobiales are found in bog peatlands only, whereas Methanosarcinales, Methanococcales, and Methanopyrales are found in fen peatland. Methanobacteriales are generally found in bogs, but some genera of Methanobacteriales occur in fens. *Methanogenium organophilium* and *Methanospirillum hungatei* are the most frequently isolated genera of Methanomicrobiales whereas *Methanobacterium bryantii* is a common member of the Methanobacteriales. *Methanosarcina mazei*, *Methanosarcina barkeri*, *Methanosarcina sicilliae*, *Methanosarcina acetivorans*, and *Methanosaeta* spp. are the important organisms identified from the order Methanosarcinales, whereas *Methanococcus jannaschii* and *Methanococcus infernus* are peatland colonizers of the order Methanococcales. The order Methanopyrales comprises *Methanopyrus kandleri*. Although all these organisms are involved in methanogenesis, their mode of action differs. The Methanomicrobiales and Methanobacteriales probably account for hydrogenotrophic methanogenesis and Methanosarcinales for acetoclastic methanogenesis. A pathway for the generation of methane by a representative of the Methanomicrobiales is presented in Fig. 9.4.

9.4.3.1 Acidophilic Methanogens

Until recently, it was widely accepted that methanogens require neutral to slightly alkaline pH conditions (pH 6.8 to 8.5) for optimum growth and methane production.

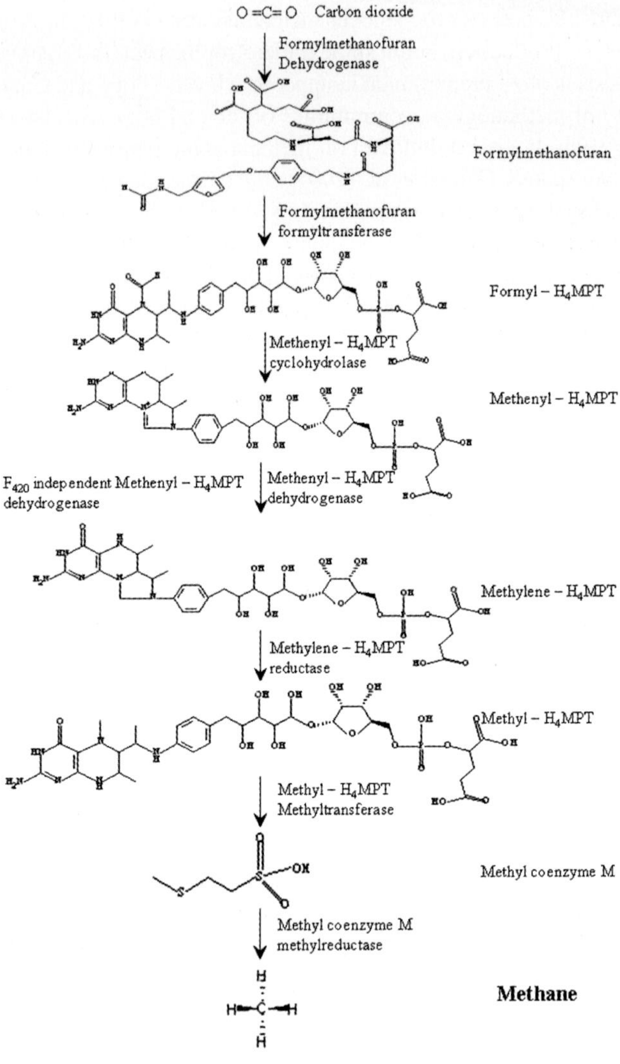

Fig. 9.4 Flow diagram of methane generation from carbon dioxide and hydrogen by *Methanobacterium thermoautotrophicum*. In an initial reaction of the pathway, formylmethanofuran dehydrogenase catalyses the reductive conversion of carbon dioxide and methanofuran to formylmethanofuran, with the required electrons being provided by molecular dihydrogen. (Used with permission: http://umbbd.msi.umn.edu/meth/meth_image_map.html; Ellis et al. 2006)

Attempts were made to isolate acid-tolerant methanogens and a methanogen identifed as *Methanobacterium uliginosum* has been reported to grow at pH 6.0 to 8.0 (König 1984). Also, *Methanobacterium espanolae* was found to grow at pH 5.5 to 6.2, which can be considered as an example of a truly acidophilic methanogen.

Certain strains of *Methanosarcina* have been shown to grow at low pH, using methanol and H_2 as the substrate (Maestrojuan and Boone 1991).

9.4.3.2 Psychrophilic Methanogens

The Archaea that are most readily isolated from naturally cold environments and that are most amenable to laboratory cultivation are methanogens. Isolates have come from an Antarctic lake (Franzmann et al. 1997), a freshwater lake in Switzerland (Simankova et al. 2001), and cold marine sediment in Alaska (Chong et al. 2002) and the Baltic Sea (Singh et al. 2005). In addition, various *Methanosarcina*, *Methanospirillum*, *Methanocorpusculum*, and *Methanomethylovorans* species have been isolated from a Boreal fen, tundra, a polluted pond in Russia, lake sediment in Switzerland, and manure digested at low temperature in a laboratory. *Methanococcoides burtonii* and *Methanogenium frigidum* were isolated from Ace Lake, Antarctica (Franzmann et al. 1997). *M. frigidum* uses H_2 and CO_2 for growth and is nonmotile; *M. burtonii* uses methyl substrates (trimethylamine and methanol) and is flagellated and motile. As *M. burtonii* is a methylotrophic methanogen, it does not compete with hydrogen-utilizing, sulfate-reducing bacteria in the environment. *M. burtonii* is cosmopolitan in cold environments, with a close relative (*Methanococcoides alaskense*; 99.8% 16S rRNA identity) isolated from Skan Bay, Alaska (Singh et al. 2005) and closely related strains identified from an Antarctic bay (Purdy et al. 2003; 99% 16S rRNA identity) and deep-sea sediment (Li et al. 1999; 98.8% 16S rRNA identity).

Although no individual micro-organism can grow at both the freezing and boiling points of water, methanogens are thermally diverse and there are species that can grow at 110°C (*Methanopyrus kandleri*), 0°C (*M. burtonii*), and all temperatures in between (Saunders et al. 2003). This illustrates that methanogenesis and the main energy and biosynthetic pathways are not restricted by growth temperature; that is, novel pathways and cellular processes are not essential for cold (or heat) adaptation.

9.4.3.3 Factors Affecting Methanogenesis

Methanogenesis is highly dependent on environmental conditions such as temperature, water table, content of organic matter, pH, and so on (Segers 1998). Temperature is known to exert a significant influence on methanogenesis and the number of methanogens in lake sediments. In culture, methanogenic bacteria metabolize best in the pH range of 6.7 to 8.0. However, very low rates of methanogenesis have been observed at pH 5.8 (Dedlysh et al. 1998; see also above, this section). Methanogens utilize sulfide or cysteine to satisfy their sulfur requirements; however, high sulfide concentrations have been shown to inhibit methanogenesis in sediments (Khan et al. 1979). Phosphate also has been found to inhibit methanogenesis in lake sediments, as has nitrate (Balderston and Payne, 1976) in salt marsh sediments. Ammonium ion may slow the conversion of acetic acid to methane at high concentrations (Wolfe and Higgins, 1979). Methanogenic substrates such as hydrogen (Winfrey et al. 1977) and acetate (Cappenberg, 1974) often stimulate methanogenesis in lake sediments. Glucose stimulated methanogenesis in some habitats (Winfrey et al. 1977), however, at glucose concentrations greater than $10\,g\,l^{-1}$, suppression of methane production occurred. Vitamins and yeast extract (Wolfe and Higgins, 1979) also have been observed to stimulate methanogenesis. Finally, sulfate has been found to inhibit methanogenesis in sediments (Winfrey et al. 1977).

9.4.4 Methanotrophs

Aerobic methanotrophic bacteria using methane as the carbon and energy source are an integral part of many natural peatland ecosystems where methane is produced and consumed. Methanotrophs maintain a balance of atmospheric methane and play a crucial role in the global methane cycle. They oxidize methane through methanol and formaldehyde to carbon dioxide and incorporate carbon into biomass at the level of formaldehyde. They typically occur at the aerobic/anaerobic interface of wet environments, for example, lake sediments, rice paddies, tundra soils, and bogs (Sundh et al. 1994; Whalen et al. 1996). The methanotrophs use methane diffusing from the anaerobic zone and often operate as a biofilter for methane. In this way they reduce and control its potentially hazardous emission to the atmosphere. Methane is one of the most effective greenhouse gases and contributes to as much as 20% of total global warming.

Methanotrophs have been divided into three major categories: type I, type II, and type X, which differ in phylogeny, chemotaxonomy, internal membrane ultrastructure, carbon assimilation pathways, and some other biochemical features (Bowman, 2006). Currently, type-I methanotrophs, that belong to the family Methylococcaceae of the Gammaproteobacteria, include six validated genera, that is, *Methylomonas, Methylobacter, Methylomicrobium, Methylosphaera, Methylosarcina,* and *Methylohalobius*. They possess lamellar stacks of intracytoplasmic membranes (ICM), assimilate formaldehyde produced from the oxidation of methane or methanol via the ribulose monophosphate (RuMP) pathway, and have DNA G+C content ranging from 45 to 55 mol%.

In contrast, type-II methanotrophs belong to the family Methylocystaceae of the Alphaproteobacteria and include four validated genera: *Methylosinus, Methylocystis, Methylocella,* and *Methylocapsa*. They generally have ICM located on the cell periphery, a higher level of G+C content in their DNA, and assimilate formaldehyde via the serine pathway.

The intermediate type-X methanotrophs belong to the Gammaproteobacteria and include *Methylococcus* and *Methylocaldum* genera, thermophilic and thermotolerant representatives of which possess ICM similar to those of type-I methanotrophs, but have a DNA G+C content of 55 to 65 mol%. They express the RuMP cycle for formaldehyde assimilation and enzymes of the serine pathway and the ribulose bisphosphate (RuBP) cycle. The traditionally recognized type-X category has now been incorporated into type I (Wise et al. 1999).

9.4.4.1 Detection of Methanotrophs

Many studies were conducted on the diversity and ecology of methane-oxidizing bacteria (MOB). Cultivation-based approaches have been shown to reflect a highly distorted picture of the original in situ community structure. Cultivation-independent, molecular approaches (mostly by cloning and sequencing, denaturing gradient gel electrophoresis (DGGE), and/or terminal restriction fragment length polymorphisms (T-RFLP)) have been based on the 16S rRNA gene, or on functional genes for the methane mono-oxygenases. Two types of such genes are known: *pmoA*, encoding the particulate methane mono-oxygenase (Lukow et al. 2000) and *mmoX*, encoding the soluble methane mono-oxygenase (Auman et al. 2000; see also below, this section). Yet another possibility is *mxaF* (McDonald and Murrell, 1997), encoding the methanol dehydrogenase, the next key enzyme in the pathway of methane metabolism. The most used marker genes are the *pmoA* and the 16S rRNA gene.

The use of *pmoA* enables us to focus on MOB and functionally related bacteria as well as to detect and identify thus far uncultured MOB. The sequence phylogeny of *pmoA* corresponds closely to that of 16S rRNA, and *pmoA* has therefore been extensively used for the cultivation-independent identification of MOB from diverse environments. The sequence of *pmoA*, encoding the 27-kDa subunit of pMMO, reflects evolutionary relationships amongst *pmoA* containing bacteria (Nguyen et al. 1998).

Studies on the diversity of MOB and the ecology of microbial methane oxidation were performed in aqueous environments (fresh, brackish, marine and underground hot spring water, and aquifers), sediments (marine and freshwater), mud, various soils (rice field, forest, arable, pasture, peat bog, landfill cover; arctic to tropical), deep-sea smokers, and shallow-sea methane seeps as well as the tissues of marine invertebrates. Soil is by far the most studied environment with respect to MOB ecology (Bourne et al. 2001). Studies have revealed a high diversity of MOB but the factors influencing competition between different MOB, determining their ecological niches and maintaining the high diversity observed, are largely unknown, except for atmospheric methane oxidation and competition between type-I and type-II MOB. The main factors in the competition between mesophilic neutrophilic type-I and type-II MOB (co-existing in many environments) are methane and oxygen partial pressures and the availability of fixed nitrogen. Type-I MOB appear to outcompete type-II strains under oxygen-rich, methane-limited conditions. Most type-II MOB and some type-I (particularly *Methylomonas*, *Methylobacter*, and *Methylococcus*) strains are capable of nitrogen fixation, thus having a competitive advantage under nitrogen-limited conditions. Some type-I MOB form desiccation-resistant or -sensitive cysts. Type-II MOB forms exospores conferring some resistance against fluctuations in nutrient supply.

9.4.4.2 Psychrotrophic Methanotrophs

Aerobic methanotrophic bacteria typically occur at the aerobic/anaerobic interface of wet environments, such as lake sediments, rice paddies, in particular land, tundra soils, and bogs (Sundh et al. 1994; Whalen et al. 1996). Tundra soils contain 15% of the biosphere's total organic carbon. Thus, the response of this large carbon reservoir to global climate warming could be important. There is increasing evidence that bacterial oxidation of methane is a vital regulator of CH_4 emission. Field and laboratory experiments on methane consumption revealed that tundra soils of Unalaska Island and the Aleutian Islands consumed methane at soil temperatures of 7°C and at CH_4 concentrations ranging from below to well above ambient levels (Whalen and Reeburgh 1990). These experiments also demonstrated that soil oxidation of atmospheric CH_4 (at a concentration of 1.7 ppm) was microbe-mediated and indicated the presence of a soil population capable of oxidizing CH_4 at concentrations tenfold lower than the ambient atmospheric concentrations.

Because the kinetic properties of methanotrophs are not consistent with their growth under atmospheric CH_4 concentration, the occurrence of any novel methanotroph having a cold active microbial methane oxidase with a high affinity for CH_4 would be of great interest. Scanning electron microscopy of tundra bog soils revealed the presence of bacterial microcolonies on organic residues sampled from the *Carex* and *Sphagnum* layers. Bacteria absorbed on the particles gave positive reactions with fluorescent antibodies targeting *Methylocystis parvus*, *Methylocystis trichosporium*, *Methylobacter bovis*, and *Methylobacter capsulatus* (Trotsenko and Khmelenina 2005; Vecherskaya et al. 1993). In this study, the number of methanotrophs in peat layers ranged from 0.1 to 22.9×10^6 cells g^{-1} of soil, thus comprising from 1 to 23% of the total bacterial population. The representatives of *Methylomonas*, *Methylobacter*, *Methylococcus*, *Methylosinus*, and *Methylocystis* genera occurred simultaneously in all soil samples. Type-I methanotrophs were more abundant than type II and members of the genus *Methylobacter* were predominant.

Micro-organisms residing in tundra soils face low temperatures, long periods in the frozen state, high water content due to poor drainage at times causing anaerobiosis, acidic pH, and a narrow range of nutrients. Such extreme conditions might favor the existence of multistress-resistant methanotrophs, which differ from those in the temperate soil community. A pure culture of a psychrophilic methanotroph that grows between 3.5 and 20°C and optimally at 5 and 10°C was first isolated from tundra soil and identified as a new species, *Methylobacter psychrophilus* (Tourova et al. 1999).

9.4.4.3 Acidophilic Methanotrophs

A novel acidophilic methanotroph was described, *Methylocella palustris*. Strains of *M. palustris* were isolated from acidic *Sphagnum* peat bogs (Dedlysh et al. 1998) and classified as type-II methanotrophic bacteria. However, phylogenetically these acidophilic strains were only moderately related to the known type-II methanotrophs and were more closely affiliated with the heterotrophic bacterium *Beijerinckia indica* subsp. *indica*. The isolation of *M. palustris* from *Sphagnum* bogs of different geographical locations (four different sites in west Siberia and European north Russia) suggests that these bacteria might be widely distributed in acidic wetlands of the northern hemisphere. However, information on the distribution and abundance of *Methylocella* in northern wetlands is still lacking. As the result of their profound distinctness from other known methanotrophs, these organisms have not been targeted by the culture-independent 16S ribosomal DNA (rDNA)-based molecular approaches developed for detection of type-I and type-II methanotrophic bacteria (Costello and Lidstrom 1999).

9.4.4.4 Methane Oxidation

Methane oxidation is divided into "low affinity" and "high affinity" categories. Typically, methanotrophs in culture oxidize high concentrations of methane (low-affinity methanotrophs), whereas in situ methane oxidation kinetics measurements indicate that there are high-affinity methanotrophs in soils. The distinction between these two populations is unclear because extant methanotrophs can also exhibit properties of high-affinity methanotrophs under certain conditions. The oxidation of methane at atmospheric levels (high affinity methane oxidation), mostly found in upland soils, appears to be mainly associated with the predominance of two uncultured MOB based on molecular analyses (Holmes et al. 1995; Pedersen 1996).

The bacterial methane oxidation pathway is catalyzed by one of the two types of the enzyme methane monooxygenase (MMO). The soluble cytoplasmic MMO (sMMO) is found in only some of these bacteria whereas the particulate membrane-bound MMO (pMMO) is present in virtually all known MOB except, perhaps, the *Methylocella* species. The pMMO consists of three membrane-associated polypeptides encoded by *pmoC*, *pmoA*, and *pmoB* (Pacheco-Oliver et al. 2002). In addition to pMMO, most type-II (*Methylosinus*, *Methylocystis*) and some type-I methanotrophs (*Methylomonas*, *Methylomicrobium*, and *Methylococcus*) possess sMMO. The enzyme from the strains belonging to *Methylosinus*, *Methylocystis*, and *Methylococcus* has been thoroughly studied and the nucleotide sequence of the sMMO gene cluster *mmoX*, *mmoY*, *mmoB*, *mmoZ*, *mmoC*, and *mmoD* appears to be highly conserved. The *pmoA* gene, encoding a 26-kDa subunit that harbors the active site of the pMMO, and the *mmoX* gene, coding for the α subunit of the sMMO hydroxylase component, can be used as appropriate gene markers for the occurrence of the enzymes in various methanotrophs.

Methanol dehydrogenase (MDH), the second enzyme involved in methane oxidation, is present in all gram-negative methylotrophs including methane and methanol utilizers. This enzyme oxidizes the methanol produced from the oxidation of methane by methane monooxygenase (MMO). The *mxaF* gene, encoding the large subunit of the enzyme, is an appropriate indicator gene for occurrence of methylotrophs in the environment (McDonald and Murrell, 1997). MDH is a pyrroloquinoline quinone-linked enzyme that carries out a key step in bacterial one-carbon (C1) metabolism because it catalyzes the oxidation of methanol to formaldehyde, the intermediate of both assimilative and dissimilative metabolism in methylotrophs. It is distinct from the alcohol dehydrogenase of gram-positive methylotrophic bacteria (de Vries et al. 1992) and methylotrophic yeasts (Williamson and Paquin, 1987).

Analysis of the predicted amino acid sequences corresponding to various structural genes for MDH, or *mxaF* genes, carried by various methanotrophs and methylotrophs revealed strong sequence conservation (McDonald and Murrell, 1997). Of the 172 amino acid residues, 47% were conserved among all 22 sequences obtained in this study. Phylogenetic analysis of these *mxaF* sequences showed that those from type-I and type-II methanotrophs form two distinct clusters and are separate from *mxaF* sequences of other gram-negative methylotrophs. Sequences of *mxaF* retrieved by PCR from DNA isolated from a blanket bog peat core sample formed a distinct phylogenetic cluster within the *mxaF* sequences of type-II methanotrophs and may originate from a novel group of acidophilic methanotrophs.

9.4.4.5 Factors Affecting Methanotrophs

The methanotrophic bacteria are affected by environmental factors such as pH, temperature, salinity, and water table level. The activity increases as aeration of the acrotelm is enhanced after water level draws down. This affects CH_4 emissions, which are lower as CH_4 consumption is higher (Svensson and Sundh 1992). In some sites, a net consumption of CH_4 has been found (Martikainen et al. 1995). This situation may occur for peatlands of temperate regions, but not necessarily for peatlands of the boreal and subarctic regions. With the higher temperatures caused by global warming, these peatland types are expected to release more CH_4 because methanogenic bacteria will be favored by the melting of the permafrost. In that sense, CH_4 emission rates will be different according to the geographical location of peatlands (Moore et al. 1998). Ambient pH from 4.3 to 5.9 favors the growth of methanotrophs. The MOB are inhibited by acetylene, which acts as a suicidal substrate, being transformed by both forms of MMO into active intermediates that in turn readily react with cell compounds containing hydroxyl and amino groups.

9.4.5 Antagonistic Bacteria

Because of their antimicrobial activity, *Sphagnum* plants were used as a natural medicine in the old Indian and Maya cultures, and as wound dressing during the First and Second World Wars (Frahm, 2001). Although they are colonized by diverse bryophilous ascomycetes, no substantial fungal diseases of these plants are known (Döbbeler 1997).

The screening of 493 bacterial isolates for antagonistic activity against fungal pathogens resulted in 237 (48%) active isolates (Opelt et al. 2007). The majority of the antagonists belonged to the genera *Serratia* (15%), *Burkholderia* (13.5%), *Staphylococcus* (13.5%), and *Pseudomonas* (10%). Interestingly, a high proportion of antagonists, for example, *Staphylococcus*, *Hafnia*, *Yersinia*, and *Pantoea*, were identified as strains that are known as facultative pathogens of humans. Thus, *Sphagnum* plants represent an ecological niche not only for diverse and extraordinary microbial populations with a high potential for biological control of plant pathogens, but also for opportunistic human pathogens.

9.4.6 Epiphytic Bacterial Communities

Associations between algae and bacteria are commonly observed in peatlands. Culture and microscopy studies have documented a number of bacterial–algal interactions. Much attention has been focused on the release of dissolved organic carbon by algal cells and its support of bacterial growth, and the surfaces of living cells may also provide microenvironmental conditions favorable for bacterial processes that otherwise could not occur under ambient water conditions (Paerl and Pinckney, 1996). Heavy bacterial colonization of algae is generally considered a sign of algal senescence, but colonization of young, active algal cells or colonies is also observed (Rosowski and Langenberg 1994) and benefits to algae of such associations have been frequently reported (Keshtacher-Liebson et al. 1995). Bacteria and algae may also compete for inorganic nutrients, and many algal taxa produce compounds that are potentially inhibitory to bacterial growth (Kellam and Walker, 1989).

Epiphytic bacterial communities within the sheath material of three filamentous green algae, *Desmidium grevillii*, *Hyalotheca dissiliens*, and *Spondylosium pulchrum* (class Charophyceae, order Zygnematales), collected from a *Sphagnum* bog, were characterized by PCR amplification, cloning, and sequencing of 16S ribosomal DNA (Fisher et al. 1998). By phylogenetic analysis, the cloned sequences were placed into several major lineages of the bacterial domain: the Bacteroidetes phylum and the Alpha-, Beta-, and Gammaproteobacteria (Fisher et al. 1998).

The representatives of the Bacteroidetes phylum mainly consisted of *Sphingobacterium heparinum*, *S. thalpophilum*, *Flavobacterium mizutaii*, *Flavobacterium ferrugineum*, *Saprospira*, *Cytophaga arvensicola*, *Flexibacter filiformis*, and *Flexibacter sancti*. Members of the Alphaproteobacteria included *Rhodospirillum rubrum*, *Rhodospirillum fulvum*, and *Azospirillum lipoferum*. Some other identified species in the Alphaproteobacteria were *Methylobacterium*, *Beijerinckia indica*, *Rhodopseudomonas acidophila*, and *Methylosinus*. Purple phototrophic bacteria belonging to the Betaproteobacteria were also observed and included *Rubrivivax gelatinosus*, *Comamonas testosteronii*, *Rhodoferax fermentans*, and *Variovorax paradoxus*. *Hydrogenoluteola thermophilus* and *Rhodocyclus*, also of the Betaproteobacteria, were isolated. *Acinetobacter lwoffii* was the only epiphytic species belonging to the Gammaproteobacteria.

The bryophytes, particularly the mosses, are a diverse group of land plants that usually colonize habitats with moist or extremely variable conditions. One of their most important features is their life cycle, which involves alternation between a diploid sporophyte and a dominant, free-living haploid gametophyte generation. Very little is known about the interaction of bryophytes with bacteria. Bacterium–host interactions can be symbiotic, commensal, or pathogenic. Therefore, bacteria associated with three bryophyte species, *Tortula ruralis*, *Aulacomnium palustre*, and *Sphagnum rubellum*, which represent typical moss species of peatlands at the southern Baltic Sea coast in Germany, was analyzed using culture-dependent and culture-independent techniques (Opelt and Berg, 2004).

A high proportion of uncultured or unidentified eubacteria were found from the surface of the mosses. *Photorhabdus luminescens, Collimonas fungivorans*, and diverse *Pseudomonas* spp. were identified. DNA analysis assigned two dominant bands obtained from *Sphagnum* to *Pseudomonas grimontii* and *Methylobacterium mesophilicum*. Species of *Acetobacter*, *Frateuria*, and *Acidocella*, bacterial genera known for their occurrence in acidic environments, were also found for *Sphagnum*. *Tortula*-specific bands were identified as *Pseudomonas aeruginosa*, *Rhodococcus erythropolis*, and *Acidovorax wohlfahrtii*.

9.4.7 Microbial Use of Nitrogen and Phosphorus

Peatlands contain up to 30% of the total organic nitrogen reserve of the world's soils, and thus have a potential to exert a significant influence on the global atmospheric budget of nitrous oxide (N_2O; Martikainen et al. 1993). Microbial N_2O production, primarily through nitrification and denitrification processes, may account for a substantial fraction of the total N loss from the peatlands (Jacks et al. 1994), either directly as N_2O or, after further reduction, as dinitrogen (N_2) (Allen et al. 1996). Large losses of N in gaseous forms are commonly related to waterlogged conditions combined with high N mineralization rates. Thus, N loss in the gaseous forms may be a significant mechanism for the removal of N from peatlands.

Denitrification is the most important N_2O-producing process in waterlogged peat soils (Velthof and Oenema, 1995), where oxygen is limited whereas nitrate and carbon are available for micro-organisms (Regina et al. 1996). Denitrification is an anaerobic microbiological process in which carbon serves as the energy source and nitrate as the electron acceptor, and mainly involves the bacterial genera *Pseudomonas, Micrococcus, Bacillus*, and *Thiobacillus*. In the process, nitrate is reduced to the gaseous nitrogen compounds N_2O and N_2. N_2O can also be formed during nitrification, that is, oxidation of NH_4^+ (Koops et al. 1997). A proportion of 50% of water-filled pore space is the optimum for N_2O production in nitrification and up to 35% of the total gaseous nitrogen loss from agricultural land is through N_2O production by nitrification. NO_2 is an intermediate product in both nitrification and denitrification processes. The NO_2 resulting from nitrification activity can in turn be reduced to N_2O and N_2 and the term "nitrifier denitrification" has been coined to account for this particular mode of nitrogen loss from soils (Koops et al. 1997).

9.5 Conclusions

Peatlands are important and unique systems for biogeochemical cycling of carbon, nitrogen, and other chemical elements. Because of the prevailing conditions in these waterlogged, mineral-poor, and biomass conservation-prone environments, metabolic processes occurring in peatlands represent particular variations of mechanisms occurring in soils and water. As such, peatlands provide a specific contribution to the maintenance and changes in planetary biological systems. Global atmospheric and climatic changes will be influenced by peatland ecology. The future of peatlands bears direct relevance to the establishment of biospheric equilibria and, in this sense, the understanding of interactive processes occurring in these wetland systems, as well as measures for their preservation and rational use, bears direct relevance to humanity's future and the form this future will take.

Acknowledgments The authors are thankful to CSIR, New Delhi for partial financial support.

References

Allen AG, Jarvis SC, Headon DM (1996) Nitrous oxide emissions from soils due to inputs of nitrogen from excreta return by livestock on grazed grassland in the U.K. *Soil Biol Biochem* 28:597–607

Anderson RL, Foster DR, Motzkin G (2003) Integrating lateral expansion into models of peatland development in temperate New England. *J Ecol* 91:68–76

Auman AJ, Stolyar S, Costello AM, Lidstrom ME (2000) Molecular characterization of methanotrophic isolates from freshwater lake sediment. *Appl Environ Microbiol* 66:5259–5266

Balderston WL, Payne WJ (1976) Inhibition of methanogenesis in salt marsh sediments and whole-cell suspensions of methanogenic bacteria by nitrogen oxides. *Appl Environ Microbiol* 32:264–269

Basiliko N, Yavitt JB, Dees PM, Merkel SM (2003) Methane biogeochemistry and methanogen communities in two northern peatland ecosystems, New York State. *Geomicrobiol J* 20:563–577

Bhatnagar K and Varma A (2006) The healthy marriage between terrestrial orchids and fungi. *J Hill Res* 19:1–12

Billett MF, Deacon CM, Palmer SM, Dawson JJC, Hope D (2006) Connecting organic carbon in stream water and soils in a peatland catchment. *J Geophys Res 111*, G02010, doi:10.1029/2005JG000065

Bourne DG, McDonald IR, Murrell, JC (2001) Comparison of *pmoA* PCR primer sets as tools for investigating methanotroph diversity in three Danish soils. *Appl Environ Microbiol* 67:3802–3809

Bowman J (2006) The methanotrophs—The families Methylococcaceae and Methylocystaceae. In: Dworkin M, Falkow S, Rosenberg E, Schleifer K-H, Stackebrandt E (eds) *The Prokaryotes, a Handbook on the Biology of Bacteria*, 3rd edn, vol 5. Springer, Berlin, Heidelberg, New York, pp 266–289

Bubier JL, Moore TR, Juggins S (1995) Predicting methane emission from bryophyte distribution in northern Canadian peatlands. *Ecology* 76:677–693

Cao M, Marshall S, Gregson K (1996) Global carbon exchange and methane emissions from natural wetlands: application of a process-based model. *J Geophys Res* 101:14399–14414

Cappenberg TE (1974) Interrelations between sulfate-reducing and methane-producing bacteria in bottom deposits of a freshwater freshwater lake. I. Field observations. *Antonie van Leeuwenhoek* 40:285–295

Chong SC, Liu Y, Cummins M, Valentine DL, Boone DR (2002). *Methanogenium marinum* sp. nov., a H_2-using methanogen from Skan Bay, Alaska, and kinetics of H_2 utilization. *Antonie van Leeuwenhoek* 81:263–270

Clymo RS, Turunen J, Tolonen K (1998) Carbon accumulation in peatland. *Oikos* 81:368–388

Costello, AM, Lidstrom ME (1999) Molecular characterization of functional and phylogenetic genes from natural populations of methanotrophs in lake sediments. *Appl Environ Microbiol* 65:5066–5074

Dalva M, Moore TR (1991) Sources and sinks of dissolved organic carbon in a forested swamp catchment. *Biogeochemistry* 15:1–19

de Vries GE, Arfman N, Terpstra P, Dijkhuizen L (1992) Cloning, expression, and sequence analysis of the *Bacillus methanolicus* C1 methanol dehydrogenase gene. *J Bacteriol* 174:5346–5353

Dedlysh SN, Panikov NS, Tiedje JM (1998). Acidophilic methanotrophic communities from *Sphagnum* peat bogs. *Appl Environ Microbiol* 64:922–929

Döbbeler P (1997) Biodiversity of bryophilous ascomycetes. *Biodivers Conserv* 6:721–738

Dunfield P, Knowles R, Dumont R, Moore TR (1993) Methane production and consumption in temperate and subarctic peat soils: Response to temperature and pH. *Soil Biol and Biochem* 25:321–326

Ellis LBM, Roe D, Wackett LP (2006) The University of Minnesota biocatalysis/biodegradation database: The first decade. *Nucl Acids Res* 34:D517–D521

Faubert P (2004) The effect of long-term water level drawdown on the vegetation composition and CO_2 fluxes of a boreal peatland in central Finland. M.Sc. Thesis, Université Laval, Québec (http://www.theses.ulaval.ca/2004/21536/21536.html)

Fisher MM, Wilcox LW, Graham, LE (1998) Molecular characterization of epiphytic bacterial communities on Charophycean green algae. *Appl Environ Microbiol* 64:4384–4389

Frahm JP (2001) *Biologie der Moose*, 1st edn. Spektrum Verlag, Berlin, Germany

Franzmann PD et al. (1997) *Methanogenium frigidum* sp. nov., a psychrophilic, H_2-using methanogen from Ace Lake, Antarctica. *Int. J. Syst. Bacteriol* 47:1068–1072

Galand PE, Fritze H, Conrad R, Yrjälä K (2005) Pathways for methanogenesis and diversity of methanogenic Archaea in three boreal peatland ecosystems. *Appl Environ Microbiol* 71:2195–2198

Garcia J-L, Patel BKC, Ollivier B (2000) Taxonomic, phylogenetic, and ecological diversity of methanogenic archaea. *Anaerobe* 6:205–226

Gorham E, Janssens JA, Glaser PH (2003) Rates of peat accumulation during the postglacial period in 32 sites from Alaska to Newfoundland, with special emphasis on northern Minnesota. *Can J Bot* 81:429–438

Holmes AJ, Owens N, Murrell JC (1995) Detection of novel marine methanotrophs using phylogenetic and functional gene probes under methane enrichment. *Microbiology* 141:1947–1955

Ingram HAP (1978). Soil layer in mires: Function and terminology. *J Soil Sci* 29:224–227

Jacks G, Joelsson A, Fleischer S (1994) Nitrogen retention in forest wetlands. *Ambio* 23:358–362

Kellam SJ, Walker JM (1989). Antibacterial activity from marine microalgae in laboratory culture. *Br Phycol J* 24:191–194

Keshtacher-Liebson E, Hadar Y, Chen Y (1995). Oligotrophic bacteria enhance algal growth under iron-deficient conditions. *Appl Environ Microbiol* 61:2439–2441

Khan AW, Trottier TM, Patel GB, Martin SM (1979). Nutrient requirement for the degradation of cellulose to methane by mixed population of anaerobes. *J Gen Microbiol* 112:365–372

König H (1984). Isolation and characterization of *Methanobacterium uliginosum* sp. nov. from a marshy soil. *Can J Microbiol* 30:1477–1481

Koops JG, van Beusichem ML, Oenema O (1997) Nitrogen loss from grassland on peat soils through nitrous oxide production. *Plant Soil* 188:119–130

Li L, Kato C, Horikoshi K (1999) Microbial diversity in sediments collected from the deepest cold-seep area, the Japan Trench. *Mar Biotechnol* 1:391–400

Lukow T, Dunfield PF, Liesack W (2000) Use of the T-RFLP technique to assess spatial and temporal changes in the bacterial community structure within an agricultural soil planted with transgenic and non-transgenic potato plants. *FEMS Microbiol Ecol* 32:241–247

Luton PE, Wayne JM, Sharp RJ, Riley PW (2002) The *mcrA* gene as an alternative to 16S rRNA in the phylogenetic analysis of methanogen populations in landfill. *Microbiology* 148:3521–3530

Maestrojuan G, Boone D (1991) Characterization of *Methanosarcina barkeri* MS^T and 227, *Methanosarcina mazei* S-6^T, and *Methanosarcina vacuolata* Z-761^T. *Int J Syst Bacteriol* 41:267–274

Martikainen PJ, Hannu N, Patrick C, Silvola J (1993) Effect of a lowered water table on nitrous oxide fluxes from northern peatlands. *Nature* 366:51–53

Martikainen PJ, Nykänen H, Alm J, Silvola, J (1995) Change in fluxes of carbon dioxide, methane and nitrous oxide due to forest drainage of mire sites of different trophy. *Plant Soil* 168–169:571–577

McDonald IR, Murrell JC (1997) The methanol dehydrogenase structural gene *mxaF* and its use as a functional gene probe for methanotrophs and methylotrophs. *Appl Environ Microbiol* 63:3218–3224

Moore TR (2001) Les processus biogéochimiques liés au carbone. In: Payette S and Rochefort L (eds) *Écologie des tourbières du Québec-Labrador*. Les Presses de l'Université Laval, Québec, pp 183–197

Moore TR, Dalva M (1993) The influence of temperature and water table position on carbon dioxide and methane emissions from laboratory columns of peatland soils. *J Soil Sci* 44:651–664

Moore TR, Roulet NT, Waddington JM (1998) Uncertainty in predicting the effect of climatic change on the carbon cycling of Canadian peatlands. *Climatic Change* 40:229–245

Nguyen HH, Elliott SJ, Yip JH, Chan SI (1998) The particulate methane monooxygenase from *Methylococcus capsulatus* (Bath) is a novel copper-containing three-subunit enzyme. *J Biol Chem* 273:7957–7966.

Opelt K, Berg G (2004) Diversity and antagonistic potential of bacteria associated with bryophytes from nutrient-poor habitats of the Baltic Sea Coast. *Appl Environ Microbiol* 70:6569–79

Opelt K, Berg C, Berg G (2007) The bryophyte genus *Sphagnum* is a reservoir for powerful and extraordinary antagonists and potentially facultative human pathogens. *FEMS Microbiol* 61:38–53

Pacheco-Oliver M, McDonald IR, Groleau D, Murrell CJ, Miguez CB (2002) Detection of methanotrophs with highly divergent pmoA genes from Arctic soils. *FEMS Microbiol Lett* 209:313–319

Paerl HW, Pinckney JL (1996) A mini-review of microbial consortia: Their roles in aquatic production and biogeochemical cycling. *Microb Ecol* 31:225–247

Payette S (2001) Les principaux types de tourbières. In: Payette S and Rochefort L (eds) *Écologie des tourbières du Québec-Labrador*. Les Presses de l'Université Laval, Québec, pp 39–89

Pedersen K (1996) Investigations of subterranean bacteria in deep crystalline bedrock and their importance for the disposal of nuclear waste. *Can J Microbiol* 42:382–391

Popp TJ, Chanton, JP (1999) Methane stable isotope distribution at a *Carex* dominated fen in north central Alberta. *Global Biogeochem Cycles* 13:1063–1077

Purdy KJ, Nedwell DB, Embley TM (2003) Analysis of the sulfate-reducing bacterial and methanogenic archaeal populations in contrasting Antarctic sediments. *Appl Environ Microbiol* 69:3181–3191.

Regina K, Nykänen H, Silvola J, Martikainen PJ (1996) Fluxes of nitrous oxide from boreal peatlands as affected by peatland type, water table level and nitrification capacity. *Biogeochemistry* 35:401–418.

Rosenberry DO, Glaser PH, Siegel DI, Weeks EP (2003) Use of hydraulic head to estimate volumetric gas content and ebullition flux in northern peatlands. *Water Resources Res* 39:3, 1066, doi:10.1029/2002WR001377

Rosowski JR, Langenberg WG (1994) The near-spineless *Trachelomonas grandis* (Euglenophyceae) superficially appears spiny by attracting bacteria to its surface. *J Phycol* 30:1012–1022

Sauer K, Thauer RK (2000) Methyl-coenzyme M formation in methanogenic archaea. *Eur J Biochem* 267:2498–2504

Saunders NFW et al. (2003) Mechanisms of thermal adaptation revealed from the genomes of the Antarctic Archaea, *Methanogenium frigidum* and *Methanococcoides burtonii*. *Genome Res* 13:1580–1588

Segers R (1998) Methane production and methane consumption: A review of processes underlying wetland methane fluxes. *Biogeochemistry* 41:23–51

Simankova MV et al. (2001) *Methanosarcina lacustris* sp. nov., a new psychrotolerant methanogenic archaeon from anoxic lake sediments. *Syst Appl Microbiol* 24:362–367

Singh N, Kendall MM, Liu Y, Boone DR (2005) Isolation and characterization of methylotrophic methanogens from anoxic marine sediments in Skan Bay, Alaska: Description of *Methanococcoides alakenese* sp. nov., and emended description of *Methanosarcina baltica*. *Int J Syst Evol Microbiol* 55:2531–2538

Sjörs H (1950) On the relation between vegetation and electrolytes in north Swedish mire waters. *Oikos* 2:241–258

Sowers KR (1995) Methanogenic *Archae*a: an overview. In: Sowers, Kr, Schreier HJ, (eds) *Archaea: A Laboratory Manual*. Cold Spring Harbor Laboratory Press, New York, pp 3–13

Sundh I, Mikkelä C, Nilsson M, Svensson BH (1995) Potential aerobic methane oxidation in a *Sphagnum*-dominated peatland - Controlling factors and relation to methane emission. *Soil Biol Biochem* 27:829–837

Sundh I, Nilsson M, Granberg G, Svensson BH (1994) Depth distribution of microbial production and oxidation of methane in northern boreal peatlands. *Microb Ecol* 27:253–265

Svensson BH, Sundh I (1992) Factors affecting methane production in peat soils. *Suo* 43:183–190

Thomas KL, Benstead J, Davies KL, Lloyd D (1996) Role of wetland plants in the diurnal control of CH_4 and CO_2 fluxes in peat. *Soil Biol Biochem* 28:17–23

Tourova TP, Omelchenko MV, Fegeding KV, Vasilieva LV (1999) The phylogenetic position of *Methylobacter psychrophilus* sp. nov. *Microbiology* 68:437–444 (Translated from Mikrobiologiya)

Trotsenko YA, Khmelenina VN (2005) Aerobic methanotrophic bacteria of cold ecosystems. *FEMS Microbiol Ecol* 53:15–26

Vecherskaya MS, Galchenko VF, Sokolova EN, Samarkin VA (1993) Activity and species composition of aerobic methanotrophic communities in tundra soils. *Curr Microbiol* 27:181–184

Velthof GL, Oenema O (1995) Nitrous oxide fluxes from grassland in the Netherlands: II. Effects of soil type, nitrogen fertiliser application and grazing. *Eur J Soil Sci* 46:541–549

Whalen SC, Reeburgh WS (1990) Consumption of atmospheric methane by tundra soils. *Nature* 346:160–162

Whalen SC, Reeburgh WS, Reimers CE (1996) Control of tundra methane emission by microbial oxidation. In: Reynolds JF, Tenhunen JD (eds) *Landscape Function: Implication for Ecosystem Response to Disturbance, a Case Study in Arctic Tundra*. Springer Verlag, Heidelberg, pp.257–274

Williams RT, Crawford RL (1984) Methane production in Minnesota peatlands. *Appl Environ Microbiol* 47:1266–1271

Williamson VM, Paquin CE (1987) Homology of *Saccharomyces cerevisiae* ADH4 to an iron-activated alcohol dehydrogenase from *Zymomonas* determinant. *Mol Gen Genet* 209:374–381

Winfrey MR, Nelson DR, Kleickis SC, Zeikus JG (1977) Association of hydrogen metabolism with methanogenesis in Lake Mendota sediments. *Appl Environ Microbiol* 33:312–318

Wise MG, McArthur JV, Shimkets LJ (1999) Methanotroph diversity in landfill soil:isolation of novel type I and type II methanotrophs whose presence was suggested by culture-independent 16S ribosomal DNA analysis. *Appl Environ Microbiol* 65:4887–4897

Woese CR (1981) Archaebacteria. *Scientific American*, 244 (6):94–106

Wolfe RS, Higgins IJ (1979) Microbial biochemistry of methane-a study in contrasts. *Int Rev Biochem* 21:267–353

Chapter 10
Subsurface Geomicrobiology of the Iberian Pyritic Belt

Ricardo Amils(✉), David Fernández-Remolar, Felipe Gómez,
Elena González-Toril, Nuria Rodríguez, Carlos Briones,
Olga Prieto-Ballesteros, José Luis Sanz, Emiliano Díaz, Todd O. Stevens,
Carol R. Stoker, the MARTE Team

10.1 Introduction

Terrestrial subsurface geomicrobiology is a matter of growing interest. On a fundamental level, it seeks to determine whether life can be sustained in the absence of radiation, whereas it also aims to develop practical applications in environmental biotechnology. Subsurface ecosystems are also intriguing exobiological models, useful for the re-creation of life on early Earth (Widdel et al. 1993) or the representation of life as it would occur in other planetary bodies (Boston et al. 1992). Subsurface ecosystems were originally reported in basalt aquifers (Stevens and McKinley 1995; Chapelle et al. 2002) and later in sedimentary aquifers, petroleum reservoirs, and alkaline and saline goldmine groundwater (Lin et al. 2006). Results obtained by deep-sea subsurface exploration initiatives are widening the scope of our knowledge in this field (D'Hondt et al. 2004). In this field there is a serious debate on whether the source of electron donors and/or acceptors is dependent on radiation-mediated reactions and also on contamination problems associated with drilling technologies, their mitigation, and control. In spite of the interest of subsurface ecosystems, information concerning microbial abundance, diversity, and sustainability is still scarce, mainly due to methodological limitations.

Among the different minerals, the metallic sulfides have the potential to be a good source of energy for subsurface chemolithotrophs. The micro-organisms that aerobically oxidize iron sulfides are well characterized (González-Toril et al. 2003), however, little is known about the possibility of subsurface chemolithoautotrophic metabolism in anoxic conditions. The Mars Analog Research and Technology Experiment (MARTE) project (Stoker et al. 2004; Fernández-Remolar et al. 2005a) outlined in this chapter was designed to search for this type of life in the nonporous volcanically hosted massive sulfide deposits (VHMS) of the Río Tinto basement at Peña de Hierro (Iberian Pyritic Belt).

Ricardo Amils
Centro de Astrobiología (INTA-CSIC), 28850 Torrejón de Ardoz, Madrid, Spain and
Centro de Biología Molecular (UAM-CSIC), Cantoblanco, 28049 Madrid, Spain
e-mail: ramils@cbm.uam.es

10.2 The Río Tinto Ecosystem

Río Tinto is an unusual ecosystem due to its size (100-km long), constant acidic pH (mean pH 2.3), high concentration of heavy metals (Fe, Cu, Zn, As, Mn, and Cr), and high level of microbial diversity, mainly eukaryotic (López-Archilla et al. 2001; Amaral-Zettler et al. 2002; Aguilera et al. 2006).

Río Tinto rises in Peña de Hierro, in the core of the Iberian Pyritic Belt (IPB), and reaches the Atlantic Ocean at Huelva. The IBP is one of the largest massive sulfidic deposits on Earth, formed as a hydrothermal deposit during the Paleozoic accretion of the Iberian Peninsula (Boulter 1996; Leistel et al. 1998). One important characteristic of Río Tinto is the high concentration of ferric iron and sulfates present in its waters, products of the bio-oxidation of pyrite, the main mineral component of the IPB system.

Pyrite, with its wide distribution on our planet, is considered an important substrate for chemolithotrophic metabolism because both of its components, sulfide and ferrous iron, can be used by sulfur- and iron-oxidizing micro-organisms as a source of energy. The first acidophilic strict chemolithotroph described, *Acidithiobacillus ferrooxidans* (formerly *Thiobacillus ferrooxidans*), was isolated from an acidic pond in a coal mine more than 50 years ago (Colmer et al. 1950). Although *At. ferrooxidans* can obtain energy by oxidizing both reduced sulfur compounds and ferrous iron, much attention was paid in the past to the sulfur oxidation reactions due to bioenergetic considerations (Amils et al. 2004)

The discovery that some strict acidophilic chemolithotrophs, such as *Leptospirillum* spp. or *Ferroplasma* spp., could grow using ferrous iron as their only source of energy, and that these micro-organisms are mainly responsible for industrial metal bioleaching processes (biomining) and the generation of acid mine drainage (AMD), has changed this perspective, shifting the interest from sulfur to iron oxidation (Golovascheva et al. 1992; Edwards et al. 1999; Malki et al. 2006).

The mechanisms by which acidophilic chemolithotrophs obtain energy by oxidizing metallic sulfides has remained controversial for many years (Ehrlich 2002). But the recent demonstration that the ferric iron present in the cell wall and in the extracellular polysaccharides of acidophilic chemolithotrophic micro-organisms is responsible for the electron transfer from the mineral substrate to the electron transport chain has clarified this issue, with important fundamental and applied consequences (Gehrke et al. 1995; Sand et al. 1995, 2001).

The differences observed during bioleaching of diverse metallic sulfides depend on the type of chemical oxidation promoted by ferric iron, which is determined by the crystallographic characteristics of the mineral. Under normal conditions of pressure and temperature, pyrite and two other sulfides, tungstenite and molybdenite, can only be oxidized by ferric iron through the so-called thiosulfate mechanism. Thiosulfate can then be further oxidized to sulfate, also by ferric iron (Sand et al. 2001). The rest of the sulfides undergo oxidation through the polysulfide mechanism. Polysulfide can be further oxidized to elemental sulfur. In this case the production of sulfate requires a subsequent biooxidation reaction promoted by sulfur-oxidizing micro-organisms (e.g., *At. ferrooxidans*). These reactions are merely chemical oxidation reactions. The critical role of iron-oxidizing micro-organisms in these processes is to maintain a high concentration of the oxidant agent, ferric iron, by means of an aerobic respiration mechanism (reaction 1):

$$Fe^{2+} + \tfrac{1}{2}O_2 + 2H^+ \rightarrow Fe^{3+} + H_2O$$

In addition, it is now well established that iron can be oxidized anaerobically in the absence of oxygen, when coupled to anoxygenic photosynthesis or to anaerobic respiration using nitrate as an electron acceptor (Widdel et al. 1993; Benz et al. 1998).

As mentioned, the most important characteristic of Río Tinto is the high concentration of ferric iron and sulfates found in its waters. Ferric iron is responsible for the maintenance of the constant pH of the water due to its buffering potential (reaction 2):

$$Fe^{3+} + 3H_2O \leftrightarrow Fe(OH)_3 + 3H^+$$

10.3 Geomicrobiology of the Iron Cycle in the Río Tinto Ecosystem

The combined use of conventional (enrichment cultures, isolation, and phenotypic characterization) and molecular microbial ecology methods (amplification of 16–18S rRNA genes and its resolution using electrophoresis in denaturating conditions (PCR-DGGE), fluorescence in situ hybridization (FISH and CARD-FISH) and molecular cloning) has led to the identification of the most representative micro-organisms of the Río Tinto basin (González-Toril et al. 2003, 2006). Eighty percent of the diversity of the water column corresponds to micro-organisms belonging to only three bacterial genera: *Leptospirillum*, *Acidithiobacillus*, and *Acidiphilium*, all members of the iron cycle (González-Toril et al. 2003). All *Leptospirillum* isolates from Río Tinto are aerobic iron oxidizers. *At. ferrooxidans* can oxidize iron aerobically and reduce it anaerobically (Malki et al. 2006). All *Acidiphilium* isolates can oxidize organic compounds using ferric iron as an electron acceptor and some isolates can do so in the presence of oxygen. Although other iron-oxidizing (*Ferroplasma* spp. and *Thermoplasma acidophilum*) or -reducing ("Ferromicrobium" spp.) micro-organisms have been detected in the Tinto ecosystem (González-Toril et al. 2003), their low numbers suggest that they play a minor role in the operation of the iron cycle, at least in the water column.

Concerning the sulfur cycle, only *At. ferrooxidans* is found in significant numbers in the water column of the Tinto ecosystem. This bacterium can oxidize both ferrous iron and reduced sulfur compounds. The oxidation of reduced sulfur compounds can be carried out aerobically and anaerobically. Some sulfate-reducing

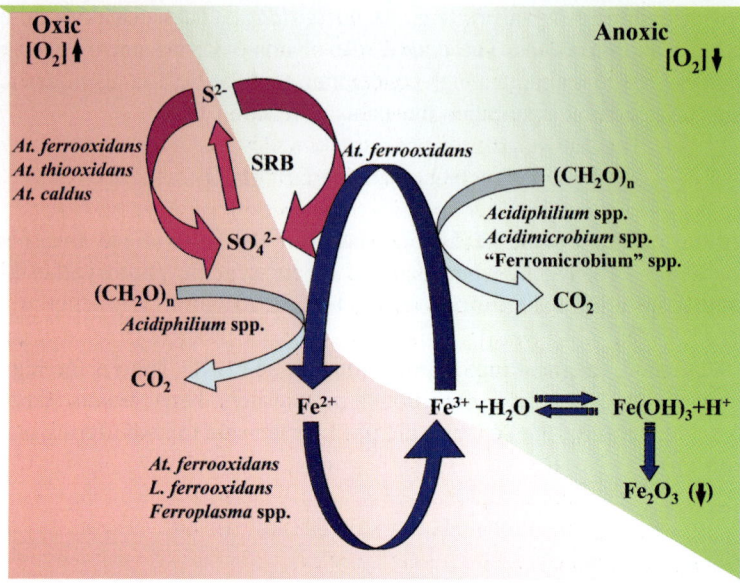

Fig. 10.1 Geomicrobiology of the Río Tinto basin ecosystem associated with the iron and sulfur cycles. In blue are indicated the metabolic reactions associated with the iron cycle, and in red those associated with the sulfur cycle, operating in aerobic (pink) or anaerobic conditions (green). Identified microbial activities are associated with the different reactions operating in the Río Tinto ecosystem. SRB: sulfate-reducing bacteria. The buffering capacity of ferric iron and the maturation of iron minerals have been placed in the model due to their relevance to the mass balance of the system

activity has been detected associated with sediments in certain parts of the river (Malki et al. 2006). Figure 10.1 shows the integrated geomicrobiological model of Río Tinto, in which the iron cycle plays a central role.

As mentioned above, besides the extreme physicochemical conditions found in the Tinto ecosystem, what makes Río Tinto a unique acidic environment is the unexpected degree of eukaryotic diversity found in its waters (López-Archilla et al. 2001; Amaral-Zettler et al. 2002; Aguilera et al. 2006). Members of the phylum Chlorophyta (*Chlamydomonas, Chlorella*, and *Euglena*) are the most frequent species followed by two filamentous algae belonging to the genera *Klebsormidium* and *Zygnemopsis*. The most acidic part of the river is inhabited by a eukaryotic community dominated by two species of the genera *Dunaliella* and *Cyanidium*, well known for their acidity and heavy-metal tolerance (Visviki and Rachlin 1993; Visviki and Santikul 2000). Among the eukaryotic decomposers, fungi are very abundant and exhibit great diversity, including yeast and filamentous forms (López-Archilla et al. 2005). The mixotrophic community is dominated by cercomonads and stramenopiles related to different genus (*Bodo, Ochroomonas, Labyrinthula*, and *Cercomonas*). The protistan consumer community is characterized by two species of ciliates, tentatively assigned to the genera *Oxytrichia* and *Euplotes*. Amoebas related to the genera *Valhkampfia* and *Naegleria* can be found in the most acidic part of the river, and one species of heliozoan belonging to the genus *Actinophyris* seems to be the most characteristic predator of the benthic food chain of the river (Aguilera et al. 2006).

Unicellular forms are not the only eukaryotes to develop in the extreme conditions of the Río Tinto basin. Different plants can be found growing in the acidic soils of the riverbanks. The strategies used by these plants to overcome the physiological problems associated with the extreme conditions of the habitat are diverse. Some are resistant to the heavy metals concentrated in the soils in which they grow (Rodríguez et al. 2007; Berazain et al. 2007). Others specifically accumulate metals in different plant tissues. Recent analysis by X-ray diffraction (XRD) and Mössbauer spectroscopy of the iron minerals found in rhizomes and leaves of *Imperata cylindrica*, an iron hyperaccumulator perennial grass growing in the Río Tinto banks, showed significant concentrations of jarosite and iron oxyhydroxides (Rodriguez et al. 2005). These results suggest that the management of heavy metals in general, and iron in particular, is much more complex and versatile in plants than has been reported to date (Schmidt 2003). Also, these results prove that multicellular complex systems can also develop in extreme conditions, like those existing in Río Tinto.

Most of the biomass of the Tinto ecosystem is located on the riverbed and the surface of the rocks, forming dense biofilms composed mainly of filamentous algae and fungi in which prokaryotic micro-organisms are trapped. Heterotrophic protists associated with these biofilms have been also detected (Aguilera et al. 2007). Significant iron mineral precipitation occurs on the surface of these negatively charged biofilms, generating iron bioformations, which grow with time following the hydrological cycles of the river. The age and depth of these iron formations strongly support the idea that Río Tinto corresponds to a natural system and not to an industrially contaminated site (van Geen et al. 1997; Davis et al. 2000; Elbaz-Poulichet et al. 2001). It is obvious that mining activity during the last 5,000 years has altered the Tinto system (Avery 1974), but evidence of its antiquity can be found in massive laminated iron beds in three iron terraces occupying different elevations above the present river. The oldest of these, Alto de la Mesa, lies 60 m above the current river level. Preliminary isotopic data indicate an age of 2 million years (My) for this formation (Fernández-Remolar et al. 2005b), although biostratigraphic considerations indicate that some altered minerals of the area (in situ gossan) may be as old as 6 My, suggesting that the IPB acidic water systems are of still older origin (Moreno et al. 2003).

The combination of bioleaching processes and high evaporation rates induce the formation of concentrated brines in the origin section of the river (Fernández-Remolar et al. 2003). Iron oxides associated with sulfates are the characteristic minerals that are formed in the modern sediments and young terraces: these are hydronium jarosite, schwermannite, copiapite, coquimbite, natronojarosite, gypsum, and other sulfate minerals. Goethite and hematite are the predominant minerals in the old terraces of the Tinto Basin (Fernández-Remolar et al. 2005b).

It is clear from these results that the main characteristics of Río Tinto are not due to acid mine drainage of exposed mineral resulting from mining activity. Our working hypothesis predicts the existence of a continuous underground reactor in which the sulfidic minerals of the IPB are the main energy source, and the river is the exhaust pipe that releases the products of the different metabolic reactions developing in the reactor. To test this hypothesis, we drilled a series of boreholes that intercepted groundwaters within the ore body in order to detect evidence of both subsurface microbial activity in the retrieved cores and potential resources to support these microbial communities in situ (MARTE project).

10.4 Subsurface Geomicrobiology of the Iberian Pyritic Belt

The main goal of the MARTE project, a collaborative effort between NASA and the Centro de Astrobiología (INTA-CSIC, Spain), was the search for subsurface microbial activity associated with the IPB. The selected study site was Peña de Hierro (Figs. 10.2 and 10.3) on the north flank of the Río Tinto Anticline, which comprises a thick volcanosedimentary succession composed of dark shales, basaltic lavas, rhiolitic materials, fine ashes and tuffites, and green/purple shales. The hydrothermal activity is recorded as complex-massive sulfide lenses or stockwork veins of pyrite and quartz, which occur at the upper part of the IPB volcanic sequence (Leistel et al. 1998).

The Peña de Hierro stratigraphy is inverted as a consequence of the Hercynian Orogenesis tectonism which produced an inverted anticline propagating along a

Fig. 10.2 View of Peña de Hierro in which the MARTE project has taken place

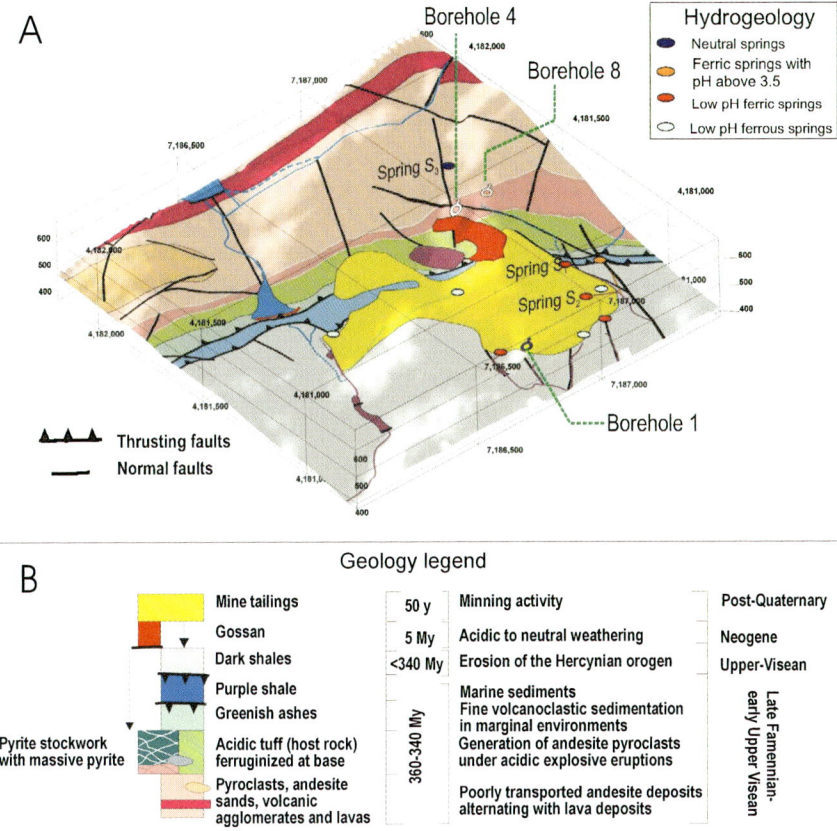

Fig. 10.3 Geological, hydrological, and stratigraphic maps of the Peña de Hierro field site. The location of boreholes and springs sampled are shown together with the stratigraphic sequence and ages of the geological units

110°N-thrusting front. This compressive structure is intersected by NNE–SSW normal fractures that are related to the generation of acidic streams in the Peña de Hierro area and which are considered the origin of the Río Tinto (Fig. 10.3). The reversed succession is topped by a gossan unit originated by the Tertiary in situ weathering of the sulfide complex.

Despite the low porosity of these rocks, the use of complementary techniques including field survey, transient electromagnetic sounding, and drilling, showed that fractures play an essential role in storing groundwater. Faults intersect the Early Carboniferous volcanic tuff-hosted pyrite bodies, which are the primary energy source for the chemolithotrophic micro-organisms found in the Río Tinto basin. The analysis of samples from drill cores, core leachates, and borehole fluids have shown distinctive microbial activity, appearing at different depths depending on environment variables in the aquifer.

10.4.1 Drilling and Sample Analysis

Drilling sites were selected using surface and subsurface techniques, including geological mapping, and surface hydrogeological and geophysical surveys (Jernsletten 2005). The well locations were selected to monitor spatial changes in microbial and hydrogeochemical processes. Coring was carried out using a commercial coring rig at three locations designated wells BH1, BH4, and BH8 (Fig. 10.3). The boreholes were continuously cored by rotary diamond-bit drilling using a Boart-Longyear (Salt Lake City, UT, USA) HQ wireline system that produced 60-mm diameter cores within a plastic liner. Water was used as drilling fluid to refrigerate the bit. Chemical tracers (NaBr) were used for controlling contamination introduced by the drilling fluids. Upon retrieval from the drill rig, cores were divided into 1-m lengths, flushed with N_2, sealed, and transported to a nearby laboratory for geomicrobiological analysis.

Samples were prepared aseptically in anaerobic conditions using an anaerobic chamber and subjected to different biological and geological analysis. Culture-independent detection of micro-organisms was done by epifluorescent microscopy after staining samples with 4′,6-diamidino-2-phenylindole dihydrochloride (DAPI) and by fluorescence in situ hybridization with specific probes (FISH and CARD-FISH) (González-Toril et al. 2006). Also a protocol has been developed for microbial biodiversity identification on subsurface cores. It is based on the amplification and molecular cloning (Sambrook and Russell 2001) of the metagenomic DNA extracted from the powdered samples (1–10 g), followed by the sequencing of the 16/18S rDNA genes of a representative number of molecular clones per sample (between 10 and 20) and, when possible, the taxonomic assignment of the clonal sequences by means of molecular phylogenetic analyses.

Chemolithotrophic enrichment cultures were performed in a minimal Mackintosh medium, with the addition of ferrous iron (20 g l^{-1}) or of a sterile rock sample, as electron donor (González-Toril et al. 2006). Anaerobic enrichments were performed as described previously, for denitrifying micro-organisms with the addition of 10 g l^{-1} $Na_2S_2O_3$ and 1 g l^{-1} KNO_3 (Stevens and McKinley 1995), for sulfate reducers according to González-Toril et al. (2006), and for methanogens according to Sanz et al. (1997).

After drilling, the wells were completed by installing PVC casings set in clean gravel packing. Underground sampling for water and gas aquifer analyses was done by the installation of multilevel diffusion samplers (MLDS) at different depth intervals. These were inserted in perforated sections of the PVC casings and allowed to equilibrate for various periods of time (between three and twelve months).

Rock leachates were produced by adding sterile anoxic water to powdered core subsample and allowing them to stand for 1 hour in anaerobic conditions before filtration and analysis. Water samples from the MLDS were filtered and the pH adjusted by adding HCl to a 0.5 N final concentration. Anion concentrations were determined by ion chromatography using a Dionex (Sunnyvale, CA, USA) 4010i system with an AS14A column, and metal concentrations were determined by ion chromatography using a CS5A column. Dissolved gases were sampled by allowing them to equilibrate across a submerged sealed polyethylene tube. Tubes were removed and analyzed within 1 hour by gas chromatography using a Carle Gas Chromatograph with a molecular sieve column and thermal conductivity detector, using purified N_2 as a carrier gas.

10.4.2 Results from 2003 and 2004 MARTE Drilling Campaigns

The characterization of the groundwater entering the ore body at Peña de Hierro was done by analyzing springs upslope. The water from these springs was aerobic, with a pH near 6 and a low ionic strength. The environment within the ore body was sampled by drilling boreholes BH4 (2003 drilling campaign) and BH8 (2004). Wells BH4 and BH8 cored around 165 m of pyrite stockwork. The lithology of borehole BH4 is shown in Fig. 10.4. The water table was encountered at nearly 90 m below the surface. From top to bottom, the lithology of borehole BH4 consists of ca. 10 m of hydrothermally altered and weathered tuff, ca. 20 m of gossanized ores composed of goethite and hematite, ca. 120 m of coarse-grained volcanogenetic tuff of ryolithic composition that contains the sulfide ore body, and ca. 10 m of chloritized volcanic tuff with disseminated pyrite and carbonate traces. The sulfide ore was a complex mixture of polymetallic sulfide minerals dominated by pyrite (Fernández-Remolar et al. 2008).

Rock leachate analyses were performed to detect contamination by drilling fluids (tracers) but also to estimate resources available to micro-organisms from the solid phase. Sulfate was abundant and is a good indicator of the degree of oxidation of the ore. Surprisingly, nitrite and nitrate were present at concentrations higher than 100 ppm in many samples. The analysis of the drilling fluid gave very low concentrations of sulfate and nitrate, and there was no correlation between bromine and anion concentration, so the presence of these anions in the core samples cannot be due to surface contamination. Both Fe(II) (average concentration 95 ppm) and Fe(III) (average concentration 22 ppm) could be leached from powdered ore samples after a 1 hour incubation with 0.5 N HCl or HCl with hydroxylamine, which is a standard method for measurement of iron availability (Chao and Zhou 1983; Lovley and Phillips 1987). Organic carbon content of the cores was near the detection limit of 0.01%. From the rock leachate experiments, we can conclude that electron acceptors for anaerobic respiration, particularly Fe(III), SO_4^{2-}, NO_2^-, NO_3^-, and carbonates, are available from the volcanically hosted massive sulfide deposits (VHMS) rock matrix.

Borehole fluids from the MLDS experiments were analyzed as a proxy for formation fluids, which could not be extracted from these low-porosity rocks. Formation water in BH4 was sampled with the MLDS from 85–105 and from 135–150 mbls (meters below surface), several months after drilling. The measured composite pH was ca. 3.5, and has remained acidic for the two sampling years after drilling. Bromide in some samples suggested that these contained between 0 and

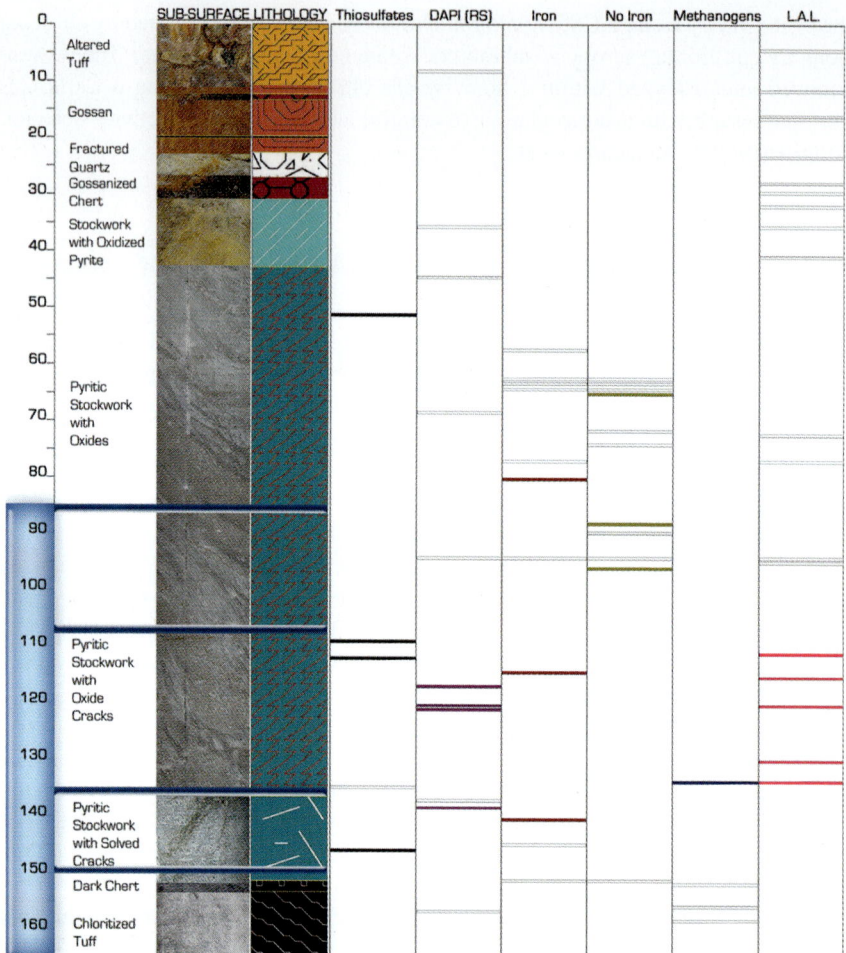

Fig. 10.4 Geomicrobiological observation within the pyrite ore body. Core lithology and locations of biological indicators for BH4. Blue shaded area at left indicates water table. Columns left to right: 1, example images of cores from each lithology; 2, lithology; 3, growth of denitrifying thiosulfate-oxidizing organisms in anaerobic chemolithotrophic enrichment cultures ; 4, detection of micro-organisms by fluorescence microscopy; 5, growth of iron-oxidizing organisms in aerobic chemolithotrophic enrichment cultures with ferrous iron; 6, growth of organisms in aerobic chemolithotrophic enrichment cultures with rock samples as source of energy; 7, growth of methanogens in enrichment cultures with added H_2; 8, positive limulus amebocyte lysate (LAL) assay. Solid lines in colums 3–8 indicate positive results in samples without detectable bromine tracer; empty lines correspond to instances with detectable bromide, indicating that some drilling fluid is present in the sample

2% of residual drilling fluid. However, this level of contamination is too low to account for the solute concentrations that were detected in the BH4 MLDS formation water. Dissolved iron ranged from 108 to 480 ppm with an average of

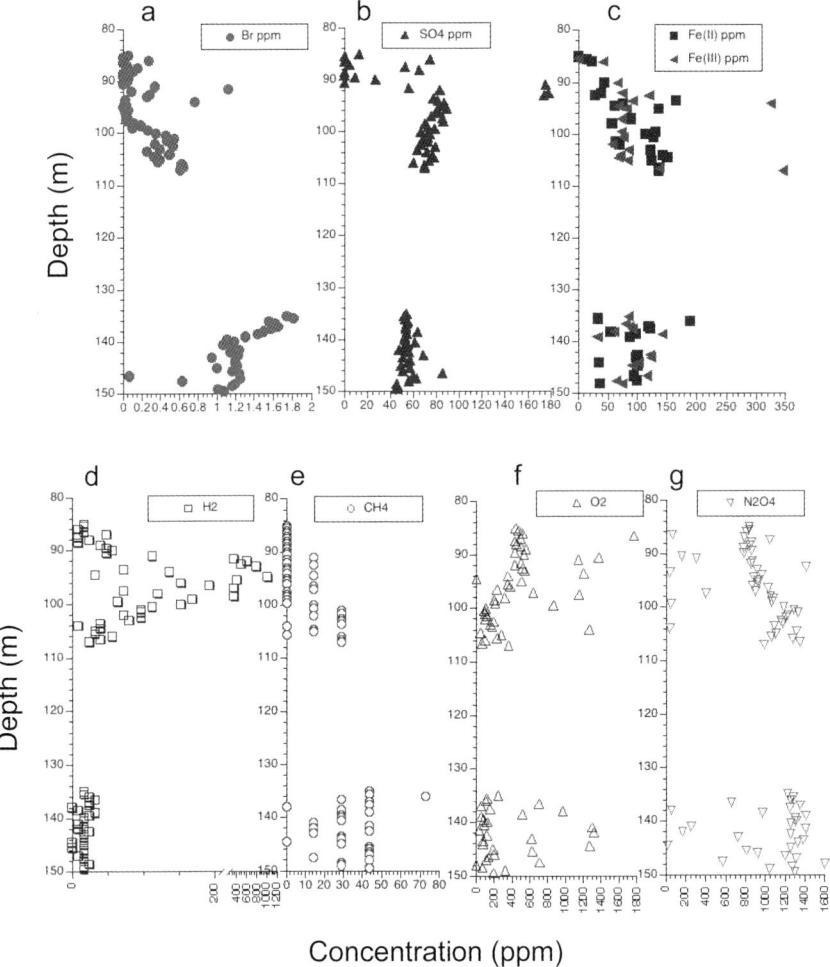

Fig. 10.5 Concentration (in ppm) of solutes in BH4 formation water sampled by MLDS. Solutes include bromide (**a**), sulfate (**b**), iron (II) and (III) (**c**), and the dissolved gases H_2 (**d**), CH_4 (**e**), O_2 (**f**), and NO_2 (**g**). For the dissolved gases, equilibrium concentrations are given

188 ppm (Fig. 10.5). The dissolved ferric to ferrous ratio ranged from 0.3 to 4.3 with an average of 1 and did not appear to correlate with total dissolved iron concentration. Sulfate concentration was relatively constant except near the water table, and was ca. 1,000-fold lower than in rock leachates. Neither nitrate nor nitrite were detected in the water. Small quantities of oxygen and NO_2 gas were present in some samples, and the two were inversely correlated. Dissolved methane was detected in many of the MLDS samples, indicating that methanogenic activity occurred throughout the ore body (Fig. 10.5).

Dissolved hydrogen concentrations averaged 25 ppm, except in a zone within the massive pyrites, just below the water table, from 90 to 100 mbls, where concentrations ranged from 100 to 1,000 ppm (Fig. 10.5). A similar pattern was observed in the second borehole, BH8, with an average H_2 concentration measured 12 months after drilling of ca. 25 ppm, and with isolated zones of higher concentration.

Electron donors available in the VHMS for microbial metabolism included Fe(II) and reduced S, which was expected, but also H_2. Laboratory experiments showed that H_2 could be produced, in concentrations similar to those in most MLDS samples, by reaction of VHMS rocks with water. Potential mechanisms for this have not yet been investigated (Wächtershäuser 1988; Drobner et al. 1990; Apps and van de Kamp 1993). We hypothesize that H_2 production supports methanogenesis throughout the wet sections of the VHMS.

Aseptically collected rock core samples, in which no tracer could be detected, were used to test for the presence or absence of micro-organisms, and to estimate their distribution. However, sampling artifacts may arise while using this protocol, inasmuch as fractured zones, which are most likely to contain microbial habitats, are also likely to be invaded by drilling fluids, thus being excluded from further analysis. It is likely that the tracer system used to monitor contamination was conservative, because many "contaminated" samples used as controls appeared to be sterile.

Micro-organisms were detected in different uncontaminated samples using culture-dependent and culture-independent methods. Distribution of micro-organisms was heterogeneous along the column, as expected in a system dominated by fracture flow. Using enrichment cultures, aerobic chemolithoautotrophs, mainly pyrite and iron oxidizers, and anaerobic thiosulfate oxidizers using nitrate as electron acceptor, sulfate reducers and methanogens were enriched from several samples.

DAPI stain and fluorescence in situ hybridization (FISH) with probes of different specificities did not yield good results due to the high level of epifluorescence exhibited by the mineral components present in the samples. The use of catalyzed reported deposition modification (CARD-FISH) improved enormously the contrast between the hybridization signal and the mineral substrate. Using this technique, we have been able to unambiguously prove the presence of active micro-organisms in different uncontaminated samples and to show that these micro-organisms occurred at extremely low density in these samples (Fig. 10.6). Those low densities revealed by in situ hybridization could explain in part the difficulties to grow them in the enrichment cultures. Higher numbers could be seen in samples from cracks (Fig. 10.7), which were normally discarded due to the presence of bromide, a signal of possible contamination by the drilling fluid.

A protocol for a direct extraction of DNA from uncontaminated samples has been developed. Preliminary results showed that amplifiable and clonable DNA could be extracted from the rock samples, leading to the detection of bacteria related to the *Acidithiobacillus* genus and sulfate-reducing bacteria in the anaerobic section of wells BH4 and BH8, together with different heterotrophic bacteria which had been enriched under strict anaerobic conditions in both wells. The combination of enrichment cultures with in situ hybridization and direct cloning-sequencing experiments should allow us to identify the subsurface microbial diversity of the IPB and to develop geomicrobiological models of the different cycles operating in the system.

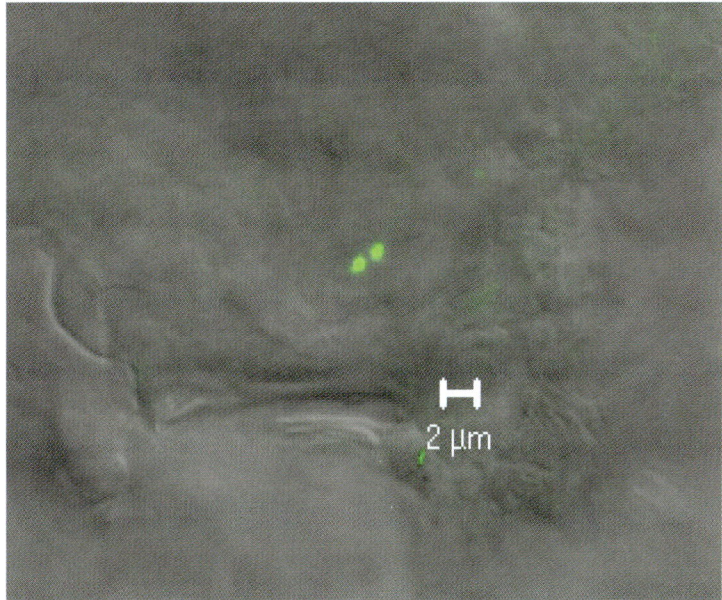

Fig. 10.6 Fluorescence in situ hybridization (CARD-FISH) of a core sample. Core sample 8,50a retrieved from an uncontaminated section hybridized with the specific probe for Bacteria, EUB 388

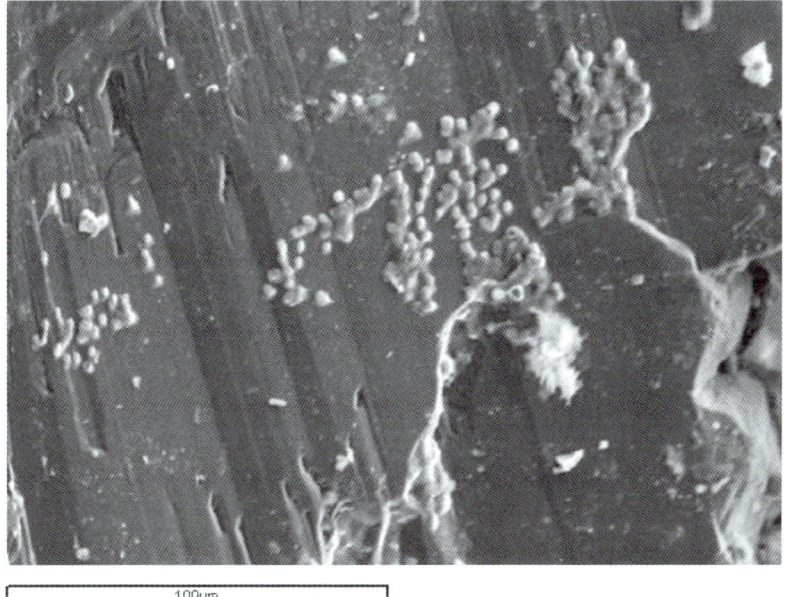

Fig. 10.7 Scanning electron microscopy of BH4 core sample 8,66a (155 mbls)

The environment down-gradient from the ore body was sampled by drilling borehole BH1. We considered that in this zone, fluids would represent the end product of subsurface interaction with the VHMS. Well BH1 cored 59 m of the younger dark shales. Core samples from BH1 consisted of carboniferous greenish shales derived from volcanic ash with fine sandy lenses and lutites bearing organic matter, and were overlaid by 7 m of mine tailings. From 7 to 12.5 mbls, shales were pale from ongoing leaching by groundwater. From 12.5 to 46 m, shales were unweathered. Below 46 m, shales were altered to a fine noncohesive black material, suggesting aqueous weathering under anoxic conditions (Fernández-Remolar et al. 2008).

As expected, sulfate and iron concentrations were lower in leachates from BH1 shales than in those from BH4 pyrites. Only small amounts of NO_3^- were detected in the leachates. Oxygen was not detected within the aquifer zone. Where present, dissolved sulfate in groundwater was in much higher concentrations than in groundwater from BH4, indicating that these waters had experienced more interaction with the ore. Neither NO_2^- nor NO_3^- were detected in water samples; however, dissolved NO_x gases were present at concentrations slightly higher than in water samples from BH4. Dissolved H_2, where detected, was at concentrations lower than in BH4 but still sufficient to make H_2 available as a microbial electron donor. Methane concentrations were several orders of magnitude higher than at BH4. These observations are consistent with the plume of groundwater representing the downstream output from reactions within the ore body.

Micro-organisms were also observed in BH1. Aerobes or denitrifiers were not detected. Sulfate reducers and methanogens were recovered in enrichment cultures and the methane concentrations that were measured near 18 and 50 mbls suggested that H_2 produced within the ore body supports these microbial activities down-gradient. At depths between 50 and 60 m, the methane-bearing water appears to mix with sulfate-bearing water. Decreasing CH_4 and H_2 was accompanied by increasing SO_4^{2-} and CO_2. Although not stoichiometric, this relationship suggests that anaerobic methane oxidation may occur in this zone.

The alteration of the sulfide ore has induced the production of different gases: CO_2, CH_4, and H_2, all of them participating in the biogeochemical cycles involved in the IPB decomposition. The observed characteristics of the underground mineralogy, dominated by iron oxyhydroxides and sulfates, resulted from the alteration of the abundant sulfides of the IPB by chemolithotrophic micro-organisms (Figs. 10.8 and 10.9). Sample analysis using scanning electron microscopy coupled with an energy-dispersive X-ray microanalysis probe (SEM-EDAX), in situ hybridization experiments, and direct cloning from deep core samples provide direct evidence of microbes metabolizing mineral substrates. As both secondary mineralogy and gas byproducts are the result of cryptic microbial communities living in the Río Tinto acidic aquifer, they can be used as potential biomarkers to explore subsurface life in deep regions.

In contrast to well-known AMD systems, the environments within and down-gradient from the Peña de Hierro VHMS appear to be anoxic, with a weakly acidic pH and evidence of methanogenic and sulfate-reducing activities. Any oxygen available from inflowing groundwater would initially be available as an electron

10 Subsurface Geomicrobiology of the Iberian Pyritic Belt

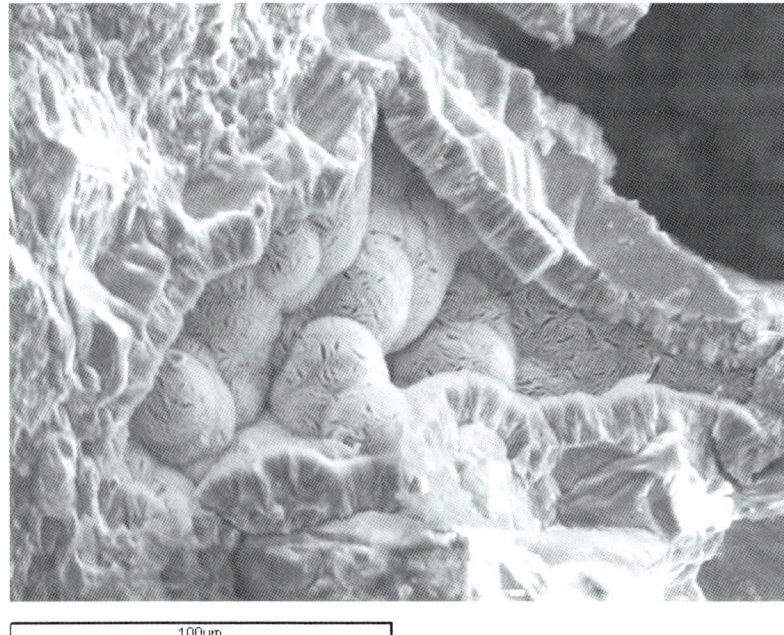

Fig. 10.8 Scanning electron microscopy analysis of BH4 core sample 8,68c (162 mbls)

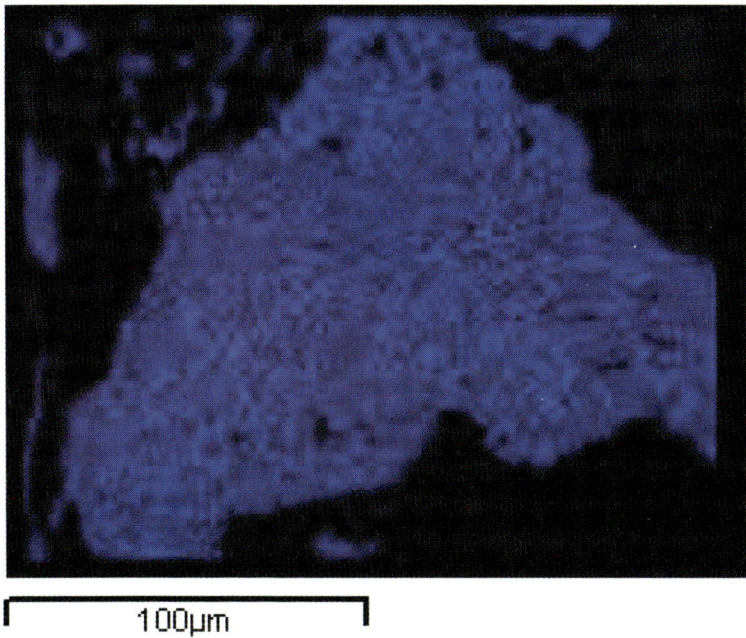

Fig. 10.9 SEM-EDAX iron mapping of Fig. 10.8

acceptor for microaerophilic micro-organisms, as suggested by enrichment cultures results, but O_2 could be also consumed by abiotic reactions (Chalk and Smith 1983; Conrad 1996). Because dissolved nitrate was not detected, quantities leached from the rock matrix are apparently consumed rapidly. Enrichment culture results suggest that some denitrifiers are present to utilize nitrate whenever it becomes available.

Some of the spring waters down-gradient from the ore body are largely acidic, high in ferric iron, and red in color, as previously described (Fernández-Remolar et al. 2003), which is typical of aerobic AMD processes. However, another group of springs found in the area (Fig. 10.3) produces anaerobic acidic waters with high concentration of ferrous iron. The origin of these iron-reduced springs remains to be determined.

10.5 Conclusions

The preliminary results from the MARTE project indicate that as groundwater enters the VHMS system, biotic and abiotic processes remove oxygen with the concomitant oxidation of iron and transient generation of acidity. Electron acceptors available for microbial metabolism include oxygen, nitrite, nitrate, sulfate, ferric iron, and inorganic carbon. Electron donors include ferrous iron, sulfide, and hydrogen gas generated by water/rock interaction. This supports a population of microaerophilic and denitrifying autotrophs. As the fluids become more reduced, methanogenesis and sulfate reduction, using hydrogen gas, become the dominant microbial processes, and the pH rises. Oxidants to drive the system appear to be supplied by the rock matrix, in contrast to conventional AMD models. These resources need only groundwater to launch microbial metabolism.

These observations confirmed the hypothesis that micro-organisms are active in the subsurface near the Río Tinto headwaters. To our knowledge, this is the first observation of a subsurface ecosystem within an undisturbed VHMS rock environment. The novel observations of H_2 and CH_4 production show that a variety of resources are available to support autotrophic microbial respiration in the subsurface both within and down-gradient from the ore body.

The Río Tinto system may be an important model for astrobiological study (Fernández-Remolar et al. 2003, 2005b; Amils et al. 2007). Recent observations of abundant layered sulfate minerals at Sinus Meridiani on Mars suggest a history of an aqueous, acidic, sulfate-rich environment (Squyres et al. 2004) that might originate from the weathering of sulfide-rich minerals (Fairén et al. 2004; Zolotov and Shock 2005). Our results suggest that such a system, if the rocks were similar to those at Peña del Hierro, could support subsurface life, even if surface conditions preclude it. In addition, we found that subsurface microbial metabolism coupled to sulfide weathering can produce large amounts of methane, which has been suggested as an atmospheric indicator of extant life on Mars (Formisano et al. 2004).

Acknowledgements MARTE Project Team (participants are listed alphabetically): Angeles Aguilera,[1] Ricardo Amils,[1,2] Carlos Briones,[1] Howard Cannon,[3] Fidel Davila,[1] Steven Dunagan,[3] Alberto G. Fairén,[2] David Fernández-Remolar,[1] Brian Glass,[3] Felipe Gómez,[1] Javier Gómez-Elvira,[1] Elena González-Toril,[1] Lawrence G. Lemke,[3] Kennda Lynch,[4] Victor Parro,[1] Olga Prieto-Ballesteros,[1] Nuria Rodríguez,[1] Todd O. Stevens,[5] Virginia Souza-Egipsy,[1] Carol R. Stoker,[3] and Jhony Zavaleta.[3] [1]Centro de Astrobiología (INTA-CSIC), Torrejón de Ardoz, Spain; [2]Centro de Biología Molecular (UAM-CSIC), U. Autónoma de Madrid, Madrid, Spain; [3] NASA Ames Research Center, Mountain View, CA, USA, [4] NASA Johnson Space Center, Houston, TX, USA, [5]Portland State University, Portland, OR, USA.

We thank the following for their contribution to this work. Drilling services were provided by INSERSA S.A., Río Tinto, Spain; fieldsites and laboratory space were provided by Fundación Río Tinto, Río Tinto, Spain; logistic support was provided by Casiano Primo at the Hotel Vazquez Díaz, Nerva, Spain. We acknowledge the technical support provided by Mercedes Moreno-Paz, Marina Postigo, María Fernández-Algar, and Moustafa Malki. Additional assistance was provided by Mary Sue Bell, James Hall, David McKay, Rachel Shelbe, and Norman Wainright. This work was supported by the NASA ASTEP program (USA), by institutional grants to the Centro de Astrobiología (INTA-CSIC) and project CGL2006-02534/BOS from the Ministerio de Educación y Ciencia.

References

Aguilera A, Manrubia SC, Gómez F, Rodríguez N, Amils R (2006) Eukaryotic community distribution and its relationship to water physicochemical parameters in an extreme acidic environment, Río Tinto (Southwestern Spain). *Appl Environ Microbiol* 72:5325–5330

Aguilera A, Souza-Egipsy V, Gómez F, Amils R (2007) Development and structure of eukaryotic biofilms in an extreme acidic environment, Río Tinto (SW, Spain). *Microb Ecol* 53:294–305

Amaral-Zettler LA, Gomez F, Zettler E, Keenan BG, Amils R, Sogin ML (2002) Eukaryotic diversity in Spain's River of Fire. *Nature* 417:137

Amils R, González-Toril E, Fernández-Remolar D, Gómez F, Aguilera A, Rodríguez N, Malki M, García-Moyano A, Fairén AG, de la Fuente V, Sanz JL (2007) Extreme environments as Mars terrestrial analogs: The Río Tinto case. *Planet Space Sci* 55:370–381

Amils R, González-Toril E, Gómez F, Fernández-Remolar D, Rodríguez N, Malki M, Zuluaga J, Aguilera A, Amaral-Zettler LA (2004) Importance of chemolithotrophy for early life on Earth: The Tinto River (Iberian Pyritic Belt) case. In: Seckbach J (ed) *Origins*. Kluwer Academic Publisher, Dordrecht, pp. 463–480

Apps J, van de Kamp P (1993). Energy gases of abiogenic origin in the Earth's crust. U.S. Geological Survey Professional Paper 1570:81–132

Avery, D (1974) *Not on Queen Victoria's Birthday*. Collins, London.

Benz M, Brune A, Schink B (1998) Anaerobic and aerobic oxidation of ferrous iron at neutral pH by chemoheterotrophic nitrate-reducing bacteria. *Arch Microbiol* 169:159–165

Berazain R, de la Fuente V, Sánchez-Mata D, Rufo L, Rodríguez N, Amils R (2007) Nickel localization on tissues of hyperaccumulator species of *Phyllanthus* L. (Euphorbiaceae) from ultramafic areas of Cuba. *Biol Trace Element Res* 115:67–86

Boston PJ, Ivanov MV, McKay CP (1992). On the possibility of chemosynthetic ecosystems in subsurface habitats on Mars. *Icarus* 95:300–308

Boulter CA (1996) Extensional tectonics and magmatism as drivers of convection leading to Iberian Pyrite Belt massive sulphide deposits? *J Geol Soc London* 153:181–184

Chalk P, Smith C (1983). Chemodenitrification. *Dev Plant Soil Sci* 9:65–89

Chao TT, Zhou L (1983). Extraction techniques for selective dissolution of amorphous iron oxides from soils and sediments. *Soil Sci Soc Am J* 47:225–232

Chapelle FH, O'Neill K, Bradley PM, Methe BA, Ciufo SA, Knobel LL, Lovley DR (2002). A hydrogen-based subsurface microbial community dominated by methanogens. *Nature* 415:312–315

Colmer AR, Temple KL, Hinkle HE (1950) An iron-oxidizing bacterium from the acidic drainage of some bituminous coal mines. *J Bacteriol* 59:317–328

Conrad R (1996) Soil microorganisms as controllers of atmospheric trace gases (H_2, CO, CH_4, OCS, N_2O and NO). *Microbiol Rev* 60:609–640

Davis Jr RA, Nelly AT, Borrego J, Morales JA, Pendon JG, Ryan JG (2000) Rio Tinto estuary (Spain): 5000 years of pollution. *Environ Geol* 39:1107–1116.

D'Hondt S et al. (2004) Distribution of microbial activities in deep subseafloor sediments. *Science* 306:22162221

Drobner E, Huber H, Wächtershäuser G, Rose D Stetter, KO (1990). Pyrite formation linked with hydrogen evolution under anaerobic conditions. *Nature* 346:742–744

Edwards KJ, Gihring TM, Banfield JF (1999). Seasonal variations in microbial populations and environmental conditions in an extreme acid mine drainage environment. *Appl Environ Microbiol* 65:3627–3632

Ehrlich, HL (2002) *Geomicrobiology*, 4th ed. Marcel Dekker

Elbaz-Poulichet F, Braungardt C, Achterberg E, Morley N, Cossa D, Beckers JM, Nomérange P, Cruzado A, Leblanc M (2001) Metal biogeochemistry in the Tinto-Odiel rivers (Southern Spain) and in the Gulf of Cadiz, a synthesis of the results of TOROS project. *Continental Shelf Res* 21:1961–1973

Fairén AG, Fernández-Remolar D, Dohm JM, Baker VR, Amils R (2004) Inhibition of carbonate synthesis in acidic oceans on early Mars. *Nature* 431:423–426

Fernández-Remolar D, Morris RV, Gruener JE, Amils R, Knoll AH (2005b) The Rio Tinto Basin, Spain: Mineralogy, sedimentary geobiology, and implications for interpretation of outcrop rocks at Meridiani Planum. *Mars Earth Planet Sci Lett* 240:149–167

Fernández-Remolar D, Prieto-Ballesteros O, Rodríguez N, Gómez F, Amils R, Gómez-Elvira J, Stoker C (2008) Underground habitats found in the Río Tinto Basin: an approach to Mars subsurface life exploration. *Astrobiol* in press

Fernández-Remolar D, Prieto-Ballesteros O, Rodríguez N, Dávila F, Stevens T, Amils R, Gómez-Elvira J, Stoker C (2005a) Rio Tinto faulted volcanosedimentary deposits as analog habitats for extant subsurface biospheres on Mars: A synthesis of the MARTE drilling project geobiology results. *Lunar and Planetary Science Conference*, Abstract 136

Fernández-Remolar D, Rodríguez N, Gómez F, Amils R (2003). Geological record of an acidic environment driven by iron hydrochemistry: The Tinto River system. *J Geophys Res* 108(E7):5080–5095, doi:1029/2002JE001918

Formisano V, Atreya S, Encrenas T, Ignatiev N, Giuranna M (2004) Detection of methane in the atmosphere of Mars. *Science* 306:1758–1761

Gehrke T, Hallmann R, Sand W (1995). Importance of exopolymers from *Thiobacillus ferrooxidans* and *Leptospirillum ferrooxidans* for bioleaching. In Jérez C, Vargas JT, Wiertz JV, Toledo H (eds) *Biohydrometallurgical Processing*, vol. 1. Universidad de Chile, Santiago, pp. 1–11

Golovacheva RS, Golyshina OV, Karavaiko GI, Dorofeev AG, Pivovarova TA, Chernykh A (1992) A new iron-oxidizing bacterium, *Leptospirillum thermoferrooxidans* sp. nov. *Microbiology* 61: 744–750

González-Toril E, Gómez F, Malki M, Amils R (2006) The isolation and study of acidophilic microorganisms. In Rainey FA, Oren A (eds) *Extremophiles. Methods in Microbiology*, vol. 35. Elsevier Academic Press, London, pp. 471–510

Gonzalez-Toril E, Llobet-Brossa E, Casamayor EO, Amann R, Amils R (2003). Microbial ecology of an extreme acidic environment, the Tinto River. *Appl Environ Microbiol* 69:4853–4865

Jernsletten JA (2005) Fast-turnoff transient electromagnetic (TEM) field study at the Mars analog site of Río Tinto, Spain. *Lunar and Planetary Science Conference*, Abstract 1014

Leistel, JM, Marcoux E, Theiblemont D, Quesada C, Sánchez A, Almodóvar GR, Pascual E, Saez R (1998). The volcanic-hosted massive sulphide deposits of the Iberian Pyrite Belt. *Mineralium Deposita* 33:2–30

Lin LH, Wang PL, Rumble D, Lippmann-Pipke J, Boice E, Pratt LM, Sherwood Lollar B, Brodie EL, Hazen TC, Andersen GL, DeSantis TZ, Moser DP, Kershaw D, Onstott TC (2006) Long-term sustainability of a high-energy, low-diversity crustal biome. *Science* 314:479–482

López-Archilla AI, González AE, Terrón MC, Amils R (2005) Diversity and ecological relationships of the fungal populations of an acidic river of Southwestern Spain: the Tinto River. *Can J Microbiol* 50:923–934

López-Archilla AI, Marín I, Amils R (2001) Microbial community composition and ecology of an acidic aquatic environment: The Tinto River, Spain. *Microb Ecol* 41: 20–35

Lovley DR, Phillips EJP (1987) Rapid assay for microbially reducible ferric iron in aquatic sediments. *Appl Environ Microbiol* 53:1536–140

Malki M, González-Toril E, Sanz JL, Gómez F, Rodríguez N, Amils R (2006) Importance of the iron cycle in biohydrometallurgy. *Hydrometallurgy* 83:223–228

Moreno C, Capitán MA, Doyle M, Nieto JM, Ruiz F, Sáez R (2003) Edad mínima del gossan de Las Cruces: Implicaciones sobre la edad de inicio de los ecosistemas extremos en la Faja Pirítica Ibérica. *Geogaceta* 33:75–78

Rodríguez N, Amils R, Jiménez-Ballesta R, Rufo L, de la Fuente V (2007) Heavy metal content in *Erica andevalensis*: An endemic plant from the extreme acidic environment of Tinto River and its soils. *Arid Land Res Manag* 21:1–15

Rodríguez N, Menéndez N, Tornero J, Amils R, de la Fuente V (2005) Internal iron biomineralization in *Imperata cylindrica*, a perennial grass: Chemical composition, speciation and plant localization. *New Phytol* 165:781–789

Sambrook J, Russell DW (2001) *Molecular Cloning: A Laboratory Manual*, 3rd edition. Cold Spring Harbor Laboratory Press, Cold Spring Harbor

Sand W, Gerke T, Hallmann R, Schippers A (1995). Sulfur chemistry, biofilm, and the (in)direct attack mechanism - A critical evaluation of bacterial leaching. *Appl Microbiol Biotech* 43:961–966

Sand W, Gehrke T, Jozsa PG, Schippers A (2001) (Bio)chemistry of bacterial leaching—Direct vs. indirect bioleaching. *Hydrometallurgy* 59:159–175

Sanz JL, Rodríguez N, Amils R (1997) Effect of chlorinated aliphatic hydrocarbons on the acetoclastic methanogenic activity of granular sludge. *Appl Microbiol Biotechnol* 47:324–328

Schmidt W (2003) Iron solutions: Acquisition strategies and signaling pathways in plants. *Trends Plant Sci* 8:188–193

Squyres S, et al. (2004) In situ evidence for an ancient aqueous environment at Meridiani Planum, Mars. *Science* 306:1709–1714

Stevens TO, McKinley JP (1995) Lithoautotrophic microbial ecosystems in deep basalt aquifers. *Science* 270:450–454.

Stevens TO, McKinley JP (2000) Abiotic controls on H_2 production from basalt-water reactions and implications for aquifer biogeochemistry. *Environ Sci Technol* 34:826–831

Stoker C, Dunagan S, Stevens T, Amils R, Gómez-Elvira J, Fernández D, Hall J, Cannon H, Zavaleta J, Glass B, Lemke L (2004) Mars analog Río Tinto Experiment (MARTE): 2003 drilling campaign to search for a subsurface biosphere at Río Tinto, Spain. *Lunar and Planetary Science Conference*, LPI contribution 1197, paper # 2025

van Geen A, Adkins JF, Boyle EA, Nelson CH, Palenques A (1997) A 120-year record of widespread contamination from mining of the Iberian Pyritic belt. *Geology* 25:291–294

Visviki I, Rachlin JW (1993) Acute and chronic exposure of *Dunaliella salina* and *Chlamydomonas bullosa* to copper and cadmium: Effects on growth. *Arch Environ Contam Toxicol* 26:149–153

Visviki I, Santikul D (2000) The pH tolerance of *Chlamydomonas applanata* (Volvocales, Chlorophyta). *Arch Environ Contam Toxicol* 38:147–151.

Wächtershäuser G (1988) Pyrite formation, the first energy source for life: A hypothesis. *System Appl Microbiol* 10:207–210

Widdel F, Schnell S, Heising S, Ehrenreich A, Assmus B, Schink B (1993) Ferrous iron oxidation by anoxygenic phototrophic bacteria. *Nature* 362:834–836

Zolotov M, Shock E (2005). Formation of jarosite-bearing deposits through aqueous oxidation of pyrite at the Meridiani Planum, Mars *Geophys Res Lett* 32:L21203. doi: 10.1029/2005GL024253

Chapter 11
The Potential for Extant Life in the Soils of Mars

Ronald L. Crawford(✉) and David A. Newcombe

11.1 Introduction

In this chapter we discuss the present state of thought on the possibility that extant life in the form of micro-organisms may exist in the soils of Mars. The Viking missions of 1976 have been the only experimental packages sent to Mars with the specific objective of searching for extant life in Martian soil samples. Landed missions since then have been geological packages that, although examining soils and rocks for minerals that might have biological origins, relied on instruments which were not designed to look specifically for living organisms. The Viking experiments provided some evidence for the possibility of life; however, the general scientific opinion (with notable exceptions; e.g., see http://mars.spherix.com/spie2/Spie2001Oxides/Spie2001-oxides.htm) has been that these experiments showed negative results (Klein 1992; Dick 2006).

The existence in Martian soils of a hypothetical reservoir of problematic "highly oxidizing" material has been suggested in many scientific articles over the past quarter century. These putative oxidizing agents are said to be capable of destroying all organic matter and thus preventing the proliferation of life. However, experimental evidence from Mariner 9, Viking, Pathfinder (Rieder et al. 1997), Kitts Peak, Mars Global Surveyor (Hynek 2004), and the rovers Spirit (Haskin et al. 2005; Gellert et al. 2004) and Opportunity (Rieder et al. 2004; Madden et al. 2004) indicate that Mars probably does not have an especially highly oxidative surface (http://mars.spherix.com/spie2/Spie2001Oxides/Spie2001-oxides.htm). The Martian regolith appears to harbor olivine-rich basaltic rock (Fig. 11.1) and materials such as the ferric sulfate mineral jarosite and gypsum. Some of these minerals appear to have been affected by water, indicating a mineralogy that is not too dissimilar from some basaltic systems that exist on Earth (Haskin et al. 2005).

Microbes are ubiquitous in the dusts on Earth. The dusts covering most of the surface of Mars thus are potential carriers of microbes across vast areas of the Martian landscape (Hagen et al. 1970). Analyses of Martian dusts by instruments

---

Ronald L. Crawford
University of Idaho, Environmental Biotechnology Institute, P.O. Box 441052, Moscow, Idaho, 83844-1052
e-mail: crawford@uidaho.edu

Fig. 11.1 Photograph of a soil in Mars' Gusev Crater where the mineral olivine was detected by instruments aboard the Mars Exploration Rover Spirit. (Photo courtesy of NASA/JPL/Cornell; released January 20, 2004.)

aboard the Mars Exploration Rovers indicate that there is nothing particularly remarkable about the chemistry of the dusts. They appear to be mostly of basaltic mineralogies, with some meteoritic material (Bandfield et al. 2003; Yen et al. 2005). Carbonates, predominately $MgCO_3$, have been spectroscopically identified at a level of 2–5% in Martian dust, and this mineral is expected to be thermodynamically stable on the surface of Mars (Quinn et al. 2006).

Goetz et al. (2005) reported results from Mossbauer spectroscopy and X-ray fluorescence of dust particles captured from the Martian atmosphere by magnets carried by the Mars Exploration Rovers. The dust collected by the magnets was found to contain magnetite and olivine and some ferric oxides, indicating basaltic origins. Magnetite appears to be the mineral responsible for the magnetic properties of the dust. Thus, Martian dusts may be reasonably hospitable to extremophilic micro-organisms such as endospore formers of the genus *Bacillus* known on Earth. Unfortunately, as is generally the case for Mars, the unique properties of this planet's atmosphere complicate such generalizations.

According to Delory et al. (2006), laboratory studies, numerical simulations, and desert field tests indicate that aeolian dust transport can generate atmospheric electricity via contact electrification or "triboelectricity." Thus, electrically charged dust generated during dust storms on Mars could potentially provide ingredients for the generation of oxidants and thereby affect the habitability of Mars dusts and even the Martian surface. One oxidant that might be produced by this mechanism from traces of water in the atmosphere is hydrogen peroxide. Atreya et al. (2006) suggest that hydrogen peroxide, or a superoxide formed from reactions catalyzed by it, might scavenge organic material from the surface soils of Mars and thereby act as a biocide. However, this impact should be minimal on microbes shielded by burial in the subsurface. Thus, theories concerning the lethality of soils on Mars, especially questions regarding the potential effects of "Martian oxidants" on possible extant Martian life, remain mostly conjecture until more direct study of the Martian surface can be performed.

Unfortunately, Mars is a very large place with many potential microbial niches. Research on Earths' harshest environments that might be considered as surrogates for Mars, such as the Atacama Desert of Chile, show that microbial life does exist, but microhabitats that harbor life are widely dispersed, are difficult to detect, and can be mostly hidden by lifeless surroundings (Warren-Rhodes et al. 2006; see Chapter 6). Viking clearly could simply have sampled such a lifeless area, missing a microniche harboring life a few centimeters, meters, or kilometers away. Thus, the Viking results cannot be considered definitive, even if it is accepted that they were negative.

Based on research performed in the most extreme locations on Earth, potential sites for life in Martian soils include: buried water ice or permafrost (Titus et al. 2003; Steven et al. 2006; Tung et al. 2005; see Chapters 7 and 12), the deep subsurface (Weiss et al. 2000; see Chapter 10), and within protected layers of rocks (Kuhlman et al. 2005; see Chapter 6). We do not examine the existing controversies about evidence for past life on Mars, such as the on-going arguments over the origin of magnetite in Martian meteorite ALH84001 (Weiss et al. 2004). Instead, we examine the question, "Is it reasonable to expect to find extant life in the soils of Mars were we to have the resources and technologies needed to look wherever we desired on or below the Martian surface?"

11.2 Mars as a Microbial Habitat

11.2.1 Mars Is an Exceedingly Harsh Environment

At first glance, by Earthly standards Mars does not constitute a favorable environment for terrestrial micro-organisms. The environmental stresses on Mars include low temperature ($-123°C$ to $25°C$), low-pressure ($\sim 600\,Pa = 6\,mbar$), high fluxes of biologically destructive, low-wavelength ultraviolet (UV) irradiation (8.4 to 67 W/m^2; Diaz and Schulze-Makuch 2006), and perhaps a highly oxidizing environment (Crawford et al. 2003; Kral et al. 2004; Benner et al. 2000). Also, inasmuch as Mars has no magnetic field and a very thin atmosphere, galactic cosmic rays and solar flare particle fluxes may impact and penetrate the surface to a greater extent than they do on Earth. Potentially lethal doses of these types of radiation may accrue on the Martian surface and even to some depth in the soil. However, such ionizing radiation does not appear to be sufficient for sterilizing soil at least in the short term (for a discussion of this topic, see http://www7.nationalacademies.org/ssb/bcmarsch3.html).

Recently, researchers analyzing data from the Electron Reflectometer (ER) on board the Mars Global Surveyor noted areas in which a dropoff in electron flux measured by the ER was observed (Mitchell et al. 2001). A surficial map of the readings corresponded to localized intense crustal magnetic fields. These fields are strong enough to stave off solar winds allowing their measurement with ER on a spacecraft 400 km above the surface and raising the question as to whether the fields may be strong enough to provide zones of reduced energetic particles and thereby moderate deleterious ionizing properties. If so, could these local zones be more favorable to microbial survival?

Even highly stress-resistant microbes such as *Deinococcus radiodurans* and the cyanobacterium *Chroococcidiopsis* do not survive very long in near-surface soils under the combined environmental stresses of low temperature, ultra-low-pressure, and intense ultraviolet irradiation (Diaz Schulze-Makuch 2006; Cockell et al. 2005). However, survival of such organisms can be greatly enhanced by the presence of liquid water and burial at depth in the soil. Thus, the subsurface of Mars is likely to be to be a far more favorable environment than the surface for survival and proliferation of micro-organisms such as those we know from Earth.

11.2.2 The Atmosphere of Mars Is Thin and Unlike That on Earth

The atmospheric composition of Mars is not Earth-like. The atmosphere of Mars contains mostly carbon dioxide (CO_2, ~95%) with some nitrogen (N_2, ~2.7%), argon (Ar, ~1.6%), oxygen (O_2, ~0.13%), and traces of water vapor (H_2O, ~0.03%; Soffen 1976; Clancy et al. 1990). There are small amounts of methane (CH_4) in the Martian atmosphere. The global average methane mixing ratio, as measured by the Planetary Fourier Spectrometer on the European Mars Express spacecraft, was 10 ± 5 parts per billion by volume (ppbv), varying between 0 and 30 ppbv over the planet (Formisano et al. 2004). There are other trace gases in the Martian atmosphere, including hydrogen (H_2) and carbon monoxide (CO) (Weiss et al. 2000; Clancy et al. 1990). The composition of the atmosphere on Mars by itself may not be detrimental to the survival and proliferation of some Earth-like microbes. For example, the cyanobacteria *Synechococcus* and *Anabaena* were shown to survive 101 kPa (100%) pCO_2 when pressure was gradually increased by 15 kPa per day, and *Plectonema* actively grew under these conditions. All of these strains grew in an anoxic atmosphere of 5 kPa pCO_2 in N_2 (Thomas et al. 2005). The composition of the gases within soils at depth on Mars is not known and may be locally different than those measured on the surface by the Viking spacecraft.

11.2.3 All Life Must Have a Source of Energy

Any form of extraterrestrial life must have a source of energy (Crawford et al. 2002, 2001; Lang et al. 2001), but this does not appear to be a problem on Mars (Irwin and Schulze-Makuch 2001; Chyba and Phillips 2002; Jakosky and Shock 1998). As Weiss et al. (2000) state, "The location and density of biologically useful energy sources on Mars will limit the biomass, spatial distribution, and organism size of any biota." On the surface there is of course adequate energy available from sunlight to support photosynthetic life forms that could use readily available carbon dioxide as a source of carbon. It is unlikely that organisms could use this radiant energy directly because of the extreme conditions on the surface (Weiss et al. 2000), unless they were able to find protected niches within the soil subsurface (Crawford et al. 2003) or in other locations such as internal layers of rocks (Kuhlman et al. 2005). Other potential sources of energy for Martian biota include hydrothermal and chemical weathering energy (Jakosky and Shock 1998; Varnes et al. 2003), thermosynthesis in the presence of a thermal gradient within ice (Muller 2003), and photochemically produced atmospheric H_2 and CO diffusing into the regolith to protected regions of the subsurface. Based on modeling experiments, the latter energy source appears sufficient to sustain a subsurface CO/H_2 oxidizing microbial community on Mars (Weiss et al. 2000).

11.2.4 Life as We Know It on Earth Requires Sources of Nitrogen and Sulfur

For life forms as exist on Earth, nitrogen is required as a nutrient for the synthesis of proteins, nucleic acids, and various metabolic cofactors. Sulfur is required for synthesis of certain amino acids and enzyme cofactors. Thus, if microbes similar to those found on Earth exist on Mars, they must be able to access sources of these two elements. As discussed above, some nitrogen (N_2, ~2.7%) is found in the atmosphere of Mars (Soffen 1976; Clancy et al. 1990). Thus, extant microbes in Martian soils might access this nitrogen supply by complex processes similar to those used by free-living nitrogen-fixing bacteria on Earth (Peters et al. 1995). Measurement of nitrogen on present-day Mars has been limited to only that present in the atmosphere, and this atmospheric nitrogen represents a small fraction of the nitrogen thought to have been received by the planet during its formation (Mancinelli and Banin 2003). As hypothesized by Mancinelli and Banin (2003), Mars soils may be like soils seen in Earth's extremely dry deserts. Such soils contain some nitrogen as nitrate salts and some as fixed ammonium bound to aluminosilicate minerals. Analyses carried out on the Martian surface by the Opportunity Rover show that rocky outcrops are rich in sulfur (Reider et al. 2004; Haskin et al. 2005). Thus, sufficient nitrogen and sulfur probably are available to support extant microbial life in the soils of Mars. Capone et al. (2006) suggest that analysis of the abundance and chemistry of nitrogen deposits will provide important clues as to the presence of life.

11.2.5 Could There Be Methanogens on Mars?

Kral et al. (2004) and others (Boston et al. 1992; McCollom 1999) suggest that Martian equivalents of Earth's methane-forming bacteria (methanogens) might be a type of micro-organism that could take advantage of the particular energy supplies available in the Martian subsurface or on other extraterrestrial locations such as Europa. As Kral et al. (2004) discuss, methanogens might be able to use a geothermal source of hydrogen for energy, which might be provided by volcanic or hydrothermal activity or the reaction of basalt and anaerobic water. For carbon they could use CO_2, which is abundant in the Martian atmosphere and soil. In addition, they would need subsurface liquid water. Kral et al. (2004) performed experiments to show that certain methanogens can grow on a Mars soil simulant when supplied with CO_2, H_2, and varying amounts of water. Thus, methanogens can grow even in relatively nutrient-poor soils, as long as liquid water is available, even if water availability is intermittent (Kendrick and Kral 2006).

Evidence is slowly accumulating from satellite-based observations such as those of the European Space Agency's Mars Express, which carries the Advanced Radar for Subsurface and Ionospheric Sounding (MARSIS) instrument, that ice and perhaps subsurface water do exist on Mars. MARSIS detected radar reflections from a subsurface base of an ice layer close to the planet's north pole, indicating that the deposit is about 1.8-km thick (Schilling 2005). Another radar instrument (SHARAD, Shallow Subsurface Radar) was launched on NASA's Mars Reconnaissance Orbiter that has been in orbit since March 10, 2006 and will greatly enhance the search for subsurface liquid water (Reichhardt 2005). SHARAD will seek liquid or frozen water at up to 1 km into Mars' crust (see also Section 11.3.1 for a discussion of recent evidence for liquid water presence).

Rother and Metcalf (2004) remind us that certain methanogens can use carbon monoxide as a growth substrate, producing methane via a pathway that involves hydrogen as an intermediate. They tested the ability of *Methanosarcina acetivorans* C2A to use CO as a growth substrate, finding it to grow on CO to high cell densities with a doubling time of only 24 h. Methane formation surpassed acetate and formate formation when the cultures entered a stationary growth phase. As discussed above, CO is available to some degree in the Martian environment (Weiss et al. 2000), so the CO/H_2 combination for support of methanogenic growth appears feasible on Mars. This possibility is further suggested by the observation that hydrogen-based methanogenic communities do occur in Earth's subsurface, providing an analogue for possible subsurface microbial ecosystems on other planets (Chapelle et al. 2002).

How likely is an extant "methanogens on Mars" scenario? As mentioned above (see Section 11.2.2), Formisano et al. (2004) recently reported the detection of methane in the Martian atmosphere at a global average methane mixing ratio of 10 ± 5 ppbv. They concluded that the source of methane could be either biogenic or nonbiogenic, including past or present subsurface micro-organisms, hydrothermal activity, or as a result of cometary impacts. Others have confirmed the presence of methane in the Martian atmosphere (Krasnopolsky et al. 2004; Mumma et al. 2004).

Onstott et al. (2006) mathematically modeled the Martian subsurface methane flux based on available data from Mars using previous information about methane sources within the South African Precambrian crust of the Witwatersrand Basin. They concluded that methane should only reach the surface if it is found as a hydrate and the hydrate is saturated in the cryosphere; otherwise they suggest it should be captured within the cryosphere. The sublimation of such a hydrate-rich cryosphere could generate the observed methane flux.

They also conclude that a microbial explanation of methane production only seems possible if there is a hypersaline environment above the hydrate stability zone. Methane as a hydrate can be derived from a number of sources, so Onstott et al. (2006) suggest that the C and H isotopic values of Martian CH_4 be analyzed by a future Mars instrument package, which would be one of the better ways to look for the possibility of an Earth-like biotic signature. Thus, there appears to be measurable methane in the Martian atmosphere, and it seems to be replenished continuously, perhaps from the subsurface. Indeed, methane survives for a relatively short time (a few hundred years) in the Martian atmosphere so it must be constantly replenished in order to maintain the observed concentrations. Its origin (biotic or abiotic) will remain an unanswered question until more analyses can be performed using instruments landed on the Martian surface or passed through its atmosphere (see also Webster 2005; Durry et al. 2004; de Bergh 1995).

11.3 Searching for Microbes in the Soils of Mars

11.3.1 Where Should Investigators Look for Extant Life on Mars?

In order to increase the likelihood of success in finding extant micro-organisms on Mars, it will be crucial to carefully consider where to look. This led Klein (1992) to state that, "Attempts to search for extant biology should be restrained until adequate new information about potential habitable microenvironments is obtained." In the last decade, considerable progress has been made toward this goal. Several locations where life might exist on Mars have been suggested (Rothschild 1990). The most commonly mentioned niches for life on Mars include those listed below. The common feature of all these environments is that they are to some degree protected from surrounding harsh conditions and/or support the availability of liquid water (Farmer et al. 1995).

- In or on rocks: This is a niche analogous to that of Earth's endolithic cyanobacteria that live inside porous sandstone rocks, protected by a thin rock crust (Friedmann and Ocampo-Friedmann 1984) or evaporates (Rothschild 1990).
- Caves: The field of cave geomicrobiology has direct relevance to studies of extant life on other planets. Caves may provide protection from surface

stresses such as UV radiation. Caves on Earth contain many unusual organisms that oxidize or reduce minerals such as manganese, iron, and sulfur (Boston et al. 2003).
- In thin mineral "varnishes": This niche is represented by the thin layer of silica and metal oxides that frequently covers rocks in virtually all of Earth's dry and cold deserts (Kuhlman et al. 2005).
- In polar ice caps: This niche would be similar to Earthly locations where snow and ice algae are found (Kohshima 2000).
- In buried water ice or permafrost (Steven et al. 2006; Tung et al. 2005).
- In possible volcanic regions: This environment would be similar to regions near deep-sea hydrothermal vents where some of Earth's chemolithoautotrophs are found (Miroshnichenko and Bonch-Osmolovskaya 2006; Gaill 1993).
- In the generic "deep subsurface" (Weiss et al. 2000).
- In areas of nitrogen salt accumulation (Mancinelli and Banin 2003; Capone et al. 2006).
- In areas maintaining an intense localized magnetic field (Mitchell et al. 2001).

Images taken by the Mars Global Surveyor Mars Orbiter Camera (MOC) suggest that liquid water existed on Mars in the recent past (Fig. 11.2A). Observations of the same Martian surface location in 1999 and then in 2005 further indicate that liquid water may have flowed on the planet's surface during the last seven years (Fig. 11.2B). Such images are very useful in planning landing locations for future life detection missions on the planet.

Ostroumov (1995) suggests that any viable micro-organism on Mars probably exists with minimum metabolism in compact zones and that these zones may contain microvolumes of unfrozen water in the Martian permafrost. Such zones, if they exist, are likely to be located deep beneath the Martian surface (Weiss et al. 2000) and may be widely dispersed and difficult to locate (Warren-Rhodes et al. 2006).

Other less obvious locations might also be considered. For example, Rothschild (1990) points out that micro-organisms can survive in salt crystals and actively metabolize while encrusted in evaporites. Evaporites may occur on Mars and can attenuate UV radiation while transmitting light that might be used for photosynthesis (light of 400–700 nm in wavelength). Thus, this author proposes that evaporites might provide a niche for extant Martian microbial communities on Mars. Ellery and Wynn-Williams (2003) also suggest that evaporites in paleolake craters might be a good place to look for life on Mars.

11.3.2 How Can Investigators Obtain Appropriate Samples of Martian Soil to Look for Extant Life?

As mentioned in the preceding section and based on what is now known about potential niches for microbial life in the soils of Mars, it appears that the best places to look will be located well below the surface of the planet (Reichhardt 2005) in

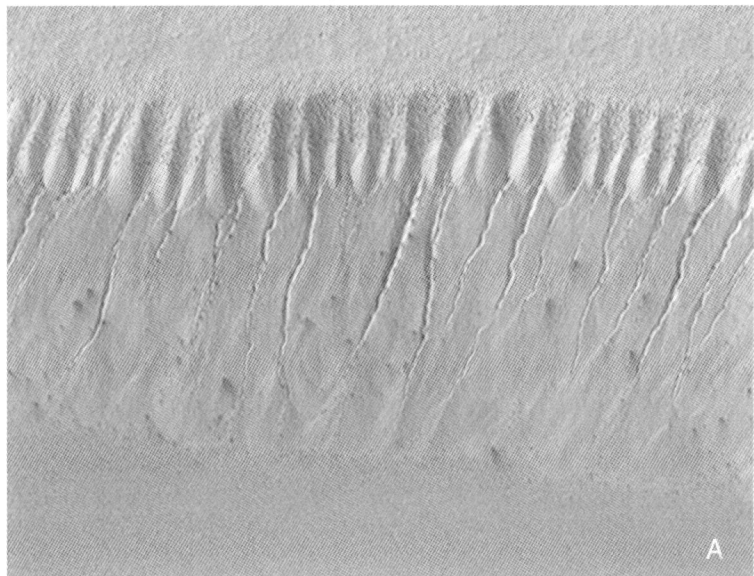

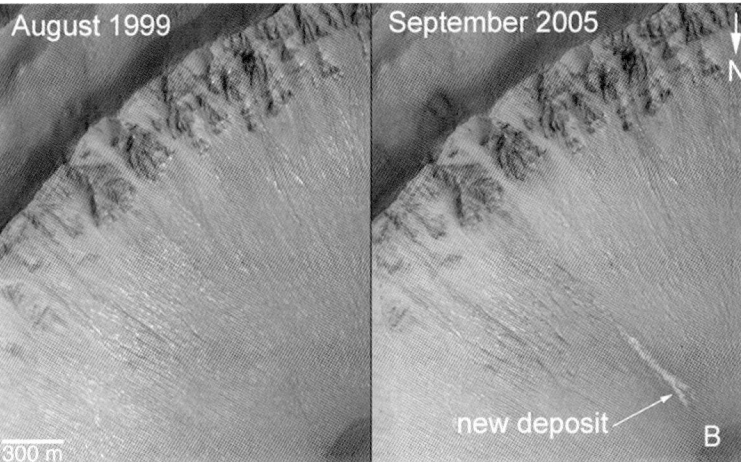

Fig. 11.2 (**A**) Gully landforms proposed to have been caused by geologically recent seepage and runoff of liquid water on Mars' south polar pitted plains (photo from the Mars Global Surveyor Mars Orbiter Camera (MOC) courtesy of NASA/JPL/Malin Space Science Systems). This image was acquired July 14, 1999 and covers an area approximately 2.8 km wide by 2.1 km high. (**B**) NASA photographs taken by the MOC have revealed bright new deposits observable in two gullies that suggest water carried sediment through them sometime during the past seven years (image released December 6, 2006)

niches that are protected from stresses such as UV light (e.g., caves) or in areas where strong magnetic fields are present that attenuate galactic cosmic rays and solar flare particle fluxes (http://www7.nationalacademies.org/ssb/bcmarsch3.html). It also will be crucial to explore sites where liquid water is likely to reside at least intermittently.

Assuming that interdisciplinary teams of scientists and engineers can choose several logical locations to search (Lobitz et al. 2001), the next problem becomes how to aseptically sample at these places. Because the best sampling locations may be multiple kilometers below the surface, the challenges will be daunting. Blacic et al (2000) analyzed some of the challenges associated with aseptic drilling and sampling on Mars to a depth of 200 m, which may not be deep enough to explore some of the most likely habitats for microbial life. Their analyses eliminated all known terrestrial drilling technologies but provided some ideas for approaches that might have promise. The NASA Ames Research Center, Honeybee Robotics, Georgia Tech, and the Mars Institute are involved in the Drilling Automation for Mars Exploration (DAME) project, focused on development of automated technologies for future drilling exploration of Mars. The team has developed a Mars-prototype drill that is lightweight, uses no lubricants, and needs relatively little power. Honeybee Robotics is building the drill that is being tested in the regolith-like material inside Haughton Crater on Devon Island in Canada's Nunavut Territory north of Ontario and Quebec. This activity involves drilling into ice layers and permafrost similar to what one might expect to find near the surface in Martian polar regions (http://www.nasa.gov/centers/ames/research/exploringtheuniverse/ai.html).

NASA is already field-testing another system designed to drill for subsurface Martian life. The drill called MARTE includes a drilling platform and suite of scientific instruments that are able to search for evidence of life in samples the robotic drill has extracted from below ground. The drill rig is about 2.4-m tall and sits on a three-legged platform about 2.1 m in diameter. The drill uses less than 150 W of power, bores using carbide-diamond cutters, and uses no drilling fluid. It makes core plugs of rock that are about 20-cm long and brings them to the surface (http://marte.arc.nasa.gov/). The drill only reaches depths of about 6 m, so if used on Mars, it would have to be employed at locations where extant life might be relatively near the surface (e.g., within caves or fields of evaporites).

One promising approach for sampling beneath the Martian surface would be through the use of a torpedolike device, or "CryoScout," that could melt its way through the ice cap on Mars (Kounaves 2003) and deliver samples to an instrument carried along or to the surface for analysis. Ellery et al. (2005) propose "Vanguard," a spacecraft that might be transported to Mars by a Mars Express-type spacecraft and would carry ground-penetrating, instrument-equipped "moles" mounted onto a rover for subsurface penetration but only to a depth of about 5 m. Each mole would take a one-way trip down a borehole and provide real-time data without the need for recovery of moles or samples.

Mancinelli (2003) suggests that a manned mission may be necessary for sampling the deep Martian subsurface. Such a mission could involve drilling 3 km or more into the Martian soil, collecting samples, and conducting preliminary analyses to select samples for return to Earth. This scenario, as others presented above, requires that the drilling equipment be sterilized prior to use; the collection, containment, and retrieval of samples would need to be conducted such that the mission crew is protected from exposure to the collected soils or their dusts. Also, Martian soils (or ice) returned to Earth must be physically and biologically isolated from the time they are collected until analysis inside a BL4-level containment facility on Earth. The presence of humans on Mars would clearly make meeting these difficult challenges more feasible. A manned mission of this magnitude is probably decades away, but scientists are already thinking about how to circumvent the many environmental and engineering challenges involved in such a mission (Ehlmann et al. 2005). Even on Earth, sampling microbial communities in the deep subsurface, without introducing contaminants from the surface or from human sources, is very difficult (see Chapter 10). Extreme measures and complex controls must be used to prove that sampling has indeed been accomplished without contamination (Lehman et al. 2001; Juck et al. 2005). Clearly, trying to sample from the deep subsurface of Mars while avoiding contamination by Earthly microbes brought along by astronauts or aboard robotic spacecraft will be an even greater challenge.

11.4 Detection of Life in Martian Soils

11.4.1 Analytical Methods Are Highly Developed for Detecting Organic Signatures of Life and Life Processes in Soil

Once investigators have decided where to look for life in Martian soils, then methods must be employed to detect any extant life present. Suggested approaches are many and varied and date from the 1960s (Levin et al. 1964; Christian et al. 1965) prior to the Viking missions to the present time. For example, Simmonds (1970) discussed pyrolysis-gas chromatography-mass spectrometry as a life detection method, and the science payload on Viking included a combined pyrolysis-gas chromatography-mass spectrometry instrument to detect thermal fragments originating from the principal classes of bio-organic matter found in living systems such as protein and carbohydrate. No such molecules were observed. This could mean that life does not exist on Mars, that the wrong location was tested, or that the detection limit of the instrument was insufficient and organic molecules were actually present but were missed or were invisible to the GC-MS used (Benner et al. 2000). Nonetheless, it is likely that chromatography-mass spectrometry will be a mainstay of extraterrestrial life detection experiments in the future because instruments have been greatly improved and miniaturized in the post-Viking era. Investigators also now know much more about Mars and what to look for as compared to the state of knowledge three decades ago (Palmer and Limero 2001; Pietrogrande et al. 2005).

A related method for chromatographic separation and detection of organic molecules extracted from soil is capillary electrophoresis, which has been developed as a useful miniaturized tool for examination of Martian soils. This technique involves chiral separations of fluorescein isothiocyanate-labeled amino acids using a microfabricated capillary electrophoresis chip, and it has been proposed to explore the feasibility of using such devices to search for extinct or extant life signs in extraterrestrial environments (Hutt et al. 1999). Also, Lang et al. (2001) used supercritical fluid extraction and both capillary electrophoresis and high-performance liquid chromatography (HPLC) equipped with diode array or electrochemical detectors as a means to detect signature biological redox compounds as an approach for the detection of molecules that could be signatures of life in extraterrestrial soils.

11.4.2 Radiorespirometry Is a Highly Useful and Sensitive Life Detection Technique

Levin et al. (1964) developed a radiorespirometric approach for life detection. This is a highly logical approach for detection of actively metabolizing heterotrophic microbes and involves the introduction of a soil sample into a microbiological medium that supports growth of a wide range of Earth micro-organisms. In this case, however, selected ingredients of the medium are labeled with radioactive ^{14}C. If these compounds are degraded (mineralized) to carbon dioxide, the CO_2 evolved will be tagged as $^{14}CO_2$, which can be detected easily with great sensitivity. In one of the classic experiments of modern times, such a radiorespirometric instrument was landed on Mars aboard Viking; $^{14}CO_2$ was in fact produced from a sample of Martian soil. However, this result is now thought by most in the scientific community to have resulted from abiotic oxidation of the substrates by strong "Mars soil oxidants" (Klein 1992), although the discussion continues (Van Dongen et al. 2005).

11.4.3 Life Might Be Detected by Observing Controlled Electron Transport Used by Living Organisms to Obtain Useful Energy

Lang et al. (2001) suggested a method for life detection based on the fact that living entities require a continual input of energy accessed through coupled oxidations and reductions (an electron transport chain). They demonstrated using Earthly soils that the identification of extracted components of electron transport chains is useful for remote detection of a chemical signature of life. The prototype instrument package these investigators developed used supercritical carbon dioxide for soil extraction, followed by chromatography or electrophoresis to separate extracted compounds, with final detection by voltammetry and tandem mass spectrometry. Later, the same group used Earth-derived soils to develop a related life detection system based on direct observation of a biological redox signature (Crawford et al. 2002).

11 The Potential for Extant Life in the Soils of Mars   237

They measured the ability of soil microbial communities to reduce artificial electron acceptors. Living organisms in pure culture and those naturally found in soil were shown to reduce 2,3-dichlorophenol indophenol (DCIP) and the tetrazolium dye 2,3-bis(2-methoxy-4-nitro-5-sulfophenyl)-2H-tetrazolium-5-carboxanilide inner salt (XTT). Uninoculated or sterilized controls did not reduce the dyes. A soil from Antarctica that was determined by chemical signature and DNA analysis to be sterile also did not reduce the dyes. Dye reduction was readily monitored by observing dye color changes upon reduction, a simple approach to implement robotically. The authors concluded that observation of dye reduction, supplemented with extraction and identification of only a few specific signature redox-active biochemicals such as porphyrins or quinones could provide a simplified means to detect a signature of life in the soils of other planets or their moons.

Direct microscopic visualization of bacteria in collected subsurface rock specimens might be another alternative for detection of Earth-like microbes. Tobin et al. (1999) developed a protocol that enables the visualization of intact microbial cells in petrographic thin sections that might be adaptable to a Mars mission. Their method avoided detaching the cells from their host mineral surfaces and avoided microbial contamination during the lapidary process. Nucleic acid stains that specifically target double-stranded DNA and RNA were utilized for *in situ* visualization of cells in surface and subsurface basalts collected in Idaho, USA. Extending this approach, the authors found that examination of samples incubated with acetic acid-UL-^{14}C using phosphor imaging allowed *in situ* visualization of ^{14}C-labeled biomass. The greatest challenge for this type of direct visualization technology will be its detection limit. Microbial distribution in Martian rocks likely would exhibit a high degree of spatial heterogeneity at the micrometer scale, and this could easily lead to failure in observing microbes that are actually present. As with many techniques discussed in this chapter, a positive result would be outstanding but a negative result still ambiguous.

Not every paper on this topic can be discussed here, but representative publications describing a variety of approaches proposed for detection of signatures of life in Martian soil are summarized in Table 11.1.

11.5 Conclusions

By Earthly standards, Mars is clearly not a favorable environment for terrestrial micro-organisms and so differs from environments across much of our own planet. The most biologically significant stresses on Mars include low temperature, high fluxes of biologically destructive radiation, possible shortages of certain nutrients such as nitrogen, and perhaps a highly oxidizing and therefore biocidal surface. In contrast, the geochemistry of the rocks and soils on Mars based on observations by orbiting and landed spacecraft over the past 35 years indicates a mineralogy not too dissimilar from some basaltic systems seen on Earth that are well colonized by micro-organisms. Also, Mars has sufficient sources of carbon and energy to support

Table 11.1 Methods suggested for the detection of extraterrestrial life

Proposed Life Detection Methods	References
Pyrolysis-gas chromatography-mass spectrometry	Simmonds (1970)
Gas chromatography-mass spectrometry	Fox (2002); Buch et al. (2003); Pietrogrande et al. (2005)
Detection of controlled electron transport and signature biological redox compounds	Sotnikov (1970); Crawford et al. (2002); Lang et al. (2001)
Fluorescence techniques	Sotnikov (1970); Kawasaki (1994)
Phosphatase activity	Kobayashi et al. (2004)
Immunological approaches	Schweitzer et al. (2005)
Raman spectroscopy	Ellery and Wynn-Williams (2003)
Remote sensing of chlorophylls	Knacke (2003)
Dichroism spectroscopy	Xu et al. (2003)
Electrochemical and polarimetric methods	Thiemann (1975); Kounaves (2003)
Observations of structural complexity	Nealson et al. (2002)
Detection of amino acids	Pollock et al. (1977); Hutt et al. (1999); Rodier et al. (2001)
Use of charge-coupled devices	Nussinov et al. (1992)
Release of heat from metabolizable substrates as measured by a microcalorimeter	Imshenetsky et al. (1976)
Determination of optical activity (turbidity) from growth on organic substrates	Imshenetskii and Evdokimova (1975)
Soil gas disequilibria	Brazhnikov et al. (1971); Kelley et al. (1975)
In situ imaging	Tobin et al. (1999)

the growth of a large variety of Earth-like micro-organisms. Based on decades of research performed in the most extreme locations on Earth, the potential for extant microbes in the soils and rocks of Mars cannot yet be excluded. Potential sites for life on Mars include: in buried water ice or permafrost, within the deep subsurface, within protected layers of rocks or in caves, in possible volcanic regions, in areas of nitrogen salt accumulation, and in areas maintaining an intense localized magnetic field that may protect life from some forms of ionizing radiation.

Based on these generalizations, it is reasonable to expect to find extant life in Martian soils if investigators were able to apply the resources and technologies needed to look wherever desired, particularly below the Martian surface. The opposite hypothesis, that Mars is sterile, likewise cannot be dismissed. Unfortunately, scientists at this point have too little information to make a definitive conclusion as to the existence of extant life on Mars. This relegates the scientific community to continued discussion and speculation until humans make the trip and look firsthand or until appropriately collected and processed samples of Martian soils and rocks are returned to Earth for careful examination. Thus, the question posed by the title of this chapter remains open for debate.

References

Atreya SK, Wong AS, Renno NO, Farrell WM, Delory GT, Sentman DD, Cummer SA, Marshall JR, Rafkin SC, Catling DC (2006) Oxidant enhancement in Martian dust devils and storms: implications for life and habitability. *Astrobiol* 6:439–450

Bandfield JL, Glotch TD, Christensen PR (2003) Spectroscopic identification of carbonate minerals in the Martian dust. *Science* 301:1084–1087

Benner SA, Devine KG, Matveeva LN, Powell DH (2000) The missing organic molecules on Mars. *Proc Natl Acad Sci USA*. 97:2425–2430

Blacic JD, Dreesen DS, Mockler T (2000) Report of conceptual systems analysis of drilling systems for 200-meter-depth penetration and sampling of the Martian subsurface. Los Alamos National Laboratory, Los Alamos, New Mexico, Technical Report Number CA-VR-00-4742

Boston PJ, Frederick RD, Welch SM, Werker J, Meyer TR, Sprungman B, Hildreth-Werker V, Thompson SL, Murphy DL (2003) Human utilization of subsurface extraterrestrial environments. *Gravit Space Biol Bull* 16:121–131

Boston PM, Ivanov MV, McKay CP (1992) On the possibility of chemosynthetic ecosystems in subsurface habitats on Mars. *Icarus* 95:300–308

Brazhnikov VV, Mukhin LM, Otrostchenko VA, Fedorova RI (1971) Gas exchange ("soil breathing") in the detection of extraterrestrial life. *Life Sci Space Res* 9:179–189

Buch A, Sternberg R, Meunier D, Rodier C, Laurent C, Raulin F, Vidal-Madjar C (2003) Solvent extraction of organic molecules of exobiological interest for *in situ* analysis of the Martian soil. *J Chromatogr A* 999:165–174

Capone DG, Popa R, Flood B, Nealson KH (2006) Follow the nitrogen. *Science* 312:708–709

Chapelle FH, O'Neill K, Bradley PM, Methe BA, Ciufo SA, Knobel LL, Lovley DR (2002) A hydrogen-based subsurface microbial community dominated by methanogens. *Nature* 415:312–315

Christian GD, Knoblock EC, Purdy WC (1965) Instruments for detection of extraterrestrial life. Anal Chem 37:29A–35A

Chyba CF, Phillips CB (2002) Europa as an abode of life. *Orig Life Evol Biosph* 32:47–68

Clancy R, Muhleman D, Berge G (1990) Global changes in the 0–70km thermal structure of the Mars atmosphere derived from 1975–1989 microwave CO spectra. *J Geophys Res* 95:14543–14554

Cockell CS, Schuerger AC, Billi D, Friedmann EI, Panitz C (2005) Effects of a simulated Martian UV flux on the cyanobacterium, *Chroococcidiopsis* sp. 029. *Astrobiol* 5:127–140

Crawford RL, Paszczynski A, Allenbach L (2003) Potassium ferrate [Fe(VI)] does not mediate self-sterilization of a surrogate Mars soil. *BMC Microbiol* 3:4–14

Crawford RL, Paszczynski A, Lang Q, Cheng IF, Barnes B, Anderson TJ, Wells R, Wai C, Corti G, Allenbach L, Erwin DP, Park J, Assefi T, Mojarradi M (2001) In search of the molecules of life. *Icarus* 154(2):531–539

Crawford RL, Paszczynski A, Lang Q, Erwin DP, Allenbach L, Corti G, Anderson TJ, Cheng IF, Wai C, Barnes B, Wells R, Assefi T, Mojarradi M (2002) Measurement of microbial activity in soil by colorimetric observation of *in situ* dye reduction: An approach to detection of extraterrestrial life. *BMC Microbiol* 2:22–29

de Bergh C (1995) Isotopic ratios in planetary atmospheres. *Adv Space Res* 15:427–440

Delory GT, Farrell WM, Atreya SK, Renno NO, Wong AS, Cummer SA, Sentman DD, Marshall JR, Rafkin SC, Catling DC (2006) Oxidant enhancement in Martian dust devils and storms: storm electric fields and electron dissociative attachment. *Astrobiol* 6:451–462

Diaz B, Schulze-Makuch D (2006) Microbial survival rates of *Escherichia* coli and *Deinococcus radiodurans* under low temperature, low pressure, and UV-Irradiation conditions, and their relevance to possible Martian life. *Astrobiol* 6:332–347

Dick SJ (2006) NASA and the search for life in the universe. *Endeavour* 30:71–75

Durry G, Amarouche N, Zeninari V, Parvitte B, Lebarbu T, Ovarlez J (2004) *In situ* sensing of the middle atmosphere with balloonborne near-infrared laser diodes. *Spectrochim Acta A Mol Biomol Spectrosc* 60:3371–3379

Ehlmann BL, Chowdhury J, Marzullo TC, Collins RE, Litzenberger J, Ibsen S, Krauser WR, DeKock B, Hannon M, Kinnevan J, Shepard R, Grant FD (2005) Humans to Mars: A feasibility and cost-benefit analysis. *Acta Astronaut* 56:851–858

Ellery A, Ball AJ, Cockell C, Dickensheets D, Edwards H, Kolb C, Lammer H, Patel M, Richter L (2005) Vanguard–A European robotic astrobiology-focused Mars sub-surface mission proposal. *Acta Astronaut* 56:397–407

Ellery A, Wynn-Williams D (2003) Why Raman spectroscopy on Mars?—A case of the right tool for the right job. *Astrobiol* 3:565–579

Farmer J, Des Marais D, Greeley R, Landheim R, Klein H (1995) Site selection for Mars exobiology. *Adv Space Res* 15:157–162

Formisano V, Atreya S, Encrenaz T, Ignatiev N, Giuranna M (2004) Detection of methane in the atmosphere of Mars. *Science* 306:1758–1761

Fox A (2002) Chemical markers for bacteria in extraterrestrial samples. *Anat Rec* 268:180–185

Friedmann EI, Ocampo-Friedmann R (1984) The Antarctic cryptoendolithic ecosystem: Relevance to exobiology. *Orig Life* 14:771–776

Gaill F (1993) Aspects of life development at deep sea hydrothermal vents. *FASEB J* 7:558–565

Gellert R, Rieder R, Anderson RC, Bruckner J, Clark BC, Dreibus G, Economou T, Klingelhofer G, Lugmair GW, Ming DW, Squyres SW, D'Uston C, Wanke H, Yen A, Zipfel J (2004) Chemistry of rocks and soils in Gusev Crater from the alpha particle x-ray spectrometer. *Science* 305:829–832

Goetz W, Bertelsen P, Binau CS, Gunnlaugsson HP, Hviid SF, Kinch KM, Madsen DE, Madsen MB, Olsen M, Gellert R, Klingelhofer G, Ming DW, Morris RV, Rieder R, Rodionov DS, de Souza PA Jr, Schroder C, Squyres SW, Wdowiak T, Yen A (2005) Indication of drier periods on Mars from the chemistry and mineralogy of atmospheric dust. *Nature* 436:62–65

Hagen CA, Hawrylewicz EJ, Anderson BT, Cephus ML (1970) Effect of ultraviolet on the survival of bacteria airborne in simulated Martian dust clouds. *Life Sci Space Res* 8:53–58

Haskin LA, Wang A, Jolliff BL, McSween HY, Clark BC, Des Marais DJ, McLennan SM, Tosca NJ, Hurowitz JA, Farmer JD, Yen A, Squyres SW, Arvidson RE, Klingelhofer G, Schroder C, de Souza PA Jr, Ming DW, Gellert R, Zipfel J, Bruckner J, Bell JF III, Herkenhoff K, Christensen PR, Ruff S, Blaney D, Gorevan S, Cabrol NA, Crumpler L, Grant J, Soderblom L (2005) Water alteration of rocks and soils on Mars at the Spirit Rover site in Gusev Crater. *Nature* 436:66–69

Hutt LD, Glavin DP, Bada JL, Mathies RA (1999) Microfabricated capillary electrophoresis amino acid chirality analyzer for extraterrestrial exploration. *Anal Chem* 71:4000–4006

Hynek BM (2004) Implications for hydrologic processes on Mars from extensive bedrock outcrops throughout Terra Meridiani. *Nature* 431:156–159

Imshenetskii AA, Evdokimova MD (1975) Determination of optical activity of the growth medium as a method for detection of extraterrestrial life. *Mikrobiol* 44:1030–1033

Imshenetsky AA, Evdokimova MD, Sotnikov GG (1976) On methods of detection of extraterrestrial life. *Life Sci Space Res* 14:345–349

Irwin LN, Schulze-Makuch D (2001) Assessing the plausibility of life on other worlds. *Astrobiol* 1:143–160

Jakosky BM, Shock EL (1998) The biological potential of Mars, the early Earth, and Europa. *J Geophys Res* 103:19359–19364

Juck DF, Whissell G, Steven B, Pollard W, McKay CP, Greer CW, Whyte LG (2005) Utilization of fluorescent microspheres and a green fluorescent protein-marked strain for assessment of microbiological contamination of permafrost and ground ice core samples from the Canadian High Arctic. *Appl Environ Microbiol* 71:1035–1041

Kawasaki Y (1994) Development of detection system of extraterrestrial microorganisms. *Biol Sci Space* 8:103–113

Kelley LM, Meyer ED, Zumberge JE, Bandurski EL, Nagy B (1975) Stereoisomeric specificity and soil gas disequilibria: Implications for Martian life detection. *Appl Microbiol* 29:229–233

Kendrick MG, Kral TA (2006) Survival of methanogens during desiccation: Implications for life on Mars. *Astrobiol* 6:546–551

Klein HP (1992) The Viking biology experiments: Epilogue and prologue. *Orig Life Evol Biosph* 21:255–261

Knacke RF (2003) Possibilities for the detection of microbial life on extrasolar planets. *Astrobiol* 3:531–541

Kobayashi K, Ito Y, Moroi A, Edazawa Y, Kaneko T, Takano Y (2004) Detection of biosphere frontier by using phosphatase activity. *Biol Sci Space* 18:144–145

Kohshima S (2000) Psycrophilic organisms in snow and ice. *Biol Sci Space* 14:353–362

Kounaves SP (2003) Electrochemical approaches for chemical and biological analysis on Mars. *Chemphyschem* 4:162–168

Kral TA, Bekkum CR, McKay CP (2004) Growth of methanogens on a Mars soil simulant. *Orig Life Evol Biosph* 34:615–626

Krasnopolsky VA, Maillard JP, Owen TC (2004) Detection of methane in the Martian atmosphere: Evidence for life? *Icarus* 172:537–547

Kuhlman KR, Fusco WG, La Duc MT, Allenbach LB, Ball CL, Kuhlman GM, Anderson RC, Erickson IK, Stuecker T, Benardini J, Strap JL, Crawford RL (2005) Diversity of microorganisms within rock varnish in the Whipple Mountains, California. *Appl Environ Microbiol* 72:1708–1715

Lang Q, Cheng IF, Wai C, Paszczynski A, Crawford RL, Barnes B, Anderson TJ, Wells R, Corti G, Allenbach L, Erwin DP, Assefi T, Mojarradi M (2001) Supercritical fluid extraction and HPLC-DAD-ECD detection of signature redox compounds from sand and soil samples. *Analyt Biochem* 301:225–234

Lehman RM, Roberto FF, Earley D, Bruhn DF, Brink SE, O'Connell SP, Delwiche ME, Colwell FS (2001) Attached and unattached bacterial communities in a 120-meter corehole in an acidic, crystalline rock aquifer. *Appl Environ Microbiol* 67:2095–2106

Levin GV, Heim AH, Thompson MF, Beem DR, Horowitz NH (1964) "Gulliver", an experiment for extraterrestrial life detection and analysis. *Life Sci Space Res* 2:124–132

Lobitz B, Wood BL, Averner MM, McKay CP (2001) Use of spacecraft data to derive regions on Mars where liquid water would be stable. *Proc Natl Acad Sci USA* 98:2132–2137

Madden ME, Bodnar RJ, Rimstidt JD (2004) Jarosite as an indicator of water-limited chemical weathering on Mars. *Nature* 431:821–823

Mancinelli RL (2003) Planetary protection and the search for life beneath the surface of Mars. *Adv Space Res* 31:103–107

Mancinelli RL, Banin A (2003) Where is the nitrogen on Mars? *Internat J Astrobiol* 2:217–225

McCollom TM (1999) Methanogenesis as a potential source of chemical energy for primary biomass production by autotrophic organisms in hydrothermal systems on Europa. *J Geophys Res* 104(E12):30729–30742

Miroshnichenko ML, Bonch-Osmolovskaya EA (2006) Recent developments in the thermophilic microbiology of deep-sea hydrothermal vents. *Extremophiles* 10:85–96

Mitchell DL, Lin RP, Mazelle C, Reme H, Cloutier PA, Connerney JEP, Acuna MH, Ness NF (2001) Probing Mars' crustal magnetic field and ionosphere with the MGS electron reflectometer. *J Geophys Res* 106:23,419–423,427

Muller AW (2003) Finding extraterrestrial organisms living on thermosynthesis. *Astrobiol* 3:555–564

Mumma MJ, Novak RE, DiSanti MA, Bonev BP, Dello Russo N (2004) Detection and mapping of methane and water on Mars. *DPS Meeting 36*, American Astronomical Society, Washington DC Abstract 26.02

Nealson KH, Tsapin A, Storrie-Lombardi M (2002) Searching for life in the universe: unconventional methods for an unconventional problem. *Int Microbiol* 5:223–230

Nussinov M D, Lysenko SV, Kozlovskii MYu, Pogodin YuS (1992) An approach to the detection of microbe life in planetary environments through charge-coupled devices. *J Br Interplanet Soc* 45:13–14

Onstott TC, McGown D, Kessler J, Lollar BS, Lehmann KK, Clifford SM (2006) Martian CH_4: Sources, flux, and detection. *Astrobiol* 6:377–395

Ostroumov V (1995) A physical and chemical characterization of Martian permafrost as a possible habitat for viable microorganisms. *Adv Space Res* 15:229–236

Palmer PT, Limero TF (2001) Mass spectrometry in the U.S. space program: Past, present, and future. *J Am Soc Mass Spectrom* 12:656–675

Peters JW, Fisher K, Dean DR (1995) Nitrogenase structure and function: A biochemical-genetic perspective. *Ann Rev Microbiol* 49:335–366

Pietrogrande MC, Zampolli MG, Dondi F, Szopa C, Sternberg R, Buch A, Raulin F (2005) *In situ* analysis of the Martian soil by gas chromatography: Decoding of complex chromatograms of organic molecules of exobiological interest. *J Chromatogr A* 1071:255–261

Pollock GE, Day R, Kinsey S, Miller SL (1977) Detection of optical asymmetry in amino acids by gas chromatography for extraterrestrial space exploration: Results of a new soil processing scheme with breadboard instrumentation. *Life Sci Space Res* 15:27–34

Quinn R, Zent AP, McKay CP (2006) The photochemical stability of carbonates on Mars. *Astrobiol* 6:581–591

Reichhardt T (2005) Mars exploration: Going underground. *Nature* 435:266–267

Rieder R, Economou T, Wanke H, Turkevich A, Crisp J, Bruckner J, Dreibus G, McSween HY Jr (1997) The chemical composition of Martian soil and rocks returned by the mobile alpha proton X-ray spectrometer: preliminary results from the X-ray mode. *Science* 278:1771–1774

Rieder R, Gellert R, Anderson RC, Bruckner J, Clark BC, Dreibus G, Economou T, Klingelhofer G, Lugmair GW, Ming DW, Squyres SW, d'Uston C, Wanke H, Yen A, Zipfel J (2004) Chemistry of rocks and soils at Meridiani Planum from the alpha particle x-ray spectrometer. *Science* 306:1746–1749

Rodier C, Vandenabeele-Trambouze O, Sternberg R, Coscia D, Coll P, Szopa C, Raulin F, Vidal-Madjar C, Cabane M, Israel G, Grenier-Loustalot MF, Dobrijevic M, Despois D (2001) Detection of Martian amino acids by chemical derivatization coupled to gas chromatography: *In situ* and laboratory analysis. *Adv Space Res* 27:195–199

Rother M, Metcalf WW (2004) Anaerobic growth of *Methanosarcina acetivorans* C2A on carbon monoxide: An unusual way of life for a methanogenic archaeon. *Proc Natl Acad Sci USA* 101:16929–16934

Rothschild LJ (1990) Earth analogs for Martian life. Microbes in evaporites, a new model system for life on Mars. *Icarus* 88:246–260

Schilling G (2005) Space science. Europe trumpets successes on Mars and Titan. *Science* 310:1598

Schweitzer MH, Wittmeyer J, Avci R, Pincus S (2005) Experimental support for an immunological approach to the search for life on other planets. *Astrobiol* 5:30–47

Simmonds PG (1970) Whole microorganisms studied by pyrolysis-gas chromatography-mass spectrometry: Significance for extraterrestrial life detection experiments. *Appl Microbiol* 20:567–572

Soffen GA (1976) Scientific results of the Viking missions. *Science* 194:1274–1276

Sotnikov GG (1970) Detection of iron-porphyrin proteins with a biochemiluminescent method in search of extraterrestrial life. *Life Sci Space Res* 8:90–98

Steven B, Leveille R, Pollard WH, Whyte LG (2006) Microbial ecology and biodiversity in permafrost. *Extremophiles* 10:259–267

Thiemann W (1975) Is the detection of optical activity in extraterrestrial samples a safe indicator for life? *Life Sci Space Res* 13:63–69

Thomas DJ, Sullivan SL, Price AL, Zimmerman SM (2005) Common freshwater cyanobacteria grow in 100% CO_2. *Astrobiol* 5:66–74

Titus TN, Kieffer HH, Christensen PR (2003) Exposed water ice discovered near the south pole of Mars. *Science* 299:1048–1051

Tobin KJ, Onstott TC, DeFlaun MF, Colwell FS, Fredrickson J (1999) *In situ* imaging of microorganisms in geologic material. *J Microbiol Meth* 37:201–213

Tung HC, Bramall NE, Price PB (2005) Microbial origin of excess methane in glacial ice and implications for life on Mars. *Proc Natl Acad Sci USA* 102:18292–18296

Van Dongen HPA, Miller JD, Levin VG, Straat PA (2005) A circadian biosignature in the labeled release data from Mars? In: Hoover RB, Levin GV, Rozanov AY, Gladstone GR (eds) *Instruments, Methods, and Missions for Astrobiology, SPIE Proceedings*, August 2005, vol. 5906, 59060C, Astrobiology and Planetary Missions. The International Society for Optical Engineering, Bellingham WA

Varnes ES, Jakosky BM, McCollom TM (2003) Biological potential of Martian hydrothermal systems. *Astrobiol* 3:407–414

Warren-Rhodes KA, Rhodes KL, Pointing SB, Ewing SA, Lacap DC, Gomez-Silva B, Amundson R, Friedmann EI, McKay CP (2006) Hypolithic cyanobacteria, dry limit of photosynthesis, and microbial ecology in the hyperarid Atacama Desert. *Microb Ecol* 52:389–398

Webster CR (2005) Measuring methane and its isotopes $^{12}CH_4$, $^{13}CH_4$, and CH_3D on the surface of Mars with *in situ* laser spectroscopy. *Appl Opt* 44:1226–1235

Weiss BP, Kim SS, Kirschvink JL, Kopp RE, Sankaran M, Kobayashi A, Komeili AM (2004) Magnetic tests for magnetosome chains in Martian meteorite ALH84001. *Proc Natl Acad Sci USA* 101:8281–8284

Weiss BP, Yung YL, Nealson KH (2000) Atmospheric energy for subsurface life on Mars? *Proc Natl Acad Sci USA* 97:1395–1399

Xu J, Ramian GJ, Galan JF, Savvidis PG, Scopatz AM, Birge RR, Allen SJ, Plaxco KW (2003) Terahertz circular dichroism spectroscopy: A potential approach to the *in situ* detection of life's metabolic and genetic machinery. *Astrobiol* 3:489–504

Yen AS, et al (2005) An integrated view of the chemistry and mineralogy of Martian soils. *Nature* 436:49–54 Erratum in: (2005) *Nature* 436(7052):881

Part III
Anthropogenic Extreme Soils

Chapter 12
Bacteriology of Extremely Cold Soils Exposed to Hydrocarbon Pollution

Lucas A.M. Ruberto, Susana C. Vazquez, and Walter P. Mac Cormack(✉)

12.1 Introduction

The study of bacterial communities present in natural ecosystems has been the object of attention of several research groups during the past 15 years, driven mainly by the development of DNA-based methodologies (Amman et al. 1995; Holben 1997). Taking advantage of these methods, it was shown that most of the individual components of natural bacterial communities are incapable of growth on standard culture media. In this sense, it was evident that more than 99% of the bacterial cells present in a soil sample were not able to grow on standard culture media (Roszak and Colwell 1987; Torsvik et al. 1990). Soil represents an extremely complex matrix. Soils considered as habitat are characterized by physical, chemical, and temporal heterogeneities at any scale from km to nm (Young and Ritz 2000). Each field, forest, tundra, or cold desert has a unique soil food web with a particular proportion of organisms, and a particular level of complexity within each group of organisms. These differences are the result of soil, vegetation, and climate factors.

Detailed studies on soil micro-organisms demonstrated that even within soil habitats deemed homogeneous at the plot scale, distribution of soil bacteria was highly structured, with different bacterial communities located in specific sites, probably in response to the heterogeneities imposed by the habitat (Franklin and Mills 2003). Although the above-cited references dealt with agricultural soils, the concept of soil heterogeneity is a general rule (Young and Crawford 2004) and must be kept in mind in all descriptive or analytical studies of soil bacterial communities. Bacterial populations located in individual soil microhabitats perform several biophysical and biochemical processes, turning the soil into the most complex and biodiverse ecosystem on Earth.

As was mentioned above, local climate conditions are among the main factors conditioning the number and type of micro-organisms present in a soil. In Antarctica (as well as in the Arctic, the Alps, the Puna, and other high-altitude soils), low

---

Walter P. Mac Cormack
Instituto Antártico Argentino, Departamento de Biología. Cerrito 1248 (C1010AAZ), Buenos Aires, Argentina
e-mail: wmac@huemul.ffyb.uba.ar

temperature represents a key selection-pressure factor restricting microbial diversity. In these extremely cold soils, only psychrotolerant and psychrophilic micro-organisms are able to survive and proliferate (Nedwell 1999; Deming 2002).

The presence of contaminant chemicals from anthropogenic origin represents an additional factor of stress that strongly affects the composition of the microbiota of contaminated soil. These changes were reported for soils and sediments from different areas of the world (Röling et al. 2002; Haack et al. 2004; Zocca et al. 2004; Katsivela et al. 2005) including Alpine (Margesin et al. 2003), Arctic (Juck et al. 2000), and Antarctic (Mac Cormack et al. 1998; Aislabie et al. 2001; Saul et al. 2005) extremely cold soils altered by hydrocarbon contamination. Although Antarctica is at the present time one of the sites least affected by human activity worldwide, the scientific and logistic stations (which are powered almost exclusively by combustion of oil derivative fuels) as well as the fishery and tourist activities (the latter having increased dramatically over the last years) have led to spatially limited hydrocarbon contamination that in some cases has been highly significant (Kennicutt II and Sweet 1992; Aislabie et al. 2004). Most of the Antarctic sites where stations are built and which are consequently at high risk of hydrocarbon pollution are restricted to those small areas remaining free of ice during the summer. Together, these areas represent less than 0.3% of the Antarctic surface, which is almost 14,000,000 km^2 in total (Fox and Cooper 1994). They occur mainly in the shore of the Antarctic Peninsula and the Ross Sea region. Humans are not the only living beings that chose these areas to settle down. On the contrary, the major part or the Antarctic micro- and macroorganisms develop in these ice-free coastal areas, including the only two vascular plants found in Antarctica and the penguins, seabirds, and several marine mammals that arrive in spring for mating and breeding. As a consequence of this, pollution in Antarctica does not affect humans only, but it is a global environmental problem that probably has a deep effect on soil food webs, which are much simpler and more fragile than those encountered in other biomes of the world. In addition to a direct effect on plants and animals, hydrocarbons may also modify the entire terrestrial food web by altering the soil microbiota, changing the type of species being present and the number of their representatives.

In the last decades, bioremediation – or the exploitation of the ability of micro-organisms (mainly bacteria) to catabolize hydrocarbons and other organic pollutants – has been considered as the most adequate tool for reducing contaminant levels and eventually clean up affected areas. Studies on variously located soils have been published, documenting the effectiveness of different bioremediation strategies for reducing soil hydrocarbon contamination (Heitzer and Sayler 1993; Hooker and Skeen 1996; Cunningham and Philp 2000; Marchal et al. 2003; Bento et al. 2005). These studies have mainly dealt with the extremely cold soils from the Arctic (Whyte et al. 1999; Eriksson et al. 2001; Mohn et al. 2001; Rike et al. 2003) and the Alps (Margesin and Schinner 1997a,b, 2001). Recently, soils from Antarctica have received some attention (Delille and Pelletier 2002; Ruberto et al. 2003; Delille et al. 2004; Snape et al. 2005). In all these cold soils, the presence of efficient hydrocarbon-degrading bacterial isolates was reported, such organisms constituting communities selected from the original soil inhabitants after a long history of contamination (Mac Cormack and Fraile 1997; Whyte et al. 1997; Thomassin-Lacroix et al. 2001;

Baraniecki et al. 2002; Margesin et al. 2003). However, the low level of metabolic activity prevailing at low temperatures, the lower evaporation rate of the volatile compounds compared with rates observed for temperate soils, and even the existence of a permafrost and an active layer under continuous cycles of freezing and thawing influence the fate of the different hydrocarbons and must be added to the list of known factors already identified as interfering with degradation of contaminating hydrocarbons in all soil types (Providenti et al. 1993).

Due to the limitations imposed on bacterial growth by the harsh climate conditions, and also to restrictions included in the Antarctic legislation (as stated in the 1959 Antarctic Treaty and its Protocol on Environmental Protection signed in Madrid in 1991, which prevent the introduction of any kind of nonindigenous organisms, including foreign micro-organisms for use in bioremediation), cleanup processes must be carried out using autochthonous bacterial strains. For this reason, a deep knowledge of the metabolic capabilities and the taxonomic position of the Antarctic micro-organisms able to proliferate in hydrocarbon-contaminated Antarctic soils is needed. Studies have been initiated only recently along those lines, with the result that little is known about the environmental impact of past human activities in the Antarctic, starting from the first settlements up to the 1980s.

The aim of this chapter is to review current information on the presence of hydrocarbons in Antarctic soils and on their effect on the composition of the natural bacterial flora. Also discussed is the available knowledge on functions of the natural Antarctic bacterial flora and the effectiveness of bioremediation strategies based on the application of external innocula to hydrocarbon-contaminated Antarctic soils. References are made to similar works carried out in other studied extremely cold soils under hydrocarbon contamination.

12.2 An Overview of the Bacteriology of Pristine Antarctic Soils

A review of the Antarctic bacteriology (as for the bacteriology of any other site) requires that recent developments in bacterial taxonomy be taken into account. In addition, soil represents an extraordinarily complex environment, so that samples taken from different sites differ in their physicochemical properties including water and organic matter content, pH, presence of plant cover, presence of guano (which is typical of ornithogenic soils), level of nutrients, and many others. Climate conditions (such as temperature range, winds, and solar radiation regime) vary considerably among sites and even for a given site at different times. All these differences make it difficult to compare richness, abundance, and other soil biodiversity determinants as they occur in different soil samples. As was remarked by Young and Crawford (2004), soil biology (including soil microbiology), soil chemistry, and soil physics are currently progressing almost as independent fields, which impedes the construction of an integrated view of soil ecology. Despite these limitations, a number of studies have examined the composition of the bacterial flora from extremely cold soils.

Antarctic and Arctic soils are generally comprised of an active layer (exposed to a broad thermal oscillation ranging between +15°C and −35°C) and a subjacent permafrost, which is defined here as the soil and/or rocks remaining below 0°C for at least two consecutive years (ACGR 1988). Permafrost is divided in an upper zone (0.5–20-m thick) with moderate temperature oscillations and a deeper permafrost, showing a stable temperature regime (−5°C to −10°C). The Arctic and most of the Antarctica permafrost is ice-cemented. Exceptions are represented by the inland valley floors and sides of the McMurdo Dry Valleys, where the low water content determines the prevalence of dry (non ice-cemented) permafrost (Bockheim 2002). Readers interested in a thorough description of the Antarctic active layer and permafrost are referred to the article by Bockheim and Hall (2002). In the present chapter, discussions are restricted to the bacterial communities colonizing the upper part of the active layer. The microbial ecology and biodiversity of permafrost has been extensively reviewed (Steven et al. 2003; see also Chapter 7).

The active layer of the Antarctic soils is eminently aerobic and, consequently, strict anaerobes are infrequent. On the contrary, peatlands and other anaerobic habitats are frequent in the Arctic, offering a wide anaerobic niche for growth of strictly anaerobic bacteria (see Chapter 9). However, the presence of strict anaerobes (*Clostridium* and *Eubacterium*) in Antarctic soils has been reported as far back as 55 years ago by Prévot and Moreau (1952). Several other *Clostridium* species were found in soils from the Antarctic Showa Station by Miwa (1975). More recently, other anaerobic bacteria were isolated from different Antarctic soils. Sulfate-reducing genera (*Desulfovibrio* and *Desulfotomaculum*) were identified from the peatlands of Signy Island (Christie 1987). In yet other studies on Antarctic permafrost and peats, anaerobic bacteria were detected by culturing (Brambilla et al. 2000) and several others were inferred using culture-independent methods (Brambilla et al. 2000; Christner et al. 2003). However, extensive phylogenetic studies of Dry Valley mineral soils mentioned in a review by Cowan and Ah Tow (2004) gave no signals of 16S rRNA sequences from anaerobic groups. In the Antarctic active layer, where soil hydrocarbon contamination occurs, the presence of anaerobic bacteria is rare and the aerobic bacteria represent almost the total fraction of the bacterial flora.

Although anaerobic pathways of hydrocarbon degradation exist (Heider et al. 1998; Shinoda et al. 2005) and could be of relevance in subsurface environments such as the permafrost table or other anaerobic contaminated terrestrial Antarctic habitats, such processes have not been studied at the present time in nature and no information about the natural interactions between hydrocarbons and strictly anaerobic bacteria is currently available. Thus, for the most part the present review concentrates on the aerobic soil bacteria.

An important number of bacterial isolates from Antarctic soils have been identified at the species level using biochemical and morphological tests. Several of these studies, carried out on vegetated soils, reported a significant fraction of coryneform bacteria. More than 30 years ago, Baker and Smith (1972) isolated species belonging to the genus *Arthrobacter, Brevibacterium, Corynebacterium, Kurthia*, and *Cellulomonas* from vegetated soils at Signy Island. The authors concluded that the microbial communities of the studied Antarctic soils were qualitatively and quantitatively similar to the microbiota of cold peat soils from other locations. Other members of the phylum Actinobacteria were frequently isolated from Antarctic soils using culture-dependent methods (Smith et al. 2000). Among Firmicutes, *Bacillus* spp. were widely reported (Johnson et al. 1978; Vishniac 1993), whereas *Pseudomonas, Flavobacterium*, and *Agrobacterium* were some of the most abundantly isolated Gracilicutes (Vishniac 1993; Zdanowski and Węgleński 2001; Smith et al. 2006). An important diversity of cyanobacteria was also reported in the terrestrial environments (Nienow and Friedmann 1993), mainly in the well-studied Dry Valleys, as these bacteria were recognized as major components of the endolithic communities. Various other bacterial species have been occasionally reported in a great number of publications dealing with the isolation and identification of soil bacteria from different Antarctic areas. Readers interested in a detailed listing of the culture-dependent isolates from Antarctic soils are referred to the reports of Hirsch et al. (1985), Line (1988), Vincent (1988), and Vishniac (1993).

Several works based on culture-dependent methods suggested that, at least in vegetated soils, most Antarctic bacteria belong to a small number of cosmopolitan genera (Vishniac 1993). However, this assumption probably was biased by the fact that those common taxa comprise a high number of strains easily culturable in simple laboratory culture media. Many other important taxa of Antarctic soil bacteria, being, if not endemic to the Antarctic soils, at least exclusive to extremely cold environments, remain undetectable by the classical culture-dependent methods. These include fastidious micro-organisms, and others that would be co-culture-dependent or that would occur in a viable but nonculturable state. The bias associated with culture-dependent methods was discussed by several authors (Franzmann 1996; Smith et al. 2006).

Most of the novel micro-organisms described in the last 15 years using the new molecular techniques belong to genera that are not exclusive to the Antarctica. As the number of studies on bacterial composition of natural environments increased, it became evident that most of the genera described for a particular Antarctic environment (soils, ice, seawater, marine sediments) also had representatives in similar but non-Antarctic extremely cold environments (such as the Arctic or deep-sea environments). Thus, adaptation to cold (psychrophilia and psychrotolerance), rather than geographic location, seems to be the main determinant of the distribution of the majority of the described genera. Finally, it must be noted that before 1988, until molecular techniques were used, no member of the Archaea domain had been recognized in Antarctic environments (Franzmann 1996).

Most of the Antarctic studies using molecular techniques have focused on bacterial communities from aquatic environments, particularly seawater, marine animals-associated bacteria, sea ice, and marine and lacustrine sediments. These studies led to the description of a great number of new Antarctic species (Bowman et al. 1997a,b; Bozal et al. 1997; Denner et al. 2001). On the other hand, fewer studies have dealt with Antarctic soil bacterial communities, so that our knowledge of the diversity of the nonculturable bacteria in these soils remains fragmentary. Unpublished data mentioned by Cowan and Ah Tow (2004) indicated that as many as 50% of the sequences retrieved from Miers Valley (Ross Desert) mineral soils are from as yet uncultured bacteria, whereas the remaining sequences belonged, either to taxa previously reported as being dominant by culture-dependent methods (27% of retrieved sequences were related to Actinobacteria and 11% to Bacteroidetes), or to other taxa with few cultivated representatives (6% of the sequences were related to Acidobacteria and 6% to Verrucomicrobia).

When a lichen-dominated lithic bacterial community of the McMurdo Dry Valleys was studied using DNA-based methods (de la Torre et al. 2003) members of the order Cytophagales (a subgroup of the Bacteroidetes lineage) dominated. Other phylotypes, including Actinobacteria (*Rhodococcus* sp., *Microsphaera* sp., *Blastococcus* sp., *Sporichthya* sp.), Alphaproteobacteria (*Sphingomonas* sp.), Gammaproteobacteria (*Actinobacter* sp.), and members of Planctomycetales were detected in low proportion. In a recent publication (Aislabie et al. 2006b), 728 clones from soil samples taken from four locations along 77°S in Victoria Land, Antarctica, were characterized by restriction fragment length polymorphism (RFLP) of the 16S rDNA. DNA sequencing showed that the ribotypes occurring more than three times grouped within the bacterial divisions Bacteroidetes, Actinobacteria, Proteobacteria, *Thermus/Deinococcus*, Acidobacteria, Firmicutes, and Cyanobacteria. A salient feature of this study was the observed dominance of a few ribotypes occurring with an abundance of 10% of the analyzed clones or more, which represents an unusually high value for surface soils. However, the dominant ribotypes changed among locations.

The presence of such over-dominant ribotypes could be a characteristic of the Antarctic soils (de la Torre et al. 2003) that is not frequently observed in soils from other latitudes, including cold soils from the Arctic (Zhou et al. 1997) or Colorado Alpine soils (Lipson and Schmidt 2004). These differences in community composition may reflect adaptations to the diverse physicochemical and biological factors present in each studied soil and mainly to the water content (Aislabie et al. 2006b). Thus, bacterial groups such as *Deinococcus* and *Rubrobacter* only prevailed in the driest sampling site (Wright Valley) due to their desiccation tolerance.

Analysis of microbial diversity from Arctic and Alpine tundra using 16S rRNA gene clone library sequencing methods showed that several of the detected bacterial groups were widely dispersed, occurring in these and also Antarctic soils. However, the relative abundance of such groups varied widely depending on the site and even on the season. Recently, Nemergut et al. (2005) have reviewed the data about Alpine and Arctic soil microbial communities, trying to relate this variability with ecological aspects of the studied systems.

12.3 Hydrocarbons in Polar Soils

The introduction of hydrocarbons in the Antarctic region results from a combination of a global input flow from low-level and long-term natural and anthropogenic sources and accidental spills. Hydrocarbons can be introduced in the Antarctic environment from local natural sources such as lichens, algae, and bacteria (Neff 1979; Cripps 1989, 1990) and even meteorites have been reported as sources of Antarctic polycyclic aromatic hydrocarbons (PAHs; Naraoka et al. 2000), although this has little relevance on the level of polycyclic aromatic hydrocarbons in Antarctic soils. However, over this baseline of hydrocarbons from biogenic sources, localized contamination events with both aliphatic and aromatic hydrocarbons, have occurred since the onset of human activity in the Antarctic continent in the early 20th century (Platt and Mackie 1980). Humans generate hydrocarbon pollution mainly through the scientific stations, and through the logistic operations supporting the pelagic fisheries and tourism. The latter, in particular, has increased rapidly starting in the mid-1980s, and has now become an important activity (Bauer 2001). In fact, the recent grounding of the Norwegian cruise ship *Nordkapp*, which occurred in February 2007 and caused a fuel spill in Deception Island, is a recent example of the pollution risks related to this growing activity.

Hydrocarbon contamination in the Arctic is a more frequent and extended problem than it is in Antarctica and has been more extensively studied. Very high levels of both aliphatic and aromatic hydrocarbons have been reported for Arctic soils (Mohn et al. 2001; Whyte et al. 2001). Alpine cold soils areas have also been reported as being affected by hydrocarbon pollution (Margesin and Schinner 2001).

Which are the current levels of hydrocarbon contamination in Antarctic soils? The main hydrocarbon source in Antarctic soils is represented by fuels (crude and fuel oils) spilled on land during the environmentally risky activities of storage-tank filling and refueling of vehicles (Fig. 12.1). Several authors have reported high

Fig. 12.1 Refueling activities in Antarctic Stations require transport of fuels from logistic ships to storage tanks by boat, helicopter, and/or pipeline. These operations are essential but imply an important environmental risk

aliphatic hydrocarbons levels caused by the mentioned activities (Gore et al. 1999; Delille 2000; Delille and Pelletier 2002). The aromatic fraction of these fuels is predominantly represented by naphthyl (2-ring) compounds and the level of PAHs with 3 or more rings (Cripps 1989) is low. However, these higher molecular-weight PAHs are the result of combustion processes of organic compounds.

Although the Antarctic Treaty and its Protocol on Environmental Protection now restrict the incineration of wastes, for decades this was a frequent practice at the stations. It generated localized highly polluted areas rich in PAHs with 3 or more rings, which were generally associated with power generators and incinerators. For example, Kennicutt II et al. (1992a) reported that subtidal sediments below an abandoned open incineration site contained combustion-derived polynuclear aromatic hydrocarbons. Soils collected at Old Palmer Station were also contaminated with these kinds of compounds. Again in a recent study (Vodopivez et al. 2007, in press), carried out at Jubany Station (King George Island, South Shetland Islands), it was found that, although all sampling sites showed very low levels of PAHs, higher values were detected in combustion-related sites (boathouse, incinerator site). However, samples from one site where a diesel oil spill occurred did not show any increase of PAHs level compared with the control site. Table 12.1 shows some of the n-alkanes and PAHs levels reported from soils and sediments near Antarctic stations.

Table 12.1 Total n-alkanes and total PAH concentration in soils and coastal sediments reported for Antarctic areas

Sampling Area	n-Alkanes in Surface Soil (ng/g dw[a])	PAHs in Surface Soil (ng/g dw)	n-Alkanes in Surface Marine Sediments (ng/g dw)	PAHs in Surface Marine Sediments (ng/g dw)	Reference
King Edward Cove, South Georgia Islands	nr[b]	nr	399	16	Mackie et al. 1978
Signy Island	1,220,000	71,000	1,731	280	Cripps 1992
Arthur Harbor	2,344,971	85,659	121,851[c]	14,491[b]	Kennicutt II et al. 1992a
Old Palmer Station	1,182,092	345,765	772,734	59,487	Kennicutt II et al. 1992b
Admiralty Bay	nr	nr	nr	32	Bicego et al. 1998
McMurdo Station	nr	88,452	nr	nr	Mazzera et al. 1999
Scott Base	nr	8,105	nr	nr	Aislabie et al. 1999
Admiralty Bay	nr	nr	nr	271	Martins et al. 2004
Potter Cove	nr	1,182	nr	1,908	Curtosi et al. 2007

[a] dw, dry weight.
[b] nr, not reported.
[c] Intertidal sediment.

Several monitorings of the Argentinean Antarctic Stations with permanent human presence have shown no significant problems of generalized contamination. However, as is the case for many stations from other countries, local manipulation of hydrocarbons has determined small restricted areas of pollution (Mac Cormack and Fraile 1997; Curtosi et al. 2007; Vodopivez et al. 2007).

The fate of hydrocarbons accumulated in any polluted soil can vary greatly depending on several factors, including the chemical characteristics of the hydrocarbons (Lundstedt 2003), ageing (Alexander 2000; Dictor et al. 2003), the soil properties (Chiou et al. 1998; da Conceição et al. 2006), the extent of the abiotic and biotic processes that promote their elimination from the contaminated site (Johnsen et al. 2005), and the enormously complex interactions which occur between them. Although an exhaustive analysis of each of the factors affecting the fate of hydrocarbons in soils exceeds the scope of this chapter, it is important to mention two of these factors that are characteristic of extremely cold soils.

The first factor is the low temperatures existing in these soils, that slow down natural biological processes. Micro-organisms show low specific affinity values (a_A) under polar temperatures, and thus become increasingly unable to sequester the substrates from their environment (Nedwell 1999). Under these conditions, even small quantities of hydrocarbons being spilled over a long time period could lead to a significant accumulation of these compounds due to the slow rate of natural biological degradation. The second determining factor is the continuous process of freezing and thawing that is maintained in high latitude areas, which results in the smallest soil particles being selectively transported to deeper layers, whereas the upper layers become enriched in larger size particles (Anderson et al. 1978). This process favors downward migration of those hydrocarbons showing the highest affinity for the smallest soil particles and modifies the distribution of the different hydrocarbons in the soil layers.

In addition, the permafrost layer could act as a low permeability barrier for the migration of hydrocarbons. Biggar et al. (1998) have reported that, under certain conditions, some contaminants can penetrate the Canadian Arctic permafrost layer, and that this was a site-specific phenomenon. However, recent studies (Curtosi et al. 2007) about distribution of PAHs in soils near Jubany Station showed the existence of a concentration gradient of PAHs, with levels being relatively low at the soil surface and increasing with depth in the active layer, reaching a maximum just below the interface between the active layer and the permafrost, and then progressively declining with depth in the permafrost (Fig. 12.2). These results show that both the freezing and thawing cycles and the permafrost barrier exert a significant effect on the distribution and fate of hydrocarbons in polar soils. Accompanying effects will be observed on the distribution of the bacterial flora able to tolerate (or degrade) such hydrocarbons.

Table 12.1 presents chronic hydrocarbon contamination events, where soils have been exposed to the pollutant for long time periods. A very different situation is created by acute contamination events, where significant amounts of hydrocarbons are spilled on a previously pristine soil. In this case, the effect produced on the microbiota, which was not adapted to the presence of contaminants, is very different from that caused by a chronic contamination. From the above discussion on factors influencing hydrocarbon mobility and other soil characteristics, it remains difficult to

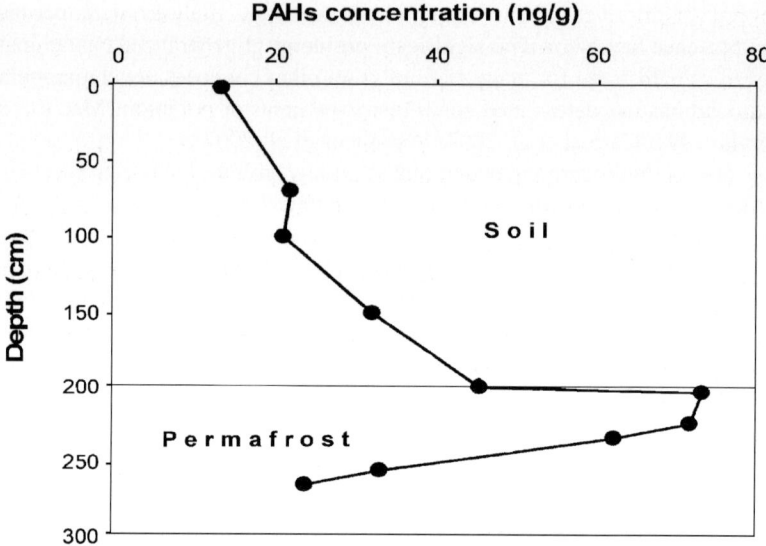

Fig. 12.2 Total PAHs concentrations in the active layer and permafrost from a sampling site near Jubany Station (South Shetland Islands, Antarctica). Samples were taken in February 2005 and patterns similar to that shown here were observed in several other sites near the station

predict the effect hydrocarbons will have on the natural bacterial flora. From here on, we present several studies dealing with the bacterial flora of hydrocarbon-contaminated Antarctic soils. This is done with the intention to extract, beyond the specificities of each particular case study, some general conclusions on the response of the bacterial communities to the presence of hydrocarbons. Similar studies performed in non-Antarctic extremely cold soils are mentioned in order to analyze whether similar responses can occur in all those soil communities under similar stress situations.

12.4 Bacteriology of Hydrocarbon-Contaminated Polar Soils

Studies on the bacteriology of hydrocarbon-contaminated Antarctic soils have been impelled mainly by the intention to apply the best hydrocarbon-adapted micro-organisms to the development of bioremediation processes of contaminated Antarctic and other extremely cold soils. Different studies have dealt with the analysis of aerobic heterotrophic (AHB) and hydrocarbon-degrading (HDB) bacterial populations, using either culture-dependent or culture-independent methods (Mac Cormack and Fraile 1997; Baraniecki et al. 2002; Delille et al. 2003; Ruberto et al. 2003; Saul et al. 2005).

It is important to distinguish the hydrocarbon-contaminated soils as a function of the acute or chronic nature of the contamination event. One possible situation is that of the pristine soil, without previous exposure to hydrocarbons, which suddenly receives a spill of a fuel such as diesel oil or jet aircraft fuel (JPl). This situation results in an "acutely contaminated soil". On the other hand, there are soils receiving small but repetitive doses of hydrocarbons. In this case, the prevailing hydrocarbons may be mainly aliphatic, aromatic, or a complex mixture of these, and the soil remains exposed to the pollutants for periods of years or decades. These soils are "chronically contaminated". From a microbiological point of view, bacteria from an acutely contaminated soil are naive with respect to hydrocarbon degradation, whereas those from a chronically contaminated soil have become adapted to the pollutant.

In general, Antarctic soils are acutely contaminated by petroleum-derived fuels, resulting in an early decrease in total bacterial counts as well as in a loss of diversity. This effect is probably caused by a toxicity effect of the hydrocarbons exerted on the nontolerant members of the bacterial community. After this acute deleterious effect of a hydrocarbon spill, evolution of population density of culturable AHB may take different courses. After crude oil contamination of nine sub-Antarctic intertidal beaches, Delille and Delille (2000) found no relevant changes in the AHB population density during a 90-day long experiment. On the contrary, when a control and contaminated soil samples from Scott Base were compared (Saul et al. 2005), AHB counts in contaminated soils were one to two orders of magnitude higher than in the control soil. A significant increase in AHB was also observed in a pristine Antarctic soil contaminated with diesel oil 28 days after contamination in a microcosm assay (Ruberto et al. 2005). In contrast with these observed differences in the evolution of AHB population density, enrichment of HDB in soils following acute contamination is general. Indeed, with respect to HDB population size, increments of several orders of magnitude were reported following acute contamination, resulting in the relative fraction of HDB among the general population rising from almost negligible levels to relatively high values. Several assays carried out by our and other research groups have shown that hydrocarbon-degrading bacteria are present even in pristine Antarctic soils and that their initially low numbers rise significantly after contamination.

Some Antarctic pristine soils showed numbers of hydrocarbon degraders extremely low or even under detection limits (Aislabie et al. 1998, 2001). However, hydrocarbon degraders represented an initial proportion of 3.2% of total AHB counts in pristine surface soils from Jubany station (Ruberto et al. 2003), and this proportion rose to 80–100% 28 days after contamination with gas-oil. Similarly, hydrocarbon degraders amounted to 0.1% of AHB counts in a 75 cm deep soil showing low levels (1.2 ppm) of PAHs (Ruberto et al. 2006), but this proportion increased to 100% 56 days after phenanthrene contamination. Also the number of HDB reported by Delille et al. (2003) for pristine soils in the vicinity of Dumond d'Urville station (Terre Adélie), which represented less than 2% of the total AHB counts, increased by several orders of magnitude after diesel and crude oil contamination.

The above-cited studies differed according to the characteristics of the soils examined and methods for evaluation of both AHB and HDB numbers. For example, in some studies the Most Probable Number technique was used, whereas others resorted to plate-count methods on various culture media and carbon sources. Such differences preclude close comparisons between the various studies. However, the fact remains that HDB are present in all the studied Antarctic soils, even in those with no previous exposure to a detectable contamination from anthropogenic origin, and that their numbers significantly increase after an acute contamination event. This conclusion applies to other cold soils, including Alpine (Margesin and Schinner 1997b) and Arctic (Thomassin-Lacroix et al. 2002) soils. The only known exception to this observation with respect to cold soils comes from analysis of soils near Scott Base, where a reduced number of AHB and an undetectable level of hydrocarbon degraders were attributed to the presence of very high (and toxic) levels of lead from leaded fuels (Aislabie et al. 1998).

Little is known about changes in bacterial diversity of Antarctic pristine soils after hydrocarbon pollution. Studying the culturable fraction of bacterial community and using biochemical techniques for identification of the isolates, Ruberto et al. (2003) found that, upon hydrocarbon contamination, a pristine soil bacterial community possessing members of *Agrobacterium, Pseudomonas, Acinetobacter, Moraxella, Flavobacterium, Bacillus, Micrococcus, Xanthomonas*, Enterobacteriaceae, and unidentified Gram-negative cocci was reduced to a "contaminated community" where only *Pseudomonas* and *Acinetobacter* members were detected. Although it considered only the culturable microbial fraction, this study suggested that bacterial communities from pristine Antarctic soils may suffer a significant reduction in diversity following exposure to hydrocarbons.

In a more recent study, Saul et al. (2005) found that control soils from Scott Base were dominated by *Fibrobacter/Acidobacterium* (20%), Actinobacteria (17%), Bacteroidetes (10%), Proteobacteria (6%), *Thermus/Deinococcus* (3%), and low-GC Gram positive bacteria (2%), this conclusion being based on the culture-independent analysis of 155 clones. However, hydrocarbon-contaminated soils, from which 367 clones were analyzed, were significantly less diverse, Proteobacteria (76%) being by far the dominant division. Although these authors found important differences when the bacterial community of the control soils was analyzed using culture-dependent methods (no members of the *Fibrobacter/Acidobacterium* division were recognized by culturing), cultural isolations from contaminated soils yielded similar results to those obtained using culture-independent methods, with Proteobacteria, mainly represented by isolates from the genera *Pseudomonas, Sphingomonas* and *Variovorax*, as the dominant group and accounting for 65% of the total AHB population.

When chronically contaminated Antarctic soils were studied, they showed a low bacterial diversity and a high proportion of hydrocarbon-degrading bacteria, an effect similar to that observed upon acute hydrocarbon contamination of pristine soils. Several hydrocarbon-degrading bacterial strains were isolated from chronically contaminated Antarctic soils. These strains belong to a number of genera comprising *Rhodococcus* (Bej et al. 2000; Ruberto et al. 2005), *Acinetobacter* (Mac Cormack and Fraile 1997), *Pseudomonas* (Panicker et al. 2002), *Sphingomonas* (Baraniecki et al. 2002), *Shewanella* (Gentile et al. 2003), and others. *Rhodococcus* seem to be dominant in Antarctic soils where aliphatic hydrocarbons are the main contaminants but not in PAHs-contaminated Antarctic soils. *Rhodococcus* species were predominant among the alkane-degrading isolates from Scott Base (Bej et al. 2000), where C9-C14 chain lengths were the most abundant *n*-alkanes. These *Rhodococcus* strains grew on a broad range of *n*-alkanes (from C6 to C20) as well as on the branched alkane pristane. No growth was observed when PAHs (toluene and naphthalene) or cyclohexane were present as the sole carbon source.

Similar results were reported by Ruberto et al. (2005) following examination of Jubany Station soils under diesel and JP fuels contamination. In this case, *Rhodococcus* strains grew on C10-C16 chain lengths *n*-alkanes and several alkane-rich fuels but failed to grow on xylene, pyrene, and phenanthrene. Only one of the three studied *Rhodococcus* strains grew on cyclohexane, and showed an important surfactant activity associated to the cell surface (Fig. 12.3). Yet another relevant characteristic of the above-mentioned strains, shared with the bulk of the hydrocarbon-degrading Antarctic strains isolated from soils, was their psychrotolerance (but no psychrophily). Although the three *Rhodococcus* strains can grow at 0–5° C, optimal growth temperature ranged from 20–30°C. This psychrotolerance property seems to be a general feature of hydrocarbon-degrading bacteria isolated from the surface of cold soils, where the temperature can drop to $-30°C$ or less during winter, whereas in summer long-term solar exposure can raise the soil

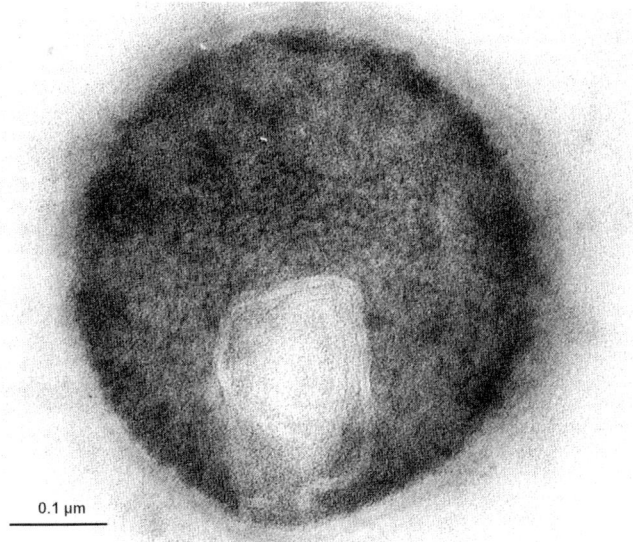

Fig. 12.3 Electron micrograph of Antarctic *Rhodococcus* ADH (Ruberto et al 2003). This strain has been isolated from a fuel-contaminated soil near Jubany Station and, as several other Antarctic hydrocarbon-degrading *Rhodococcus* strains, can use crude oil, a number of light fuels, and a broad range of alkanes as a carbon source. When grown on hydrocarbons, strain ADH has an important surfactant activity and frequently shows membrane complexes (as can be observed in the lower part of the image), probably associated with hydrocarbon metabolism

temperature to 20°C or more (Margesin and Schinner 1997a; Whyte et al. 1997; Deppe et al. 2005). Under these conditions, psychrotolerant bacteria possess a crucial adaptive advantage over the strict psychrophiles. The prevalence of *Rhodococcus* and other Actinobacteria in fuel-contaminated cold soils has also been confirmed by culture-independent methods in the case of Arctic soils (Juck et al. 2000), but not of Alpine soils (Margesin et al. 2003).

A search for *alk* genes has shown that homologues of two alkane hydroxylase systems are present in contaminated as well as uncontaminated Antarctic soils (Whyte et al. 2002), suggesting that alkane-metabolizing members of the *Rhodococcus* genus are common inhabitants of both pristine and hydrocarbon-contaminated soils. It seems that *Rhodococcus* represents one of the main components of the HDB in pristine Antarctic soils and that the spill of alkane-rich fuels generates a selective pressure that allows members of this genus to prevail in a chronic-contamination situation. Another genus that was reported as a relevant group in some pristine soils is *Acinetobacter* (Ruberto et al. 2003). Although some species of this genus were isolated from fuel-contaminated Antarctic soils (Mac Cormack and Fraile 1997), their prevalence in Antarctic contaminated sites seems to be low (Whyte et al. 2002). On the contrary, *Acinetobacter* sp. was reported to be significantly enriched from contaminated Alpine soils (Margesin et al. 2003).

The metabolic versatility of *Pseudomonas* and related organisms from the Gammaproteobacteria is well known (Palleroni 1995), with various representatives exhibiting the capacity for utilization of a wide range of organic compounds as the carbon and energy source, including aliphatic and aromatic hydrocarbons. As was mentioned in Section 12.2, *Pseudomonas* has been frequently found in Antarctic (as well as in Arctic and Alpine) pristine soils. Although their ability to metabolize aliphatic hydrocarbons in low-temperature environments is well documented (Stallwood et al. 2005), it is in PAHs-contaminated environments, including cold soils, that *Pseudomonas* and related genera clearly became dominant. Isolates of these genera have frequently been reported to use PAHs as a substrate (Whyte et al. 1997; Aislabie et al. 2000). Through phylogenetic analysis, these studies also showed that *Pseudomonas* strains isolated from Antarctica and the Arctic were closely related and fell in the same cluster.

Recently, the composition of the Antarctic bacterial consortium M10, isolated from a contaminated Antarctic soil by enrichment cultures on PAHs (phenanthrene, anthracene, fluorene, and dibenzothiophene) was determined (Mestre 2006). All the culturable strains characterized using 16S rDNA sequencing and biochemical techniques were *Pseudomonas*, *Stenotrophomonas*, and *Burkholderia*. In addition to this culture-based analysis, direct cloning of total DNA from M10 consortium followed by 16S rDNA sequencing was performed, with very similar results except for the detection of a minor member of the Bacteroidetes.

A similar phylogenetic study recently performed by Ma et al. (2006) with 22 PAH-degrading bacterial strains isolated from Antarctic soils with naphthalene and phenanthrene as the sole carbon source resulted in the identification of 21 *Pseudomonas* and one *Rahnella* sp., again pointing to the important role played by *Pseudomonas* in PAHs-contaminated cold environments. As the *ndo* (naphthalene dioxigenase) genes of the Antarctic *Pseudomonas* showed no differences with *ndo* genes from mesophilic strains, it was proposed that these genes were acquired by the Antarctic *Pseudomonas* through horizontal gene transfer from exogenous strains. Margesin et al. (2003), who also observed a sharp increase in the percentage of genotypes containing degradative genes similar to those of *Pseudomonas putida* in contaminated Alpine soils compared to pristine controls, attributed this increase to the r-type strategy followed by pseudomonads, which are fast colonizers and grow rapidly on nutrient-rich materials, such as hydrocarbons. This being the case, it may be that at a later stage of the contamination process, when the readily degraded compounds are consumed, other bacterial groups, more adapted to low nutrient concentrations and showing a K-type strategy (such as *Rhodococcus* species), occupy this ecological niche and replace *Pseudomonas* and other fast-growing bacteria.

12.5 Bioremediation in Extremely Cold Soils

The existence of natural hydrocarbon-degrading bacteria in pristine Antarctic soils and their increase in numbers after exposure to these contaminants suggest that bioremediation may help reduce hydrocarbon contamination in these and other cold soils. As polar areas are remote and transferring the affected soils to more temperate areas is very expensive and frequently impracticable, the treatment of affected soils on site (i.e., on or near the site where the contamination event occurs) seems to be the best option.

As was mentioned above (see Section 12.1), bioremediation techniques for cleaning hydrocarbon-contaminated soils have been successfully applied to the cleaning of temperate soils all around the world. Although extremely cold soils might not be an exception, a number of factors, and primarily low temperatures, tend to limit microbial activity in these environments. Antarctic micro-organisms are adapted to survive and grow, not merely at low temperatures, but also in the presence of temperature oscillations and the consequent freeze–thaw cycles resulting from temperatures rising well above 0°C during the short summer period (see Section 12.4). For example, the soil at Jubany station, located in the North part of the Antarctic Peninsula (where a large part of the Antarctic stations are established), can reach 15–20°C during a sunny day in the summer, but the temperature can then drop to values below 0°C at night. A strictly psychrophilic micro-organism could not survive these temperature fluctuations, whereas psychrotolerant strains harbor an adaptive advantage under these particular conditions.

Although the prevailing low temperatures determine a low bacterial metabolic rate and reduce evaporation of volatile compounds, a significant abiotic loss of hydrocarbons has been reported when fuels are spilled on Antarctic soils. Ruberto et al. (2003) reported that at least 54% of an initial level of 14,380 ppm from an acutely diesel-contaminated soil was lost in 10 days due to volatilization and stripping. A lesser but relevant abiotic elimination of diesel oil (16–23%) was found by Margesin and Schinner (1997a) during a laboratory assay where an acute spill was simulated on Alpine soils. However, despite the abiotic loss of the light components of fuels suddenly spilled on soils, a significant fraction of the hydrocarbons remain in the soil many years after the contamination event (Gore et al. 1999). This fact suggests that, in many cases, biodegradation rate of the natural Antarctic microbiota is too low to eliminate the contaminants.

An unequal spatial distribution of micro-organisms and contaminants and a retardation of substrate diffusion in the soil matrix can strongly limit hydrocarbon bioavailability (Harms and Bosma 1997). Enell et al. (2005) have shown this retardation process (particularly relevant with PAHs) to be temperature-dependent and found that the desorption rate of several PAHs from contaminated soil decreased by a factor of 11–12 when temperature declined from 23 to 7°C. Thus, the low temperatures at which the extremely cold soils are exposed could exacerbate the substrate diffusion limitation. The extent of biodegradation also depends on the chemical structure (Heitkamp and Cerniglia 1987) and the physical state of the compounds (Wodzinski and Coyle 1974). Several other factors limiting biodegradation of hydrocarbons in polar soils, such as pH, nutrient availability, and soil moisture, were reviewed in a recent and complete paper by Aislabie et al. (2006a).

There is no general consensus about the best strategy for bioremediation of Antarctic (and others extremely cold) soils. Although natural attenuation proved to reduce hydrocarbon levels in recently contaminated sub-Antarctic (Delille and Pelletier 2002) and Alpine soils (Margesin and Schinner 2001), the time required for this process and the remaining levels of hydrocarbons suggest that other strategies such as biostimulation and bioaugmentation represent valid alternatives to improve the speed and extent of the biodegradation in Antarctic soils.

It seems that chronically contaminated areas, where natural microbiota is adapted to the hydrocarbons, need no additional bacterial inocula for a successful biodegradation. A chronically contaminated soil (containing 12,000 ppm of diesel and JP1 fuels) from Marambio Station was dispensed in separate 3-kg quantities and placed in metal trays (Fig. 12.4a) to compare the effects of biostimulation of the autochthonous microbiota with inorganic N and P sources and of bioaugmentation with different hydrocarbon-degrading bacterial strains and consortia (Ruberto et al. 2004). After 45 days, the biostimulated autochthonous bacterial flora was as efficient for hydrocarbon degradation as the best bioaugmented system. We have obtained similar results working with chronically contaminated soils from Jubany Station and using 1 m^2 land plots as an experimental model (Fig. 12.4b). The same conclusion was obtained by other researchers working in the Arctic (Whyte et al. 1999; Thomassin-Lacroix et al. 2002). In the Alps, Margesin and Schinner (1997b) found that bioaugmentation of diesel oil-contaminated soils enhanced biodegradation rates only slightly and temporarily. Hence, bioaugmentation of chronically contaminated cold soils would serve only to reduce the length of the initial acclimation period by the natural microbiota, and would not increase the extent of the biodegradation processes. In any given situation, an exhaustive time–cost balance analysis should be made to determine the best bioremediation strategy.

Fig. 12.4 Different microcosm systems used as experimental models in studies on bioremediation of Antarctic soils. (**a**) Metallic trays with 3 kg of soil; (**b**) 1 m² land plots; (**c**) small flasks with 20 g of soil; (**d**) larger flasks with 250 g of soil

The possibility remains that the relatively low impact of bioaugmentation is due to what is called by Thompson et al. (2005) "the challenge of the strain selection." These authors argue that selection of strains has been frequently done on the basis of their catabolic competence but not according to many other essential features that are crucial for the function and persistence of the biological model in the target habitat. In this sense, recent results using molecular techniques to detect the presence and progress of two bacterial consortia used for bioaugmentation of Antarctic soils (Vazquez et al. 2007), suggest that consortia do not survive the prevailing environmental conditions and are outcompeted by the natural microbiota, when they are inoculated at densities similar to those shown by the culturable fraction of the soil community (10^6–10^7 colony-forming units (CFU)/g dry soil). Further studies using higher inoculum levels (10^9 CFU/g or more) must be done to determine to what extent the inoculation of chronically contaminated soils improves the hydrocarbon degradation activity. An infrequent case where inocula had significant stimulatory effects on bioremediation in field experiments was reported by Mohn et al. (2001), using biopiles for on-site bioremediation of chronically hydrocarbon-contaminated tundra soils. This stimulatory effect of bioaugmentation was significant during the first summer after treatment (39–53 days), but no difference with the fertilized systems was observed one year after treatment.

A very different situation arises upon contamination of a previously pristine soil. In this case, the low numbers of hydrocarbon-degrading bacteria suggest the need for inoculation (bioaugmentation) with previously isolated hydrocarbon-degrading bacteria. However, contradictory results have been obtained in such systems. A significant improvement in degradation efficiency was obtained when an Antarctic soil acutely contaminated with diesel oil was inoculated with indigenous hydrocarbon-degrading bacteria (Ruberto et al. 2005). A positive effect of bioaugmentation was also obtained when a soil showing low levels (1.2 ppm) of PAHs was exposed to 1,744 ppm of phenanthrene (Ruberto et al. 2006) in the presence of fish meal (as N and P source) and a surfactant (Brij®700) as a bioavailability enhancer. Bioaugmentation with a *Pseudomonas* strain (Stallwood et al. 2005) also enhanced the bioremediation rate in an uncontaminated Antarctic soil exposed to 1% w/w Polar Blend marine diesel oil.

All of the above-mentioned studies used microcosms, performed in flasks filled with small quantities of soil, as an experimental model (see Fig. 12.4c,d). Although few studies have been conducted along those lines, the currently available literature on artificially contaminated Antarctic soils indicates that bioaugmentation could be a valid strategy when a fuel is accidentally spilled on pristine soils. Under this situation, inoculation could effectively reduce the lag period intervening before the HDB counts rise and, thus, improve the extent of pollutant elimination. This acceleration of bioremediation processes would be highly desirable in Antarctica and the Arctic, where a significant biodegradation activity is only possible during the short summer period.

Whereas the usefulness of bioaugmentation strategies still remains controversial, the requirement of N and P addition seems obvious for Antarctic soils, where the levels of these nutrients are generally low. Although it is true that many hydrocarbon-degrading bacteria are well adapted to oligotrophic conditions prevailing in soils (Johnsen et al. 2005), the presence of high levels of carbon source (represented by the pollutants) leads to an unbalanced C:N:P ratio which limits bacterial growth. As a fraction of the hydrocarbons is used to generate bacterial biomass, additional levels of N and P are required to support this biomass increase in polluted soils.

As was mentioned above (see Section 12.4), a surge in bacterial growth and activity is often observed in the first stages of hydrocarbon biodegradation processes in cold soils. In these cases, both the natural bacterial flora and the bacterial inocula used for bioaugmentation require additional amounts of N and P. Some studies with Antarctic soils used commercial slow-release fertilizers, such as Inipol EAP-22, for biostimulation. This product, which is a microemulsion containing nitrogen (urea) and phosphorus (tri(laureth-4)-phosphate) encapsulated within oleic acid, efficiently enhanced biodegradation of alkanes in soils from Kerguelen Islands when added at a C:N:P ratio of 100:12:1.1. However, no effect of the fertilizer on the PAHs degradation rate was detected (Delille et al. 2004). In addition, the stimulating effect of the fertilizer was stronger on a desert soil having low levels of nutrients than on a vegetated soil. When used for bioremediation of an Alpine soil, a commercial water-soluble N-P-K fertilizer enhanced biodegradation of diesel oil at an N:P ratio of 20:1 (Margesin and Schinner 2001). Biostimulation was also effective when N and P were added as inorganic salts ($(NH_4)_2SO_4$ and K_2HPO_4) on flask-microcosms prepared with oil-contaminated Antarctic soils from South Orkney Islands (Stallwood et al. 2005). The effectiveness of biostimulation was also demonstrated with Arctic soils using diverse nutrient sources as urea and diammonium phosphate (Thomassin-Lacroix et al. 2002) or a water-soluble fertilizer (Braddock et al. 1997).

However, some results highlight the care that must be taken when N and P are added as "enhancers" of the bioremediation processes. Ruberto et al (2003) found that when N and P were added to diesel oil-contaminated Antarctic soils as inorganic salts ($NaNO_3$ and Na_2HPO_4) at a C:N:P ratio of 100:12:3, an initial inhibition of bacterial growth, and hence a slower hydrocarbon degradation activity, was observed. It is possible that the relatively high amounts of nutrients used in this study (1,800 mg N/kg soil and 500 mg P/kg soil) were inhibitory to the microbiota. This hypothesis is in keeping with the results obtained by Braddock et al. (1997) who found that 50–100 mg N/kg enhanced microbial hydrocarbon degradation in soils from Alaska, whereas 200 mg N/kg were inhibitory. Negative effects of $(NH_4)NO_3$ and K_2HPO_4 were also observed in non-cold soils by Trindade et al. (2002) at C:N ratios of 100:10 and 100:5 and a C:P ratio of 100:5.

Soil texture strongly influences the optimum level of N for hydrocarbon degradation. Indeed, sandy soils (which represent the majority of Antarctic soils and have a lower water-holding capacity than the silt- and clay-rich soils), are more prone to inhibition by inorganic nitrogen (Walworth and Reynolds 1995). Also, depending on their structure, hydrocarbons might require different levels of nutrients for their degradation. This relationship was suggested by the observation by Breedveld and Sparrevik (2000) that 3-ring and 4-ring PAHs were degraded to different extents in Norwegian cold soils. Finally, a comparison of available results from various studies suggests that what might be considered as a "high level" of nutrients depends on a number of factors, including the tolerance of the microbiota involved, the physical characteristics of the soil, and the type of N and P source used.

For these reasons, it is difficult to define the adequate level of nutrients required by a soil bacterial community for an optimum degradation activity. If only the C:N:P ratio is taken into account, the highly contaminated soils presenting levels as high as 10,000 ppm of TPH or more would require the addition of N and P to levels that could be toxic or inhibitory for the microbiota, if the aim were to reach a C:N:P ratio of 100:10:1. In recent years, the use of complex organic materials as slow-release nutrient sources has been successfully tested. Biostimulation of the autochthonous microbiota using fish meal improved phenanthrene removal from Jubany Station soils (Ruberto et al. 2006). Dry fish compost also was reported by Pelletier et al. (2004), working with soils from the Kerguelen Archipelago at the northern limit of the Antarctic ocean, as an efficient fertilizer as well as an excellent solid carrier for bioremediation additives (nutrients and surfactants). The use of such low-cost complex organic matrixes is also being explored in several other regions for bioremediation of temperate soils (Rahman et al. 2002; Molina-Barahona et al. 2004).

12.6 Major Challenges for the Near Future

One of the major challenges for the near future is not related exclusively to the cold soils or to their exposure to the contaminants. It is a challenge inherent to the complexity of the soil matrix in general. As stated in Section 12.2, at the present time there exists no theory linking the dynamics of the bacterial communities to biodiversity and function in terms of the different microenvironments existing in the soil (Young and Crawford 2004). Progress along those lines will require the combined efforts of ecologists, microbiologists, molecular biologists, soil physicists, and mathematicians, to fully account for the complexity of the problem. Of course, such multidisciplinary effort would also benefit our understanding of the bacterial communities of the cold soils as well as their functioning and evolution upon pollutant exposure.

As regards hydrocarbons in cold soils, in spite of the valuable contributions of several research groups, many important aspects remain poorly studied, especially with respect to Antarctic soils. Although much information has been gathered from other areas, little is known about the effect of hydrocarbons of different chemical compositions on the Antarctic soil ecosystems, either in the short or in the long term. Due to the characteristics of Antarctic food webs, with scarce terrestrial primary producers, low diversity of macroorganisms, and seasonal migration of most consumers located at the highest trophic levels to more temperate regions, Antarctic ecosystems are particularly sensitive to perturbation. Hence, alteration of the resident bacterial community by hydrocarbons could have enormous effects on the function of the sensitive Antarctic ecosystems, these effects occurring up to the highest trophic levels.

Several questions remain unanswered in relation to the fate of hydrocarbons in extremely cold soils. The presence of the permafrost introduces a unique feature that changes the distribution and migration of hydrocarbons in an unknown way. Recent results obtained by us in Jubany Station (Curtosi et al. 2007; see also Section 12.3) showed that hydrocarbons accumulate in the permafrost table. Conclusions from these studies on the fate of hydrocarbons in extremely cold soils should be validated for many other sites in Antarctica and the Arctic by applying well-planned monitoring programs. It will lead us to estimate the consequences of the global warming that probably will affect the structure of permafrost. If the permafrost table, retaining a significant concentration of PAHs, is melted and the compounds washed out to the near marine basin, a massive flow of PAHs may ensue, with unpredictable ecological consequences.

We have only scant knowledge of the metabolism of the different hydrocarbon compounds in cold soils. Because different micro-organisms can metabolize the same hydrocarbon compound using different pathways and producing different metabolites (Sutherland et al. 1995), deeper studies on the biochemical fate of hydrocarbons in nature should be conducted. The issue is more than strictly academic, as some metabolites can result in being highly toxic to the natural biota. For this reason, the potential of the soil microbiota to produce any particular intermediate compound bears relevance on the foreseeable progress of hydrocarbon degradation and the fate of the individual components spilled on the soil.

In relation to the bioaugmentation strategies, one of the main limiting steps is the lack of practical tools for investigating the fate of the bacterial inoculum. This problem is made more complicated still when a consortium (and not a single bacterium) is used and when the consortium has been obtained from the treated location. Because allochthonous micro-organisms cannot be used for bioremediation in Antarctica, to monitor the fate of the inoculated micro-organisms throughout the processes represents a real challenge. It will be necessary to find solutions to this problem before bioaugmentation can be considered as a valuable alternative. Related unsolved problems concern the immobilization of the inoculum and strategies for inoculum application to soils. Immobilization has been analyzed recently in order to extend the permanence in the soil of the inoculated hydrocarbon-degrading bacteria and hence to enhance removal of the contaminants (Cunningham et al. 2004). However, these studies were carried out at a laboratory scale and not using extremely cold soils. Thus, there is currently no information about in situ bioaugmentation of cold soils with immobilized bacteria.

Inoculation strategies that improve hydrocarbon removal are needed for the Antarctic soils. It has been reported that repeated inoculation enhances hydrocarbon mineralization rates (Schwartz and Scow 2001). However, studies using Antarctic soil micro-organisms exposed to harsh climatic conditions are needed to define the adequate inoculation strategy for in situ bioremediation involving bioaugmentation. Progress along those lines would contribute to address the problem pointed out by Aislabie et al. (2004): to date, no consensus has been reached on remediation guidelines for hydrocarbon contamination cleanup protocols for the Antarctic.

12.7 Conclusions

Extremely cold soils have a natural microbiota composed mainly of psychrotolerant micro-organisms adapted to survive under the prevailing extreme environmental conditions. Additional stress represented by the presence of hydrocarbons of anthropogenic origin tends to increase the heterotrophic bacterial counts and markedly stimulate the hydrocarbon-degrading bacterial fraction. However, contaminated soils present lesser bacterial richness and diversity in comparison to pristine controls. Currently, bioremediation strategies are considered as one of the best alternatives to restore the restricted Antarctic areas where hydrocarbons represent a real contamination problem. As was observed in other areas of the world, biostimulation of the natural bacterial flora seems to be the adequate choice for reducing the level of hydrocarbons from chronically contaminated cold soils.

On the other hand, bioaugmentation currently represents a controversial strategy that seems to be useful only in soil with no previous exposure to the pollutants or with low populations of hydrocarbon-degrading bacteria. For Antarctica, as human activities (particularly scientific research, fisheries, and tourism) increase, new and deep multidisciplinary knowledge is needed to face the problem of hydrocarbon contamination and to realize the goals stated by the Antarctic treaty and its Protocol; in particular, it is intended that Antarctica shall continue forever to be used exclusively for peaceful purposes and that its status as a special conservation area will be preserved.

References

ACGR (Associate Committee on Geotechnical Research) (1988) Glossary of permafrost and relative ground ice terms. Permafrost Subcommittee, National Research Council of Canada, Technical Memorandum 142

Aislabie J, Balks M, Astori N, Stevenson G, Symons R (1999) Polycyclic aromatic hydrocarbons in fuel-oil contaminated soils, Antarctica. *Chemosphere* 39:2201–2207

Aislabie J, Foght J, Saul D (2000) Aromatic-hydrocarbon degrading bacteria isolated from soil near Scott Base, Antarctica. *Polar Biol* 23:183–188

Aislabie J, Saul D, Foght J (2006a) Bioremediation of hydrocarbon-contaminated polar soils. *Extremophiles* 10:171–179

Aislabie JM, Balks MR, Foght JM, Waterhouse EJ (2004) Hydrocarbon spills on Antarctic soils: effects and management. *Environ Sci Technol* 38:1265–1274

Aislabie JM, Chhour KL, Saul DJ, Miyauchi S, Ayton J, Paetzold RF, Balks MR (2006b) Dominant bacteria in soils of Marble Point and Wright Valley, Victoria Land, Antarctica. *Soil Biol Biochem* 38:3041–3056

Aislabie JM, Fraser R, Duncan S, Farrell RL (2001) Effects of oil spills on microbial heterotrophs in Antarctic soils. *Polar Biol* 24:308–313

Aislabie JM, McLeod M, Fraser R (1998) Potential of biodegradation of hydrocarbons in soil from the Ross Dependency, Antarctica. *Appl Microbiol Biotechnol* 49:210–214

Alexander M (2000) Ageing, bioavailability and overestimation of risk from environmental pollutants. *Environ Sci Technol* 34:4259–4265

Amman RI, Ludwig W, Schleifer KH (1995) Phylogenetic identification and in situ detection of individual microbial cells without cultivation. *Microbiol Rev* 59:143–169

Anderson DM, Pusch R, Penner E (1978) Physical and thermal properties of frozen ground. In: Andersland OB, Anderson DM (eds) *Geotechnical Engineering for Cold Regions*, vol 2. McGraw-Hill, New York, pp 37–102

Baker JH, Smith DS (1972) The bacteria in an Antarctic peat. *J Appl Bacteriol* 35:589–596

Baranięcki CA, Aislabie J, Foght JM (2002) Characterization of *Sphingomonas* sp. Ant 17, an aromatic hydrocarbon-degrading bacterium isolated from Antarctic soil. *Microb Ecol* 43:44–54

Bauer TG (2001) *Tourism in the Antarctic: Opportunities, Constraints and Future Prospects*. Haworth Hospitality Press, New York, 275 pp

Bej AK, Saul D, Aislabie J (2000) Cold-tolerant alkane-degrading *Rhodococcus* species from Antarctica. *Polar Biol* 23:100–105

Bento FM, Camargo FAO, Okeke BC, Frankenberger WT (2005) Comparative bioremediation of soils contaminated with diesel oil by natural attenuation, biostimulation and bioaugmentation. *Bioresource Technol* 96: 1049–1055

Bicego MC, Zanardi E, Ito RG, Webber RR (1998) Hydrocarbons in surface sediments of Admiralty Bay, King George Island, Antarctica. *Pesquisa Antártica Brasileira* 3:15–21

Biggar KW, Haidar S, Nahir M, Jarrett PM (1998) Site investigations of fuel spill migration into permafrost. *J Cold Region Eng* 12:84–104

Bockheim JG (2002) Landform and soil development in the McMurdo Dry Valleys, Antarctica: A regional synthesis. *Arct Antarct Alpine Res* 34:308–317

Bockheim JG, Hall KJ (2002) Permafrost, active-layer dynamics and periglacial environments of continental Antarctica. *South African J Sci* 98:82–90

Bowman JP, McCammon SA, Brown JL, Nichols PD, McMeekin TA (1997b) *Psychroserpens burtonensis* gen. nov., sp. nov., and *Gelidibacter algens* gen. nov., sp. nov., psychrophilic bacteria isolated from Antarctic lacustrine and sea ice habitats. *Int J Syst Bacteriol* 47:1040–1047

Bowman JP, Nichols DS, McMeekin TA (1997a) *Psychrobacter glacincola* sp. nov., a halotolerant, psychrophilic bacterium isolated from Antarctic sea ice. *Syst Appl Microbiol* 20:209–215

Bozal N, Tudela E, Rossello-Mora R, Lalucat J, Guinea J (1997) *Pseudoalteromonas antarctica* sp. nov., isolated from an Antarctic coastal environment. *Int J Syst Bacteriol* 47:345–351

Braddock JF, Ruth ML, Catterall PH, Walworth JL, McCarthy KA (1997) Enhancement and inhibition of microbial activity in hydrocarbon-contaminated Arctic soils: Implications for nutrient-amended bioremediation. *Environ Sci Technol* 31:2078–2084

Brambilla E, Hippe H, Hagelstein A, Tindall BJ, Stackebrandt E (2000) 16S rDNA diversity of cultured and uncultured prokaryotes of a mat sample from Lake Fryxell, McMurdo Dry Valleys, Antarctica. *Extremophiles* 5:23–33

Breedveld GD, Sparrevik M (2000) Nutrient-limited biodegradation of PAH in various soil strata at a creosote contaminated site. *Biodegradation* 11:391–399

Chiou CT, McGroddy SE, Kile DE (1998) Partition characteristics of polycyclic aromatic hydrocarbons on soils and sediments. *Environ Sci Technol* 32:264–269

Christie P (1987) Nitrogen in two contrasting Antarctic bryophyte communities. *J Ecol* 75:73–94

Christner BC, Kvitko BH, Reeve JN (2003) Molecular identification of Bacteria and Eukarya inhabiting an Antarctic cryoconite hole. *Extremophiles* 7:177–183

Cowan DA, Ah Tow L (2004). Endangered Antarctic environments. *Annu Rev Microbiol* 58:649–690

Cripps GC (1989) Problems in the identification of anthropogenic hydrocarbons against natural background in the Antarctic. *Antarct Sci* 1:307–312

Cripps GC (1990) Hydrocarbons in the seawater and pelagic organisms of the Southern Ocean. *Polar Biol* 10:393–402

Cripps GC (1992) The extent of hydrocarbon contamination in the marine environment from a research station in the Antarctic. *Mar Pollut Bull* 25:9–12

Cunningham CJ, Ivshina IB, Lozinsky VI, Kuyukima MS, Philp JC (2004) Bioremediation of diesel-contaminated soil by microorganisms immobilised in polyvinyl alcohol. *Int Biodeter Biodeg* 54:167–174

Cunningham CJ, Philp JC (2000) Comparison of bioaugmentation and biostimulation in *ex situ* treatment of diesel contaminated soil. *Land Contam Reclamation* 8:261–269

Curtosi A, Pelletier E, Vodopivez C, Mac Cormack WP (2007) Distribution pattern of polycyclic aromatic hydrocarbons in soil, surface marine sediments and suspended particulate matter in the seawater column near Jubany Station (Antarctica). *Berichte zur Polarforschung* (Reports on Polar and Marine Research) in press

da Conceição M, Alvim-Ferraz M, Albergaria JT, Delerue-Matos C (2006) Soil remediation time to achieve clean-up goals. I. Influence of soil water content. *Chemosphere* 62:853–860

de la Torre JR, Goebel BM, Friedmann EI, Pace NR (2003) Microbial diversity of cryptoendolithic communities from the McMurdo Dry Valleys, Antarctica. *Appl Environ Microbiol* 69:3858–3867

Delille D (2000) Response of Antarctic soil bacterial assemblages to contamination by diesel fuel and crude oil. *Microb Ecol* 40:159–168

Delille D, Coulon F, Pelletier E (2004) Effects of temperature warming during a bioremediation study of natural and nutrient-amended hydrocarbon-contaminated sub-Antarctic soils. *Cold Reg Sci Technol* 40:61–70

Delille D, Delille B (2000) Field observations on the variability of crude oil impact on indigenous hydrocarbon-degrading bacteria from sub-Antarctic intertidal sediments. *Mar Environ Res* 49:403–417

Delille D, Pelletier E (2002) Natural attenuation of diesel-oil contamination in a sub-Antarctic soil (Crozet Island). *Polar Biol* 25:682–687

Delille D, Pelletier E, Delille B, Coulon F (2003) Effects of nutrients enrichment on the bacterial assemblage of Antarctic soils contaminated by diesel or crude oil. *Polar Rec* 39:1–10

Deming JW (2002) Psychrophiles and polar regions. *Curr Opin Microbiol* 5:301–309

Denner EBM, Mark B, Busse H-J, Turkiewicz M, Lubitz W (2001) *Psychrobacter proteolyticus* sp. nov., a psychrotrophic, halotolerant bacterium isolated from the Antarctic krill Euphausia superba Dana, excreting a cold-adapted metalloprotease. *Syst Appl Microbiol* 24:44–53

Deppe U, Richnow H-H, Michaelis W, Antranikian G (2005) Degradation of crude oil by an Arctic microbial consortium. *Extremophiles* 9:461–470

Dictor MC, Berne N, Mathieu O, Moussay A, Saada A (2003) Influence of ageing of polluted soils on bioavailability of phenanthrene. *Oil Gas Sci Technol* 58:481–488

Enell A, Reichenberg F, Ewald G, Warfvinge P (2005) Desorption kinetics studies on PAH-contaminated soil under varying temperatures. *Chemosphere* 61:1529–1538

Eriksson M, Ka JO, Mohn WM (2001) Effects of low temperature and freeze-thaw cycles on hydrocarbon biodegradation in Arctic tundra soil. *Appl Environ Microbiol* 67:5107–5112

Fox AJ, Cooper PR (1994) Measured properties of the Antarctic ice sheet derived from the SCAR Antarctic digital database. *Polar Rec* 30:201–206

Franklin RB, Mills AL (2003) Multi-scale variation in spatial heterogeneity for microbial community structure in an eastern Virginia agricultural field. *FEMS Microbiol Ecol* 44:335–346

Franzmann PD (1996) Examination of Antarctic prokaryotic diversity through molecular comparisons. *Biodiv Conserv* 5:1295–1305

Gentile G, Bonasera V, Amico C, Giuliano L, Yakimov MM (2003) *Shewanella* sp. GA-22, a psychrophilic hydrocarbonoclastic antarctic bacterium producing polyunsaturated fatty acids. *J Appl Microbiol* 95:1124–1133

Gore DB, Revill AT, Guille D (1999) Petroleum hydrocarbons ten years after spillage at a helipad in Bunger Hills, East Antarctica. *Antarct Sci* 11:427–429

Haack SK, Fogarty LR, West TG, Alm EW, McGuire JT, Long DT, Hyndman DW, Forney LJ (2004) Spatial and temporal changes in microbial community structure associated with recharge-influenced chemical gradients in a contaminated aquifer. *Environ Microbiol* 6:438–448

Harms H, Bosma TNP (1997) Mass transfer limitation of microbial growth and pollutant degradation. *J Ind Microbiol Biotechnol* 18:97–105

Heider J, Spormann AM, Beller HR, Widdel F (1998) Anaerobic bacterial metabolism of hydrocarbons. *FEMS Microbiol Rev* 22:459–473

Heitkamp MA, Cerniglia CE (1987) Effect of chemical structure and exposure on the microbial degradation of polycyclic aromatic hydrocarbons in freshwater and estuarine ecosystems. *Environ Toxicol Chem* 6: 535–546

Heitzer A, Sayler GS (1993) Monitoring the efficacy of bioremediation. *Trends Biotechnol* 11:334–343

Hirsch P, Gallikowski CA, Friedmann EI (1985) Microorganisms in soil samples from Linnaeus Terrace southern Victoria Land: Preliminary observations. *Antarct J U S* 20:183–186

Holben WE (1997) Isolation and purification of bacterial community DNA from environmental samples. In: Hurst CJ, Knudsen GR, McInerney MJ, Stetzenbach LD, Walter MV (eds) *Manual of Environmental Microbiology*. ASM Press,Washington DC, pp 431–436

Hooker BS, Skeen RS (1996) Intrinsic bioremediation: An environmental restoration technology. *Curr Opin Biotechnol* 7:317–320

Johnsen AR, Wick LY, Harms H (2005) Principles of microbial PAHs-degradation in soil. *Environ Pollut* 133:71–84

Johnson RM, Madden JM, Swafford JR (1978) Taxonomy of Antarctic bacteria from soils and air primarily of the McMurdo Station and Victoria Land dry valley region. *Antarct Res Ser* 30:35–64

Juck D, Charles T, Whyte LG, Greer CW (2000) Polyphasic microbial community analysis of petroleum hydrocarbon-contaminated soils from two northern Canadian communities. *FEMS Microbiol Ecol* 33:241–249

Katsivela E, Moore ERB, Maroukli D, Strömpl C, Pieper D, Kalogerakis N (2005) Bacterial community dynamics during *in-situ* bioremediation of petroleum waste sludge in landfarming sites. *Biodegradation* 16:169–180

Kennicutt II MC, McDonald TJ, Denoux GJ, McDonald SJ (1992a) Hydrocarbon contamination on the Antarctic Peninsula. I. Arthur Harbor - Subtidal sediments. *Mar Pollut Bull* 24:499–506

Kennicutt II MC, McDonald TJ, Denoux GJ, McDonald SJ (1992b) Hydrocarbon contamination on the Antarctic Peninsula II. Arthur Harbor inter and subtidal limpets (*Nacella concinna*). *Mar Poll Bull* 24:506–511.

Kennicutt II MC, Sweet ST (1992) Hydrocarbon contamination on the Antarctic Peninsula III. The Bahia Paraiso – Two years after the spill. *Mar Pollut Bull* 25:303–306

Line MA (1988) Microbial flora of some soils of Mawson Bays and the Vestfold Hills, Antarctica. *Polar Biol* 8:421–427

Lipson DA, Schmidt SK (2004) Seasonal changes in an alpine bacterial community in the Colorado Mountains. *Appl Environ Microbiol* 70:2867–2879

Lundstedt S (2003) Analysis of PAHs and their transformation products in contaminated soil and remedial processes. PhD Thesis, Umeå University, Sweden, 56 pp

Ma Y, Wang L, Shao Z (2006) *Pseudomonas*, the dominant polycyclic aromatic hydrocarbon-degrading bacteria isolated from Antarctic soils and the role of large plasmids in horizontal gene transfer. *Environ Microbiol* 8:455–465

Mac Cormack WP, Fraile ER (1997) Characterization of a hyrocarbon degrading psychrotrophic Antarctic bacterium. *Antarct Sci* 9:150–155

Mac Cormack WP, Rios Merino L, Fraile R (1998) Bacterial hydrocarbon degradation in Antarctica. *Berichte zur Polarforschung* 299:304–315

Mackie PR, Platt HM, Hardy R (1978) Hydrocarbons in the marine environment II. Distribution of *n*-alkanes in the fauna and environment of the sub-Antarctic island of South Georgia. *Estuarine Coastal Mar Sci* 6:301–313

Marchal R, Penet S, Solano-Serena F, Vandercasteele JP (2003) Gasoline and diesel oil biodegradation. *Oil Gas Sci Technol* 58:441–448

Margesin R, Labbé D, Schinner F, Greer CW, Whyte LG (2003) Characterization of hydrocarbon-degrading microbial populations in contaminated and pristine Alpine soils. *Appl Environ Microbiol* 69:3085–3092

Margesin R, Schinner F (1997a) Efficiency of indigenous and inoculated cold-adapted soil microorganisms for biodegradation of diesel oil in Alpine soils. *Appl Environ Microbiol* 63:2660–2664

Margesin R, Schinner F (1997b) Bioremediation of diesel-oil-contaminated Alpine soils at low temperatures. *Appl Microbiol Biotechnol* 47:462–468

Margesin R, Schinner F (2001) Bioremediation (natural attenuation and biostimulation) of diesel-oil-contaminated soil in an Alpine glacier skiing area. *Appl Environ Microbiol* 67:3127–3133

Martins CC, Bícego MC, Taniguchi S, Montone RC (2004) Aliphatic and polycyclic aromatic hydrocarbons in surface sediments in Admiralty Bay, King George Island, Antarctica. *Antarct Sci* 16:117–122

Mazzera D, Hayes T, Lowenthal D, Zielinska B (1999) Quantification of polycyclic hydrocarbons in soil at McMurdo Station, Antarctica. *Sci Total Environ* 229:65–71

Mestre MC (2006) Caracterización bioquímica y molecular de los componentes cultivables de un consorcio bacteriano antártico capaz de degradar hidrocarburos. Diplomat Thesis, Facultad de Ciencias Exactas y Naturales, Universidad de Buenos Aires, 85 pp

Miwa T (1975) Clostridia in the soil of the Antarctic. *Jap J Medical Sci Biol* 28:201–213

Mohn WW, Radziminski CZ, Fortin M-C (2001) On site bioremediation of hydrocarbon-contaminated Arctic tundra soils in inoculated biopiles. *Appl Microbiol Biotechnol* 57:242–247

Molina-Barahona L, Rodríguez-Vázquez R, Hernández-Velasco M, Vega-Jarquín C, Zapata-Pérez O, Mendoza-Cantú A, Albores A (2004) Diesel removal from contaminated soils by biostimulation and supplementation with crop residues. *Appl Soil Ecol* 27:165–175

Naraoka H, Shimoyama A, Harada K (2000) Isotopic evidence from an Antarctic carbonaceous chondrite for two reaction pathways of extraterrestrial PAH formation. *Earth Planet Sci* Lett 184:1–7

Nedwell DB (1999) Effect of low temperature on microbial growth: Lowered affinity for substrates limits growth at low temperature. *FEMS Microbiol Ecol* 30:101–111

Neff JM (1979) *Polycyclic Aromatic Hydrocarbons in the Aquatic Environment: Sources, Fates and Biological Effects*. Applied Science, Essex, England, 262 pp

Nemergut DR, Costello EK, Meyer AF, Pescador MY, Weintraub MN, Smidt SK (2005) Structure and function of Alpine and Arctic soil microbial communities. *Res Microbiol* 156:775–784

Nienow JA, Friedmann EI (1993) Terrestrial lithophytic (rock) communities. In: Friedmann EI (ed) *Antarctic Microbiology*. Wiley-Liss, New York, pp 343–412

Palleroni NJ (1995) Microbial versatility. In: Young LY, Cerniglia CE (eds) *Microbial Transformation and Degradation of Toxic Organic Chemicals*. Wiley-Liss, Ney York, pp 3–25

Panicker G, Aislabie J, Saul D, Bej AK (2002) Cold tolerance of *Pseudomonas* sp 30–3 isolated from oil-contaminated soil, Antarctica. *Polar Biol* 25:5–11

Pelletier E, Delille D, Delille B (2004) Crude oil bioremediation in sub-Antarctic intertidal sediments: chemistry and toxicity of oiled residues. *Mar Environ Res* 57:311–327

Platt HM, Mackie PR (1980) Distribution and fate of aliphatic and aromatic hydrocarbons in Antarctic fauna and environment. *Helgoland Mar Res* 33:236–245

Prévot AR, Moreau M (1952) Recherches sur les bactéries anaérobes de la Terre Adélie (prélevées par la première expédition antarctique française). *Ann Inst Pasteur* 82:13–18

Providenti MA, Lee H, Trevors JT (1993) Selected factors limiting the microbial degradation of recalcitrant compounds. *J Ind Microbiol Biotechnol* 12:379–395

Rahman KSM, Banat IM, Thaira J, Thayumanavan T, Lakshmanaperumalsamy P (2002) Bioremediation of gasoline contaminated soil by a bacterial consortium amended with poultry litter, coir pith and rhamnolipid biosurfactant. *Bioresource Technol* 81:25–32

Rike AG, Haugen KB, Børrensen M, Engene B, Kolstad P (2003) In situ biodegradation of petroleum hydrocarbons in frozen Arctic soils. *Cold Reg Sci Technol* 37:97–120

Röling WFM, Milner MG, Jones DM, Lee K, Daniel F, Swannell RJP, Head IM (2002) Robust hydrocarbon degradation and dynamics of bacterial communities during nutrient-enhanced oil spill bioremediation. *Appl Environ Microbiol* 68:5537–5548

Roszak DB, Colwell RR (1987) Survival strategies of bacteria in the natural environment. *Microbiol Rev* 51:365–379

Ruberto L, Vazquez SC, Curtosi A, Mestre MC, Pelletier E, Mac Cormack WP (2006) Phenanthrene biodegradation in soils using an Antarctic bacterial consortium. *Biorem J* 10:191–201

Ruberto L, Vazquez SC, Lobalbo A, Mac Cormack WP (2004) Biorremediación de suelos contaminados con hidrocarburos utilizando bacterias antárticas sicrotolerantes. Actas del V° Simposio Argentino y 1° Latinoamericano sobre Investigaciones Antárticas On line publication in www.dna.gov.ar/CIENCIA/SANTAR04/CD/PDF/CVAUTR.htm, Cód 206BH, 5 pp

Ruberto L, Vazquez SC, Lobalbo A, Mac Cormack WP (2005) Psychrotolerant hydrocarbon-degrading *Rhodococcus* strains isolated from polluted Antarctic soils. *Antarct Sci* 17:47–56

Ruberto L, Vazquez SC, Mac Cormack WP (2003) Effectiveness of the natural bacterial flora, biostimulation and bioaugmentation on the bioremediation of a hydrocarbon contaminated Antarctic soil. *Int Biodeter Biodeg* 52:115–125

Saul DJ, Aislabie JM, Brown CE, Harris L, Foght JM (2005) Hydrocarbon contamination changes the bacterial diversity of soil from around Scott Base, Antarctica. *FEMS Microbiol Ecol* 53:141–155

Schwartz E, Scow KM (2001) Repeated inoculation as a strategy for the remediation of low concentration of phenanthrene in soil. *Biodegradation* 12:201–207

Shinoda Y, Akagi J, Uchihashi Y, Hiraishi A, Yukawa H, Yurimoto H, Sakai Y, Kato N (2005) Anaerobic Degradation of aromatic compounds by *Magnetospirillum* strains: Isolation and degradation genes. *Biosci Biotechnol Biochem* 69:1483–1491

Smith JJ, Ah Tow L, Stafford W, Cary C, Cowan DA (2006) Bacterial diversity in three different Antarctic cold desert mineral soils. *Microb Ecol* 51:413–421

Smith MC, Bowman JP, Scott FJ, Line MA (2000) Sublithic bacteria associated with Antarctic quartz stones. *Antarct. Sci* 12:177–184

Snape I, Harvey PM, Ferguson SH, Rayner JL, Revill AT (2005) Investigation of evaporation and biodegradation of fuel spills in Antarctica. I. A chemical approach using GC-FID. *Chemosphere* 61:1485–1494

Stallwood B, Shears J, Williams PA, Hughes KA (2005) Low temperature bioremediation of oil-contaminated soil using biostimulation and bioaugmentation with a *Pseudomonas* sp. from maritime Antarctica. *J Appl Microbiol* 99:794–802

Steven B, Léveillé R, Pollard WH, Whyte LG (2003) Microbial ecology and biodiversity of permafrost. *Extremophiles* 10:259–267

Sutherland JB, Rafii F, Kahn AA, Cerniglia CE (1995) Mechanisms of polycyclic aromatic hydrocarbon degradation. In: Young LY, Cerniglia CE (eds) *Microbial Transformation and Degradation of Toxic Organic Chemicals*. Wiley-Liss, New York, pp 269–306

Thomassin-Lacroix EJM, Eriksson M, Reimer KJ, Mohn WW (2002) Biostimulation and bioaugmentation for on-site treatment of weathered diesel fuel in Arctic soil. *Appl Microbiol Biotechnol* 59:551–556

Thomassin-Lacroix EJM, Yu Z, Eriksson M, Reimer KJ, Mohn WW (2001) DNA-based and cultured-based characterization of hydrocarbon-degrading consortium enriched from Arctic soil. *Can J Microbiol* 47:1107–1115

Thompson IP, van der Gast J, Ciric L, Singer AC (2005) Bioaugmentation for bioremediation: the challenge of strain selection. *Environ Microbiol* 7:909–915

Torsvik V, Goksoyr J, Daae FL (1990) High diversity in DNA of soil bacteria. *Appl Environ Microbiol* 56:782–787

Trindade P, Sobral LG Rizzo AC, Leite SGF, Lemos JLS, Milloili VS Soriano AU (2002) Evaluation of the biostimulation and bioaugmentation techniques in the bioremediation process of petroleum hydrocarbon contaminated soils. *Proceedings of the 9th International Petroleum Environmental Conference, IPEC (Integrated Petroleum Environmental Consortium)*, Albuquerque, NM, 15 pp

Vázquez SC, Nogales B, Ruberto L, Hernandez E, Christie-Oleza J, Bosch R, Lalucat J, Mac Cormack WP (2007) Bacterial community dynamics during bioremediation of diesel-oil contaminated Antarctic soil in a mesocosm assay. Manuscript in preparation

Vincent WF (1988) *Microbial Ecosystems of Antarctica*. Cambridge University Press, Cambridge, 304 pp

Vishniac HS (1993) The microbiology of Antarctic soils. In: Friedmann EI (ed) *Antarctic Microbiology*. Wiley-Liss, New York, pp 297–341

Vodopivez C, Mac Cormack WP, Villamil E, Curtosi A, Pelletier E, Smichowski P (2007) Evidence of pollution with hydrocarbons and heavy metals at the surroundings of Jubany Station. *Berichte zur Polarforschung* (Reports on Polar and Marine Research). in press

Walworth JL, Reynolds CM (1995) Bioremediation of petroleum contaminated cryic soil: effects of phosphorous, nitrogen and temperature. *J Soil Contam* 4:299–310

Whyte LG, Bourbonniere L, Bellerose C, Greer CW (1999) Bioremediation assessment of hydrocarbon-contaminated soils from the high Arctic. *Biorem J* 3:69–79

Whyte LG, Bourbonniere L, Greer CW (1997) Biodegradation of petroleum hydrocarbons by psychrotrophic *Pseudomonas* strains possessing both alkane (*alk*) and naphthalene (*nah*) catabolic pathways. *Appl Environ Microbiol* 63:3719–3723

Whyte LG, Goalen B, Hawari J, Labbé D, Greer CW, Nahir M (2001) Bioremediation treatability assessment of hydrocarbon-contaminated soils from Eureka, Nunavut. *Cold Reg Sci Technol* 32:121–132

Whyte LG, Schultz A, van Beilen JB, Luz AP, Pellizari V, Labbé D, Greer CW (2002) Prevalence of alkane monooxygenase genes in Arctic and Antarctic hydrocarbon-contaminated and pristine soils. *FEMS Microbiol Ecol* 41:141–150

Wodzinski RS, Coyle JE (1974) Physical state of phenanthrene for utilization by bacteria. *Appl Microbiol* 27:1081–1084

Young IM, Crawford JW (2004) Interactions and self-organization in the soil-microbe complex. *Science* 304:1634–1637

Young IM, Ritz K (2000) Tillage, habitat space and function of soil microbes. *Soil Tillage Res* 53:201–213

Zdanowski MK, Weglenski P (2001) Ecophysiology of soil bacteria in the vicinity of Henryk Arctowski Station, King George Island, Antarctics. *Soil Biol Biochem* 33:819–829

Zhou J, Davey ME, Figueras JB, Rivkina E, Gillichinsky D, Tiedje JM (1997) Phylogenetic diversity of a bacterial community determined from Siberian tundra soil DNA. *Microbiol* 143:3913–3919

Zocca C, Di Gregorio S, Visentini F, Vallini G (2004) Biodiversity amongst cultivable polycyclic aromatic hydrocarbon-transforming bacteria isolated from an abandoned industrial site. *FEMS Microbiol Lett* 238:375–382

Chapter 13
Microbiology of Oil-Contaminated Desert Soils and Coastal Areas in the Arabian Gulf Region

Samir Radwan

13.1 Introduction

Deserts are of global distribution; they cover considerable areas of all continents, with the exception of Europe. Desert soils are poor in organic substances and water, and are usually subjected to rather high temperature in summer and chilling in winter, and to extensive light. In spite of their extreme character, desert soils usually accommodate communities of micro-organisms including actinomycetes, cyanobacteria and other bacteria, fungi, protozoa, and phototrophic microalgae. Many of such micro-organisms live naturally under stress, and must possess special adaptive mechanisms in order to survive and propagate (see Chapter 2). Desert micro-organisms appear to be limited in their physiological activities due to low availability of certain nutrients, according to Liebig's "law of the minimum" (Liebig 1840).

Sometimes, microbial activities in the desert soil are arrested according to Shelford's "law of tolerance" (Shelford 1913) saying that there are maxima and minima for environmental factors above and below which micro-organisms cannot survive. Nutrient starvation seems to be one of the most serious problems desert soil micro-organisms have to overcome. Apparently, primary producers such as microalgae and cyanobacteria are the major sources of organic materials in the poor desert soils. Roszak and Colwell (1987) identified among bacteria a number of survival approaches against starvation, that fall within two strategies (Jannasch 1967), namely the potential for growth at low nutrient levels, and the potential for entering into dormancy.

Another possible mechanism for survival of starving bacteria is cell size reduction through the phenomenon of multiple division, thus producing the so-called ultramicrobacteria (Novitsky and Morita 1976, 1977, 1978; Morita 1982). The smaller the bacterial cell, the larger is its surface-to-volume ratio, and consequently the greater is its potential for accumulating diluted nutrients from the surroundings.

Samir Radwan
Department of Biological Sciences, Kuwait University, P.O. Box 5969, Safat 13060, Kuwait
e-mail: radwan@kuc01.kuniv.edu.kw

Ultramicrobacteria are widely distributed in marine waters (Button et al. 1993). They may be rather dormant and consequently, relatively resistant to environmental stresses (Roszak and Colwell 1987). Some starving bacteria with a depleted amino acid pool exhibit the so-called "stringent response" (Neidhardt et al. 1990) which ultimately reduces the protein synthesis rate by inhibiting rRNA synthesis.

In addition to stresses exerted on desert micro-organisms by factors such as nutrient deficiency, drought, and heat, additional stress may arise due to pollution of the same desert soil areas with crude oil. This is particularly true for oil-producing countries such as the Arabian Gulf countries. Such desert soil areas have been polluted for many ages with small amounts of hydrocarbon vapors naturally volatilizing from the deep oil reservoirs. However, the pollution problems associated with modern and intensive oil production and transport occur on a much expanded scale.

In addition, coastal regions of oil-producing countries are particularly exposed to oil pollution which of course exerts stress on the indigenous coastal microflora.

The objective of this chapter is to shed light on changes in the indigenous microflora of desert and coastal regions in response to pollution. Emphasis is put on the desert and coastal regions of the Arabian Gulf for several reasons. Oil-utilizing micro-organisms of the Gulf area have now been the subject of study in our laboratory for more than 15 years. This area contains ancient oil, and has produced oil intensively for decades. The Gulf oil spill associated with and resulting from the Iraqi occupation of Kuwait, from August 2, 1990, to February 26, 1991, is so far the greatest in the history of mankind.

13.2   The Gulf Oil Spill

Shortly before their withdrawal from Kuwait in February 1991, the Iraqi forces deliberately blew up the Kuwaiti wells, amounting to more than 700 in the desert. It took the Kuwaiti authorities about seven months to get the resulting fires under control. During that period, crude oil kept gushing, and thus about 300 so-called "oil lakes" (Fig. 13.1) formed, covering in total about $50\,km^2$ of the Kuwaiti Desert. There are estimates (McKinnon and Vine 1991) that such lakes used to contain about 22 million barrels of oil, but 18 million barrels have been recovered and exported, and 3 million remain as pollutants. Oil penetrated between 40 to 60 cm deep into the sand. The total volume of polluted desert soil is estimated to be $20 \times 10^6\,m^3$, and still contains highly viscous to solid crude.

The Gulf water body also received a share of the oil pollution. The Iraqi forces on January 19, 1990 deliberately released crude oil from the Mina Al-Ahmady oil terminal directly into the water. According to different estimates, the amount of oil released in the course of three successive days ranged between half a million and twelve million barrels (McKinnon and Vine 1991). The slick was 16-km long, that is, several times the size of the famous *Exxon Valdez* spill in Alaska. Most of the oil was transported counterclockwise by the water currents to the south. Most of the crude then became sedimented in the intertidal zone along more than 700 km of the western Gulf coast, leaving the water and the subtidal zone almost free of oil sediments.

Fig. 13.1 One of the small oil lakes in the Kuwaiti desert (February 1993)

13.3 Composition of Crude Oil

Crude oil consists chemically of four major constituents: saturates, aromatics, asphaltenes, and resins (Leahy and Colwell 1990). Saturated hydrocarbons, including normal alkanes with chains of up to 44 carbon atoms, branched alkanes, and cycloalkanes (naphtenes), are the major constituents of the crude, making up between 40 and 60% of the total weight. Aromatic hydrocarbons with from one to six benzene or substituted benzene rings follow the saturates in quantitative importance and amount to roughly 20% of the crude weight. Asphaltenes, which include tar, are very high-molecular weight hydrocarbons, which are used as road paving materials. Resins are crude constituents that contain, in addition to carbon and hydrogen, sulfur and oxygen. The chemistry of asphaltenes and resins is still not yet completely known; both constituents make up 1 to 5% of light oils and up to 25% of heavy oils, which correspondingly contain lower proportions of saturates and aromatics.

13.4 General Description of Oil-Utilizing Micro-Organisms

Several reports have been published on oil-utilizing micro-organisms, for example, Klug and Markovetz (1971), Levi et al. (1979), Einsele (1983), Radwan and Sorkhoh (1993), Van Hamme et al. (2003), Rosenberg (2006), and Widdel et al. (2006). It is important to note that such micro-organisms are normal indigenous soil and water inhabitants, and that most of them can consume conventional carbon sources. Their defining characteristic, which makes them capable of utilizing hydrocarbons as substrate, is that they possess the so-called mono-oxygenase and/ or dioxygenase enzyme systems. Such systems catalyze the introduction of oxygen atoms from molecular oxygen into aliphatic and aromatic hydrocarbon molecules producing the corresponding alcohols that in turn become further oxidized to aldehydes, and ultimately acids. The resulting acids are then biodegraded by β-oxidation, producing acetyl CoA that can be further metabolized (for reviews see Rehm and Reiff 1981; Fukui and Tanaka 1981; Boulton and Ratledge 1984).

The capacity to utilize hydrocarbons is widely distributed among conventional micro-organisms including prokaryotes and eukaryotes. Bacterial genera reported to attack hydrocarbons include *Acinetobacter, Micrococcus, Vibrio, Azospirillum* (Roy et al. 1988), *Aeromonas, Alcaligenes, Chromobacterium, Flavobacterium, Klebsiella, Pseudomonas* (Klug and Markovetz 1971), *Bacillus* (Loginova et al. 1981; Sorkhoh et al. 1993), *Arthrobacter, Brevibacterium, Corynebacterium, Rhodococcus, Mycobacterium, Nocardia* and other nocardioforms (Egorov et al. 1986), and *Streptomyces* (Barabas et al. 1995). Yeast genera capable of utilizing hydrocarbons include (for review see Radwan and Sorkhoh 1993) *Candida, Dabayomyces, Endomyces, Leucosporidium, Lodderomyces, Metschnikowia, Pichia, Rhodosporidium, Rhodotorula, Saccharomycopsis, Schwanniomyces, Selenotila, Sporidiobalus, Sporobolomyces, Torulopsis, Trichosporon,* and *Wingea*. Filamentous fungi with hydrocarbon utilization potential include *Absidia* (Hoffman and Rehm 1978), *Aspergillus, Aureobasidium, Beauveria* (Davies and Westlake 1979), *Botrytis, Cephalosporium, Cladosporium, Corellospora, Canninghamella, Dendyphiella* (Kirk and Gordon 1988), *Fusarium, Hormodendrum* (Lin et al. 1971a,b), *Lulworthia* (Kirk and Gordon 1988), *Mortierella, Mucor, Penicillium, Phialophora, Phoma* (Davies and Westlake, 1979), *Scedosporium* (Ornodera et al. 1989), *Scoleobasidium* (Davies and Westlake 1979), *Sporotrichum, Varicosporina,* and *Verticillium* (Kirk and Gordon 1988). In addition, there are reports that phototrophic bacteria such as *Rhodospirillum* and *Rhodopseudomonas* (Cerniglia et al. 1980b); cyanobacteria such as *Oscillatoria* (Cerniglia et al. 1980a), *Microcoleus,* and *Phormidium* (Al-Hasan et al. 1994, 1998); microalgae such as *Chlamydomonas* and *Chlorella* (Ellis 1977); and the phytoflagellate *Euglena* (Ellis 1977) can oxidize aliphatic and/or aromatic hydrocarbons.

None of the micro-organisms listed above can consume all of the crude oil constituents; and each organism has the potential for utilization of only a limited range of compounds. Yet, collectively, all crude constituents from the gaseous low molecular weight (van Ginkel et al. 1987; Ornodera et al. 1989) up to the medium and high molecular weight (Demanova et al. 1980) compounds, including asphaltenes, can be attacked by micro-organisms.

Growth on oil and hydrocarbons is associated in some micro-organisms with certain unique morphological and/or cytological features. One of the most frequent features is the appearance of cytoplasmic hydrocarbon inclusions in actinomycetes (Barabas et al. 1995) and other bacteria (e.g., Scott and Finnerty 1966; Atlas and Heintz 1973; Kennedy and Finnerty 1975), and also in filamentous fungi (Cundell et al. 1976; Koval and Redchitz 1978; Redchitz 1980). The picocyanobacteria *Synechococcus* and *Synechocystis* exhibit much wider interthylakoid spaces in the presence of oil and hydrocarbons than in the absence of these compounds (Al-Hasan et al. 2001). Some micro-organisms produce dense intraplasmic membranes (Kennedy and Finnerty 1975; Ivshina et al. 1982) and volutin inclusions (Redchitz and Koval 1979; Ivshina et al. 1982). *Penicillium* grows in shaken cultures in the presence of hydrocarbons as hollow mycelial balls enclosing hydrocarbon droplets, whereas in hydrocarbon-free media the balls are solid (Cundell et al. 1976).

13.5 Oil-Utilizing Micro-Organisms in the Arabian Gulf Desert Soils

Our group in Kuwait, working for more than 15 years on oil and hydrocarbon-utilizing micro-organisms indigenous to the desert and marine environments of the Arabian Gulf, has collected a wealth of useful information on this subject. This information may help in understanding the composition of oil-utilizing microflora indigenous to other desert and coastal areas similar to those of the Gulf area.

The predominant indigenous oil-utilizing bacteria in the Kuwaiti desert belong to *Micrococcus, Pseudomonas, Bacillus, Arthrobacter*, and the group of nocardioforms, particularly the genus *Rhodococcus* (Sorkhoh et al. 1990, 1995; Radwan et al. 1997). The genus *Streptomyces* is the predominant oil-utilizing actinomycete in the Kuwaiti desert (Barabas et al. 1995, 2000; Radwan et al. 1998b).

The oil-utilizing fungal flora of the Kuwaiti desert comprises predominantly the genera *Aspergillus, Penicillium, Fusarium*, and *Mucor*. Members of other genera are also found (Sorkhoh et al. 1990). The Gulf region is characterized by a rather long, dry, and very hot summer. Therefore, it may be expected that the desert in this region may accommodate thermophilic hydrocarbon-utilizing micro-organisms. The analysis of 38 Kuwaiti Desert soil samples polluted with crude oil revealed the occurrence of 3.7×10^3 to 1.1×10^7 cells of thermophilic (with an optimal temperature of 55°C) oil-utilizing bacteria per g of soil, all of which were identified as *Bacillus stearothermophilus* (now *Geobacillus stearothermophilus*) (Sorkhoh et al. 1993). The isolation of hydrocarbon-utilizing thermophiles is not unexpected inasmuch as Loginova et al. (1981) observed the growth of obligate thermophilic bacteria in a medium with paraffin. Similarly, Zarilla and Perry (1984) reported on *Thermoleophilum album* as a novel bacterium obligate for thermophily and utilizing *n*-alkane substrates.

In the course of our studies on oil-utilizing micro-organisms in the Kuwaiti Desert soils, we noticed that oil-polluted areas generally contained higher numbers of such organisms than pristine areas. However, pristine desert soil was never free of oil-utilizing bacteria. In particular, in both pristine and contaminated Kuwaiti Desert areas, the soil fraction in direct contact with the desert plant roots (or rhizosphere) represented microenvironments enriched in oil-utilizing micro-organisms. The rhizospheres of desert plants growing in the Kuwaiti Desert were found to contain more hydrocarbon-utilizing micro-organisms than the soil farther away from the roots (Radwan et al. 1995b, 1998a). These plants included *Senecio glaucus*, *Cyperus cenglomeratus*, *Launaea mucronata*, *Picris babylonica*, and *Salsola imbricata*. It was observed that some plants, although growing in black, oil-polluted, desert soil areas, possessed white clean roots rich in oil-utilizing bacteria (Radwan et al. 1995b). Not only the rhizospheres of wild plants, but also those of legume crops (e.g., *Vicia faba* and *Lupinus albus*) were richer in hydrocarbon-utilizing bacteria than nonrhizosphere soil. The rhizosphere effect (which is measured as the ratio of the number of micro-organisms in the rhizosphere soil to the number of micro-organisms in nonrhizosphere soil (Anderson et al. 1993)) was much more pronounced for plants growing in oil-polluted than in pristine soils. The most prevalent hydrocarbon-utilizing bacteria in the rhizospheres of the above plants were *Cellulomonas flavigena*, *Rhodococcus erythropolis*, and *Arthrobacter* spp.

There is experimental evidence for self-cleaning of oily desert soil in Kuwait through the activities of the indigenous hydrocarbon-utilizing microflora. The total amounts of extractable alkanes from heavily polluted soil cores in a Kuwaiti oil field exposed to the open air were quantitatively determined once every two weeks through a whole year (Radwan et al. 1995a). It was found that the total amount of extractable alkanes remained fairly constant during the dry hot summer months, but decreased during the rainy months reaching, after one year, slightly more than one half of the amount at zero time. This result demonstrates the self-cleaning capacity of the Kuwaiti Desert soil and the essential role of moisture in this process. The loss of alkanes could not be attributed to simple physical volatilization, because loss occurred at a slower pace during the hot summer. On the contrary, self-cleaning was faster during the rainy period of the year, suggesting that it was mainly occurring through biological processes.

13.6 Micro-Organisms in the Gulf Coastal Areas

One of the interesting observations our group made during trips to the oily coastal regions in the early 1990s was the appearance of mats of an intense blue-green color at the top of oil sediments in the Saudi Arabian Gulf coasts, about 300 km south of Kuwait (Sorkhoh et al. 1992). Those mats are frequent in the Gulf coasts, even in the nonpolluted areas (Golubic 1992). Later, we made similar observations along the oily Kuwaiti coasts (Fig. 13.2). Strikingly, the mats were tightly associated with oil, and oil-free coastal areas were also free of those mats. It was also noted that all forms of higher life on the oily coasts were dramatically inhibited by the oil sediments. Animal inhabitants were absent or dead, and their coastal underground tunnels were full of crude. The mats appeared to be at that time the only, or one of the few living things in the oily coasts. The microbiological analysis of mat samples we collected from the Saudi research station of Jubail revealed that they consisted mainly of photosynthetic and heterotrophic prokaryotes. The phototrophs included filamentous cyanobacteria, such as *Microcoleus*, *Phormidium*, *Spirulina*, and others, in addition to some eucaryotes, mainly diatoms. The filamentous

Fig. 13.2 Microbial mats on the top of oil sediments along the Kuwaiti Coast of the Arabian Gulf

cyanobacteria adhered together through excreted mucilage forming the mat matrix. The heterotrophic bacteria associated with those mats included millions of oil-utilizing bacteria per g fresh mat.

In this context, blue-green mats, also comprising the cyanobacterium *Microcoleus*, have also been recorded in the pristine coast of Abu Dhabi (Golubic 1992). The hydrocarbon-utilizing bacteria associated with the mats (Sorkhoh et al. 1995) consisted of nocardioforms (63%), belonging mostly to the genus *Rhodococcus*, the genera *Bacillus* (21%), and *Arthrobacter* (13%), in addition to *Pseudomonas* as a minor bacterium. Actinomycetes belonging to the genus *Streptomyces* and filamentous fungi belonging to the genera *Aspergillus* and *Penicillium* were also identified, but were much less frequent than the other heterotrophs. *Microcoleus* consortia comprising heterotrophic bacteria capable of biodegrading oil have also been described by Garcia De Oteyza et al. (2004). The microbial consortium in the blue-green mats is active in self-cleaning of the oily coasts of the Gulf. Successive visits to the lightly polluted coasts along Kuwait revealed that the oil sediments gradually vanished. Today, those coasts have become absolutely oil-free, and their mats have disappeared.

The mats harbor a number of advantages as valuable biological systems in self-cleaning of oily coasts. Heterotrophic oil-utilizing bacteria are naturally immobilized within the mats, thus avoiding being washed out into the open sea. Furthermore, such bacteria are adequately aerated with oxygen produced by the photosynthetic partners in the mats. It is known that oxygen could be a limiting factor in microbial biodegradation of hydrocarbons (for a review, see Radwan and Sorkhoh 1993). In addition, some of the cyanobacterial partners in the mats are probably nitrogen fixers (see also Steppe et al. 1996); nitrogen fertilization is known to enhance microbial degradation of hydrocarbons (Radwan et al. 1995c). In addition, there is experimental evidence for direct hydrocarbon oxidation by cyanobacteria and algae (Cerniglia et al. 1980a,b; Al-Hasan et al. 1994, 1998; Raghukumar et al. 2001; Todd et al. 2002).

Along the Gulf coast and also probably elsewhere, littoral materials from intertidal zones are associated with much higher numbers of oil-degrading micro-organisms than inshore and offshore water samples (Radwan et al. 1999). This may indicate that the coasts have a better potential for oil biodegradation than the water body. Oil-utilizing bacteria found associated with coastal materials along the Arabian Gulf belonged to the genera *Acinetobacter*, *Micrococcus*, and the group of nocardioforms (Radwan et al. 1999). Interestingly, those bacteria are more frequent in association with littoral animate materials, such as microbial mats and epilithic biomass, than in association with inanimate materials, such as sand, stonelets, and gravel particles. However, gravel particles coated with blue-green biofilms are also rich in oil-utilizing bacteria (Radwan and Al-Hasan 2001).

In coastal and offshore waters of the Gulf, oil-utilizing bacteria are found preferentially associated with macroalgae (Radwan et al. 2002), fish (Radwan et al. 2007a), and picoplankton (Radwan et al. 2005a), rather than free-living. Thus, microbial consortia in biofilms along the coasts and various associations in the water body seem to play a major role in self-cleaning of the oily marine ecosystems.

13.7 Extremophilic Oil-Utilizing Micro-Organisms

There is an information gap regarding the extremophilic oil-utilizing micro-organisms in desert, coastal, and marine environments. Mention has already been made of the few studies on thermophilic hydrocarbon-utilizing bacteria (Loginova et al. 1981

Zarilla and Perry 1984; Sorkhoh et al. 1990; see Section 13.5). However, there are no studies on hyperthermophilic hydrocarbon-utilizing bacteria.

A significant number of oil-polluted ecosystems are characterized by rather high alkalinities and/or salinities. These include estuaries, beaches, salt marshes, inland lakes, rockpools, desert rain pools, and others. Furthermore, billions of gallons of wastewaters with high contents of salts and waste organics are generated by industry, and disposed of in the environment. An example of such industrial activities is the production, transport, and refining of crude oil, normally associated with the generation of a large volume of oily salt water that displays a wide range of alkalinities and/or salinities (Flynn et al. 1996; Roe et al. 1996). This makes oil pollution difficult to treat using conventional microbial strains, whose cell membranes may be disrupted and enzymes denatured (Kargi and Dincer 2000; Woolard and Irvine 1994). To bioremediate oily alkaline and saline environments without costly pretreatment, alkaliphilic and halophilic micro-organisms should be used (Díaz et al. 2002).

With these facts in mind, recent work performed in our laboratory (Sulaiman, 2006; Al-Awadhi et al. 2007) concerned the isolation and identification of alkaliphilic and halophilic oil-utilizing bacteria from the Arabian Gulf coasts of Kuwait. The results showed that animate coastal materials such as epilithic biomass and cyanobacterial mats were associated with considerable numbers of alkaliphilic and halophilic oil-utilizing bacteria. Inanimate material, such as coastal sand and gravel particles, as well as coastal waters contained much fewer numbers, if any, of these bacteria. The alkaliphilic oil-utilizing bacteria were found to belong to the genera *Marinobacter*, *Micrococcus*, *Dietzia*, *Bacillus*, *Oceanobacillus*, and *Citricoccus*. The halophilic oil-utilizing bacteria were found to belong to the genera *Marinobacter*, *Georgenia*, *Microbacterium*, *Stappia*, *Bacillus*, *Isoptericola*, and *Cellulomonas*. All isolates had a good potential for hydrocarbon degradation, and consequently may be suitable tools for self-cleaning and bioremediation for oily alkaline and salty regions.

13.8 Microbial Consortia and Associations

In previous sections of this review, mention has been made of the frequent occurrence of microbial associations probably involved in oil biodegradation in the Gulf environments (see also Grötzschel et al. 2002; Abed et al. 2002; Abed and Köster 2005; Sanchez et al. 2005). Such associations were recorded in the desert and coastal environments as well as in the water body of the Gulf. In the present section, more light is shed on those associations. Although it should be expected that oil-biodegradation in oily desert soil is mediated by "hypothetical" microbial consortia comprising bacteria and fungi, with the possibility of establishment of cometabolic and syntrophic strategies, there is only little information on the role of individual partners in such consortia.

The rhizospheric microflora associated with roots of desert plants may offer unique opportunities in this respect, as in this case the benefits of the association to the oil-utilizing microbes are obvious. In the rhizosphere, oil-utilizing micro-organisms find proper conditions for propagation and activity. These micro-organisms can cover their vitamin requirements from exudates excreted by the root tissues. Indeed, more than 90% of oil-utilizing bacteria need vitamins for optimal growth and activity (Radwan and Al-Muteirie 2001). Furthermore, plants are known to aerate their rhizospheres by pumping air down into the soil. Oxygen was reported to be a limiting factor for the microbial attack on hydrocarbon molecules (Rehm and Reiff 1981; Fukui and Tanaka 1981). The roots, particularly of legumes, may also provide oil-utilizing bacteria with nitrogen fixation potential, thus enriching soil with compounds reported to enhance microbial hydrocarbon biodegradation (Atlas 1981; Leahy and Colwell 1990).

The coastal and aquatic marine environments appear to contain more interesting oil-degrading microbial consortia than those of the desert soil. In most of those associations there are phototrophic partners and heterotrophic oil-utilizing partners. This is true for coastal epilithic biomass (Radwan et.al. 1999), coastal cyanobacterial mats (Radwan et al. 1999), coastal gravel particles coated with picocyanobacteria and other phototrophs (Radwan and Al Hasan 2001), macroalgae coated with bacterial biofilms (Radwan et al. 2002), and picocyanobacteria associated with oil-utilizing bacteria at the water surface (Radwan et al. 2005a). In all these consortia, the conditions are suitable for growth and activity of oil-utilizing bacteria. Thus, oxygen becomes available as a byproduct of photosynthesis, and nitrogenous compounds as a result of nitrogen fixation by some cyanobacteria. Furthermore, as mentioned above, in many of these associations the bacteria are protected against dispersal in the open sea.

Another interesting association in coastal and offshore waters is that of oil-utilizing bacteria in biofilms coating fish surfaces and gills and gut linings (Radwan et al. 2007a). All test samples of ten types of the Arabian Gulf fish and two of farm fish were found to accommodate rather high numbers of oil-utilizing bacteria, with 10^5 to 10^7 cells being found per cm^2 of fish surface and per gram of gills and guts. Such numbers were much higher than in the surrounding Gulf water which contained only 10^2–10^3 bacteria per ml. Such bacteria belonged to the genera *Acinetobacter*, *Micrococcus*, *Bacillus*, *Rhodococcus*, and other nocardioforms. It should be expected that ships, boats, and other vehicles navigating in the Gulf (and other open waters) are also probably coated by biofilms rich in oil-utilizing micro-organisms which apparently play a role in self-cleaning of the polluted aquatic environments.

13.9 Responses of Micro-Organisms to Environmental Variables

The numbers and activities of oil-utilizing micro-organisms in oily desert soil and coastal regions are affected by a number of environmental factors.

13.9.1 Organic Matter Content

There is a lack of information on the effect of desert soil organic matter on oil-utilizing micro-organisms. Desert soils are characterized by their very low content of organic substances. Therefore, it may be assumed that the organic oil pollutants would specifically favor oil-utilizing micro-organisms. They certainly do, and we have observed this through our studies on the Gulf desert microflora. How oil-utilizing micro-organisms would be affected, should an oily desert soil sample receive conventional organic substances, should also be examined. This question is relevant, because most of the oil-utilizing micro-organisms also have the potential for utilizing conventional organic carbon sources. A partial answer to this question lies in the fact that hydrocarbon-utilizing micro-organisms are more frequent in the rhizospheric than in the nonrhizospheric soils (Radwan et al. 1998a). It is well established that plant roots permanently excrete into the soil organic exudates, for example, sugars, vitamins, and amino acids. Such exudates probably stimulate the oil-utilizing micro-organisms in the rhizospheric soils.

A more direct answer to the interesting question of conventional substrate utilization is provided by the results of one of our studies (Radwan et al. 2000). It was found that fertilizing an oily sample of desert soil with a mixture of glucose and peptone resulted in enhanced oil attenuation in that sample. The magnitude of the stimulation effect was too great to be attributed solely to nitrogen fertilization by the added peptone. Soil fertilization with KNO_3 containing an equivalent amount of nitrogen to that in peptone brought about a much lower oil attenuation value than that obtained with peptone. Glucose/peptone addition to a clean desert soil sample resulted in a dramatic increase of the total numbers of oil-utilizing micro-organisms in that sample. After 13 days, the micro-organisms had depleted all the added glucose and peptone and their numbers decreased. In the oily desert soil sample, glucose/peptone addition also increased the numbers of oil-utilizing micro-organisms. Yet, after the depletion of glucose and peptone, the numbers of oil-utilizing micro-organisms remained high and enhanced oil attenuation was recorded. It was thus concluded that easily utilizable conventional carbon and nitrogen sources in desert soil favor the oil-utilizing microflora, and consequently oil attenuation in such soils.

A question may be raised here regarding the origin of easily utilizable organic matter in the poor desert and coastal soils. It is well known that the primary producers in the various ecosystems of our planet are the phototrophic (and chemolithotrophic) organisms. In this respect, coastal areas, being wet all or most of the time, benefit from supporting algae and cyanobacteria. The latter produce and liberate easily utilizable organic matter via photosynthesis. On the other hand, dry desert soils seem to support such phototrophs only temporarily, following rainy periods. Therefore, their organic substance contents should be low, and dependent on the frequency of precipitation.

Certain organic compounds, such as carboxylic acids, alcohols, and aldehydes, inhibit or even kill most micro-organisms. Interestingly, hydrocarbon-utilizing micro-organisms permanently produce such compounds from their hydrocarbon substrates as metabolic intermediates and release some of them into the environment (for a review see Radwan and Sorkhoh 1993).

13.9.2 Temperature

Micro-organisms in the Gulf desert soils are naturally subjected to a rather wide range of temperatures. During daytime the surface desert soil in the Arabian Gulf environment may reach or exceed 70°C. In winter nights the temperature may fall below the freezing point. Although there are a few reports in the literature on thermophilic hydrocarbon-utilizing bacteria (Loginova et al. 1981; Zarrilla and Perry 1984; Sorkhoh et al. 1993), our experience indicates that hydrocarbon-utilizing micro-organisms in desert soils and coastal regions of the Gulf are predominantly mesophilic with optima at 30–35°C. We have isolated only one thermophilic species, *Bacillus stearothermophilus*, from desert soils, and this organism was not markedly predominant in any of the numerous desert soil samples analyzed (Sorkhoh et al. 1993). We also failed to isolate obligate psychrophilic hydrocarbon-utilizing micro-organisms from desert soil samples.

It appears that desert micro-organisms are provided with survival mechanisms at both high and low temperatures. Our knowledge of such mechanisms is still far from clear. The production of "dormant" units such as endospores, cysts, and others, is one of such mechanisms; yet it is limited only to a few species of bacteria and fungi. However, diminishing metabolic activity in vegetative cells to a minimum level may probably be an effective strategy for survival at unsuitable temperatures. This strategy might be more easily applied at suboptimal than at superoptimal temperatures. The moisture content may be an important interfering factor here. Moist heat is known to be more effective in killing micro-organisms than dry heat. Dry proteins need higher temperatures for denaturation than wet proteins. In view of the fact that the highest temperatures in the desert environment are reached during the long "dry" summer, a preliminary understanding of probable surviving mechanisms of mesophilic micro-organisms at temperatures much above their optima could be gained through examination of dry heat effects on proteins.

Wet coastal regions appear to be more protected against the high summer temperatures than the dry desert areas.

13.9.3 Hydrogen Ion Concentration

Our routine measurements revealed that pristine and oily desert soils of the Gulf were rather neutral. This was also true for coastal soil samples except for a tendency of these soils to become slightly alkaline. It is known that soil pH affects the dissociation of the carboxyl and amino groups of proteins and thus the microbial enzymatic activities. For optimal activity, enzymes have to be in a certain state of dissociation. Furthermore, the pH affects the solubility and consequently the availability of many nutrients, for example, phosphate and ammonium (see Atlas and Bartha 1998). It has been mentioned (see Section 13.4) that hydrocarbon-utilizing micro-organisms in the Gulf desert soils and coastal areas comprise bacteria and fungi. Bacteria predominantly prefer neutral environments and are usually sensitive to acidity. Most fungi also prefer a neutral pH value, but are tolerant of acidity.

These facts mean that the microbiological attack on hydrocarbons in the Gulf desert soils and coastal areas can occur at a wide range of hydrogen ion concentrations, yet with an optimum at pH 7. In view of the well-known fact that fungi are commonly slower in growth and activity than bacteria, it should be expected that acidity would slow down the hydrocarbon biodegradation in the polluted soils. On the other hand, the coastal areas of the Gulf contain limited numbers of alkaliphilic hydrocarbon-utilizing bacteria (Al-Awadhi et al. 2007). This also means that extremely alkaline areas polluted with oil could be enriched with alkaliphilic hydrocarbon-utilizing bacteria that would sustain the oil biodegradation process.

Hydrocarbon biodegradation in oily soils proceeds in the pH range of 4.5 to 11.5 with optima between pH 6.5 and 8 (Daylan et al. 1990). Our group found that the irrigation of oil desert soil samples with sewage-effluent as source of water and nitrogen inhibited alkane-biodegradation due to increasing soil acidity (Radwan et al. 1995c). However, liming relieved this inhibition.

13.9.4 Moisture and Aeration

Desert soils are characterized by extremely low moisture content most of the year. In the Arabian Gulf region, precipitation is rather rare and occurs only during the short winter. The dry period may exceed nine months in the year. On the other hand, coastal areas are submerged with sea water during tidal and wave movements, and suffer from drought only temporarily. There is an inverse proportion between the soil moisture content and the soil degree of aeration. Thus, the Gulf desert soils are well aerated most of the time. Microbial activities in dry soils are substantially enhanced by increasing the water content, but only until the latter starts to fill the soil air space. When water logging occurs, growth and metabolism of aerobic micro-organisms become inhibited. Moisture content of soil is optimal for residing micro-organisms at 50–75% of its water holding capacity.

Oil-utilizing micro-organisms are predominantly aerobic, and the first step in the microbial attack on hydrocarbon substrates involves the introduction of one (for alkanes) or two (for aromatic rings) oxygen atoms into the substrate molecule (Rehm and Reiff 1981; Boulton and Ratledge 1984; Buehler and Schindler 1984; Singer and Finnerty 1984). There are reports on anaerobic biodegradation of hydrocarbons, but such a process is so slow that it is considered negligible in nature (Ratledge 1978; Atlas 1981; Aeckersberg et al. 1991). Experimental results showed that the microbiological degradation of hydrocarbons in Kuwaiti Desert soil samples was insignificant during the long dry period of the year, but was resumed actively during rainy months (Radwan et al. 1995c). This result demonstrates that water is a major limiting factor controlling the microbiological hydrocarbon degradation in desert soils. Water in soil is also needed for microbial motility, microbial spacial proliferation and substrate transport (Smiles 1988).

Salinity reduces water activity and thus, water availability to micro-organisms and other living beings. However, oil-utilizing bacteria in the desert soils and coastal areas seem to be adapted to the range of salinities common in their natural habitats. We have noticed that terrestrial oil-degrading bacteria operate optimally in the absence of any added sodium chloride, whereas coastal and sea water isolates needed about 3.5% (w/v) sodium chloride for optimal activity.

13.9.5 Inorganic Nutrients

Desert soils are commonly quite poor in organic matter but usually contain adequate amounts of most of the inorganic nutrients needed by micro-organisms. Under certain conditions, however, soil micro-organisms may show increased requirements for some specific inorganic nutrients. Oil pollution is a typical example of such a condition. Hydrocarbons are consumed by micro-organisms as carbon and energy sources. In order for hydrocarbon-utilizing micro-organisms to synthesize proteins, nucleic acids, and other organic nitrogenous compounds, and to enhance their energy metabolism, they require additional amounts of nitrogen and phosphorus (Atlas and Bartha 1972; Gibbs 1975; Gibbs et al. 1975). It has been estimated that 60 mg N and 6 mg P are needed for the metabolism of one gram of hydrocarbons (Kant et al. 1985).

Experimental results in our laboratory showed that the microbiological hydrocarbon degradation in oily Kuwaiti Desert soil samples and coastal areas was enhanced by nitrate but not by phosphate fertilization (Radwan et al. 1995c). This result indicates that such soil samples contain insufficient concentrations of nitrogen, but adequate concentrations of phosphorus. Some authors suggested the use of oleophilic nitrogen and phosphorus fertilizers for enhancing oil biodegradation especially in the oily marine ecosystem (Atlas and Bartha 1972; Atlas 1977). Hydrocarbon-utilizing bacteria predominantly have the potential for using both inorganic and organic nitrogen sources (Radwan et al. 1995a,b, 2000). Unpublished results in our laboratory indicate that some of such bacteria, such as *Bacillus* spp., are capable of atmospheric nitrogen fixation, and that the symbiotic nitrogen-fixing nodule bacteria (*Rhizobium* and *Bradyrhizobium* spp.) can utilize hydrocarbons as the sole sources of carbon and energy (see also Prantera et al. 2002).

Oil-polluted desert soils and coastal areas become simultaneously enriched with heavy metals which are known to inhibit micro-organisms at rather low concentrations (Gadd 1990). However, micro-organisms are provided with defense mechanisms against heavy metal toxicity. These mechanisms include the reduction of transport of such metals across the cell envelope, their complexing and subsequent precipitation outside the cell, and compartmentalization inside the cell (Atlas and Bartha 1998).

13.9.6 Surfactants

Many, but not all, hydrocarbon-utilizing micro-organisms produce surfactants extracellularly in order to emulsify or pseudosolubilize these water-insoluble substrates prior to their uptake (Desai and Banat 1997; Cameotra and Makkar 1998; Makkar and Cameotra 1998, 2002; Banat et al. 2000). Such biosurfactants comprise low molecular weight compounds such as trehalose lipids, rhamnolipids, surfactin, polyol lipids, and fatty acids, and high molecular weight compounds such as emulsan, liposan, mannan, and lipoproteins. Apparently, biosurfactants and chemical surfactants change the physical nature of oil, but do not eliminate it from the environment.

There are contradicting reports on the effect of such compounds on the microbiological degradation of oil substrates, as both enhancing and inhibitory effects were recorded (Tumeo et al. 1994; Bai et al. 1997; Lang and Wullbrandt 1999). The inhibitory effect was attributed to surfactant toxicity, preferential metabolism of the surfactants over the hydrocarbons, and/or interference with the membrane uptake process (Efroymson and Alexander 1991; Rouse et al. 1994; Mulligan et al. 2001). Several authors emphasize the need for more study before surfactant-enhanced bioremediation approaches may be suggested (Leavitt and Brown 1994; Van Eyk 1994; Korda et al. 1997; Bolba et al. 1998).

13.10 Bioremediation Strategies

Bioremediation has been defined as the technology in which microbial activities are implemented to mineralize and remove xenobiotic pollutants from the environment (Atlas and Pramer 1990). During the 1990s a number of books and review articles were published on this subject (e.g., Hinchee and Olfenbuttel 1991a,b; Riser-Roberts 1992; Rosenberg 1993, Alexander 1994; Stoner 1994; Atlas 1995; Radwan et al. 1995c). Out of the xenobiotic pollutants, the hydrocarbon contaminants were among the first to receive close attention (Mueller et al. 1989; Song et al. 1990; Hinchee and Olfenbuttel 1991a,b). Bioremediation technology involves two basic approaches: biostimulation, or enhancing the activity of indigenous micro-organisms capable of degrading the pollutants, and bioaugmentation, or seeding the environment with pollutant-degrading micro-organisms.

The current section of this review deals with bioremediation strategies for cleaning oily desert and coastal areas of the Arabian Gulf region. However, before examining such strategies, it may be useful to refer to earlier bioremediation attempts from other parts of the world. One such attempt was the partial bioremediation of about 800,000 gallons of oily wastewater in the bilge tanks of the *Queen Mary* moored in Long Beach Harbor, CA (Applied Biotreatment Association 1989). There were also attempts to clean oily terrestrial environments in refineries and tank farms using bioremediation technology (Zitrides 1990). Even oil spills in the open sea were subjected to bioremediation attempts. In the summer of 1990, the Norwegian tanker *Mega Borg* discharged 100,000 barrels of crude oil into the Caribbean, and the Texan Company Alpha Environmental applied a mixture of oil-degrading micro-organisms in an attempt to clean up the resulting spill (McKinnon and Vine 1991). In the spring of 1989, the *Exxon Valdez* tanker released about 11 million gallons of crude oil into Prince William Sound in Alaska, thus heavily contaminating more than 1,000 miles of the Alaskan coasts. Because physical removal of oil by simple washing of the coasts was of only limited effectiveness, it was suggested to fertilize contaminated areas with nutrient that might enhance the potential of the indigenous microbial flora for biodegrading the oil (Pritchard and Costa 1991; Bragg et al. 1992, 1994). Reportedly, the rates of oil biodegradation were enhanced three to five times when contaminated coasts were fertilized with 4–8 kg nitrogen per $10\,m^2$, and the time required for oil removal was thus reduced from 10–20 years to only 2–3 years.

There are no large-scale attempts to bioremediate in situ oily desert soils and coastal areas in the Gulf. Yet, results from basic studies make it possible to design cleaning protocols. As mentioned above, bioremediation involves two major approaches, seeding with oil-degrading micro-organisms and fertilization with nutrients enhancing the indigenous microflora. As far as seeding is concerned, there are commercial mixtures of micro-organisms available in the market for in situ application (Applied Biotreatment Association 1989, 1990). However, experience in our laboratory indicates that seeding should not be the approach of choice for bioremediating oily environments in the Gulf (Radwan 1990; Radwan et al. 1997). Bioremediation should depend on indigenous strains, which are adapted to the prevailing environmental conditions. Oil spills create specific conditions for enhancing indigenous oil-degrading strains. In one of our studies (Radwan et al. 1997), we found that after a simulated oil spill, and for 28 weeks, desert soil samples became steadily enriched with one specific, indigenous, oil-degrading *Arthrobacter* strain, KUCC201. Other indigenous oil-utilizing bacteria, including other *Arthrobacter* strains either remained unchanged at low numbers or steadily disappeared. Conversely, seeding the 24-week-old polluted samples with local or foreign oil-degrading isolates resulted in dramatic decreases in the numbers of the predominant, indigenous, oil-degrading *Arthrobacter* strain, KUCC201. We concluded that seeding is probably a useless, or even harmful approach for bioremediation. In this context, genetically engineered oil-degrading micro-organisms have been created for decades (Hartmann et al. 1979; Reineke and Knachmuss 1979). However, the introduction of such strains into the environment is obviously hazardous and requires governmental regulation (Halvorson et al. 1985).

The above discussion indicates that the approach of choice for bioremediating oily environments should rely on the enhancement of activity of indigenous hydrocarbon-utilizing micro-organisms. This may be achieved, in particular, by fertilization with nutrients such as nitrogen.

As far as the oily desert soil areas are concerned, management should primarily include irrigation and nitrogen fertilization. The Gulf desert soils receive intermittent precipitation only during the short winter, therefore self-cleaning of oily areas by indigenous micro-organisms ceases most of the year (Radwan et al. 1995). Mere irrigation with water activates the micro-organisms in the summer, and fertilization with KNO_3 enhances the cleanup process considerably (Radwan et al. 1995). In view of these facts, a phytoremediation approach depending on rhizosphere technology may be suggested as the biotechnology of choice for bioremediating oily deserts.

Our group studied the role of rhizospheric bacteria in the attenuation of hydrocarbons in oily desert soils. As mentioned before, roots of desert plants and crop plants growing in pristine and oily soils were found to be densely colonized with oil-utilizing bacteria (Radwan et al. 1995a, 1998). Our group offered experimental evidence that cropping is a successful practice for cleaning up oily desert soils (Radwan et al. 2000). The crop we used was broad beans (*Vicia faba*), because the plants were able to tolerate up to 10% crude oil in soil. Common cropping practices such as irrigation and mechanical managements improve soil moisture content and aeration. Obviously, such practices provide better conditions for the growth and activities of soil micro-organisms. The roots of legume plants, including *Vicia faba*, carry nodules that fix molecular nitrogen, thus providing a natural and economical route for nitrogen fertilization. We found that the amounts of extractable hydrocarbons recovered from oily desert soils supporting *V. faba*, were less than those recovered from the uncultivated soil controls. A group of rhizospheric bacteria known as plant growth-promoting rhizobacteria (PGPR), predominantly pseudomonads, has been described to enhance plant growth when inoculated into the root area (Polonenko et al. 1987; Zhang et al. 1996, 1997).

We have found that inoculation of *V. faba* roots with PGPR enhanced the growth of this plant in oily desert soil samples, and increased the phytoremediation potential of this crop for the oily soil (Radwan et al. 2005b). In this context, we also found that nodule bacteria and PGPR have the potential for biodegradation of hydrocarbons (Radwan et al. 2007b). Thus, the rhizosphere microenvironments provide conditions optimal for activity of hydrocarbon-degrading bacteria. The rhizospheres are well aerated by the oxygen pumped down through the roots which may be an important factor because oxygen is a limiting factor for hydrocarbon biodegradation. The rhizospheres of legume crops are enriched with compound nitrogen fixed by the nodule bacteria, and assimilable nitrogen is also a limiting factor for hydrocarbon biodegradation. In addition, the rhizospheres are enriched with exuded vitamins and organic substances that have been recorded to enhance growth and activities of oil-utilizing bacteria in culture (Radwan et al. 2000; Radwan and Al-Muteirie 2001).

The main technical problem involved in bioremediation of oily coastal regions is that any added microbial inocula or nutrient fertilizers can potentially be washed out into the open sea during wave and tidal movements. On the other hand, such environments do not seem to suffer much from drought. In view of these facts, we believe that the most promising approach for bioremediating oily coastal areas in the Gulf region is self-cleaning via oil-utilizing micro-organisms in biofilms (see Section 13.6) along the coasts. Those biofilms occur in cyanobacterial mats and epilithic algal growth, and coat gravel particles, sand grains, and other animate and inanimate coastal materials. In such biofilms, micro-organisms are firmly immobilized, and thus are not readily washed out into the open water. Furthermore, in cyanobacterial mats and epilithic biomass, such micro-organisms are provided with oxygen and other products of photosynthesis such as organic nutrients including nitrogenous compounds and vitamins. Our regular field trips along the oil-polluted Kuwaiti coasts after the war revealed that the coastal oil sediments were predominantly covered with cyanobacterial mats. Month after month, we noticed that the oil sediment layers became steadily thinner until they completely disappeared together with their mat covers three to five years after the spill. Thus, it became obvious that such natural phenomena are efficient in self-cleaning of the coasts in the Gulf region.

13.11 Conclusions

Desert soils and coastal areas of the Arabian Gulf contain indigenous hydrocarbon-utilizing micro-organisms. In desert soils, those micro-organisms are found in the sand and the rhizospheres of desert plants, whereas in coastal areas they occur mainly in association with inanimate (e.g., sand, gravel) and animate (e.g., blue-green mats, epilithic algal growth) materials. The predominant desert soil bacteria belong to the genera *Micrococcus*, *Pseudomonas*, *Bacillus*, *Arthrobacter*, and to the group of nocardioforms. The predominant fungi belong to the genera *Aspergillus*, *Penicillium*, *Fusarium*, and *Mucor*. Predominant micro-organisms in the rhizospheres are the genera *Cellulomonas*, *Rhodococcus*, and *Arthrobacter*. The major actinomycete in the desert soils is *Streptomyces*.

In general, organisms from the same genera occur in coastal areas, where they are mainly associated with blue-green mats in which the predominant cyanobacterial partners are the genera *Microcoleus* and *Phormedium*. Oil-pollution results in dramatic increases of the frequencies of these organisms. Micro-organisms in the desert soils suffer mainly from drought and lack of nutrients, particularly fixed nitrogen. Therefore, bioremediation of oily desert soils should involve activating indigenous micro-organisms via watering and fertilization with nitrogenous and easily utilizable carbon compounds. Phytoremediation, particularly by planting oily areas with legume crops whose roots are loaded with nodule bacteria, may be the approach of choice for cleaning oily desert soil. In coastal areas, which are naturally rich in blue-green mats, self-cleaning occurs at a satisfactory rate.

Acknowledgements Thanks are due to Mrs. Samar Salamah and Mrs. Majeda Khanafer for their valuable help during the manuscript preparation. Some of the unpublished results of the author in this chapter have been obtained through Research Projects No.SL07/03 and RS01/04, supported by Kuwait University.

References

Abed RMM, Köster J (2005) The direct role of aerobic heterotropic bacteria associated with cyanobacteria in the degradation of oil compounds. *Int Biodeterior Biodegrad* 55:29–37

Abed RMM, Safi NMD, Köster J, de Beer D, El-Nahhal Y, Rullkötter J, Garcia-Pichel F (2002) Microbial diversity of a heavily polluted microbial mat and its community changes following degradation of petroleum compounds. *Appl Environ Microbiol* 68:1674–1683

Aeckersberg F, Bak F, Widdel F (1991) Anaerobic oxidation of saturated hydrocarbons to CO_2 by a new type of sulfate-reducing bacterium. *Arch Microbiol* 155:5–14

Al-Awadhi H, Sulaiman RHD, Mahmoud HM, Radwan SS (2007) Alkaliphilic and halophilic hydrocarbon-utilizing bacteria from Kuwaiti coasts of the Arabian Gulf. *Appl Microbiol Biotechnol.* In press

Alexander M (1994) *Biodegradation and Bioremediation*, Academic Press, San Diego

Al-Hasan RH, Al-Bader DA, Sorkhoh NA, Radwan, SS (1998). Evidence for *n*-alkane consumption and oxidation by filamentous cyanobacteria from oil-contaminated coasts of the Arabian Gulf. *Mar Biol* 130: 521–527

Al-Hasan RH, Khanafer M, Elyas M, Radwan SS (2001) Hydrocarbon accumulation by picocyanobacteria from the Arabian Gulf. *J Appl Bacteriol* 91:533–540

Al-Hasan RH, Sorkhoh NA, Al-Bader D, Radwan SS (1994) Utilization of hydrocarbons by cyanobacteria from microbial mats on oily coasts of the Gulf. *Appl Microbiol Biotechnol* 41:615–619

Anderson TA, Guthrie EA, Walton BT (1993) Bioremediation in the rhizosphere. Plant roots and associated microbes in clean and contaminated soil. *Environ Sci Technol* 27:2630–2636

Applied Biotreatment Association (1989) *Case History Compendium.* Applied Biotreatment Association, Washington DC

Applied Biotreatment Association (1990) *The Role of Biotreatment of Oil Spills.* Applied Biotreatment Association, Washington DC

Atlas RM (1977) Stimulated petroleum biodegradation. *Crit Rev Microbiol* 5:371–386

Atlas RM (1981) Microbial degradation of petroleum hydrocarbons: An environmental perspective. *Microbiol Rev* 45:180–209

Atlas RM (1995) Bioremediation. *Chem Eng News*, April 3:32–42

Atlas RM, Bartha R (1972) Degradation and mineralization of petroleum in seawater. Limitation by nitrogen and phosphorus. *Biotech Bioeng* 14:309–318

Atlas RM, Bartha R (1998) *Microbial Ecology, Fundamentals and Applications*, 4th edn, Benjamin/Cummings, California

Atlas RM, Heintz CE (1973) Ultrastructure of two species of oil-degrading marine bacteria. *Can J Microbiol* 19:43–45

Atlas RM, Pramer D (1990) Focus on bioremediation. *ASM News* 56:7

Bai GY, Brusseau MI, Miller RM (1997) Biosurfactant-enhanced removal of residual hydrocarbon from soil. *J Cont Hydrol* 25:157–170

Banat IM, Makkar RS, Cameotra SS (2000) Potential commercial applications of microbial surfactants. *Appl Microbiol Biotechnol* 53:495–508

Barabas G, Penyige A, Szabo I, Vargha G, Damianovich S, Matko J, Szollosi J, Radwan SS, Matyus A, Hirano T (2000) Hydrocarbon uptake and utilization by *Streptomyces* strains. In: Wise DL, Trantolo DJ (eds) *Remediation of Hazardous Wastes Contaminated Soils*, 2nd edn. Marcel Dekker, New York, pp 291–309

Barabas G, Sorkhoh NA, Fardoon F, Radwan SS (1995) *n*-alkane utilization by oligocarbophilic actinomycete strains from oil-polluted Kuwaiti desert soil. *Actinomycetol* 9:13–18

Bolba MT, Al-Awadhi N, Al-Daher R (1998) Bioremediation of oil-contaminated soil: Microbiological methods for feasibility assessment and field evaluation. *J Microbiol Methods* 32:155–164

Boulton CA, Ratledge C (1984) The physiology of hydrocarbon-utilizing microorganisms. In: Wiseman A (ed) *Topics in Fermentation and Enzyme Technology*, vol 9, Ellis Horwood, Chichester, pp 11–77

Bragg JR, Prince RC, Harner EJ, Atlas RM (1994) Effectiveness of bioremediation for the Exxon Valdez oil spill. *Nature* 368:413–418

Bragg JR, Prince RC, Wilkinson JB, Atlas RM (1992) Bioremediation for shoreline cleanup following the 1989 Alaskan oil spill. Exxon Co, Houston, TX

Buehler M, Schindler J (1984) Aliphatic hydrocarbons. In: Rehm H-J, Reed G (eds) *Biotechnology: A Comprehensive Treatise*, vol 6a, Verlag Chemie, Weinheim, pp 329–385

Button DK, Schut F, Quang P, Martin R, Robertson BR (1993) Viability and isolation of marine bacteria by dilution culture: Theory, procedures and initial results. *Appl Environ Microbiol* 59:881–891

Cameotra SS, Makkar RS (1998) Synthesis of biosurfactants in extreme conditions. *Appl Microbiol Biotechnol* 50:520–529

Cerniglia CE, Gibson DT, van Baalen C (1980a) Oxidation of naphthalene by cyanobacteria and microalgae. *J Gen Microbiol* 116:495–500

Cerniglia CE, van Baalen C, Gibson DT (1980b) Metabolism of naphthalene by the cyanobacterium *Oscillatoria* sp. strain JCM. *J Gen Microbiol* 116:485–494

Cundell AM, Mueller WC, Traxier RW (1976) Morphology and ultrastructure of a *Penicillium* sp. grown on *n*-hexadecane or peptone. *Appl Environ Microbiol* 31:408–414

Davies JS, Westlake DWS (1979) Crude oil utilization by fungi. *Can J Microbiol* 25:146–156

Daylan U, Harder H, Hoepner Th (1990) Hydrocarbon biodegradation in sediments and soils. A systematic examination of physical and chemical conditions. II. pH-values. *Erdoel Kohle Erdgas Petrochem* 43:337–342

Demanova NF, Davydov ER, Golobov AD (1980) Assimilation of *n*-alkanes with a varying length of the carbon chain by the yeast *Candida guilliermondii*. *Prikl Biokhim Mikrobiol* 26:5–12

Desai JD, Banat IM (1997) Microbial production of surfactants and their commercial potential. *Microbiol Mol Biol Rev* 61:47–64

Díaz MP, Boyd KG, Grigson SJW, Burgess, JG (2002) Biodegradation of crude oil across a wide range of salinities by an extremely halotolerant bacterial consortium MPD-M, immobilized onto polypropylene fibers. *Biotech Bioeng* 79:145–153

Efroymson RA, Alexander M (1991) Biodegradation by an *Arthrobacter* species of hydrocarbons partitioned into an organic solvent. *Appl Environ Microbiol* 57:1441–1447

Egorov NS, Koronelli TV, Milko ES, Stepanova RA, Rozynov B, Pletenko MG (1986) Comparison of lipid composition among *Rhodococcus rubropertinctus* R,S and M variants. *Microbiologiya* 55:227–230

Einsele A (1983) Biomass from higher *n*-alkanes. In: Rehm H-J, Reed G (eds) *Biotechnology- A Comprehensive Treatise in Eight Volumes*, vol 3, Verlag Chemie, Weinheim, pp 43–81

Ellis BE (1977) Degradation of phenolic compounds by freshwater algae. *Plant Sci Lett* 8:213–216

Flynn S, Butler J, Vance I (1996) Produced water composition, toxicity and fate. In: Reed M, Johnsen S (eds) *Produced Water 2. Environmental Issues and Migration Technologies*. Plenum Press, New York, pp. 69–80

Fukui S, Tanaka A (1981) Metabolism of alkanes by yeasts. *Adv Biochem Eng* 19:217–237

Gadd GM (1990) Metal tolerance. In: Edwards C (ed) *Microbiology of Extreme Environments*. McGraw-Hill, New York, pp 178–210

Garcia De Oteyza T, Grimalt JO, Diestra E, Solé T, Esteve I (2004) Changes in the composition of the polar and apolar crude oil fractions under the action of *Microcoleus* consortia. *Appl Microbiol Biotechnol* 66:226–232

Gibbs CF (1975) Quantitative studies on marine biodegradation of oil. I. Nutrient limitation at 14°C. *Proc Roy Soc London* 188:61–82

Gibbs CF, Pugh KB, Andrews AP (1975) Quantitative studies on marine biodegradation of oil. II. Effect of temperature. *Proc Roy Soc London* 188:83–94

Golubic S (1992) Microbial mats of Abu Dhabi. In: Margulis L, Olendzenski L (eds) *Environmental Evolution: Effects of the Origin and Evolution of Life on Planet Earth*, MIT Press, Cambridge, MA, pp. 103–130

Grötzschel S, Köster J, Abed RMM, de Beer D (2002) Degradation of petroleum model compounds immobilized on clay by a hypersaline microbial mat. *Biodegradation* 13:273–282

Halvorson HO, Pramer D, Rogul M, eds (1985) *Engineered Organisms in the Environment: Scientific Issues*. American Society of Microbiology, Washington DC

Hartmann J, Reineke W, Knackmuss HJ (1979) Metabolism of 3-chloro,4-chloro, and 3,5-dichlorobenzoate by a pseudomonad. *Appl Environ Microbiol* 37:421–428

Hinchee RE, Olfenbuttel RE, eds (1991a). *In Situ Bioreclamation: Applications and Investigations for Hydrocarbon Contaminated Site Remediation*. Butterworth-Heinemann, Boston

Hinchee RE, Olfenbuttel RE, eds (1991b). *On Site Bioreclamation: Processes for Xenobiotic And Hydrocarbon Treatment*. Butterworth-Heinemann, Boston

Hoffman B, Rehm H-J (1978) Degradation of *n*-alkanes by mucorales IV. Lipid formation and fatty acid composition of *Absidia spinosa*, *Cunninghamella echinulata* and *Mortierella isabellina* grown on glucose and some *n*-alkanes. *Eur J Appl Microbiol Biotechnol* 5:189–195

Ivshina IB, Nesterenko OA, Glazacheva LE, Shekhovtsev VP (1982) Facultative gas assimilating *Rhodococcus rhodochrous* studied by electron microscope. *Mikrobiologiya* 51:477–481

Jannasch HW (1967) Growth of marine bacteria at limiting concentrations of organic carbon in seawater. *Limnol Oceanog* 12:264–271

Kant U, Kiesewetter K, Michaelsen M, Hoepner T (1985) Estimation of hydrocarbon biodegradation velocities in tidal sediments under standard conditions. Final report to the commission of European Communities, Oldenburg, Germany

Kargi F, Dincer AR (2000) Use of halophilic bacteria in biological treatment of saline wastewater by fed-batch operation. *Water Environ Res* 72:170–174

Kennedy RS, Finnerty WR (1975) Microbial assimilation of hydrocarbons. 1. The fine structure of hydrocarbon-oxidizing *Acinetobacter* sp. *Arch Microbiol* 10:75–83

Kirk PW, Gordon AS (1988) Hydrocarbon degradation by filamentous marine higher fungi. *Mycologia* 80:776–782

Klug MJ, Markovetz AJ (1971) Utilization of aliphatic hydrocarbons by microorganisms. *Adv Microb Physiol* 5:1–43

Korda A, Santas P, Tenente A, Santas R (1997) Petroleum hydrocarbon bioremediation: Sampling and analytical techniques, in situ treatment and commercial microorganisms currently used. *Appl Microbiol Biotecnol* 48:677–686

Koval EZ, Redchitz TI (1978) Fatty inclusions in the mycelium of aspergilli grown under surface cultivation on media with hydrocarbons. *Mikrobiol Zh* 40:736–740

Lang S, Wullbrandt D (1999) Rhamnose lipids-biosynthesis, microbial production and application potential. *Appl Microbiol Biotechnol* 51:22–32

Leahy JG, Colwell RR (1990) Microbiological degradation of hydrocarbons in the environment. *Microbiol Rev* 54:305–315

Leavitt ME, Brown KI (1994) Biostimulation versus bioaugmentation-three case studies. In: Hinchee, RE, Alleman BC, Hoeppel RE, Miller RN (eds) *Hydrocarbon Bioremediation*, Lewis, Boca Raton, FL, pp 72–79

Levi ID, Shennan JL, Ebbon GP (1979) Biomass from liquid *n*-alkanes. In: Rose AH (ed) *Microbial Biomass*. Academic Press, New York, London, pp 361–491

Liebig J (1840) *Chemistry in its Application to Agriculture and Physiology*. Taylor and Walton, London

Lin HT, Iida M, Iizaka H (1971a) Formation of organic acids and ergosterol from *n*-alkanes by fungi isolated from aircraft fuel. *J Ferment Technol* 49:206–212

Lin HT, Iida M, Iizaka H (1971b) Formation of organic acids and ergosterol from *n*-alkanes by fungi isolated from oil fields in Japan. *J Ferment Technol* 49:771–777

Loginova LG, Bogdanova TI, Seregina IM (1981) Growth of obligate thermophilic bacteria in a medium with paraffin. *Mikrobiologiya* 50:49–54

Makkar RS, Cameotra SS (1998) Production of biosurfactant at mesophilic and thermophilic conditions by a strain of *Bacillus subtilis*. *J Ind Microbiol Biotechnol* 20:48–52

Makkar RS, Cameotra SS (2002) An update on the use of unconventional substrates for biosurfactant production and their new applications. *Appl Microbiol Biotechnol* 58:428–434

McKinnon M, Vine P (1991) *Tides of War*. IMMEL, London

Morita RY (1982) Starvation-survival of heterotrophs in the marine environment. *Adv Microb Ecol* 6:171–198

Mueller JG, Chapman PJ, Pritchard PH (1989) Creosote-contaminated sites: Their potential for bioremediation. *Environ Sci Technol* 23:1197–1201

Mulligan CN, Yong RN, Gibbs BF (2001) Surfactant enhanced remediation of contaminated soil: a review. *Eng Geol* 60:371–380

Neidhardt EC, Ingraham JL, Schaechter M (1990) *Physiology of the Bacterial Cell*. Sinauer, Sunderland, MA

Novitsky JA, Morita RY (1976) Morphological characterization of small cells resulting from nutrient starvation of a psychrophilic marine vibrio. *Appl Environ Microbiol* 32:617–632

Novitsky JA, Morita RY (1977) Survival of a psychrophilic marine vibrio under long-term nutrient starvation. *Appl Environ Microbiol* 33:635–641

Novitsky JA, Morita RY (1978) Possible strategy for the survival of marine bacteria under starvation conditions. *Mar Biol* 48:289–295

Ornodera M, Endo Y, Ogasawa N (1989) Oxidation of gaseous hydrocarbons by a gaseous hydrocarbon assimilating mold, *Scedosporium* sp. A-4. *Agric Biol Chem* 53:1947–1951

Polonenko DR, Scher FM, Kloepper JW, Singleton CA, Laliberte M, Zaleska I (1987) Effects of root colonizing bacteria on nodulation of soybean roots by *Bradyrhizobium japonicum*. *Can J Microbiol* 33:498–503

Prantera MT, Drozdowicz A, Gomes-Leite S, Soares-Rosado A (2002) Degradation of gasoline aromatic hydrocarbons by two N_2-fixing soil bacteria. *Biotechnol Lett* 24:85–89

Pritchard PH, Costa CF (1991) EPA's Alaska oil spill bioremediation project. *Environ Sci Technol* 25:372–379

Radwan SS (1990) Gulf oil spill. *Nature* 350:456

Radwan SS, Al-Awadhi H, El-Nemr IM (2000) Cropping as a phytoremediation practice for oily desert soil with reference to crop safety as food. *Int J Phytoremed* 2:383–396

Radwan SS, Al-Awadhi H, Sorkhoh NA, El-Nemr I (1998a) Rhizospheric hydrocarbon-utilizing microorganisms as potential contributors to phytoremediation for oily Kuwaiti desert. *Microbiol Res* 153:247–251

Radwan SS, Al-Hasan RH (2001) Potential application of coastal biofilm-coated gravel particles for treating oily waste. *Aqu Microb Ecol* 23:113–117

Radwan SS, Al-Hasan RH, Al-Awadhi H, Salamah S, Abdullah HM (1999) Higher oil biodegradation potential at the Arabian Gulf coast than in the water body. *Mar Biol* 135:741–745

Radwan SS, Al-Hasan RH, Mahmoud HM, Eliyas M (2007a) Oil-utilizing bacteria associated with fish from the Arabian Gulf. *J Appl Microbiol* (in press)

Radwan SS, Al-Hasan RH, Salamah S, Al-Dabbous S (2002) Bioremediation of oily sea water by bacteria immobilized in biofilms coating macroalgae. *Int Biodeter Biodeg* 50:55–59

Radwan SS, Al-Hasan RH, Salamah S, Khanafer M (2005a) Oil-consuming microbial consortia floating in the Arabian Gulf. *Int Biodeter Biodeg* 56:28–33

Radwan SS, Al-Mailem D, El-Nemr I, Salamah S (2000) Enhanced remediation of hydrocarbon contaminated desert soil fertilized with organic carbons. *Int Biodet Biodeg* 46:129–132

Radwan SS, Al-Muteirie AS (2001) Vitamin requirements of hydrocarbon utilizing soil bacteria. *Microbiol Res* 155:301–307

Radwan SS, Dashti N, El-Nemr IM (2005b) Enhancing the growth of *Vicia faba* plants by microbial inoculation to improve their phytoremediation potential for oily desert areas. *Int J Phytoremed* 7:19–32

Radwan SS, Dashti N, El-Nemr IM, Khanafer M (2007b) Hydrocarbon utilization by nodule bacteria and plant growth promoting rhizobacteria. *Int J Phytoremed* (in press)

Radwan SS, et al. (1998b) Hydrocarbon uptake by *Streptomyces*. *FEMS Microbiol Lett* 169:87–94

Radwan SS, Sorkhoh NA (1993) Lipids of n-alkane-utilizing microorganisms and their application potential. *Adv Appl Microbiol* 39:29–90

Radwan SS, Sorkhoh NA, Al-Hasan RH (1995a) Self-cleaning and bioremediation potential of the Arabian Gulf. In Cheremisinoff P (ed) *Encyclopedia of Environmental Control Technology*, vol 9, Gulf, Houston, pp 901–924

Radwan SS, Sorkhoh NA, El-Nemr I (1995b) Oil-biodegradation around roots. *Nature* 376:302

Radwan SS, Sorkhoh NA, El-Nemr I, El-Desouky AF (1997) A feasibility study on seeding as a bioremediation practice for the oily Kuwaiti desert. *J Appl Microbiol* 83:353–358

Radwan SS, Sorkhoh NA, Fardoun F, Al-Hasan RH (1995c) Soil managements enhancing hydrocarbon biodegradation in the polluted Kuwaiti desert. *Appl Microbiol Biotechnol* 44:265–270

Raghukumar C, Vipparty V, David JJ, Chandramohan D (2001) Degradation of crude oil by cyanobacteria. *Appl Microbiol Biotechnol* 57:433–436

Ratledge C (1978) Degradation of aliphatic hydrocarbons. In: Watkinson I (ed) *Developments in Biodeterioration of Hydrocarbons*, vol 1, Applied Science, Essex, pp 1–45

Redchitz TI (1980) Fatty incorporations in *Aspergillus* mycelium during submerged cultivation in media with hydrocarbons. *Microbiol Zh* 42:596–600

Redchitz TI, Koval EZ (1979) Formation of volutin inclusions in the mycelium of aspergilli growing on media with hydrocarbons. *Mikrobiol Zh* 41(1):34–39

Rehm H-J, Reiff I (1981) Mechanisms and occurrence of microbial oxidation of long-chain alkanes. *Adv Biochem Eng* 19:175–216

Reineke W, Knackmuss H-J (1979) Construction of haloaromatic-utilizing bacteria. *Nature* 277:385–386

Riser-Roberts E (1992) *Bioremediation of Petroleum Contaminated Sites*. CRC Press, Boca Raton, FL

Roe TI, Johnsen S, The Norwegian Oil Industry association (1996) Discharges of produced water to the North Sea. In: Reed M, Johnsen S (eds) *Produced Water 2. Environmental Issues and Migration Technologies*. Plenum Press, New York, pp. 13–25

Rosenberg E (1993) *Microorganisms to Combat Pollution*. Kluwer Academic, Dordrecht

Rosenberg E (2006) Hydrocarbon-oxidizing bacteria. In: Dworkin M, Falkow S, Rosenberg E, Schleifer K-H, Stackebrandt E (eds) *The Prokaryotes, a Handbook on the Biology of Bacteria*, 3rd edn, vol 2. Springer, Berlin, pp 564–577

Roszak DB, Colwell RR (1987) Survival strategies of bacteria in the natural environment. *Microbiol Rev* 51:365–379

Rouse JD, Sabatini DA, Suffita JM, Harwell JH (1994) Influence of surfactants on microbial degradation of organic compounds. *Crit Rev Microbiol* 24:325–370

Roy I, Shukla SK, Mishra AK (1988) *n*-Dodecane as a substrate for nitrogen fixation by an alkane-utilizing *Azospirillum* sp. *Curr Microbiol* 16:303–309

Sanchez O, Diestra E, Esteve I, Mas J (2005) Molecular characterization of an oil-degrading cyanobacterial consortium. *Microb Ecol* 50:580–588

Scott GL, Finnerty WR (1966) Characterization of intracytoplasmic hydrocarbon inclusions from the hydrocarbon-oxidizing *Acinetobacter* species. *J Bacteriol* 127:481–489

Shelford VE (1913) *Animal Communities in Temperate America*. University of Chicago, Chicago

Singer ME, Finnerty WR (1984) Microbial metabolism of straight-chain and branched alkanes. In: Atlas RM (ed) *Petroleum Microbiology*, Macmillan, New York, pp 1–59

Smiles DE (1988) Aspects of the physical environment of soil organisms. *Biol Fertil Soils* 6:204–215

Song HG, Wang X, Bartha R (1990) Bioremediation potential of terrestrial fuel spills. *Appl Environ Microbiol* 56:652–656

Sorkhoh N, Al-Hassan R, Radwan S, Höpner T (1992) Self-cleaning of the Gulf. *Nature* 359:109

Sorkhoh NA, Al-Hasan RH, Khanafer M, Radwan SS (1995) Establishment of oil-degrading bacteria associated with cyanobacteria in oil polluted soil. *J Appl Bacteriol* 78:194–199

Sorkhoh NA, Ghannoum MA, Ibrahim AS, Stretton RJ, Radwan SS (1990) Crude oil and hydrocarbon degrading strains of *Rhodococcus rhodochrous* isolated from soil and marine environments in Kuwait. *Environ Pollut* 65:1–17

Sorkhoh NA, Ibrahim AS, Ghannoum MA, Radwan SS (1993) High-temperature hydrocarbon degradation by *Bacillus stearothermophilus* from oil-polluted Kuwaiti desert. *Appl Microbiol Biotechnol* 39:123–126

Steppe TF, Olson JB, Paerl HW, Litaker RW, Belnap J (1996) Consortial N_2 fixation: A strategy for meeting nitrogen requirements of marine and terrestrial cyanobacterial mats. *FEMS Microbiol Ecol* 21:149–156

Stoner DL (1994) *Biotechnology for the Treatment of Hazardous Waste*. Lewis, Boca Raton, FL

Sulaiman R (2006) Alkalinity and salinity loving (tolerant) oil-utilizing microorganisms from Kuwait coasts. A thesis submitted for the M.Sc. degree in microbiology, Kuwait University, Kuwait

Todd SJ, Cain RB, Schmidt S (2002) Biotransformation of naphthalene and diaryl ethers by green microalgae. *Biodegradation* 13:229–238

Tumeo M, Brandock J, Venator T, Rog S, Owens D (1994) Effectiveness of biosurfactants in removing weathered crude oil from subsurface beach material. *Spill Sci Technol Bull* 1:53–59

Van Eyk J (1994) Venting and bioventing for the in situ removal of petroleum from soil. In: Hinchee, RE, Alleman BC, Hoeppel RE, Miller RN (eds) *Hydrocarbon Bioremediation*, Lewis, Boca Raton, FL, pp 243–251

Van Ginkel GG, Welten HGJ, de Bont JAM (1987) Oxidation of gaseous and volatile hydrocarbons by selected alkene-utilizing bacteria. *Appl Environ Microbiol* 53:2903–2907

Van Hamme JD, Singh A, Ward O (2003) Recent advances in petroleum microbiology. *Microbiol Molec Biol Rev* 67:503–549

Widdel F, Boetius A, Rabus R (2006) Anaerobic biodegradation of hydrocarbons including methane. *The Prokaryotes, a Handbook on the Biology of Bacteria*, 3rd edn, vol 2. Springer, Berlin, pp 1028–1049

Woolard CR, Irvine RL (1994) Biological treatment of hypersaline wastewater by a biofilm of halophilic bacteria. *Water Environ Res* 66:230–235

Zarilla KA, Perry JJ (1984) *Thermoleophilum album* gen. nov. and sp. nov., a bacterium obligate for thermophily and *n*-alkane substrates. *Arch Microbiol* 137:286–290

Zhang F, Dashti N, Hynes RK, Smith D (1996) Plant growth promoting rhizobacteria and soybean [*Glycine max* (L.) Merr] nodulation and nitrogen fixation at suboptimal root zone temperatures. *Ann Bot* 77:453–459

Zhang F, Dashti N, Hynes RK, Smith D (1997) Plant growth-promoting rhizobacteria and soybean [*Glycine max* (L.) Merr] growth and physiology of suboptimal root zone temperatures. *Ann Bot* 79:243–249

Zitrides TG (1990) Bioremediation comes of age. *Pollut Eng* XXII:59–60

Chapter 14
Microbial Communities in Fire-Affected Soils

Christopher Janzen and Tammy Tobin-Janzen(✉)

14.1 Introduction

Of all the ways in which human activities can affect soil environments, fires are perhaps the most dramatic. Whether set deliberately or accidentally, the worldwide impact that such fires have is extensive, including the loss of human and animal lives as well as economic and ecological damage (UNECE et al. 2000; UNECE and FAO 2001; Davidenko and Eritsov 2003; Kudoh 2005; FAO 2005). The United Nations Food and Agricultural Organization (2005) estimated that 350 million hectares burn annually, and that approximately 90% of those fires are of human origin. The frequency and severity of surface fires have also increased in many parts of the world due to changes in climate and land management practices (Houghton et al. 1992; Renkin and Despain 1992; Glantz 1996; Neary et al. 1999; Westerling et al. 2006). Large fires tend to draw media attention, particularly when they impinge on densely inhabited or well-known wildlife areas. Their rapid spread, extreme temperatures, and the barren landscapes they leave behind can alter the surrounding ecosystem for years, decades, or even permanently. It is no surprise, therefore, that a premium has historically been placed on extinguishing fires as they happen, rather than on studying their ecological significance. Only recently have scientists begun to understand the critical roles that some fires play in sustaining natural environments, and on the parts that soil micro-organisms play in that process.

Two basic types of fires are discussed in this chapter: surface fires and underground fires. Surface fires include both wildfires and prescribed fires and their effects are predominantly 'top down'. That is, the fire source is aboveground, and the heat from the fire, although it can be intense enough to sterilize the surface soil, may not penetrate more than a few centimeters below the soil surface. Surface fire effects often consist of a patchwork of severely affected sites interspersed with less affected areas, the pattern of which is dictated by the availability of fuel, topography, and other factors.

Tammy Tobin-Janzen
Biology Department, Susquehanna University, Selinsgrove, PA17870, USA
e-mail: tobinjan@susqu.edu

Underground fires, by way of contrast, have a 'bottom-up' impact. They are typified by the large coalmine fires found in the Jharia Coalfield of India (Agarwal et al. 2006), throughout much of northern China (Zhou et al. 2006), and throughout the coal-mining regions of the United States, most notably in Centralia, Pennsylvania (Trifonoff 2000). In these fires, steam and gases carrying vaporized combustion products from the burning coal below migrate, or 'vent', upward through soil fractures, cooling as they rise. As the gases from these 'anthracite smokers' cool, their dissolved chemicals either escape to the atmosphere or condense into the surrounding soils. Thus, underground fires are distinguished by subsurface soil temperatures that are generally much hotter than surface temperatures, and by soil chemical changes that tend to be clustered around actively venting soil fractures. In many ways, these fires resemble geothermal environments as closely as they do environments affected by surface fires. In this chapter, we use coalmine fires as the model system for underground fires in general.

Regardless of whether a fire has a top-down or a bottom-up impact, it alters many surface and subsurface soil properties as it migrates through an area. These properties include the availability of water, the structure and composition of soil particles, and the availability of nutrients. Sulfur, nitrogen, phosphorus, and carbon species, in particular, can experience dramatic shifts when pre- and post-fire environments are compared.

The extent and severity of these fire-induced changes is determined by many factors including the duration of the fire, the intensity of the fire, the rate of the fire's spread, the topography of the burn area, the initial soil composition, the soil moisture content, fuel characteristics, weather conditions, the interval between fire events, and the frequency of fire events. However, the greatest of these is duration. A fast-moving aboveground fire of even moderate intensity does not persist long enough to transfer enough heat to the soils to cause substantial belowground chemical changes. Slow-moving fires cause severe and complex changes that can be long lasting or permanent. Fire temperatures can range from 50 to >1,500°C. Heat release can range from 2.11 kJ/kg of fuel to 2.1 MJ/kg. The effect of fire intensity is related to how well the energy produced by the fire is transferred to the soil. A slow-moving intense fire will have a greater impact than a fast-moving fire of the same intensity. Likewise, a slow-moving fire of low intensity may have as great an impact as a fast-moving fire due to the greater length of time the fire will be transferring the generated heat to the soil.

Ultimately, it is both the intensity and persistence of fires, as well as the resulting changes to the surface and subsurface soil properties that determine the fire's impact on resident microbial communities. Although these communities are likely to play critical roles in both post-fire biogeochemical nutrient cycling and in the ultimate recovery of fire-affected environments, they have remained quite poorly studied. Most prokaryotic research, in particular, has focused on studies of post-fire microbial biomass, microbial metabolic assays, and culture-dependent assays (Ahlgren 1974; Dunn et al. 1979; Bissett and Parkinson 1980; Klopatek and Klopatek 1987; Klopatek et al. 1990; Fritze et al. 1993; Vazquez et al. 1993; Acea and Carballas 1996; Hernandez et al. 1997; Ross et al. 1997; Prieto-Fernández et al. 1998; Choromanska and DeLuca 2002; Andersson et al. 2004; Treseder et al. 2004; Naumova 2005; Giai and Boerner 2007). Only recently have culture-independent methods such as 16S rRNA gene sequencing, terminal restriction fragment length polymorphism (T-RFLP) analysis, phospholipids fatty acid analysis (PLFA), and denaturing gradient gel electrophoresis (DGGE) been combined with more traditional methods to give a more complete picture of the effects that fires have on specific prokaryotic communities, and on the roles that those communities play in the post-fire environment (Baath et al. 1995; Pietikainen et al. 2000; Jaatinen et al. 2004; Tobin-Janzen et al. 2005; Yeager et al. 2005; Izzo et al. 2006).

14.2 The Effects of Surface Fires on Microbial Environments

Because the fire-induced thermal, chemical, and physical changes to the soil environment directly drive microbial community responses, it is important to understand these factors before turning to a more complete discussion of microbial ecology. The soil changes that most clearly affect microbial community responses depend on the intensity of the fire, and correspond to changes in the availability of water, nitrogen, sulfur, phosphorus, and carbon.

14.2.1 Fire Intensity

The intensity of a surface fire, and its resulting impact on the microbial communities in underlying soils, varies according to the total amount of heat produced by the fire, the mode of transfer of that heat to the underlying soils, and the speed with which the fire progresses through an area. The total heat produced by the fire is governed by the nature of the fuel source and the fuel load. For example, grassland fires (Raison 1979; Ross et al. 1997; Neary et al. 1999) tend to have lower fuel loads and thus generate lower fire temperatures than fires in tropical woodlands or conifer forests. Because microbial mortality generally begins at around 50°C, but varies tremendously from species to species, the total heat produced by the fire is of paramount importance in determining the post-fire microbial community composition.

The modes of energy transfer from a surface fire to the soil are radiation, convection, conduction, vaporization/condensation, and mass transport (Neary et al. 1999). Electromagnetic radiation from the fire can play a significant role in energy transfer at the onset of a burn. Movement of air masses by convection can transfer heat from the active fire to nearby areas, heating nearby soils as well as spreading the fire. Conduction of heat by direct contact between hot or burning fuel and cooler soil can be important especially where fuel loads are heavy or where fuels themselves are massive such as with slash piles. The high specific heat capacity of water (4.184 J $g^{-1}°C^{-1}$) means that a substantial amount of energy is required to vaporize the water in soil. The enthalpies of heating and vaporization of soil moisture tend to mitigate any rise in soil temperature until all the water is vaporized. Indeed, the soil temperature cannot rise above 95°C until vaporization is complete (Campbell et al. 1994). The vaporized water carries latent heat through soils faster and deeper than other modes of energy transport such as radiation or conduction. Thus, whereas moist soils initially suffer lower temperatures, the overall impact of the fire may be more rapid and affect soil to a greater depth. The transfer of energy via mass transport generally has little to no effect on soil temperatures.

Once all of the moisture in soils under a fire has vaporized, soil temperatures typically rise to 200–300°C. However, under severe or slow-moving fires, surface soil temperatures can rise to as high as 700°C (DeBano et al. 1998). The depth-temperature profile for a soil under a fire is determined by intensity and duration of the fire as well as the moisture content of the soil. With fast-moving or low severity fires, belowground soil temperatures generally do not exceed 100–150°C at 5 cm depth and demonstrate no heating below 30 cm (Agee 1973; DeBano 2000). A slow-moving or intense fire will clearly have a more significant effect on both the maximum soil temperatures observed and the depth at which temperatures are raised.

14.2.2 Available Moisture

Available soil moisture levels are not only decreased initially as a direct result of vaporization, but may also suffer long-term reductions. At moderate temperatures, incomplete burning of organic matter can result in the formation of a hydrophobic coating in the mineral components of soil resulting in increased repellency and decreased soil permeability. This repellency can result in a persistent decrease in soil moisture in the post-burn soils or in greater run-off and erosion. Despite these tendencies, situations in which soil moisture has increased (Klock and Hevey 1976; Haase 1986) or remained the same (Campbell et al. 1977; Milne 1979) after fire have also been documented, and thus the overall impact that post-fire soil moisture has on microbial communities is varied (Letey 2001).

14.2.3 Available Nutrients

In an aboveground fire, the most dramatic and well-understood changes to available nutrients involve the volatilization, chemical transformation, and biogeochemical cycling of nitrogen, carbon, phosphorus, and sulfur. Although changes can and do occur to trace minerals such as As, Ca, K, Na, Fe, and Al (Grier 1975; Feller 1982; Macadam 1989; DeBano et al. 1998; Neary et al. 1999; Arocena and Opio 2003) the effects that these changes have on microbial communities remain mostly unstudied, and thus are not discussed in this chapter.

14.2.3.1 Nitrogen

At moderate temperatures, dead partially combusted plant and microbial biomass can be easily oxidized resulting in an increase in inorganic nitrogen immediately post burn (Diaz-Ravina et al. 1996). More extreme temperatures can result in volatilization of nitrogen (Giovannini et al. 1990). Total nitrogen decreases slightly as the temperature rises from 25 to 220°C but dramatically decreases between 220 and 460°C (Giovannini et al. 1990). Ammonium concentrations steadily increase up to 220°C but, like total nitrogen, drop quickly as the temperature rises above 220°C until very little remains at 460°C. It is presumed that this increase in NH_4^+-N is due to mineralization of organic nitrogen. Nitrate concentrations are initially unaffected by fire, even moderate to intense fires. However, NO_3^--N is produced by biochemical nitrification of ammonium in the time following the fire resulting in NO_3^--N concentrations that can be significantly higher in the weeks or years post-fire (Covington et al. 1991).

Therefore, although total nitrogen may decrease, the bioavailable forms of nitrogen, nitrate and ammonium, may be at elevated levels for several years following a burn. Ammonium tends to adsorb onto the mineral soil and become immobilized. Unless regrowth of vegetation occurs soon after the fire, the nitrate can be easily lost to leaching, thus depleting the total nitrogen for long periods. Conversely, if regrowth is rapid, soil organic nitrogen concentrations can rapidly recover to pre-fire levels (Adams and Attiwill 1984; Weston and Attiwill 1996).

14.2.3.2 Carbon

Fire severity as described by the maximum sustained surface temperature under a fire affects the degree of consumption of litter and soil organic matter. A low severity burn where the soil temperatures do not exceed 250°C can cause partial scorching of litter. As depth in the soil increases, the insulating ability of the soil mitigates the heating of soils. Thus, with surface temperatures at or below 250°C, soil temperatures down to 2.5 cm rarely exceed 100°C and at 5 cm the temperature is typically below 50°C. The lower temperatures at depth result in only partial distillation of organic matter above 2.5 cm and little to no effect below 5 cm. A moderate burn characterized by surface temperatures up to 400°C results in temperatures at 2.5 and 5 cm of up to 175 and 50°C, respectively. These temperatures result in very significant charring of litter, some charring of organic matter to 2.5 cm and the start of distillation of organic matter above 5 cm. In a severe burn, surface temperatures exceed 675°C, resulting in complete combustion of the litter. The soil down to 2.5 cm can experience heating to 190°C with concomitant charring or consumption of large portions of the organic matter. Even at depths of 5 cm, soils are being heated to 75°C resulting in significant distillation of volatile organic matter and some charring (DeBano et al. 1977).

14.2.3.3 Phosphorus

Organic phosphorus is readily converted to inorganic forms of phosphate. Even in low severity fires where temperatures do not exceed 200°C, the concentration of organic phosphorus decreases markedly. In an artificial heating experiment, organic phosphorus was completely depleted from soils above 220°C (Giovannini et al. 1990). Organic carbon is easily lost, however, total phosphorus remains fairly constant due to its low volatility. The organic phosphate is instead converted to orthophosphate, the predominant form of bioavailable phosphorus (Cade-Menun et al. 2000). The concentrations of inorganic and available phosphorus peak at soil temperatures of approximately 450°C (Giovannini et al. 1990).

Other factors play a role in determining how much of the inorganic phosphorus produced by a burn is available. Bioavailability peaks at a pH of around 6.5 (Sharpley 2000). The ash produced in a fire tends to move the soil pH higher toward this value. However, in calcareous soils phosphate can complex very strongly to calcium, resulting in removal of phosphorus from the available pool. In acidic soils, phosphate can bind with other soil metals such as iron, manganese, and aluminum. In summation, aboveground fires easily convert most or all of the organic phosphorus to inorganic phosphorus, thus initially increasing the bioavailability. How long the increase in bioavailable phosphorus persists depends on soil pH, the composition of the mineral soil, and the rate of uptake of available phosphorus by recolonizing vegetation.

14.2.3.4 Sulfur

In an artificial heating study, Badia and Marti (2003) found that a slight steady increase in total sulfur occurred upon heating gypsiferous soils up to 500°C and upon incorporation of ash into the soil, but that no change in total sulfur occurred in calcareous soils. In another study, Castelli and Lazzari (2002) found that total sulfur peaked a year after a controlled burn in both grass-covered and shrub-covered soils. After a second controlled burn three years after the first burn, the total sulfur in the soils under grass cover dropped back to the original (before the first burn) levels below 1 cm depths. In the same study, available sulfur under shrub cover immediately jumped significantly, but then dropped back to pre-burn levels within two years. The available sulfur under grass cover was not affected. In both cases, the available sulfur did not change with the second controlled burn. These two studies indicate that the sulfur chemistry of soils is affected by aboveground fires in manners similar to other nutrients.

14.3 The Effects of Underground Fires on Microbial Environments

Underground coalmine fires have very different effects on soil chemistry than those caused by aboveground fires. The reasons stem from three major factors. The heat source is below the surface, the duration of the belowground fires can be considerably longer than surface fires, and the composition of the coal fuel is quite different from that found in typical prescribed or wild aboveground fires. These factors lead to unique physical and chemical properties and processes.

14.3.1 Fire Intensity

In an aboveground fire, the highest temperatures the soils experience are at the air–surface interface. Conversely, the lowest temperature of an affected soil is at the air–surface interface with a belowground fire. In other words, the soil temperature gradient in a belowground fire is inverted from that observed in aboveground fires.

In addition, the impact of the fire is not evenly distributed spatially. The primary modes of energy transfer from the fire to the surface and near-surface soils are convection and conduction. Convection of hot combustion gases takes the paths of least resistance, following the small fractures and faults in the subsurface. The gases tend to escape to the atmosphere in vents. The temperature of the surrounding soils is highest near the vent as a result of the transfer of thermal energy from the hot gases to the surrounding soils. These hot soils are then able to transfer heat away from the vent via conduction. Eventually, this heat is slowly lost to the atmosphere. Thus, the surface temperature decreases with increasing distance from the vent. In addition to heat brought to the surface by convection, the hot subsurface rock and mineral matter can transfer heat to the surface directly by conduction. This conduction results in surface regions that are quite hot while displaying no outward signs of venting of hot gases. Indeed, the authors have measured surface temperatures above the Centralia Pennsylvania mine fire exceeding 400°C in nonventing areas (Tobin-Janzen, unpublished results). The underground environment (including both structure and composition) is not uniform. Therefore, the heat transferred by conduction is spatially uneven. The end result of the modes of heat transfer is that the surface temperatures above an underground fire are distributed in an irregular pattern.

Another significant difference between aboveground and belowground fires is the duration. Aboveground fires persist at a certain location for hours to weeks. The available fuel is consumed quickly and when gone, no further burning is possible. Stated another way, aboveground fires are fuel limited. The rate of progression of belowground fires is much slower, and can range from ten meters per year to several hundred meters per year. Even at the fastest rates, underground fires progress more slowly than aboveground fires. This slow movement means that an area on the surface that becomes affected by the underground fire may experience these effects for years. Indeed, certain areas in Centralia have been affected continuously for decades (Trifonoff 2000).

The principal reason for the slow movement is a shortage of oxygen. That is, belowground fires are oxidant limited. Oxygen is replenished through the fractures in the subsurface structures. Indeed, weather systems affect the progress of the fire. Low-pressure weather systems result in more venting of gases whereas little venting takes place during high-pressure systems. Another consequence of this shortage of oxygen is that the underground fire is a reducing fire. Thus, the reduced and volatile forms of nitrogen (ammonium) and sulfur ($S°$ and H_2S) are formed and transported to the surface.

14.3.2 Available Moisture

In general, as the hot combustion products of the fire rise through the mineral and soil column, those components with low volatilities begin to condense. One simple manifestation of this phenomenon is the formation of a very moist semi-liquid layer beneath active vents. When the hot gas encounters cooler soil, water condenses creating a very hot mud between 0.5 and 1 m beneath the surface. (Tobin-Janzen, unpublished results). Above this level, the soil moisture varies tremendously, and the soil can even be quite dry, more closely resembling soils affected by aboveground heat sources (Tobin-Janzen et al. 2005).

14.3.3 Available Nutrients

The chemical composition of coal is markedly different from the organic fuel burned in aboveground fires. Coal is a highly variable fossilized form of ancient plant matter that has been significantly altered by exposure to elevated temperature and pressure. In addition to the organic components, coal contains significant inorganic materials. When burned, the inorganic components can melt and volatilize. Indeed, these inorganic phases pose technological problems for the use of coal as a clean energy source. The combustion products, including the inorganic phases, can percolate to the surface and near-surface environments, where they have been shown to condense into a variety of minerals, including elemental sulfur, downeyite (SeO_2), orpiment (As_2S_3), laphamite $(As_2(Se,S)_3)$, ammonium chloride, gypsum $(CaSO_4 H_2O)$, and mullite $(Al_6Si_2O_{13}$; Finkelman and Mrose 1977; Lapham et al. 1980; Dunn et al. 1986). Indeed, the authors have employed X-ray powder diffraction to identify pure ammonium chloride and needlelike elemental sulfur crystallizing on the surface of the ground above the Centralia Pennsylvania mine fire (Janzen, unpublished results). Although the rate of deposition of these compounds is not high, the impact caused by a long and slow-burning underground coalmine fire has the potential to be significant. Also, as described above, the distribution of combustion gases and volatile chemical species at the surface is not even. Rather, it is a function of the subsurface structure. Thus localized areas of high chemical concentrations, temperature, and soil moisture are interspersed with areas of lower concentrations and more moderate temperatures.

14.3.3.1 Nitrogen

Tobin-Janzen et al. (2005) studied the inorganic nitrogen (NH_4^+-N and NO_3^--N) levels in surface soils overlying the Centralia, Pennsylvania coalmine fire over a two-year period. During that time, the mine fire front progressed through the sample site, and surface soil temperatures increased an average of 16.1°C. Many of the hottest areas showed elevated ammonium (as high as 13.79 mg/kg) and nitrate (as high as 103.1 mg/kg) levels, however, these concentrations were not correlated with absolute temperature values, nor to proximity to an active vent, as some areas with elevated temperature, close proximity to an active vent, or both, also had inorganic nitrogen concentrations that were comparable to those in nearby unaffected soils. Rather, inorganic nitrogen levels were related to more complex environmental trends, such as the length of time an area had been affected by the mine fire and whether the overall soil temperatures were rising or falling in the area due to the movement of the mine fire front.

14.3.3.2 Sulfur

The sulfur-containing gases percolating up from the burning coal have a profound effect on the soil chemistry of the overlying soils, and thus represent one of the most striking differences between underground mine fire-affected soils and those affected by surface fires. For example, in unpublished work, the total sulfur concentrations are clearly affected by the underground fire (Fig. 14.1). The concentrations are significantly higher near the vent. Interestingly, the sulfur concentrations are not highest directly at the vent. Rather, they increase to a maximum approximately 0.5 m

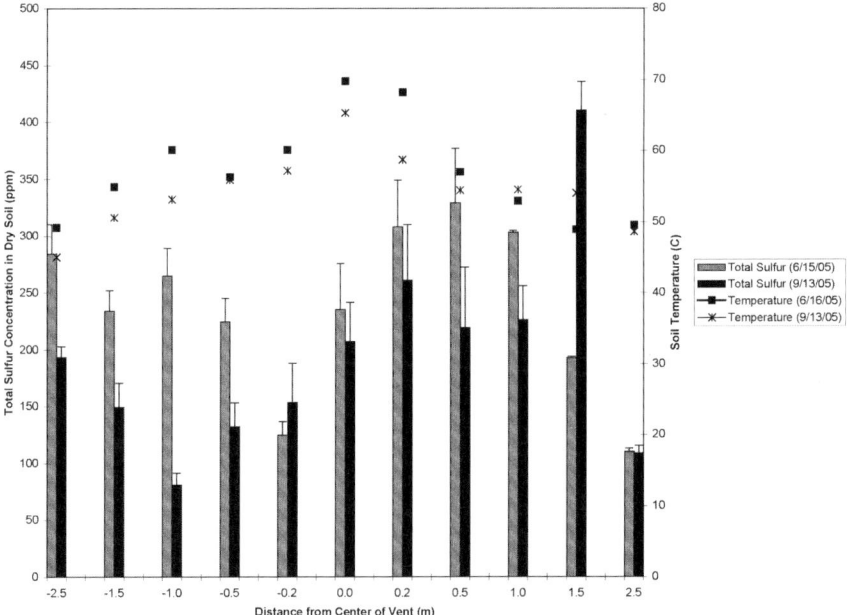

Fig. 14.1 Spatial distribution of total sulfur around an active vent above the Centralia, Pennsylvania mine fire. The concentrations of total sulfur were measured in dry soil on two days during the summer of 2005. The results are correlated with soil temperature at a depth of 5 cm and distance from the vent. Triplicate analyses were performed on each sample and the error bars represent the 95% confidence intervals

from the vent, then decrease with decreasing distance from the vent. At greater distances from the vent the concentrations of total sulfur again begin to drop due to the depletion of volatile sulfur components of the gases at greater distances from the source. Similar results have been observed for elemental sulfur (S^0).

Using thermodynamic principles, Stracher has proposed a stability relationship between the partial pressure of S_2 gas and temperature (Stracher 1995). He developed a model predicting which phase, orthorhombic or S_2 gas, would be stable under a given set of conditions. Depending on the partial pressure of S_2, the temperature at which the solid form of elemental sulfur becomes the favored species varies. At very low partial pressures of S_2, sulfur prefers to be solid at low temperatures. As the pressure rises, the solid form becomes favored at higher temperatures. Thus, at the vent where temperatures are highest, the gaseous form is predominant. Moving a short distance from the vent, the temperature is lower, allowing sulfur to solidify. Farther from the vent, the sulfur has already been depleted in the effluent gas, thus no further deposition is possible.

14.4 Microbial Communities in Fire-Affected Soils

The response of microbial communities to fire varies primarily with the nature of the fire. The fire may have a top-down or a bottom-up profile, its intensity and duration may vary, as also vary the chemical, physical, and biological changes it generates in the surrounding environment. Not surprisingly, the fire's intensity and persistence tend to be the most important factors in governing post-fire microbial community numbers and diversity. Nevertheless, other factors, such as the release of organic toxins, including polychlorinated dibenzo-p-dioxins (PCDDs), dibenzofurans (PCDFs), and polycyclic (or polynuclear) aromatic hydrocarbons (PAHs), into the surrounding soils can dramatically and negatively affect microbial biomass and diversity (Kim et al. 2003). Conversely, the increase of inorganic nitrogen and sulfur levels in fire-affected soils can allow microbes capable of exploiting these molecules to thrive in many post-fire environments, particularly if they are able to withstand the fire's high temperatures or to recolonize fire-affected areas quickly as they cool.

14.4.1 Fire Intensity and Frequency

The fire's intensity, as determined by its total heat, the transfer of that heat to the under- or overlying soils, and the migration rate of the fire, is generally the most important single factor driving microbial community responses in most environments (Neary et al. 1999; Hart et al. 2005; Tobin-Janzen et al. 2005). Heat from the fire is by itself often high enough to sterilize soils, with surface temperatures above mine fires exceeding 730°C (Lapham et al. 1980), and those in surface fires capable of exceeding 675°C. Even nonsterilizing heat levels can directly lyse bacterial cells, and reduce their reproductive capabilities (Covington and DeBano 1990). As a result, total microbial numbers as well as microbial community diversity tend to decrease dramatically immediately post-fire (Dunn et al. 1979; Fritze et al. 1993; Prieto-Fernández et al. 1998; Tobin-Janzen et al. 2005), although some exceptions to this rule have been observed (Newman et al. 2003).

Soil moisture, which varies with both above- and belowground fire conditions as previously described, can intensify the fire's impact on microbial cells. Moist heat is much more efficient at killing soil micro-organisms than dry heat, with threshold values for fungal and bacterial survival estimated at 80°C and 60°C for fungi, and 120°C and 100°C for bacteria in dry and moist soils, respectively (Dunn and DeBano 1977; Dunn et al. 1985). Thus, increased soil moisture can be responsible for augmenting a fire's short-term impacts on microbial communities. Conversely, decreased soil moisture can exacerbate a fire's long-term impacts. Summer droughts, for example, have been proposed to be responsible for the slow recovery of bacterial and fungal community biomass following both surface and underground fires (Cilliers et al. 2005; Tobin-Janzen et al. 2005; Yeager et al. 2005). Finally, the condensing steam from underground mine fires moistens the overlying soils, and may actually help to increase microbial survival in fire-affected soils, even during summer droughts (Tobin-Janzen et al. 2005).

The migration rate of a fire likewise determines its impact on soil microbial communities. Low intensity, rapidly moving surface fires do not generally exceed surface temperatures of 250°C, and microbes at depths as shallow as 25 mm generally survive. By contrast, in high-intensity surface fires with surface temperatures exceeding 675°C, microbes as deep as 50 mm underground experience selective die-off (Neary et al. 1999). By comparison to surface fires, underground fires tend to move very slowly, and their effects can last for decades or longer. The Centralia mine fire, which started in 1962, is expected to burn for at least another 100 years, and the Jharia coalmine fire has been burning in India since 1916.

Thus, the long-term effects on soil microbial communities in these areas, although yet unstudied, are expected to be extensive. Over a one-year period, our laboratory (Tobin-Janzen et al. 2005) used T-RFLP analysis of domain Bacteria 16S rRNA to ascertain that the complexity of microbial communities decreased uniformly as temperatures in the site increased. Furthermore, these changes were not correlated with other environmental factors such as soil pH, soil moisture, inorganic nitrogen, or total sulfur, but rather with general trends in the mine fire progression itself. It will be of interest to determine what happens to these microbial communities as the impact time stretches from months to decades.

Another important manner in which the intensity of surface fires and underground fires differ from each other is in the bottom-up nature of underground fires. Because the fire source is belowground, subsurface temperatures become progressively more elevated as soil depth increases. Although the subsurface microbial communities below 50 cm have not yet been extensively assayed in our laboratory, our unpublished observations suggest that as temperatures increase with depth, the total microbial biomass, as determined by the amount of bacterial DNA that can be extracted from the soil samples, decreases as well. Despite this decrease, we have successfully amplified domain Bacteria 16S rRNA genes from soil samples at depths of up to 50 cm, and at temperatures of up to 87°C.

The overall impact of a fire's intensity on microbial communities varies tremendously with the species composition of that community. Pietikainen et al. (2000) demonstrated that bacteria tend to be more resistant to fire-induced heat than fungi. However, not all bacteria are similarly resistant, nor are all fungi similarly susceptible. In fact, Izzo et al. (2006) used greenhouse experiments to demonstrate that certain species of ectomycorrhizal fungi may actually compete more successfully for roots in the presence of fire, and *Neurospora* ascospores germinate in response to heat or fire (Emerson 1948; Jesenska et al. 1993; Pandit and Maheshwari 1996).

Heterotrophic bacteria often show the sharpest declines in numbers following fires, presumably as a result of the combustion of readily available food sources. By contrast, autotrophic bacterial numbers can actually increase as nutrients are released from their organic forms. In our work above the Centralia coalmine fire, for example, autotrophs are frequently dominant members of the bacterial communities present in fire-affected soils. However, *Nitrobacter* are generally more heat-sensitive than heterotrophic bacteria, (Dunn and DeBano 1977; Dunn et al. 1985), and actinomycetes are generally more heat-resilient than other culturable heterotrophs (Cilliers et al. 2005). Recent experiments have shown that many of the most common bacteria in fire-affected soils include endospore formers (Ahlgren 1974; Moseby et al. 2000; Yeager et al. 2005). This finding is not surprising, as these endospore formers probably survive short-term intense heat as endospores, and then rapidly germinate once soil moisture, temperature, and nutrient levels have returned to sustaining levels. Ultimately, thermophilic bacteria are the most resilient of all microbial species, and are often preferentially selected by fire conditions. Our team and others have demonstrated that they often make up the dominant populations in fire-affected soils (Norris et al. 2002; Tobin-Janzen et al. 2005; Yeager et al. 2005).

14.4.2 Nitrogen-Cycling Bacteria

Available nitrogen is often a limiting nutrient in pre-fire environments (Allen et al. 2002; Yeager et al. 2005). However, both surface and underground fires can release inorganic nitrogen as a result of the combustion of organic molecules. This nitrogen initially exists as NH_4^+-N, but is quickly oxidized to NO_3^--N (DeLuca and Zouhar 2000; Tobin-Janzen et al. 2005; Yeager et al. 2005). Soil bacteria are probably critical components in this oxidation, and yet they have remained poorly studied in post-fire environments until recently. Yaeger et al. (2005) used molecular analysis of *nif*H and *amo*A genes to study nitrogen-fixing and nitrifying bacteria following a forest fire. They demonstrated that although there was a decrease in overall microbial biomass, including that of nitrogen-cycling bacteria, the nitrogen-fixing community actually became more diverse within a month after the fire. By contrast, a single ammonia-oxidizing type, belonging to *Nitrosospira* spp. cluster 3A, dominated in the post-fire soils.

Our laboratory has similarly demonstrated the presence of ammonia-oxidizing bacteria, as determined by the presence of *amo*A genes, in hot mine-fire affected soils with elevated NH_4^+-N and NO_3^--N, and at temperatures of up to 60°C (Tobin-Janzen et al. 2005). Furthermore, we have isolated spore-forming *Geobacillus* that can reduce nitrate to N_2O at 60°C (Kauffman and Tobin-Janzen 2005). Thus, nitrogen-cycling bacteria appear to be important components of both surface and underground mine fire environments, where they are most likely exploiting the elevated nitrogen levels, and where they could be contributing to greenhouse emissions in the fire areas.

14.4.3 Sulfur-Cycling Bacteria

Perhaps the most interesting difference between environments affected by belowground versus surface fires is the high levels of sulfur that can condense into the soils above the former. In this regard, subsurface fires tend to more closely resemble fumaroles and other similar geothermal environments, in which surface soils are affected by hot gases containing varying levels of H_2O, CO_2, SO_2, H_2S, H_2, N_2, CO, and CH_4 (for further details, see Delmelle et al. 2000; Giggenbach and Sheppard 1989).

Janzen et al. (unpublished results) have recently studied the sulfur levels surrounding an active vent above the Centralia coalmine fire, where they demonstrated that the highest levels of sulfur did not necessarily correlate with the highest soil temperatures, but rather with temperatures that favored condensation of the sulfur into the soils, rather than its loss to vaporization (Fig. 14.1). In order to determine if sulfur-metabolizing bacteria were capitalizing on the high levels of sulfur in these soils, our laboratory collected soil samples from the same boreholes, and identified total bacterial community members using 16S rRNA gene sequencing.

In Fig. 14.2, it can be seen that the bacterial community members of this fire-affected soil are in fact enriched for thermophilic bacteria, including thermophilic sulfur metabolizers, nitrogen-cycling bacteria, and endospore-forming bacteria. The community itself was only modestly diverse, with the dominant bacterial groups including Actinobacteria, Alphaproteobacteria, Deltaproteobacteria, Betaproteobacteria, along with *Acidothermus, Clostridium,* and *Geobacillus* species. These results are in good agreement with the trends seen in other hot, fire and geothermally affected soil environments (Norris et al. 2002; Yeager et al. 2005).

14.5 Conclusions

Although the micro-organisms capable of surviving and thriving in fire-affected areas are poorly studied at this time, studying their life cycles and environmental effects is likely to produce important discoveries. Microbes found in fire-affected

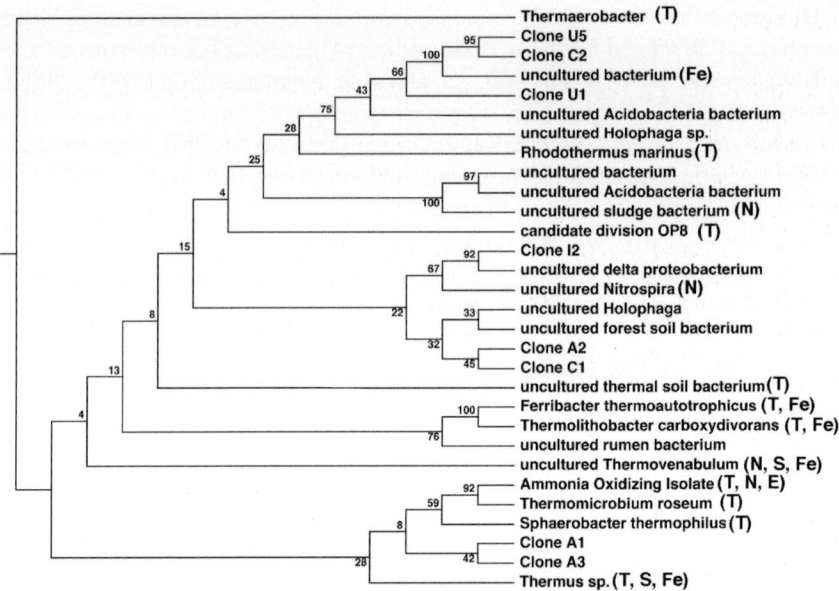

Fig. 14.2 Dendrogram of 16S rRNA gene sequences obtained from the field samples surrounding the active vent depicted in Fig. 14.1. Samples designated 'Clone C' and 'Clone I' originated from the boreholes one meter to the left and right of the active vent, respectively. Samples designated Clone A and Clone U came from an affected surface site (50°C) located outside of the study area depicted in Fig. 14.1. Taxa containing known thermophiles (T), endospore-formers (E), and nitrogen (N), sulfur (S), and iron (Fe) cycling bacteria are indicated. The dendrogram was generated using nearest neighbor analysis and the CLC Combined Workbench Program (CLC Bio). Bootstrap values from 100 resamplings are shown above each internal node. *Thermaerobacter* 16S rRNA was used as the outgroup

areas most likely play critical roles in the biogeochemical cycling of nitrogen and in rhizosphere ecology, and thus can be expected to play important roles in the recovery of fire-affected ecosystems. Thermophilic bacteria, in general, have already provided biotechnology with some of its most powerful tools, and thermophilic sulfur-metabolizing bacteria are currently at the heart of several studies geared toward producing more efficient methods of bioleaching metals and desulfurizing flue gases (Huber and Stetter 1998; Kaksonen et al. 2006). Inasmuch as these bacteria appear to be enriched in hot mine-fire affected soils, continued research into their biology, environmental roles, and metabolism will be particularly interesting. Finally, not all fire-associated bacteria are benign. Iron and sulfur-metabolizing bacteria catalyze the rate-limiting step responsible for acidification of drainage effluents in coal mining environments, and the nitrogen-cycling bacteria identified in this study and others may also be responsible for releasing the greenhouse gases NO and N_2O into the environment. Thus, further study of these interesting extremophiles holds the promise of beneficial technological advances in a variety of compelling areas.

References

Acea MJ, Carballas T (1996) Changes in physiological groups of microorganisms in soil following wildfire. *FEMS Microbiol Ecol* 20:33–39

Adams MA, Attiwill PM (1984) Role of *Acacia* spp. in nutrient balance and cycling in regenerating *Eucalyptus regnans* F. Muell. Forests. I. Temporal changes in biomass and nutrient content. *Aust J Bot* 32:205–215

Agarwal R, Singh D, Chauhan DS, Singh KP (2006) Detection of coal mine fires in the Jharia coal field using NOAA/AVHRR data. *J Geophys Eng* 3:212–218

Agee JK (1973) Prescribed fire effects on physical and hydrologic properties of mixed-conifer forest floor and soil. Report 143. University of California Resources Center

Ahlgren IF (1974) The effect of fire on soil organisms. In: Kozlowski, TT, Ahlgren, CE (eds) *Fire and Ecosystems*. Academic Press, New York, pp 47–72

Allen CD, Savage M, Falk DA, Suckling KF, Swetnam TW, Schulke T, Stacey PB, Morgan P, Hoffman M, Klingel JT (2002) Ecological restoration of Southwestern ponderosa pine ecosystems: A broad perspective. *Ecol Appl* 12:1418–1433

Andersson M, Michelsen A, Jensen M, Kjoller A (2004) Tropical savannah woodland: effects of experimental fire on soil microorganisms and soil emissions of carbon dioxide. *Soil Biol Biochem* 36:849–858

Arocena JM, Opio C (2003) Prescribed fire-induced changes in properties of sub-boreal forest soils. *Geoderma* 113:1–16

Baath E, Frostegard A, Pennanen T, Fritze H (1995) Microbial community structure and pH response in relation to soil organic matter quality in wood-ash fertilized, clear-cut or burned coniferous forest soils. *Soil Biol Biochem* 27:229–240

Badia D, Marti C (2003) Plant ash and heat intensity effects on chemical and physical properties of two contrasting soils. *Arid Land Res Manag* 17:23–41

Bissett J, Parkinson D (1980) Long-term effects of fire on the composition and activity of the soil microflora of a subalpine, coniferous forest. *Can J Bot* 58:1704–1721

Cade-Menun BJ, Berch SM, Preston CM, Lavkulich LM (2000) Phosphorus forms and related soil chemistry of Podzolic soils on northern Vancouver Island. II. The effects of clear-cutting and burning. *Can J For Res* 30:1726–1741

Campbell GS, Jungbauer JD, Bidlake WR, Hungerford RD (1994) Predicting the effect of temperature on soil thermal conductivity. *Soil Sci* 158:307–313

Campbell RE, Baker MB, Ffolliott PF, Larson FR, Avery CC (1977) Wildfire effects on a ponderosa pine ecosystem: An Arizona case study. USDA For Serv Res Pap RM-191

Castelli LM, Lazzari MA (2002) Impact of fire on soil nutrients in central semiarid Argentina. *Arid Land Res Manag* 16:349–364

Choromanska U, DeLuca TH (2002) Microbial activity and nitrogen mineralization in forest mineral soils following heating: Evaluation of post-fire effects. *Soil Biol Biochem* 34:263–271

Cilliers CD, Botha A, Esler KJ, Boucher C (2005) Effects of alien plant management, fire and soil chemistry on selected soil microbial populations in the Table Mountain National Park, South Africa. *S Afr J Bot* 71:211–220

Covington WW, DeBano LF (1990) Effects of fire on pinyon-juniper soils. In: Krammes, J.S., (ed. Effects of Fire Management of Southwestern Natural Resources. USDA For Serv Gen Tech Rep RM-191, pp 78–86

Covington WW, DeBano LF, Huntsberger TG (1991) Soil nitrogen changes associated with slash pile burning in pinyon-juniper woodlands. *For Sci* 37:347–355

Davidenko EP, Eritsov A (2003) The fire season in 2002 in Russia. Report of the Aerial Forest Fire Service, Avialesookhrana. *Int For Fire News* 28:15–17

DeBano LF (2000) The role of fire and soil heating on water repellency in wildland environments: A review. *J Hydrol* 231:195–206

DeBano LF, Dunn PH, Conrad CE (1977) Fire's effect on physical and chemical properties of chaparral soils. In: Mooney, HA, Conrad, CE (eds.) *Proceedings of the Symposium on the*

Environmental Consequences of Fire and Fuel Management in Mediterranean Ecosystems, Palo Alto, CA. USDA For Serv Gen Tech Rep WO-3. pp 65–74

DeBano LF, Neary DG, Ffolliott PF (1998). *Fire Effects on Ecosystems*. Wiley, New York

Delmelle P, Bernard A, Kusakabe M, Fischer TP, Takano B (2000) Geochemistry of the magnatic-hydrothermal system of Kawah Ijen volcano, East Java, Indonesia. J Volcanol Geotherm Res 97:31–53

DeLuca TH, Zouhar KL (2000) Effects of selection harvest and prescribed fire on the soil nitrogen status of ponderosa pine forests. *For Ecol Manag* 138:263–271

Diaz-Ravina M, Prieto A, Baath E (1996) Bacterial activity in a forest soil after heating and organic amendments measured by the thymidine and leucine incorporation technique. *Soil Biol Biochem* 28:419–426

Dunn PH, Barro SC, Poth M (1985) Soil moisture affects survival of microorganisms in heated chaparral soil. *Soil Biol Biochem* 17:143–148

Dunn PH, DeBano LF (1977) Fire's effects on biological and chemical properties of chaparral soils. In: Mooney, HA, Conrad, CE (eds) *Proceedings of the Symposium on the Environmental Consequences of Fire and Fuel Management In Mediterranean Ecosystems*, Palo Alto, CA. USDA For Serv Gen Tech Rep WO-3, pp 75–84

Dunn PH, DeBano LF, Eberlein GE (1979) Effects of burning on chaparral soils: II Soil microbes and nitrogen mineralization. *Soil Sci Soc Am J* 43:509–514

Dunn PJ, Peacor DR, Criddle AJ, Finkelman RB (1986) Laphamite, an arsenic selenide analogue of orpiment, from burning anthracite deposits in pennsylvania. *Mineral Mag* 50:279–282

Emerson M (1948) Chemical activation of ascospore germination in *Neurospora crassa*. *J Bacteriol* 55:327–330

FAO (2005) *State of the World's Forests 2005*. Food and Agriculture Organization of the United Nations. Rome

Feller MC (1982) The ecological effects of slash burning with particular reference to British Columbia: A literature review. Province of British Columbia Ministry of Forests Land Management Report No. 13

Finkelman RB, Mrose ME (1977) Downeyite, the first verified natural occurance of SeO_2. *Am Mineral* 62:316–320

Fritze H, Pennanen T, Pietikäinen J (1993) Recovery of soil microbial biomass and activity from prescribed burning. *Can J For Res* 23:1286–1290

Giai C, Boerner REJ (2007) Effects of ecological restoration on microbial activity, microbial functional diversity, and soil organic matter in mixed-oak forests of southern Ohio, USA. *Appl Soil Ecol* 35:281–290

Giggenbach WF, Sheppard DS (1989) Variations in the temperature and chemistry of white Island fumarole discharges 1972–85 New Zealand Geological Survey Bulletin 103:119–126

Giovannini G, Lucchesi S, Giachetti M (1990) Effects of heating on some chemical parameters related to soil fertility and plant growth. *Soil Sci* 149:344–350

Glantz MH (1996) *Currents of Change: El Niño's Impact on Climate and Society*. Cambridge University Press, Cambridge

Grier CC (1975) Wildfire effects on nutrient distribution and leaching in a coniferous ecosystem. *Can J For Res* 5:599–607

Haase SM (1986) Effect of prescribed burning on soil moisture and germination of southwestern ponderosa pine seed on basaltic soils. USDA For Serv, Rocky Mt For Range Exp Stn Res Note RM-462

Hart SC, DeLuca TH, Newman GS, MacKenzie MD, Boyle SI (2005) Post-fire vegetative dynamics as drivers of microbial community structure and function in forest soils. *For Ecol Manag* 220:166–184

Hernandez T, Garcia C, Reinhardt I (1997) Short-term effect of wildfire on the chemical, biochemical and microbiological properties of Mediterranean pine forest soils. *Biol Fertil Soils* 25:109–116

Houghton JH, Callander BA, Varney SK (1992) Climate change 1992: The supplementary report to the IPCC Scientific Assessment Press Syndicate of the University of Cambridge

Huber H, Stetter KO (1998) Hyperthermophiles and their possible potential in biotechnology. *J Biotechnol* 64:39–52

Izzo A, Canright M, Bruns TD (2006) The effects of heat treatments on ectomycorrhizal resistant propagules and their ability to colonize bioassay seedlings. *Mycol Res* 110:196–202

Jaatinen K, Knief C, Dunfield PF, Yrjala K, Fritze H (2004) Methanotrophic bacteria in boreal forest soil after fire. *FEMS Microbiol Ecol* 50:195–202

Jesenska Z, Pieckova E, Bernat D (1993) Heat resistance of fungi from soil. *Int J Food Microbiol* 19:187–192

Kaksonen AH, Plumb JJ, Robertson WJ, Spring S, Schumann P, Franzmann PD, Puhakka JA (2006) Novel thermophilic sulfate-reducing bacteria from a geothermally active underground mine in Japan. *Appl Environ Microbiol* 72:3759–3762

Kauffman C, Tobin-Janzen T (2005) Isolation and identification of soil microbial populations in Centralia, Pennsylvania. *J Penn Acad Sci* 78:117

Kim E-J, Oh J-E, Chang Y-S (2003) Effects of forest fire on the level and distribution of PCDD/Fs and PAHs in soil. *Sci Total Environ* 311:177–189

Klock GO, Hevey JD (1976) Soil-water trends following wildfire on the Entiat Experimental Forest. *Ann. Proc. Tall Timber Fire Ecol Conf* 15:193–200

Klopatek CC, DeBano LF, Klopatek JM (1990) Impact of fire on the microbial processes in pinyon-juniper woodlands: Management implications. In: Effects of Fire Management of Southwestern Natural Resources. USDA For Serv Gen Tech Rep RM-191, pp197–205

Klopatek CC, Klopatek JM (1987) Mycorrhizae, microbes, and nutrient cycling processes in pinyon-juniper systems. Everett, R.L. (ed. *Proceedings of the Pinyon-Juniper Conference*, January 13–16, 1986 Reno, NV, USDA For Serv Gen Tech Rep INT-215, pp360–367

Kudoh J (2005) Report of the View of Northeast Asia Forest Fire from Cosmos International Symposium Center for Northeast Asian Studies (CNEAS), Tohoku University, Sendai, Miyagi, Japan, 17–18 January 2005

Lapham DM, Barnes JH, Downey WF, Finkelman RB (1980) Minerology associated with burning anthracite deposits of eastern Pennsylvania Pennsylvania Geological Survey Mineral Research Report M 78.

Letey J (2001) Causes and consequences of fire-induced soil water repellency. *Hydrol Process* 15:2867–2875

Macadam A (1989) Effects of prescribed fire on forest soils. BC Ministry of Forests Research Report 89001-PR, Victoria

Milne MM (1979) The effects of burning root trenching and shading on mineral soil nutrients in southwestern ponderosa pine. M.Sc Thesis, Northern Arizona University, Flagstaff, AZ

Moseby AH, Burgos J, Reed J, Tobin-Janzen T (2000) Isolation and identification of soil bacteria from the Centralia mine fire area. *J Penn Acad Sci* 73:150

Naumova NB (2005) Biomass and activity of soil microorganisms after a surface fire in a pine forest. *Eurasian Soil Sci* 38:870–873

Neary DG, Klopatek CC, DeBano LF, Ffolliott PF (1999) Fire effects on belowground sustainability: A review and synthesis. *For Ecol Manag* 122:51–71

Newman GS, Hart SC, Guido D, Overby S (2003) Wildfire effects on soil microbial activity and community-level physiological profiles in a ponderosa pine ecosystem. *ESA 2003 Annual Meeting*, August 3–8, 2003 Savannah, GA

Norris TB, Wraith JM, Castenholz RW, McDermott TR (2002) Soil microbial community structure across a thermal gradient following a geothermal heating event. *Appl Environ Microbiol* 68:6300–6309

Pandit A, Maheshwari R (1996) Life-history of *Neurospora intermedia* in a sugar cane field. *J Biosci* 21:57–79

Pietikainen J, Hiukka R, Fritze H (2000) Does short-term heating of forest humus change its properties as a substrate for microbes? *Soil Biol Biochem* 32:277–288

Prieto-Fernández A, Acea MJ, Carballas T (1998) Soil microbial and extractable C and N after wildfire. *Biol Fertil Soils* 27:132–142

Raison RJ (1979) Modification of the soil environment by vegetation fires, with particular reference to nitrogen transformations: A review. *Plant Soil* 51:73–108

Renkin RA, Despain DG (1992) Fuel moisture, forest type and lightning caused fire in Yellowstone National Park. *Can J For Res* 22:37–45

Ross DJ, Speir TW, Tate KR, Feltham CW (1997) Burning in a New Zealand snow-tussock grassland: Effects on soil microbial biomass and nitrogen and phosphorous availability. *NZ J Ecol* 21:63–71

Sharpley A (2000) Phosphorous availability. In: Sumner ME (ed) *Handbook of Soil Science*. CRC Press, Boca Raton, FL, pp D18–D38

Stracher GB (1995) The anthracite smokers of eastern Pennsylvania: P_{S_2}-T stability diagram by TL analysis. *Math Geo* 27:499–511

Tobin-Janzen T, Shade A, Marshall L, Torres K, Beblo C, Janzen C, Lenig J, Martinez A, Ressler D (2005) Nitrogen changes and domain bacteria ribotype diversity in soils overlying the Centralia, Pennsylvania, underground coal mine fire. *Soil Sci* 170:1–10

Treseder KK, Mack MC, Cross A (2004) Relationships among fires, fungi, and soil dynamics in Alaskan boreal forests. *Ecol Appl* 14:1826–1838

Trifonoff KM (2000) The mine fire in Centralia, Pennsylvania. *Penn Geo* 38:3–24

UNECE, FAO (2001) *Forest Fire Statistics 1998–2000*

UNECE, FAO, Organization IL, Center GFM (2000) Baltic exercise for fire information and resources exchange. BALTEX FIRE 2000, 5–9 June 2000 Kuopio, Finland

Vazquez FJ, Acea MJ, Carballas T (1993) Soil microbial populations after wildfire. *FEMS Microbiol Ecol* 13:93–103

Westerling AL, Hidalgo HG, Cayan DR, Swetnam TW (2006) Warming and earlier spring increase western US forest wildfire activity. *Science* 313:940–943

Weston CJ, Attiwill PM (1996) Clearfelling and burning effects on nitrogen mineralization and leaching in soils of old-age *Eucalyptus regnans* forests. *For Ecol Manag* 89:13–24

Yeager CM, Northup DE, Grow CC, Barns SM, Kuske CR (2005) Changes in nitrogen-fixing and ammonia-oxidizing bacterial communities in soil. *Appl Environ Microbiol* 71:2713–2722

Zhou F, Ren W, Wang D, Song T, Li X, Zhang Y (2006) Application of three-phase foam to fight an extraordinarily serious coal mine fire. *Int J Coal Geo* 67:95–100

Chapter 15
Endophytes and Rhizosphere Bacteria of Plants Growing in Heavy Metal-Containing Soils

Angela Sessitsch(✉) and Markus Puschenreiter

15.1 Introduction: Heavy Metal Contamination of Soils

As a consequence of industrialization during the last centuries, the heavy metal concentration of soils has increased worldwide (Adriano 2001). Hot spots of soil contamination are located in areas of large industrial activities, where surrounding agricultural areas are affected by atmospheric deposition of heavy metals. Also, agricultural practices, such as the application of sewage sludge or phosphate fertilisers, has led to increased metal concentration in soils (Puschenreiter et al. 2005a).

Metal contamination of soils may also derive from geogenic sources. These natural metalliferous soils are the classical habitats for metal-accumulating plants and may be divided into four different main groups, depending on the parent rock materials (see Section 15.2). Most of these naturally contaminated soils are quite infertile, which is particularly true for Ni-rich serpentine soils (which are characterized by low NPK levels and a low Ca:Mg ratio; Baker et al. 2000). Indigenous soil contamination may be restricted to either very small spots of only a few square meters or may affect large areas of several square kilometers such as those found in Cornwall and Devon (UK), where up to 700 km^2 are contaminated with As (Mitchell and Barr 1995).

15.2 Heavy Metal Tolerance and Hyperaccumulation in Plants

Heavy metal tolerance mechanisms were summarized by Schat et al. (2000) and by Clemens (2001). Briefly, the simplest heavy metal tolerance strategy, termed "avoidance", consists in limited uptake into the plant body. However, a clear evidence for this mechanism was only found for some arsenite-tolerant plants. In most metal-tolerant higher plants, heavy metals are strongly retained in root

---

Angela Sessitsch
Austrian Research Centers GmbH, Dept. of Bioresources, A-2444 Seibersdorf, Austria
e-mail: angela.sessitsch@arcs.ac.at

tissues. Regardless of whether heavy metals are mainly accumulated in roots or in shoots, internal tolerance mechanisms are the basis for efficient detoxification of the metals. This internal detoxification is based on (i) sequestration of the metals, that is, transport to cell components not involved in physiological processes (vacuole, cell wall); and (ii) complexation with metal-binding peptides, that is, metallothioneines and phytochelatins.

Hyperaccumulator plants are able to take up large amounts of metals in their aerial tissues without showing any symptoms of toxicity. Plants accumulating >1,000 mg kg^{-1} of Cu, Co, Cr, Ni, or Pb, or >10,000 mg kg^{-1} of Mn or Zn have been defined as hyperaccumulator species (Baker and Brooks 1989). An extended definition was provided by Baker and Whiting (2002), who claimed that the shoot/root or leaf/root ratio has to be >1, indicating a clear partitioning of metals to the shoots.

More than 400 plant species have been identified as hyperaccumulators, of which 75% are Ni hyperaccumulators growing on ultramafic soils (Baker et al. 2000). Hyperaccumulation of heavy metals is found throughout the whole plant kingdom in temperate as well as tropical climates, but is typically restricted to endemic species growing on mineralised soils and related rock types. The most important types of metalliferous soil hosting hyperaccumulator plants are (1) serpentine soils (enriched in Ni, Cr, Co), (2) "calamine" soils (enriched in Cd, Pb, Zn), (3) Se-rich soils, and (4) Cu- and Co-containing soils (Reeves and Baker 2000). Related to the soil habitat, the hyperaccumulators are classified into accumulators of (1) Ni, (2) Cd, Pb, Zn, (3) Se, (4) Co, Cu, and (5) other elements (Al, As, Cr, Mn, Tl; Reeves and Baker 2000). Some examples for metal hyperaccumulating plants are shown in Table 15.1.

Very effective translocation and storage processes have been described in metal hyperaccumulating plants (e.g., Lasat and Kochian 2000; Salt and Krämer 2000; Assunção et al. 2003). Metal sequestration into the vacuole or the apoplast and the complexation of metals by organic acids, amino acids, or specific high-affinity ligands are the basis for the high metal tolerance of hyperaccumulators (Salt and Krämer 2000). To accumulate and be sequestered in plants, heavy metals must be taken up,

Table 15.1 The highest observed foliar concentrations of As, Cd, Co, Cu, Ni, Pb, and Zn in metal-hyperaccumulating plants (Baker et al. 2000)

Element	Species	Highest Observed Foliar Concentration (mg kg^{-1} dry weight)
As	Pteris vittata*	23,000
Cd	Thlaspi caerulescens	1,000
Co	Haumaniastrum robertii	10,200
Cu	Aeollanthus biformifolius	13,700
Ni	Sebertia acuminata	260,000**
Pb	Thlaspi rotundifolium	8,200
Zn	Thlaspi caerulescens	35,000

* Ma et al. (2001).
** Found in the blue-green latex.

for which high expression of metal transporter genes and active root proliferation towards contaminated soil spots have been reported as the main processes (Assunção et al. 2003). Additionally, some evidence for heavy metal mobilisation processes in the rhizosphere was reported (Puschenreiter et al. 2003, 2005b).

15.3 Rhizosphere Bacteria and Endophytes

Roots supply inorganic nutrients and water to the rest of the plant, whereas shoots fix carbon through photosynthesis and transport organic carbon compounds to the roots. The roots excrete a significant proportion of the transported carbon into the surrounding soil environment, which is biologically and biochemically influenced by the living root, known as the rhizosphere. The rhizosphere is a dynamic environment and hosts a wide variety of micro-organisms. A schematic presentation of the distribution of microbes at the rhizoplane is shown in Fig. 15.1. Exudates released by plant roots and associated microbes may significantly mobilize heavy metals and thus increase their bioavailability (Wenzel et al. 1999). In addition, the apoplast of plants is commonly colonised by a wide range of bacterial endophytes that do

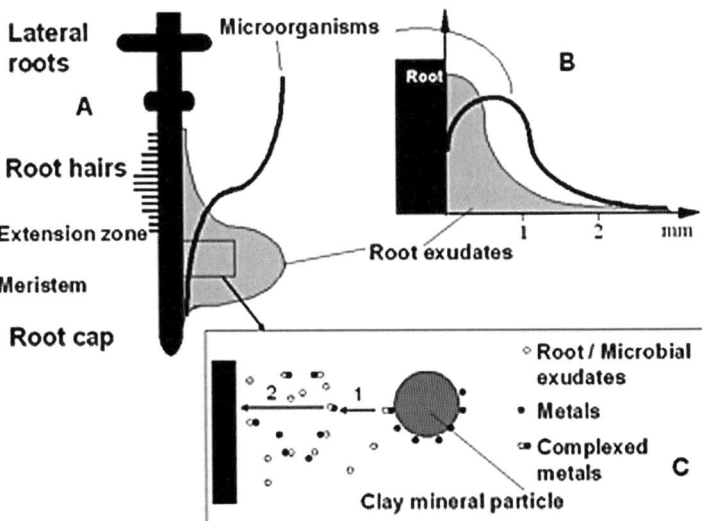

Fig. 15.1 Distribution patterns of microbial populations and root exudates in the rhizosphere, (**A**) along the rhizoplane and (**B**) perpendicular to the rhizoplane; (**C**) mobilisation of mineral nutrients and heavy metals in the rhizosphere from the soil solid phase (e.g., clay minerals) by complexation with root and/or microbial exudates. After mobilisation (1), the complexed nutrients/metals are transported (2) to the root surface by mass flow and diffusion. (Compiled from Römheld 1991; Marschner 1995; Wenzel et al. 1999)

not exhibit pathogenicity (Hallmann et al. 1997; Sturz and Nowak 2000; Reiter et al. 2002; Sessitsch et al. 2002; Idris et al. 2004).

Some plant-associated micro-organisms are detrimental to plant health because they compete with the plant for nutrients or cause disease. However, a wide range of bacteria have a beneficial effect on plants. Some of these support plant resistance against pathogens by either producing antibiotic substances or by inducing plant defenses. Other bacteria are able to stimulate plant growth by increasing the supply of nutrients to the plants or by producing plant growth hormones (Lugtenberg et al. 1991).

Plant growth-promoting bacteria (PGPB) can directly promote plant growth by production of bacterial metabolites that positively affect the plant (Mahaffee and Kloepper 1994). Although the mechanisms still remain unclear, some reports indicate the production of phytohormones (de Freitas et al. 1990; Frommel et al. 1991) as being responsible for plant growth promotion. Many plant-associated bacteria synthesize the plant hormone and growth regulator indole-3-acetic acid (IAA) (Costacurta et al. 1994; Patten and Glick 1996), and enhanced production of IAA by a rhizosphere bacterium can improve plant growth (Beyeler et al. 1999). In addition, the production of cytokinin by rhizobacteria has also been suggested to enhance plant growth (Timmusk et al. 1999). The enzyme 1-aminocyclopropane-1-carboxylic acid (ACC) deaminase has been isolated from some bacteria (Campbell and Thompson 1996; Shah et al. 1998). This enzyme has no function in bacteria, but cleaves ACC, the precursor of ethylene in plants, and thus modulates ethylene levels, which can be particularly high under stress conditions. This enzymatic reaction contributes to plant growth promotion (Glick et al. 1997; Burd et al. 1998).

Many PGPB suppress phytopathogens by mechanisms such as antibiosis or by inducing systemic resistance. A variety of biocontrol strains simply outcompete phytopathogens by efficiently colonizing plants (Dekkers et al. 2000; Chin-A-Woeng et al. 2000). In addition, the production of siderophores plays an important role in iron competition and has been identified as a mechanism contributing to biocontrol activity (Penyalver et al. 2001). Furthermore, it has been suggested that siderophores might be involved in the induction of systemic resistance (ISR) (De Meyer and Höfte 1997; Maurhofer et al. 1998).

Bacterial endophytes have been defined as "bacteria, which for all or part of their life cycle invade the tissues of living plants and cause unapparent and asymptomatic infections entirely within plant tissues, but cause no symptoms of disease" (Wilson 1995). Endophytes colonize a similar ecological niche as plant pathogens and may gain entry into plants by a number of mechanisms. Host entry points include tissue wounds (Agarwhal and Schende 1987; Lamb et al. 1996), stomata (Roos and Hattingh 1983), lenticels (Scott et al. 1996), and germinating radicles (Gagné et al. 1987). Bacteria also may invade intact plants by penetrating root hair cells (Huang 1986) or by producing cell wall-degrading enzymes (Huang 1986; Quadt-Hallmann et al. 1997). Endophytes including Proteobacteria, Firmicutes, and the Bacteroidetes phylum (Hallmann et al. 1997) have been studied mainly by cultivation-based methods (Bell et al. 1995; Stoltzfus et al. 1998; Sturz et al. 1998; Sessitsch et al. 2004). Results of recent endophyte analyses by cultivation-independent, 16S rRNA-based approaches indicated that individual plants host a broad phylogenetic range of endophytic bacteria

(Chelius and Triplett, 2001; Sessitsch et al. 2002; Idris et al. 2004). By 16S rRNA gene analysis, it has been also demonstrated that biotic and abiotic stress may affect potato endophyte communities (Reiter et al. 2002; Sessitsch et al. 2002). Several studies suggested that the plant growth-promoting and biocontrol potential of endophytic bacteria is high as compared to that of rhizosphere microbes (van Buren et al. 1993; Reiter et al. 2002; Sessitsch et al. 2004). The mechanisms for biocontrol and growth promotion by an endophyte may be similar to those exhibited by rhizosphere bacteria.

15.4 Diversity of Bacterial Communities Associated with Heavy Metal-Tolerant Plants

Heavy metals affect the growth and activity of micro-organisms in soils mainly through destruction of the integrity of cell membranes, protein denaturation, and functional disturbance (Leita et al. 1995). Studies of the impact of heavy metals upon bacterial diversity in soils have shown mostly a negative influence (Hirsch et al. 1993; Smit et al. 1997; Sandaa et al. 2001; Moffett et al. 2003; Hinojosa et al. 2005). Other factors such as pH, temperature, moisture, and organic matter content may interfere with metal toxicity (Giller et al. 1998). It is well known that bacteria and fungi isolated from polluted habitats are tolerant of higher levels of metals than those isolated from unpolluted areas (Bååth et al. 1989; Doelman et al. 1994; Huysman et al. 1994; Mertens et al. 2006). After the addition of metals, metal tolerance is increased in bacterial communities by the death of sensitive species and subsequent competition and adaptations of surviving bacteria (Díaz-Roviña et al. 1996). Horizontal transfer of plasmids containing resistance genes may greatly contribute to the adaptation process. It has been shown that microbial biomass and enzyme activities are more sensitive to heavy metal contamination than is species diversity (Kandeler et al. 2000).

Concentrations and bioavailabilities of heavy metals, and therefore their toxicity to the rhizosphere microflora, might be altered in the rhizosphere. Roots may absorb heavy metals to a certain extent. In addition, root exudates may either complex metals, making them unavailable to microbes, or enhance metal release, for example, by altering the pH. Contrasting results regarding the ability of root excretions to mobilize heavy metals have been reported, showing either an increased mobility (Morel et al. 1986; Bernal and McGrath 1994; Cieslinski et al. 1998; Fitz et al. 2003; Wenzel et al. 2003; Puschenreiter et al. 2005b) or no effect (Whiting et al. 2001a; Zhao et al. 2001; Amir and Pineau, 2003).

Due to their importance for practical applications, plant–microbe interactions involving plant species with some relevance to phytoremediation have been the object of particular attention. Kunito et al. (2001) compared the characteristics of bacterial communities in the rhizosphere of *Phragmites* with those of nonrhizosphere soil in a highly Cu-contaminated area near a copper mine in Japan. *Phragmites* is an important plant for phytoremediation applications, but does not hyperaccumulate heavy metals. Higher bacterial numbers were detected in the rhizosphere, which may be due to the lower Cu concentrations or to the availability of root exudates.

Nevertheless, the percentage of highly resistant strains was higher in the rhizosphere than in nonrhizosphere soil. Cu toxicity was found to be lower in rhizosphere soil. The study by Kunito et al. (2001) also indicated that Cu toxicity lowered the frequency of r-strategists (bacteria capable of rapid growth and utilization of resources) as these are more sensitive to toxic substances (Kozdroj 1995).

Rhizosphere and nonrhizosphere isolates behaved very differently regarding their doubling time and exopolymer production (Kunito et al. 2001). Exopolymers produced by bacteria were shown to strongly bind heavy metals (Bitton and Freihofer 1978), leading to the formation of organic–metal complexes, which are difficult to degrade (Francis et al. 1992; Hattori et al. 1996; Huysman et al. 1994). Furthermore, heavy metal concentrations induce the production of exopolymer production (Chao and Chen 1991; Kidambi et al. 1995) and metal resistance due to exopolymer production has been shown (Bitton and Freihofer 1978). Kunito et al. (1997) reported a dominance of Cu-resistant *Bacillus* spp. in the rhizosphere of *Phragmites*, whereas nonrhizosphere soil was dominated by *Methylobacterium* spp.

In the rhizosphere of heavy metal hyperaccumulating plants, higher proportions of resistant bacteria were found (Schlegel et al. 1991; Mengoni et al. 2001). Mengoni et al. (2001) sampled soils from three Ni-containing, serpentine sites in Italy and found an increasing number of resistant strains in increasing proximity to the Ni-hyperaccumulating plant, *Alyssum bertolonii*. At all sites, the culturable Ni-resistant rhizosphere community was dominated by *Pseudomonas*, whereas soil samples contained a high number of Ni-resistant *Streptomyces* spp. In general, isolates obtained in this study showed co-resistance to Cr and Co, although other resistance combinations (e.g., Ni, Zn, and Cu) or single resistance were also found, indicating independent evolution of heavy metal resistance determinants. Mengoni et al. (2004) analyzed the same sites by cultivation-independent analysis, for which soil samples at different distances to the *Alyssum bertolonii* roots were examined. Results showed that the plant rather than the locality shaped the microbial community structure. In agreement with the cultivation-dependent analysis, proteobacteria were predominantly found in the rhizosphere.

As was the case for the *Alyssum bertolonii* rhizosphere (see above, this section), a high percentage of proteobacteria was found in the rhizosphere of the Ni-hyperaccumulator *Thlaspi goesingense* by cultivation-independent analysis. In addition, *T. goesingense* hosted members of the phylum *Holophaga/Acidobacterium*, high-GC Gram positive bacteria, members of the Bacteroidetes phylum and Verrucomicrobia (Idris et al. 2004). Cultivation of bacteria on Ni-containing medium resulted mostly in the isolation of *Methylobacterium* spp., an alpha-proteobacterial genus, as well as *Rhodococcus* spp. and *Okibacterium* spp., belonging to the Actinobacteria (or Gram-positive bacteria with a high G+C content; Idris et al. 2004).

Methylobacteria were found to be dominant as endophytic colonizers of *Thlaspi goesingense* (Idris et al. 2004). Although *Methylobacterium extorquens* and *Methylobacterium mesophilicum* were both identified in the rhizosphere as well as inside the plant, different strains were found suggesting that these habitats provide distinct growth conditions for micro-organisms. One strain isolated from

Thlaspi goesingense shoots fell into a newly described species, *Methylobacterium goesingense* (Idris et al. 2006). *M. extorquens*, *M. mesophilicum*, and *M. goesingense* strains were highly resistant to Ni, but were also shown to be resistant against different combinations of heavy metals, indicating the independent evolution of resistance traits. Despite the fact that heavy metal determinants are frequently located on plasmids, horizontal transfer of plasmids between *Methylobacterium* spp. isolated from *Thlaspi goesingense* shoots and rhizosphere was found to be unlikely (Idris et al. 2006).

A large number of heavy metal tolerating methylobacteria were also isolated from shoots of the Zn-hyperaccumulator *Thlaspi caerulescens*, but they were not found in association with roots (Lodewyckx et al. 2002). Striking was the high diversity of endophytes colonizing *Thlaspi goesingense*, as revealed by cultivation-independent analysis (Idris et al. 2004). Bacteria belonging to all major bacterial phyla could be found, including Acidobacteria. The phylum *Holophaga/Acidobacterium* has been found to be dominant in many soils worldwide (e.g., Sessitsch et al. 2001) and Acidobacteria may also colonize the rhizosphere (e.g., Idris et al. 2004; Mengoni et al. 2004). However, usually these bacteria do not colonize the apoplast of plants.

In the highly toxic environment represented by the apoplast of a Ni-hyperaccumulating plant, the diversity of endophytes was higher than that usually observed among endophytes from other plants (e.g., Rasche et al. 2006; Reiter and Sessitsch 2006). Lodewyckx et al. (2002) compared root and rhizoplane isolates with endophytes isolated from *Thlaspi caerulescens* shoots. Similar species were found in both compartments. However, shoot endophytes showed higher resistance to Zn and Cd than strains isolated from roots and rhizoplane. Most isolates from the Zn hyperaccumulators were affiliated with the genera *Methylobacterium* and *Sphingomonas*, which were also found to be represented in high numbers in shoots of the Ni hyperaccumulator *Thlaspi goesingense* (Idris et al. 2004) as well as in Zn accumulating willows (Kuffner et al. unpublished results).

Cultivation-independent 16S rDNA-based analysis of bacteria has revolutionized our understanding of bacterial diversity, as micro-organisms can be investigated irrespectively of their culturability. However, DNA-based analysis does not give any information on the activity of cells, and even dead cells may be detected insofar as their DNA has not been destroyed by nuclease activity. As metabolically active cells usually contain a higher amount of ribosomes than resting or dormant cells, Gremion et al. (2003) used 16S rDNA as well as 16S rRNA-based analysis to characterize bacterial diversity in heavy metal-contaminated bulk soil and in the rhizosphere of *Thlaspi caerulescens*. DNA-based analysis indicated the dominance of proteobacteria, acidobacteria, and planctomycetes in the rhizosphere. A minority of bacteria were affiliated with the high-GC Gram-positive bacteria (actinobacteria) and verrucomicrobia.In contrast, analysis of the metabolically active population showed that members of the Rubrobacteria subdivision, belonging to the high-GC Gram-positive bacteria, were highly dominating.

15.5 The Influence of Associated Micro-Organisms on Plant Heavy Metal Tolerance and Uptake

The analysis of 40 ultramafic soil samples from New Caledonia identified a high positive correlation between bioavailability of heavy metals and microbial activity, suggesting a possible role of micro-organisms in the release of heavy metals (Amir and Pineau 2003). These findings were confirmed by an experiment in which autoclaved soil was inoculated with a small portion of the same nonheated soil, leading to the release of metals (Amir and Pineau 2003). The effect of pH was found to be marginal, whereas the presence of a heavy metal-tolerant plant as well as the addition of compost stimulated the metal mobilization process. Amir and Pineau (2003) suggested that the rhizosphere effect on metal release is due to the stimulation of chemoorganotrophic micro-organisms rather than to the direct effect of root secretions.

The effect of root exudates on metal mobilization was reported by Bernal and McGrath (1994). Amir and Pineau (2003) did not exclude the possibility that roots would absorb a part of the released metals. Several studies have shown that rhizosphere bacteria may contribute to the plant metal tolerance and increase metal uptake. Stimulation of plant growth and root biomass was most probably responsible for the increased heavy metal uptake by *Brassica juncea* in the presence of heavy metal-tolerant bacteria (Salt et al. 1999). de Souza et al. (1999a) showed that rhizosphere bacteria are necessary to achieve optimum rates of selenium accumulation and volatilization by Indian mustard. In this study, the tested rhizosphere bacteria increased root hair production of Indian mustard and enhanced metal uptake was observed. However, from the different experiments performed the authors concluded that stimulation of root hair production was not responsible for the enhanced accumulation of Se. A heat-labile compound was shown to enhance Se accumulation in axenic plants, but it was not clear whether it was produced by rhizosphere bacteria or by bacterized roots (de Souza et al. 1999a). The authors postulated that the heat-labile compound stimulated the selenate transporter in plants.

Whiting et al. (2001b) studied the effect of rhizosphere bacteria on Zn accumulation by the hyperaccumulating plant *Thlaspi caerulescens* and the nonaccumulator *T. arvense*. They showed that bacteria facilitated biomass production and Zn uptake of the hyperaccumulator, whereas *T. arvense* was not affected. Furthermore, significantly higher microbial numbers were detected in the rhizosphere of *T. caerulescens* than in that of *T. arvense*, which was probably a result of the higher root biomass of the accumulating species. Nevertheless, it was shown that the increased uptake of Zn was not a result of increased root surface area. Data suggested that the bacteria enhanced the availability of water-soluble Zn in the soil, which overcame a major rate-limiting step for Zn uptake by *T. caerulescens* in soils with low concentrations of labile Zn (Whiting et al. 2001b). Similarly, rhizobacteria were considered as highly important for the mobilization of nickel in soil and for its uptake by the hyperaccumulating plant *Alyssum murale* (Abou-Shanab et al. 2003). Our own studies revealed a strong effect on accumulated heavy metal contents in willows due to plant growth-promoting bacteria (Kuffner et al. unpublished results).

Apart from plant growth promotion, which may be achieved by a range of bacterial activities such as the production of hormones or the provision of nutrients, micro-organisms may improve the stress tolerance of the plant. This may be achieved by the enzyme ACC deaminase leading to a reduction of stress-induced ethylene levels in the plant (Burd et al. 1998, 2000; Glick 2004). Furthermore, bacteria may stimulate the production of metal transporters in plants (de Souza et al. 1999a,b). Heavy metal mobilization was proposed to be due to the action of bacterial siderophores (Lodewyckx et al. 2002; Abou-Shanab et al. 2003). These compounds show high affinity for ferric iron but also form complexes with bivalent heavy metal ions (Evers et al. 1989) that can be assimilated by the plant. Furthermore, heavy metals have been shown to stimulate the production of bacterial siderophores (van der Lelie et al. 1999).

In addition to these direct effects, bacterial siderophores can indirectly alleviate heavy metal toxicity by increasing the supply of iron to the plant (Burd et al. 1998, 2000). Our own studies showed that siderophore production is a frequently found trait among bacteria associated with heavy metal accumulating plants (Idris et al. 2004; Kuffner et al. unpublished results). However, those siderophore-producing bacteria which were tested for their potential to support heavy metal uptake did not show the expected effects. On the other hand, several strains that did not produce siderophores mobilized high amounts of heavy metals, probably due the production of other secondary metabolites (Kuffner et al. unpublished results). In addition to the potential effect of siderophores it has been suggested that bacterial exopolymers may complex heavy metals leading to reduced availablity for plants (Diels et al. 1995; Kunito et al. 2001).

15.6 Potential Applications of Improved Plant–Microbe Interactions

Various studies demonstrating that plant-associated micro-organisms greatly contribute to the mobilization and accumulation of heavy metals as well as to the stress resistance of plants suggest that appropriate strains may be inoculated in order to further improve phytoextraction applications. Initial studies have been performed by Whiting et al. (2001b), who inoculated *Thlaspi caerulescens* seeds grown in sterile as well as in nonsterile soil with rhizosphere bacteria belonging to the species *Microbacterium saperdae*, *Pseudomonas monteilii*, and *Enterobacter cancerogenes*. Bacteria enhanced Zn uptake in sterile soils. However, in nonsterile soils the effect of the bacterial inoculant strains was masked by the natural soil microflora. This indicates that several native strains have the capability to enhance heavy metal mobilization. When inoculated, superior strains, which are promising as inoculants for enhancing phytoextraction, usually encounter high numbers of highly adapted micro-organisms and therefore have to show high competitive abilities. This is particularly true for rhizosphere bacteria, whereas endophytes usually encounter less competition.

Endophytic bacteria with good colonization potential were endowed with heavy metal resistance using genetic methods. The *ncc-nre* nickel resistance system of *Cupriavidus metallidurans* (previously *Ralstonia metallidurans*) was efficiently expressed in two endophyte strains of the species *Burkholderia cepacia* and *Herbaspirillum seropedicae* (Lodewyckx et al. 2001). When inoculated onto *Lupinus luteus* L. plants, the Ni-resistant *B. cepacia* strain induced a significant increase of Ni concentrations in roots but not in shoots. Similarly, the Ni-resistant *Herbaspirillum* strain enhanced Ni uptake of *Lolium perenne* plants; however, the wild-type bacterial strain showed the same effect, indicating that nickel resistance was not responsible for the effect observed with the recombinant *Herbaspirillum* strain (Lodewyckx et al. 2001). Recombinant strains of the rhizosphere bacterium *Pseudomonas aureofaciens* were engineered by either adding the arsenite resistance operon or the citrate synthase gene of *Pseudomonas aeruginosa* PA01 (Sizova et al. 2004). The latter genetically modified strain mobilized arsenic in soils, whereas the former was highly resistant to arsenite and arsenate. Both strains increased the survival of sorghum (*Sorghum saccharatum* L.) plants and led to highly increased levels of plant-accumulated arsenic.

Genetic engineering of micro-organisms and/or plants for improved phytoextraction has been also proposed by Sauge-Merle et al. (2003). As phytochelatins (PCs) bind heavy metals by complex formation, the authors expressed PC synthase of *Arabidopsis thaliana* in *Escherichia coli*. Significant increases of cellular heavy metal content were found and the authors suggest using genes of the PC biosynthetic pathway for the design of bacterial strains or higher plants with increased abilities to accumulate toxic metals.

15.7 Conclusions

Rhizosphere and endophytic bacteria are associated with plants showing tolerance to heavy metals and may modulate this tolerance. Endophytic colonizers may possess a better potential at influencing plant activities, as compared to rhizospheric micro-organisms. The structure of plant-associated bacterial communities is influenced by heavy-metal soil contamination, with resistant strains being selected. Rhizosphere and endophytic bacteria may increase the capacity of the plant to hyperaccumulate heavy metals by a variety of direct and indirect mechanisms. Direct mechanisms include enhanced heavy-metal mobilization and an alleviation of heavy-metal toxicity to the plant by improving iron nutrition. Indirect mechanisms comprise plant growth promotion and improved stress tolerance.

These effects of rhizosphere and endophytic micro-organisms suggest that specific strains with good activity and colonization potential would be useful in enhancing phytoextraction applications. In particular, the application of genetically engineered plant-associated micro-organisms may be a promising approach for phytoremediation of soils contaminated with heavy metals. However, the performance of these micro-organisms under natural conditions has to be investigated in detail. Although these strains are likely to be superior in terms of heavy metal resistance and mobilization, they might face competition problems similar to promising natural strains. Furthermore, biosafety aspects have to be considered and their release depends on national legislation. Addressing the issue of persistence and competition capacity of inoculant strains, while developing their potential for plant growth promotion, stress resistance, and heavy metal accumulation, represent promising strategies for improving current phytoremediation techniques.

References

Abou-Shanab RA, Angle JS, Delorme TA, Chaney RL, van Berkum P, Moawad H, Ghanem K, Ghozlan HA (2003). Rhizobacterial effects on nickel extraction from soil and uptake by *Alyssum murale*. *New Phytol* 158: 219–224

Adriano DC (2001) *Trace Elements in Terrestrial Environments: Biogeochemistry, Bioavailability and Risk of Metals*. 2nd edn., Springer, New York

Agarwhal S, Shende ST (1987) Tetrazolium reducing microbes inside the roots of *Brassica* species. *Curr Sci* 56:187–188

Amir H, Pineau R (2003) Release of Ni and Co by microbial activity in New Caledonian ultramafic soils. *Can J Microbiol* 49:288–293

Assunção AGL, Schat H, Aarts MGM (2003) *Thlaspi caerulescens*, an attractive model species to study heavy metal hyperaccumulation in plants. *New Phytol* 159: 351–360

Bååth E (1989) Effects of heavy metals in soil on microbial processes and populations: A review. *Water Air Soil Pollut* 47:335–379

Baker AJM, Brooks RR (1989) Terrestrial higher plants which accumulate metallic elements - A review of their distribution, ecology and phytochemistry. *Biorecovery* 1: 81–126

Baker AJM, McGrath SP, Reeves RD, Smith JAC (2000) Metal hyperaccumulator plants: A review of the ecology and physiology of a biological resource for phytoremediation of metal-polluted soils. In: Terry N, Bañuelos GS (eds) *Phytoremediation of Contaminated Soil and Water*, CRC Press, Boca Raton, FL, pp 85–107

Baker AJM, Whiting SN (2002) In search of the Holy Grail – A further step in understanding metal hyperaccumulation? *New Phytol* 155: 1–4

Bell CR, Dickie GA, Harvey WLG, Chan JWYF (1995) Endophytic bacteria in grapevine. *Can J Microbiol* 41: 46–53

Bernal MP, McGrath SP (1994) Effects of pH and heavy metal concentrations in elemental composition of *Alyssum murale* and *Raphanus sativus* L. *Plant Soil* 166: 83–92

Beyeler M, Keel C, Michaux P, Haas D (1999) Enhanced production of indole-3-acetic acid by a genetically modified strain of *Pseudomonas fluorescens* CHA0 affects root growth of cucumber, but does not improve protection of the plant against *Phytium* root rot. *FEMS Microbiol Ecol* 28:225–233

Bitton G, Freihofer V (1978) Influence of extracellular polysaccharide on the toxicity of copper and cadmium toward *Klebsiella aerogenes*. *Microb Ecol* 4:119–125

Burd GI, Dixon DG, Glick BR (1998) A plant growth promoting bacterium that decreases nickel toxicity in seedlings. *Appl Environ Microbiol* 64: 3663–3668

Burd GI, Dixon DG, Glick BR (2000). Plant growth promoting bacteria that decrease heavy metal toxicity in plants. *Can J Microbiol* 46: 237–245

Campbell BG, Thompson JA (1996) 1-Aminocyclopropane-1-carboxylate deaminase genes from *Pseudomonas* strains. *FEMS Microbiol Lett* 138:207–210

Chao WL, Chen CLF (1991) Role of exopolymer and acid-tolerance in the growth of bacteria in solutions with high copper ion concentrations. *J Gen Appl Microbiol* 37: 363–370

Chelius MK, Triplett EW (2001) The diversity of archaea and bacteria in association with the roots of *Zea mays* L. *Microb Ecol* 41:252–263

Chin-A-Woeng TFC, Bloemberg GV, Mulders IHM, Dekkers LC, Lugtenberg BJJ (2000) Root colonization by phenazine-1-carboxamide-producing bacterium *Pseudomonas chlororaphis* PCL1391 is essential for biocontrol of tomato foot and root rot. *Mol Plant-Microbe Interact* 13:1340–1345

Cieslinski G, van Rees KCJ, Szmigielska AM, Krishnamurti GSR, Huang PM (1998) Low-molecular-weight organic acids in the rhizosphere soils of durum wheat and their effect on cadmium bioaccumulation. *Plant Soil* 203: 109–117

Clemens S (2001) Molecular mechanisms of plant metal tolerance and homeostasis. *Planta* 212: 475–486

Costacurta A, Keijers V, Vanderleyden J (1994) Molecular cloning and sequence analysis of an *Azospirillum brasiliense* indole-3-pyruvate decarboxylase gene. *Mol Gen Genet* 243:463–472

de Freitas RJ, Germida JJ (1990) A root tissue culture system to study winter wheat-rhizobacteria interactions. *Appl Microbiol Biotechnol* 33:589–595

Dekkers LC, Mulders IHM, Phoelich CC, Chin-A-Woeng TFC, Wijfjes AHM, Lugtenberg BJJ (2000) The *sss* colonization gene of the tomato-*Fusarium oxysporum* f. sp. *radicislycopersici* biocontrol strain *Pseudomonas fluorescens* WCS365 can improve root colonization of other wild-type *Pseudomonas* spp. bacteria. *Mol Plant-Microbe Interact* 13:177–1183

De Meyer G, Höfte M (1997) Salicylic acid produced by the rhizobacterium *Pseudomonas aeruginosa* 7NSK induces resistance to leaf infection by *Botrytis cinerea* on bean. *Phytopathology* 87:588–593

de Souza MP, Chu D, Zhao M, Zayed AM, Ruzin SE, Schichnes D, Terry N (1999a) Rhizosphere bacteria enhance selenium accumulation and volatilization by Indian mustard. *Plant Physiol* 119.565–573

de Souza MP, Huang CPA, Chee N, Terry N (1999b) Rhizosphere bacteria enhance the accumulation of selenium and mercury in wetland plants. *Planta* 209: 259–263

Díaz-Roviña M, Bååth E (1996) Development of metal tolerance in soil bacterial communities exposed to experimentally increased metal levels. *Appl Environ Microbiol* 62:2970–2977

Diels L, Dong QH, van der Lelie D, Baeyens W, Mergeay M (1995) The *czc* operon of *Alcaligenes eutrophus* CH34: From resistance mechanism to the removal of heavy metals. *J Ind Microbiol* 14: 142–153

Doelman P, Janson E, Michels M, van Til M (1994) Effects of heavy metals in soil on microbial diversity and activity as shown by the sensitivity-resistance index, an ecologically relevant parameter. *Biol Fert Soils* 17:177–184

Evers A, Hancock RD, Martell AE, Motekaitis RJ (1989) Metal ion recognition in ligands with negatively charged oxygen donor groups. Complexation of Fe(III), Ga(III), In(III), Al(III) and other highly charged metal ions. *Inorg Chem* 28: 2189–2195

Fitz WJ, Wenzel WW, Zhang H, Nurmi J, Stipek K, Fischerova Z, Schweiger P, Köllensperger G, Ma LQ, Stingeder G (2003) Rhizosphere characteristics of the arsenic hyperaccumulator *Pteris vittata* L. and monitoring of phytoremoval efficiency. *Environ Sci Technol* 37: 5008–5014

Francis AJ, Dodge CJ, Gillow JB (1992) Biodegradation of metal citrate complexes and implications for toxic-metal mobility. *Nature* 356:140-142

Frommel MI, Novak J, Lazarovits G (1991) Growth enhancement and developmental modification of *in vitro* potato (*Solanum tuberosum* ssp. *tuberosum*) as affected by nonfluorescent *Pseudomonas* sp. *Plant Physiol* 96:928–936

Gagné S, Richard C, Rousseau H, Antoun H (1987) Xylem-residing bacteria in alfalfa. *Can J Microbiol* 33:996–1005

Giller KE, Witter E, McGrath SP (1998) Toxicity of heavy metals to micro-organisms and microbial processes in agricultural soils: A review. *Soil Biol Biochem* 30:1389–1414

Glick BR (2004) Teamwork in phytoremediation. *Nature Biotechnol* 22:526–527

Glick BR, Liu C, Ghosh S, Dumbroff EB (1997) Early development of canola seedlings in the presence of the plant growth-promoting rhizobacterium *Pseudomonas putida* GR12-2. *Soil Biol Biochem* 29:1233–1239

Gremion F, Chatzinotas A, Harms H (2003) Comparative 16S rDNA and 16S rRNA sequence analysis indicates that *Actinobacteria* might be a dominant part of the metabolically active bacteria in heavy metal-contaminated bulk and rhizosphere soil. *Environ Microbiol* 5: 896–907

Hallmann J, Quadt-Hallmann A, Mahaffee WF, Kloepper JW (1997) Bacterial endophytes in agricultural crops. *Can J Microbiol* 43: 895–914

Hattori H (1996) Decomposition of organic matter with previous cadmium adsorption in soils. *Soil Sci Plant Nutr* 42: 745–752

Hinojosa MB, Carreira JA, García-Ruíz R, Dick RP (2005) Microbial response to heavy metal polluted soils. *J Environ Qual* 34:1789–1800

Hirsch PR, Jones MJ, McGrath SP, Giller KE (1993) Heavy metals from past applications of sewage sludge decrease the genetic diversity of *Rhizobium leguminosarum* biovar *trifolii* populations. *Soil Biol Biochem* 25:1485–1490

Huang J (1986) Ultrastructure of bacterial penetration in plants. *Annu Rev Phytopathol* 24:141–157

Huysman F, Verstraate W, Brookes PC (1994) Effect of manuring practises and increased copper concentrations on soil microbial populations. *Soil Biol Biochem* 26:103–110

Idris R, Kuffner M, Bodrossy L, Puschenreiter M, Monchy S, Wenzel WW, Sessitsch A (2006) Characterization of Ni-tolerant methylobacteria associated with the hyperaccumulating plant *Thlaspi goesingense* and description of *Methylobacterium goesingense* sp. nov. *Syst Appl Microbiol* 29:634–644

Idris R, Trivonova R, Puschenreiter M, Wenzel WW, Sessitsch A (2004) Bacterial communities associated with flowering plants of the Ni-hyperaccumulator *Thlaspi goesingense. Appl Environ Microbiol* 70:2667–2677

Kandeler E, Tscherko D, Bruce KD, Stemmer M, Hobbs PJ, Bardgett RD, Amelung W (2000) Structure and function of the soil microbial community in microhabitats of a heavy metal polluted soil. *Biol Fertil Soils* 32:390–400

Kidambi SP, Sundin GW, Palmer DA, Chakrabarty AM, Bender CL (1995) Copper as a signal for alginate synthesis in *Pseudomonas syringae* pv. *syringae. Appl Environ Microbiol* 61:2172–2179

Kozdroj J (1995) Microbial responses to single or successive soil contamination with Cd or Cu. *Soil Biol. Biochem* 27:1459–1465

Kunito T, Nagaoka K, Tada N, Saeki K, Senoo K, Oyaizu H, Matsumoto S (1997) Characterization of Cu-resistant bacterial communities in Cu-contaminated soils. *Soil Sci Plant Nutr* 43:709–717

Kunito T, Saeki K, Nagaoka K, Oyaizu H, Matsumoto S (2001) Characterization of copper-resistant bacterial community in rhizosphere of highly copper-contaminated soil. *Eur J Soil Biol* 37:95–102

Lamb TG, Tonkyn DW, Kluepfel DA (1996) Movement of *Pseudomonas aureofaciens* from the rhizosphere to aerial plant tissue. *Can J Microbiol* 42:1112–1120

Lasat MM, Kochian LV (2000) Physiology of Zn hyperaccumulation in *Thlaspi caerulescens*. In: Terry N, Bañuelos GS (eds) *Phytoremediation of Contaminated Soil and Water*, CRC Press, Boca Raton, FL, pp. 159–169

Leita L, de Nobili M, Muhlbachova G, Mondini C, Marchiol L, Zerbi G (1995) Bioavailability and effects of heavy metals on soil microbial biomass survival during laboratory conditions. *Biol Fertil Soils* 19:103–108

Lodewyckx C, Mergeay M, Vangronsveld J, Clijsters H, van der Lelie D (2002) Isolation, characterization, and identification of bacteria associated with the zinc hyperaccumulator *Thlaspi caerulescens* subsp. *calaminaria. Int J Phytoremed* 4:101–115

Lodewyckx C, Taghavi S, Mergeay M, Vangronsveld J, Clijsters H, van der Lelie D (2001) The effect of recombinant heavy metal-resistant endophytic bacteria on heavy metal uptake by their host plant. *Int J Phytoremediat* 3:173–187

Lugtenberg BJJ, de Weger LA, Bennett JW (1991) Microbial stimulation of plant growth and protection from disease. *Curr Opin Biotechnol* 2:457–464

Ma LQ, Komar KM, Tu C Zhang W, Cai Y, Kennelley ED (2001) A fern that hyperaccumulates arsenic. *Nature* 409:579

Mahaffee WF, Kloepper JW (1994) Applications of plant growth-promoting rhizobacteria in sustainable agriculture. In: Pankhurst CE, Edwards C, Jeans K (eds) *Soil Biota: Management in Sustainable Farming Systems*, CSIRO, Australia, pp 23–31

Marschner H (1995) *Mineral Nutrition of Higher Plants*. 2nd edn. Academic Press, London

Maurhofer M, Reimmann C, Schmidli-Sacherer P, Heeb S, Haas D, Défago G (1998) Salicylic acid biosynthetic genes expressed in *Pseudomonas fluorescens* strain P3 improve the induction of systemic resistance in tobacco against tobacco necrosis virus. *Phytopathology* 88:678–684

Mengoni A, Barzanti A, Gonnelli C, Gabrielli R, Bazzicalupo M (2001) Characterization of nickel-resistant bacteria isolated from serpentine soil. *Environ Microbiol* 3:691–698

Mengoni A, Grassi E, Brazanti A, Biondi EG, Gonnelli C, Kim CK, Bazzicalupo M (2004) Genetic diversity of microbial communities of serpentine soil and of rhizosphere of the Ni-hyperaccumulator plant *Alyssum bertolonii*. *Microb Ecol* 48:209–217

Mertens J, Springael D, De Troyer I, Cheyns K, Wattiau P, Smolders E (2006) Long-term exposure to elevated zinc concentrations induced structural changes and zinc tolerance of the nitrifying community in soil. *Environ Microbiol* 8:2170.2178

Mitchell P, Barr D (1995) The nature and significance of public exposure to arsenic: A review of its relevance to South West England. *Environ Geochem Health* 17:57–82

Moffett BF, Nicholson FA, Uwakwe NC, Chambers BJ, Harris JA, Hill TCJ (2003) Zinc contamination decreases the bacterial diversity of agricultural soil. *FEMS Microbiol Ecol* 43:13–19

Morel JL, Mench M, Guckert A (1986) Measurement of Pb^{2+}, Cu^{2+} and Cd^{2+} binding with mucilage exudates from maize (*Zea mays* L.) roots. *Biol. Fertil. Soils* 2: 29-34

Patten CL, Glick BR (1996) Bacterial biosynthesis of indole-3-acetic acid. *Can J Microbiol* 42:207–220

Penyalver R, Oger P, López MM, Farrand SK (2001) Iron-binding compounds from *Agrobacterium* spp.: biological control strain *Agrobacterium rhizogenes* K84 produces a hydroxamate siderophore. *Appl Environ Microbiol* 67:654–664

Puschenreiter M, Horak O, Friesl W, Hartl W (2005a) Low-cost agricultural measures to reduce the heavy metal transfer into human food chain – A review. *Plant Soil Environ* 51:1–11

Puschenreiter M, Schnepf A, Molina Millán I, Fitz WJ, Horak O, Klepp J, Schrefl T, Lombi E, Wenzel WW (2005b) Changes of Ni biogeochemistry in the rhizosphere of the hyperaccumulator *Thlaspi goesingense*. *Plant Soil* 271:205–218

Puschenreiter M, Wieczorek S, Horak O, Wenzel WW (2003) Chemical changes in the rhizosphere of metal hyperaccumulator and excluder *Thlaspi* species. *J Plant Nutr Soil Sci* 166:579–584

Quadt-Hallmann A, Benhamou N, Kloepper JW (1997) Bacterial endophytes in cotton: mechanisms of entering the plant. *Can J Microbiol* 43:557–582

Rasche F, Velvis H, Zachow C, Berg G, van Elsas JD, Sessitsch A (2006) Impact of transgenic potatoes expressing antibacterial agents on bacterial endophytes is comparable to effects of wildtype potatoes and changing environmental conditions. *J Appl Ecol* 43:555–566

Reeves RD, Baker AJM (2000) Metal-accumulating plants. In: Raskin I, Ensley BD (eds) *Phytoremediation of Toxic Metals*, John Wiley & Sons, New York, pp 193–229

Reiter B, Pfeifer U, Schwab H, Sessitsch A (2002) Response of endophytic bacterial communities in potato plants to infection with *Erwinia carotovora* ssp. *atroseptica*. *Appl Environ Microbiol* 68:2261–2268

Reiter B, Sessitsch A (2006) The bacterial microflora in association with the wildflower *Crocus albiflorus*. *Can J Microbiol* 52:1–10

Römheld V (1991) The role of phytosiderophores in acquisition of iron and other micronutrients in graminaceous species: An ecological approach. *Plant Soil* 130:127–134

Roos IMM, Hattingh MJ (1983) Scanning electron microscopy *Pseudomonas syringae* pv. *morsprunorum* on sweet cherry leaves. *Phytopathol Z* 108:18–25

Salt DE, Benhamou N, Leszczyniecka M, Raskin I (1999) A possible role for rhizobacteria in water treatment by plant roots. *Int J Phytoremediat* 1:67–69

Salt DE, Krämer U (2000) Mechanisms of metal hyperaccumulation in plants. In: Raskin I, Ensley BD (eds) *Phytoremediation of Toxic Metals: Using Plants to Clean Up the Environment*. Wiley, New York, pp 231–246

Sandaa R-A, Torsvik V, Enger Ø (2001) Influence of long-term heavy metal contaminations on microbial communities in soil. *Soil Biol Biochem* 33:287–295

Sauge-Merle S, Cuiné S, Carrier P, Lecomte-Pradines C, Luu D-T, Peltier G (2003) Enhanced toxic metal accumulation in engineered bacterial cells expressing *Arabidopsis thaliana* phytochelatin synthase. *Appl Environ Microbiol* 69:490–494

Schat H, Llugany M, Bernhard R (2000) Metal-specific patterns of tolerance, uptake, and transport of heavy metals in hyperaccumulating and nonhyperaccumulating metallophytes. In: Terry N, Bañuelos GS (eds) *Phytoremediation of Contaminated Soil and Water*, CRC Press, Boca Raton, FL, pp 174–188

Schlegel HG, Cosson JP, Baker AJM (1991) Nickel-hyperaccumulating plants provide a niche for nickel-resistant bacteria. *Bot Acta* 104:18–25

Scott RI, Chard JM, Hocart MJ, Lennard JH, Graham DC (1996) Penetration of potato tuber lenticels by bacteria in relation to biological control of blackleg disease. *Potato Res* 39:333–344

Sessitsch A, Reiter B, Berg G (2004) Endophytic bacterial communities of field-grown potato plants and their plant-growth-promoting and antagonistic abilities. *Can J Microbiol* 50:239–249

Sessitsch A, Reiter B, Pfeifer U, Wilhelm E (2002) Cultivation-independent population analysis of bacterial endophytes in three potato varieties based on eubacterial and *Actinomycetes*-specific PCR of 16S rRNA genes. *FEMS Microbiol Ecol* 39:23–32

Sessitsch A, Weilharter A, Gerzabek MH, Kirchmann H, Kandeler E (2001) Microbial population structures in soil particle size fractions of a long-term fertilizer field experiment. *Appl Environ Microbiol* 67:4215–4224

Shah S, Li J, Moffatt BA, Glick BR (1998) Isolation and characterization of ACC deaminase genes from two different plant growth-promoting rhizobacteria. *Can J Microbiol* 44:833–843

Sizova OI, Lyubun EV, Kochetkov VV, Validov SZ, Boronin AM (2004) Effect of wild and genetically modified rhizosphere bacteria *Pseudomonas aureofaciens* on the accumulation of arsenic by plants. *Appl Biochem Microbiol* 40:67–70

Smit E, Leeflang P, Wernars K (1997) Detection of shifts in microbial community structure and diversity in soil caused by copper contamination using amplified ribosomal DNA restriction analysis. *FEMS Microbiol Ecol* 23:249–261

Stoltzfus JR, So R, Malarvithi PP, Ladha JK, de Bruijn FJ (1998) Isolation of endophytic bacteria from rice and assessment of their potential for supplying rice with biologically fixed nitrogen. *Plant Soil* 194:25–36

Sturz AV, Christie BR, Matheson BG (1998) Associations of bacterial endophyte populations from red clover and potato crops with potential for beneficial allelopathy. *Can J Microbiol* 44:162–167

Sturz AV, Nowak J (2000) Endophytic communities of rhizobacteria and the strategies to create yield enhancing associations with crops. *Appl Soil Ecol* 15:183–190

Timmusk S, Nicander B, Granhall U, Tillberg E (1999) Cytokinin production by *Paenibacillus polymyxa*. *Soil Biol Biochem* 31:1847–1852

Van Buren AM, Andre C, Ishimaru CA (1993) Biological control of the bacterial ring rot pathogen by endophytic bacteria isolated from potato. *Phytopathology* 83:1406

van der Lelie D, Corbisier P, Diels L, Gilis A, Lodewyckx C, Mergeay M, Taghavi S, Spelmans N, Vangronsveld J (1999). The role of bacteria in the phytoremediation of heavy metals. In: Terry N, Banuelos G (eds) *Phytoremediation of Contaminated Soil and Water*. Lewis, London, UK

Wenzel WW, Bunkowski M, Puschenreiter M, Horak O (2003) Rhizosphere characteristics of indigenously growing nickel hyperaccumulator and tolerant plants on serpentine soil. *Environ Pollut* 123:131–138

Wenzel WW, Lombi E, Adriano DC (1999) Biogeochemical processes in the rhizosphere: Role in phytoremediation of metal-polluted soils. In: Prasad NMV, Hagemeyer J (eds) *Heavy Metal Stress in Plants - From Molecules to Ecosystems*. Springer Verlag, Heidelberg, pp. 273–303

Whiting SN, de Souza SP, Terry N (2001b) Rhizosphere bacteria mobilize Zn for hyperaccumulation by *Thlaspi caerulescens*. *Environ Sci Technol* 35:3144–3150

Whiting SN, Leake JR, McGrath SP and Baker AJM (2001a) Assessment of Zn mobilization in the rhizosphere of *Thlaspi caerulescens* by bioassay with non-accumulator plants and soil extraction. *Plant Soil* 237:147–156

Wilson D (1995) Endophyte – The evolution of a term, and clarification of its use and definition. *Oikos* 73:274–276

Zhao FJ, Hamon RE, McLaughlin MJ (2001) Root exudates of the hyperaccumulator *Thlaspi caerulescens* do not enhance metal mobilization. *New Phytol* 151:613–620

Chapter 16
Interactions of Fungi and Radionuclides in Soil

John Dighton(✉), Tatyana Tugay, and Nelli Zhdanova

16.1 Introduction

Following the development of nuclear weapons and the subsequent evolution of nuclear energy-generating industries, there has been considerable concern regarding the safe storage of radionuclide waste. Widescale release, in the aftermath of nuclear detonations or as the result of malfunction of atomic energy plants and reprocessing facilities, has also been a preoccupation. The International Commission on Radiological Protection recommendations on the ecological aspects of radionuclide release were discussed by Coughtree (1983), in which Heal and Horrill (1983) summarized element transfers within terrestrial ecosystems, highlighting the importance of organic soil horizons and their microbial communities as potential accumulators of both nutrient elements and radionuclides. This was a significant step forward from initial discussions of the impact of radionuclide fallout on ecosystems, where the involvement of fungi in regulating radionuclide movement was limited to one sentence in a paragraph describing radionuclide accumulation in organic horizons of forest soils, which may be related to fungal biomass (Osburn 1967). Now, in a more recent model of radiocesium migration in forest ecosystems, Avila and Moberg (1999) place fungal activity in the pivotal point of the diagonal of their interaction matrix, as one of the important biotic regulators of radionuclide movement in soils.

Concerns over global climate change and the limitations of conventional fuels is making the world consider alternative energy sources other than fossil fuels. One of these alternatives, nuclear energy, is already in use in many countries, although its expansion has been shadowed by the explosion of the reactor at Chernobyl, Ukraine in 1986. It is likely, however, that there will be future expansion of nuclear activity, which may lead to an increased possibility of local or widespread nuclear contamination of terrestrial systems. It is, therefore, essential that we have adequate knowledge of the behavior of radionuclides in a variety of environments and enough

---

John Dighton
Rutgers University Pinelands Field Station, New Lisbon, New Jersey 08064, USA
e-mail: dighton@camden.rutgers.edu

understanding of the biotic regulators of radionuclide movement to assist us in limiting potential damage and effecting remediation as efficiently as possible.

It is probable that future studies along those lines will point to fungi as an important component of environmental systems involving radionuclides. Indeed, Steinera et al. (2002) open their introduction by saying 'Fungi are one of the most important components of forest ecosystems, since they determine to a large extent the fate and transport processes of radionuclides (…)', and fungi have been suggested as being potentially important agents for bioremediation (Gray 1998; Skladany and Metting 1992). A brief review of the interactions between radionuclides and fungi (Zhdanova et al. (2005b) calls for greater understanding of the nature of these interactions, particularly from a molecular standpoint and from the point of view of using fungal and plant–fungal interactions in remediation of polluted systems.

16.2 Fungi as Regulators of Radionuclide Movement

16.2.1 Role of Saprotrophic Fungi

Saprotrophic fungi are involved in the decomposition of dead organic residues (plant and animal remains) in soil. In order to assimilate the end products of extracellular enzymatic activity, fungi present an enormous surface area of hyphae to their surrounding environment. As such, these organisms are well fitted to absorb other elements, including radionuclides, from the soil environment. Microbial immobilization of radionuclides was identified by Witkamp and Barzansky (1968), who showed that microbial communities on decomposing leaf litter and cellophane accumulated almost three times the amount of radiocesium immobilized by sterile equivalents. Witkamp (1968) measured radiocesium incorporation into the fungus *Trichoderma viride* from fresh or highly decomposed leaf litter and wood. He showed that the concentration factor in fungi decreased as the leaf litter source of cesium aged (2.36, 0.46, and 0.23 for litters of 4, 16, and 48 month-old leaf litter). The highest concentration factor (4.31) was found from freshly fallen wood. This pioneering work provided basic evidence to suggest that fungi could be significant ecosystem components in regulating movement of radionuclides in the environment.

Grassland soil saprotrophic fungi have been shown to have great potential for uptake and immobilization of radiocesium fallout (Olsen et al. 1990; Dighton et al. 1991). Assuming an average influx rate of 134 nmol Cs g^{-1} dry weight of mycelium (determined from laboratory uptake studies) and an estimate of hyphal biomass of up to 6 g dry weight m^{-2} of soil (determined by hyphal length measurements of field collected samples; Dighton and Terry 1996), Dighton et al. (1991) estimated that the fungal community of upland grass ecosystems in northern England would be able to take up between 350 and 804 nmol Cs m^{-1} h^{-1}. These fungi might have accumulated a large percentage of the total Chernobyl fallout in the United Kingdom as the radiocesium concentrations in soil pore water were reportedly at the micromolar level (Oughton 1989). Once incorporated into microbial (fungal and bacterial) biomass, some 40–80% of radionuclide (Cs) is retained by the fresh soil during leaching, in comparison to either irradiated (Guillitte et al. 1994) or autoclaved soil (Sanchez et al. 2000), where the microbiota has been killed. Thus, retention is thought to be due mainly to microbial immobilization of radionuclides. This is in contrast to sandy mineral soils where biomass accumulation is low and surface adsorption is the main form or radionuclide binding, only an additional 0.005% of immobilized ^{137}Cs being extracted from these soils following sterilization (Cawse 1983).

Zhdanova et al. (1990) compared ^{90}Sr accumulation by conidia or mycelium. They showed greater accumulation by mycelia and differences between unpigmented and pigmented fungal species. For light pigmented species of the family Moniliaceae ^{90}Sr accumulation varied between 18 and 335 Bq g^{-1}, whereas for dark pigmented mycelia of the Dematiaceae accumulation ranged from 20 to 1,510 Bq g^{-1}; which is up to an order of magnitude higher than for light pigmented fungi.

16.2.2 Mechanisms of Fungal Immobilization of Radionuclides

Binding of radionuclides to cell wall ion exchange sites and by potassium replacement in cytoplasm are two mechanisms proposed for radionuclide accumulation by fungi. Potassium replacement is, however, species-specific with suggestions that Rb and Cs replace K in the filamentous fungus *Fusarium solani* (Das 1991), whereas only Rb, but not Li, Na, or Cs, could replace K in the yeast *Candida utilis* (Aiking and Tempest 1977). Connolly et al. (1998) demonstrated that the wood decay fungus *Resinicium bicolor* could absorb strontium from strontianite sand and redeposit it in calcium oxalate crystals after translocation through mycelial cords. Here, strontium behaves similarly to calcium in fungal metabolism and the research suggests a pathway for translocation of the radionuclide ^{90}Sr within decomposer fungi.

Remediation of contaminated areas is usually approached by phytoremediation (Raskin and Ensley 2000). However, the propensity of fungal hyphae to adsorb and absorb radionuclides provides an alternative means to effect radionuclide cleanup. The usefulness of fungal mycelia in environmental cleanup has been suggested by White and Gadd (1990) who developed air-lift bioreactor systems containing live cultures of fungi for biosorption of radiothorium. *Rhizopus arrhizus* and *Aspergillus niger* were found to be more efficient absorbers than *Penicillium italicum* and *Penicillium chrysogenum*. Fungi have also been shown to be useful adsorbers of heavy metals and radionuclide contaminants of industrial effluents, where their dead mycelia have been used as filters (Tobin et al. 1984; Singleton and Tobin 1996). In live mycelia, the ability to immobilize radionuclides is species dependent. Mahmoud (2004) showed that *Alternaria alternata* had a greater uptake capacity than *Aspergillus pulverulens* or *Fusarium verticilloides* for both ^{60}Co and ^{137}Cs. Greatest uptake was attributed to high melanin content of the fungal cell walls and could account for between 45 and 60% of the radionuclide uptake into the fungal hyphae. Clint et al. (1991) also showed significant differences in Cs uptake capacity between species of both mycorrhizal and saprotrophic basidiomycetes. As a result, it is often difficult to generalize on the efficiency of radionuclide immobilization exhibited by fungi per se.

Soil saprotrophic microfungi have been shown to grow into and decompose carbon-based radioactive debris from the Chernobyl reactor (Zhdanova, et al. 1991). *Cladosporium cladosporioides* and *Penicillium roseopurpureum* were shown to overgrow these 'hot particles' that contained less than 1,147 Bq of γ-activity and destroy them within 50 to 150 days. Fungal accumulation of radionuclide released from intact hot particles was shown to be greater for ^{152}Eu than for ^{137}Cs (Zhdanova et al. 1991), but similar when the particles were ground. Thus, the capacity to acquire radionuclides in these fungi is determined by an interaction between the physical nature of the radioactive source, fungal species, and, presumably, enzymatic potential. Indeed, it is the immobilization of radionuclides into fungal biomass that is reportedly causing the retention and accumulation of radiocesium in upper soil horizons. Hence, immobilization by fungi reduces the downward movement of radionuclides that accompanies leaching (Rommelt et al. 1990; Guillitte et al. 1990). Drissner et al. (1998) point out that most ^{137}Cs in spruce forest soil is in the Ah layer, located just under the leaf litter, due to the high abundance of fungal mycelia in the surficial organic horizons of forest soils (Baeza et al. 2002). This surface accumulation prevents leaching loss of radionuclides and enhances uptake by plants through mycorrhizal associations. Retention by fungi in soil is likely, as fungi compete more strongly than plants for Cs^+, accumulating 10 to 150 times more of this element (Avery 1996).

In addition to immobilization, fungal mycelia are known to translocate materials within the fungal body. Translocation of ^{14}C and ^{32}P through hyphal systems of *Rhizopus, Trichoderma*, and *Stemphylium* species occurs by diffusion (Olsson and Jennings 1991). The rate of carbon translocation is influenced by source–sink relations within the fungal thallus to provide directional flow to the building phases of the hyphae (Olsson 1995). Jennings (1990) showed that the absorption of phosphorus by rhizomorphs of *Phallus impudicus* and *Mutinus caninus* was effected by two transport systems.

In contrast to the diffusion of C and P, translocation of ^{137}Cs through hyphae of *Armillaria* spp. and *Schizophyllum commune* has been shown to be slower than diffusion, suggesting a possible mechanism for accumulation, probably due to cell wall binding of the radionuclide (Gray et al. 1995, 1996). Gray et al. (1996) also demonstrated preferential movement of radiocesium to developing fruitbodies, which were acting as nutrient sinks. This has been confirmed by work of Baeza et al. (2002), who showed translocation of ^{85}Sr and ^{134}Cs from substrate to mycelium, and then to fruitbody in *Pleurotus eringii*. These observations support the finding of Dighton and Horrill (1988) and Yoshida and Muramatsu (1994), who observed that radiocesium accumulation in basidiomycete fungi could be high and long-lived. For example, Dighton and Horrill (1988) found that up to 92% of the radiocesium in mycorrhizal basidocarps in the United Kingdom was derived from pre-Chernobyl sources of fallout; this would be the result of accumulation during the previous 30 years or so. Measures of influx and accumulation of radiocesium into hyphae of a range of fungal species suggested that saprotrophic species had higher accumulation than mycorrhizal species (on a per weight basis; Clint et al. 1991). Hence, this high rate of radiocesium immobilization in United Kingdom upland grassland by saprotrophic fungal mycelia could also have accounted for a high proportion of the immobilization of Chernobyl fallout radiocesium (Dighton et al. 1991).

16.2.3 Role of Endomycorrhizae

Mycorrhizae are a symbiotic association between plant roots and a variety of fungi. The benefits of this association are seen in about 95% of all vascular plant species and consist in improved water and nutrient acquisition, and also defense against root herbivory and root pathogens. The endomycorrhizae are associations where most of the symbiotic fungal tissue is located within the host plant tissue, along with extraradical hyphae extending from the root surface into the soil to increase the absorptive surface area. In this section, arbuscular mycorrhizae (associations between a small number of fungal species belonging to the phylum Glomeromycota, and a huge number of grasses, forbs, and tropical tree species) and ericoid mycorrhizae (associations between even fewer fungal species from the phylum Ascomycota and plants in the Ericaceae, Empetraceae, and Epacridaceae) are considered. Ectomycorrhizae are considered below (see Section 16.2.4). These are associations between many fungal species from the Basidiomycota and the Ascomycota and a smaller number of plant (tree) species. In this association, most of the symbiotic fungal biomass occurs outside the host root tissue as a fungal sheath and emanating extraradical hyphae. For further information on mycorrhizae the reader is referred to texts such as Smith and Read (1997).

Haselwandter and Berreck (1994) reviewed the role of arbuscular mycorrhizae in plant uptake of radionuclides and found that the current information is somewhat conflicting. They cite an example of arbuscular mycorrhizal inoculation of sweet clover and Sudan grass, which showed slight, and statistically insignificant increases in uptake of ^{137}Cs and ^{60}Co by mycorrhizal plants. In a later study Berreck and Haselwandter (2001) showed that mycorrhizal colonization of the grass *Agrostis tenuis* by *Glomus mosseae* significantly decreased the Cs content of shoots. Increased competition between K and Cs in soil, by increasing K concentration with fertilizer, led to further suppression of Cs uptake.

Jackson et al. (1973) found similar differences between host plant species, where arbuscular mycorrhizal colonization of roots of soybeans by *Glomus mosseae* significantly increased ^{90}Sr uptake from soil, whereas mycorrhizal symbioses in the grass *Festuca ovina* reduced the uptake of radiocesium into shoots. This protective effect is also shown by de Boulois et al. (2005) where uptake of radiocesium by transformed carrot root in association with the mycorrhizal fungus *Glomus* is strongly correlated to fungal hyphal length. Hyphae have a greater uptake capacity than roots, resulting in preferential hyphal accumulation. This effect, however, is likely to be plant species-dependent, as suggested by Joner et al. (2004), who saw no enhanced mycorrhizal uptake of ^{134}Cs into clover, eucalyptus, or maize, but enhanced uptake of ^{65}Zn by mycorrhized maize. Roséna et al. (2005) also showed increased ^{137}Cs uptake by leek, but not by ryegrass, when colonized by arbuscular mycorrhizae.

In keeping with the information summarized above, suggesting a reduced plant uptake in the mycorrhizal condition, Clint and Dighton (1992) showed that influx of radiocesium into mycorrhizal heather plants (*Calluna vulgaris*) with ericoid mycorrhizae was lower than that into nonmycorrhizal plants. However, the internal redistribution of Cs within mycorrhizal plants allowed a greater proportion of the Cs taken up in mycorrhizal plants to be translocated to shoots compared to nonmycorrhizal plants. Strandberg and Johansson (1998) also showed that mycorrhizal heather had 18% higher shoot accumulation of ^{137}Cs than nonmycorrhizal plants. Similar enhanced translocation of radiocesium into shoots of arbuscular mycorrhizal *Festuca ovina*, but not into clover (*Trifolium repens*), were shown by Dighton and Terry (1996). However, in a more recent study, Berreck and Hasselwandter (2001) showed a decrease in the Cs translocated to the shoots of *Agrostis tenuis* in the presence of arbuscular mycorrhizae.

Elevated levels of radionuclide occur in roots as compared to shoots in mycorrhizal plants, which suggests that the endomycorrhizal fungi accumulate radiocesium in the fungal tissue in a similar manner to that with which ectomycorrhizae accumulate heavy metals (Denny and Wilkins 1987a,b; see also Section 16.2.4). This concept has been reviewed for arbuscular mycorrhizal symbioses by Berreck and Haselwandter (2001), who investigated the impact of potassium fertilization as a method to reduce uptake of cesium by the mycorrhizal grass (*Agrostis tenuis*). They showed that mycorrhizal development in plant roots reduced Cs uptake by the plant at moderate nutrient levels in the soil. This further suggests that the mechanism of protection is due to sequestration of Cs in the extraradical hyphae of the endomycorrhizal fungus and a reduced translocation into the host plant. They also demonstrated that, for this fungal–plant interaction, there was no benefit of adding potassium to reduce Cs uptake.

Thus, it would appear that endomycorrhizal fungi perform a plant protective effect by immobilizing radionuclides in fungal components associated with roots. However, there is some evidence to suggest that any radionuclide that enters the plant is translocated to aboveground parts with greater efficiency in the mycorrhizal condition than in nonmycorhizal plants.

16.2.4 Role of Ectomycorrhizae

Ectomycorrhizae are different from arbuscular mycorrhizae in that in many instances the symbiosis occurs with basidiomycete fungi. These produce relatively large fruitbodies (or mushrooms) as well as associated belowground mycelia. Thus, in addition to accumulation of radionuclides in belowground mycelia, there is translocation to mushrooms which are a source of accumulation up food chains by consumption or which could be harvested for site remediation. Hence, the earlier work of Haselwandter (1978), Eckl et al. (1986), Haselwandter et al. (1988), and Byrne (1988), which showed that lichens and mushroom-forming fungi took up and accumulated radionuclides in their fruiting structures, became an important stimulus to investigate fallout radionuclide accumulation in these structures following the Chernobyl explosion. The literature is replete with records of radionuclide concentrations following the Chernobyl accident. A few examples of accumulation levels reported are given in Table 16.1 and others can be found in articles by Guillitte et al. (1987), Elstner et al. (1987), Oolbekkink and Kuyper (1989), Watling et al. (1993), Guillitte et al. (1994), Yoshida and Muramatsu (1994), Mietelski et al. (1994), and Muramatsu et al. (1991) to name a few.

Although a variety of radionuclides were released from Chernobyl, most of the surveys relate to the radiocesium content of mushrooms. Haselwandter and Berreck (1994) reviewed the accumulation of radionuclides in fungal fruitbodies after the Chernobyl disaster and found that in addition to Cs, fungi have been shown to take up ^{7}Be, ^{60}Co, ^{90}Sr, ^{95}Zr, ^{95}Nb, ^{100}Ag, ^{125}Sb, ^{144}Ce, ^{226}Ra, and ^{238}U (see Chapter 6 in Dighton 2003).

The Chernobyl explosion released ^{137}Cs and ^{134}Cs in a 2:1 ratio, which has been used as fingerprint to distinguish ^{137}Cs from pre- and post-Chernobyl sources. Using this fingerprint, Dighton and Horrill (1988) showed that a large proportion

Table 16.1 Selected data on radionuclide accumulation in fruitbodies of a number of fungal species

Fungus	Isotope	Concentration (kBq kg^{-1} dry wt)	Source
Cortinarius praestans	^{137}Cs	0.5	Byrne (1988)
Laccaria amethystine	^{137}Cs	43	Byrne (1988)
Cortinarius armillatus	^{137}Cs	44	Byrne (1988)
Agaricus spp.	^{110}Ag	0.05	Byrne (1988)
Lycoperdon spp.	^{110}Ag	0.56	Byrne (1988)
Various	Various	0.09–947	Haselwandter and Berreck (1994)
Various	^{137}Cs	0–33 (low level contamination)	Grodzinskaya et al. (1995)
Various	^{137}Cs	1400–700 (high level contamination)	Grodzinskaya et al. (1995)
Various	^{137}Cs	0.3–20	Mietelski et al. (1994)
Various	^{137}Cs	300–1800	Mietelski et al. (1994)
Various	^{90}Sr, $^{239+240}$Pu	0–0.004	Mietelski et al. (1994)
Lactarius rufus	^{137}Cs	1.8–7.2	Dighton and Horrill (1988)
Inocybe longicystis	^{137}Cs	8.7–14.2	Dighton and Horrill (1988)
Xercomus badius	^{40}K	0.3–6.7	Malinowska et al. (2006)
Xercomus badius	^{210}Pb	0.007–0.03	Malinowska et al. (2006)
Various	^{137}Cs	0–2.86	Kirschner and Dalliant (1998)
Various	^{210}Pb	0.001–0.36	Kirschner and Dalliant (1998)

(up to 92%) of pre-Chernobyl ^{137}Cs was accumulated in the fruitbodies of two ectomycorrhizal fungal species. Similar figures (13 to 69%) for pre-Chernobyl accumulation of radiocesium in mushrooms were presented by Byrne (1988) and Giovani et al. (1990). This information suggests that fungi could be long-term accumulators and retainers of radionuclides in the environment. Laboratory studies showed that there was a wide range of rates of uptake and incorporation of radiocesium into fungal mycelia grown in liquid culture, with the three saprotrophic basidiomycetes, *Mycena polygramma, Cystoderma amianthinum,* and *Mycena sanguinolenta* having the highest rates of biomass accumulation as compared to many ectomycorrhizal species (Clint et al. 1991). Accumulation varied between about 100 to 250 nmol g^{-1} dry weight of fungus per hour. However, if the uptake is expressed on a hyphal surface area basis, the ranking of species differs and the uptake values range from 0.1 to 2.5 nmol m^{-2} hyphal area per hour. This suggests that acquisition of radionuclides by fungi may be influenced by the growth conditions in soil and consequential foraging strategy of the mycelium (Rayner 1991; Ritz 1995).

The ability of ectomycorrhizal basidiomycete fungal species to accumulate radionuclides could, in theory, be an important component in the restoration of radionuclide-contaminated terrestrial ecosystems (Gray 1998). The formation of large and harvestable fruiting structures (mushrooms) provides a potential means of removal of radionuclides that have been accumulated within. Mycorrhizal fungi have been shown to be a major component of the radionuclide accumulation (as evaluated by radiocesium concentration measurements) in a boreal coniferous ecosystem in Sweden (Guillitte et al. 1994), although its relative contribution to the total standing biomass of the forest is probably not large (Vogt et al. 1982; Fogel and Hunt 1983).

16.3 Effects of Radionuclides on Fungi

16.3.1 Isolation of Fungi from Contaminated Areas

Although they did not cite radiation levels, Durrell and Shields (1960) isolated 41 fungal species from soil in the Nevada Test Site (USA) where atomic weapons testing had occurred. Under equally extreme circumstances, the ability of fungi to survive high levels of radioactivity has been shown by the isolation of microfungi from the walls of the nuclear reactor room at the Chernobyl Atomic Electric Station (ChAES) under conditions of 1.5 to 800 mR h^{-1} in the presence of alpha, beta, and gamma-emitting ^{239}Pu, $^{240+241}$Pu, ^{241}Am, ^{244}Cm, and ^{137}Cs (Zhdanova et al. 2000). Thirty-seven species of 19 genera were isolated with the frequency of isolation being greater at higher than at lower levels of radioactivity. Table 16.2 presents a list of fungal species isolated from within and around Chernobyl and the radiation dose at site of isolation.

16 Soil Fungi and Radionuclides

Table 16.2 List of fungal species isolated from within and around the Chernobyl nuclear reactor in relation to radiation exposure at site of isolation and directional growth (radiotropism[a]) towards a collimated beam of ionizing radiation

Species (and Designation) of Isolate	Substrate	Radioactivity of Substrate (mR h^{-1})	Tropism
Penicillium steckii Zaleski 2 (Pst2)	Soil, "hot" particles	20	Positive
Penicillium hirsutum Dierckx (Phir1)	Soil, "hot" particles	20	Positive
Cladosporium cladosporioides (Fres.) de Vries (Ccl4)	Soil 10 km zone ChAES	30	Positive
Penicillium lanosum Westl. (Plan10)	The 4th block ChAES	20	Positive
Aspergillus versicolor (Vuill.) Tiraboschi (Aver43)	The 4th block ChAES	55	Positive
Aspergillus versicolor (Vuill.) Tiraboschi (Aver54)	The 4th block ChAES	98	Positive
Cladosporium sphaerospermum (Penz.) (Cshp35)	The 4th block ChAES	35	Positive
Cladosporium sphaerospermum (Penz.) (Csph70)	The 4th block ChAES	40	Positive
Hormoconis resinae (Lindau) v. Arx & de Vries f. resinae (Hres 30)	The 4th block ChAES	40	Positive
Aspergillus versicolor (Vuill.) Tiraboschi (Aver55)	The 4th block ChAES	250	Positive
Aspergillus versicolor (Vuill.) Tiraboschi (Aver57)	The 4th block ChAES	500	Positive
Cladosporium sphaerospermum (Penz.) (Csph21)	The 4th block ChAES	105	Positive
Hormoconis resinae (Lindau) v. Arx & de Vries f. resinae (Hres52)	The 4th block ChAES	120	Positive
Hormoconis resinae (Lindau) v. Arx & de Vries f. resinae (Hres61)	The 4th block ChAES	200	Positive
Hormoconis resinae (Lindau) v. Arx & de Vries f. resinae (Hres21)	The 4th block ChAES	300	Positive
Penicillium aurantiogriseum Dierckx 60 (P aur 60)	The 4th block ChAES	380	Positive
Aspergillus versicolor (Vuill.) Tiraboschi (Aver101)	The 4th block ChAES	100,000	No tropism
Aspergillus versicolor (Vuill.) Tiraboschi (Aver102)	The 4th block ChAES	100,000	No tropism
Penicillium spinulosum Thom (Pspin85)	The 4th block ChAES	30,000	No tropism
Penicillium spinulosum Thom (Pspin87)	The 4th block ChAES	100,000	No tropism
Hormoconis resinae (Lindau) v. Arx & de Vries f. resinae (Hres11)	The 4th block ChAES	1500	No tropism
Hormoconis resinae (Lindau) v. Arx & de Vries f. resinae (Hres76)	The 4th block ChAES	10,000	No tropism
Hormoconis resinae (Lindau) v. Arx & de Vries f. resinae (Hres77)	The 4th block ChAES	30,000	No tropism
Cladosporium herbarum (Pers. : Fr.) Lk (Cher80)	The 4th block ChAES	100 000	No tropism

(continued)

Table 16.2 (continued)

Species (and Designation) of Isolate	Substrate	Radioactivity of Substrate (mR h^{-1})	Tropism
Aspergillus versicolor (Vuill.) Tiraboschi (Aver222)	Control	Control	No tropism
Aspergillus versicolor (Vuill.) Tiraboschi (Aver429)	Control	Control	No tropism
Aspergillus versicolor (Vuill.) Tiraboschi (Aver432)	Control	Control	No tropism
Cladosporium sphaerospermum (Penz.) (Csph3925)	Control	Control	No tropism
Cladosporium sphaerospermum (Penz.) (Csph2538)	Control	Control	No tropism
Cladosporium cladosporioides (Fres.) de Vries (Ccl4061)	Control	Control	No tropism

ª For a description of radiotropism, see Section 16.3.3.

These microfungi have been isolated from walls, reinforced concrete framing, and wooden structures of the fourth reactor room of ChAES. Between 1997 and 2004, 155 samples in 49 inner locations of the reactor room have been isolated. The radioactivity levels of the surveyed rooms varied widely and have been assigned to four groups of radioactivity: (I) weak: 0.1–100 mR h^{-1}; (II) average: 101–500 mR h^{-1}; (III) high: 501–5,000 mR h^{-1}; and (IV) superhigh: above 5,000 mR h^{-1}. Identification of the fungal cultures revealed 58 species from 25 genera, some of which are listed in Table 16.2 in relation to the radiation dose at source of isolation. The frequency of isolation of two fungal genera (*Aspergillus* and *Cladosporium*) in relation to radiation exposure and date is given in Fig. 16.1 (Zhdanova et al. 2005a).

The highest numbers of fungal species (13–27) were obtained from a group of rooms with the lowest level of radiation (zone I) and these numbers declined in areas of higher radiation to 5–11 and 6–11 fungal species in zones II and III, respectively, excluding species that were isolated only once. Species of genera *Penicillium*, *Fusarium*, *Chrysosporium*, *Scopulariopsis*, *Hyalodendron*, *Verticillium*, and *Mucor* made up to 70–80% of all isolated species. Less frequently isolated species (10–50% frequency) included *Aspergillus versicolor*, *Aspergillus niger*, *Hormoconis resinae* (*Cladosporium chlorocephalum*), *Cladosporium sphaerospermum*, *Cladosporium herbarum*, *Cladosporium cladosporioides*, *Alternaria alternata*, *Aureobasidium pullulans*, *Penicillium aurantiogriseum*, *Penicillium spinulosum*, and *Acremonium strictum*.

Durrell and Shields (1960) noted that a high proportion (25%) of fungal species isolated form the Nevada Test Site contained melanin or other pigments in their hyphae. Similarly, the proportion of melanin-containing species (>40%) among all

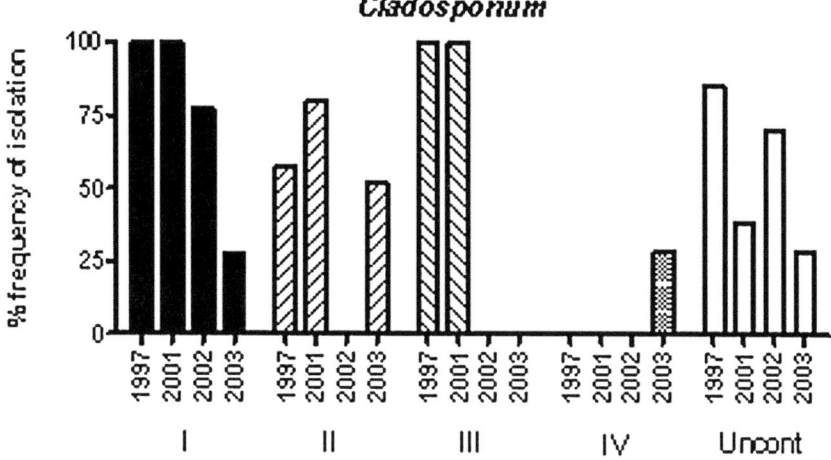
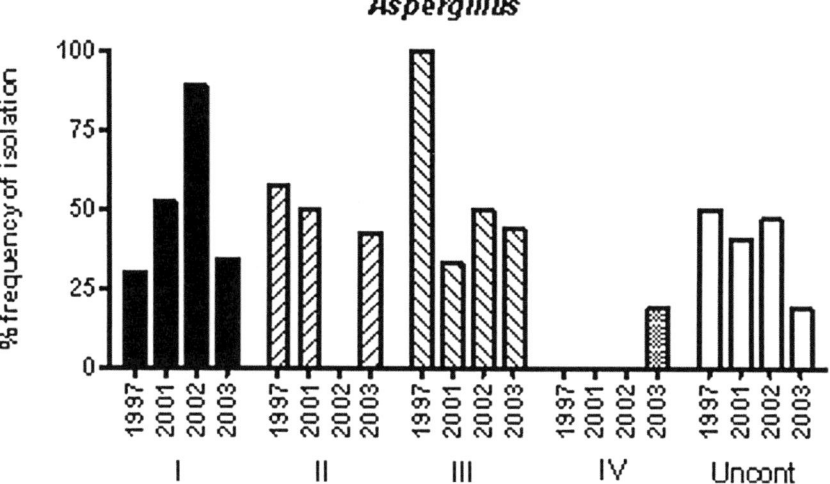

Fig. 16.1 Frequency of occurrence of *Aspergillus* and *Cladosporium* in the fourth reactor room of ChAES over time. Frequency was calculated as the percentage of number of times the species was isolated from all isolations collected between 1997 and 2003. Results from isolations in the reactor room were divided according to the four categories of radiation exposure (I = 0.1–100 mR h^{-1}; II = 101–500; III = 501–5000; and IV >5000 mR h^{-1}) and are accompanied by results from an uncontaminated site (Uncont)

fungal groups isolated from the radioactively contaminated rooms of the Chernobyl reactor room noticeably exceeded the ratio found in other, clean environments. The most frequently occurring pigmented species belonged to genera of *Cladosporium* (*C. sphaerospermum, C. herbarum*), *Hormoconis resinae, Alternaria alternata,* and *Aureobasidium pullulans*. The greatest quantity of species belonged to the genus

Penicillium (17 species), the majority of which were encountered sporadically, with only one instance of isolation from rooms of high and extremely high levels of radioactivity (*Penicillium aurantiogriseum* and *P. spinulosum*).

16.3.2 Influence of Ionizing Radiation on Fungal Communities

Monitoring the fungal communities in the region around Chernobyl during the last 18 years has resulted in the isolation and culturing of about 2,000 strains of 180 species of 92 genera (Zhdanova et al. 2000). The dynamics of change of soil microfungal communities during this time inside a 10-km radius to the zone of alienation has allowed us to reveal in more detail the processes of natural rehabilitation acting in the highly radionuclide-polluted soils of this region. Initial studies showed that soil fungal communities were altered by the intense radiation doses (Zhdanova et al. 1995) leading to simpler community structure and a dominance of melanin-containing (pigmented) fungal species at higher levels of radioactivity.

Durrell and Shields (1960) suggested that fungal pigmentation could provide some resistance to radiation in the same way as it does to UV light. Their data on enhanced spore survival under UV light due to the presence of pigmentation suggest a possible evolutionary adaptation to life in light-exposed situations. Tugay (2006) and Tugay et al. (2006b) suggest that this adaptation may also enhance fungal spores survival in the presence of ionizing radiation. Fungal survival within the Chernobyl reactor room is not entirely due to spore resistance to radiation as active mycelia also exist under high levels of ionizing radiation. Observations on diversity of fungal species revealed a significant decline in diversity in communities within the reactor room at extreme radiation levels (Group IV, above $5,000\,\text{mR}\,\text{h}^{-1}$) compared to uncontaminated areas.

Within this diverse assemblage of fungi, the frequency of isolation of radio-resistant, melanin-containing species characteristically increased to 40–60%, as compared to a 5–7% occurrence in clean soils of the Ukraine. Melanization of soil microfungi in the contaminated zone around ChAES was abundant during 1986–1988, but has subsequently declined during 1997–1998 to reach zero occurrence in 1999–2000 (Zhdanova et al. 2000).

Ecological studies have characterized fungal bioindicator traits (melanization and species composition) of high, middle, and low levels of radioactive contamination (Zhdanova et al. 1990, 1991, 1994, 1995, 2000). Zhdanova et al. (1995) showed that both *Chaetomium aureum* and *Paecilomyces lilacinus* were indicators of high levels of radionuclide contamination of soil in woodland ecosystems. On the other hand, *Acremonium strictum* and *Arthrinium phaeospermum* were bioindicators of soils at middle level of pollution, and *Myrothecium roridum* and *Metarhizium anisopliae* were bioindicators of low contaminated soils. Representatives of *Chaetomium aureum* and *Paecilomyces lilacinus* were constantly present in areas of high and average levels of pollution around the villages of Kopachi, Novo-Shepelichi, Chistogalovka, Prypyat', and Chernobyl until 2000. But their maximum occurrence was between 1990 and 1995. During subsequent years a decrease in frequency of occurrence of high pollution indicator species and an increase in frequency of bioindicators of middle and low level radioactive pollution was observed from 1987 up to 2004 as radiation levels decreased by two to four orders of magnitude. This suggests that melanin may be involved in the protection of fungal tissue against damage by radionuclide emissions as pigmented fungal species occur with greater frequency in polluted sites (Zhdanova et al. 1995). This may be a protective mechanism similar to that seen in melanin-containing lichen fungi that protect the fungal symbionts against ultraviolet light (Zhdanova et al. 1978; Gauslaa and Solhaug 2001).

Genetic changes of the fungi isolated from the fourth reactor rooms during 1997–1998 have been revealed by means of uniplex polymerase chain reaction (U-PCR) analysis of populations of two species of fungi, *Alternaria alternata* and *Cladosporium sphaerospermum*. Less intraspecific genetic variability (estimated on the basis of gel electrophoresis patterns of U-PCR products from two universal primers) was found within populations in the radioactively contaminated zones than in populations outside. This measure of reduced intraspecific genetic variability is referred to as 'radioresistance'. For *A. alternata*, a correlation was observed between the radioresistance of the isolates (deduced from the structure of their PCR patterns) and the level of radioactivity measured in their respective site of origin. A similar situation was observed with populations of *C. sphaerospermum*. These results show that in the elevated radioactive conditions of the fourth reactor rooms, surviving fungal strains have increased radioresistance and their populations show less genome diversity than those from noncontaminated sites (Burlakova et al. 2001). These findings offer opportunities for identification of specific genetic markers associated with increased radioresistance (Mironenko et al. 2000).

16.3.3 Influence of Ionizing Radiation on Fungal Growth

Ionizing radiation has a detrimental effect on the growth of many cells and organisms. It is, therefore, no surprise that the presence of ionizing radiation may be selectively inhibitory, causing changes in fungal community composition (see Section 16.3.2) and resulting in a restricted fungal flora (see Section 16.3.1). It is, however, surprising to think that ionizing radiation could have a stimulatory effect on growth of certain fungi.

Tugay et al. (2006a) investigated 13 fungal strains, isolated from sites with different level of radioactive pollution. Of these fungi, 9 (69%) showed increased germination in the presence of ionizing radiation produced by ^{137}Cs, and 6 (46%) were stimulated by ^{121}Sn. Of these, 5 strains (48%) also showed activation of conidia germination under exposure to gamma (^{121}Sn) and mixed (gamma + beta) radiation (^{137}Cs), and 5 strains (48%) showed activation upon exposure to only one of the tested sources of radiation. Of the various fungi isolated from inner locations of the damaged reactor room with a level of radioactivity of 30,000 mR h^{-1}, only *Hormoconis resinae* 77 showed inhibition of conidia germination under exposure to both sources of radiation. None of the fungi isolated from radioactively clean locations showed radiostimulation of conidia germination under the influence of radiation. On the other hand, 92% of fungal strains isolated from radioactively contaminated sites showed enhanced conidia germination under exposure to at least one type of ionizing radiation. Further examples of spore germination responses to irradiation are shown in Table 16.3.

Based on preliminary observations of fungi growing over and decomposing radioactive hot particles (Zhdanova 1991), there appeared to be preferential growth of some fungal species towards the particle. It was suggested that this oriented growth was the result of an attraction of the fungi towards the ionizing radiation (or radiotropism) rather than towards the food base (Vember et al. 1999; Zhdanova et al. 1994; Tugay 2006; Tugay et al. 2003, 2004, 2005). This phenomenon of growth of fungal hyphae towards a source of ionizing radiation was further studied by Zhdanova et al. (2004). In these experiments, sources of radioactive emissions at approximately 2×10^8 Bq, namely ^{32}P (beta emissions) and ^{109}Cd (gamma emissions) were contained in a cuboid lead irradiation box, to provide collimated beams of ionizing radiation through 1-mm diameter holes drilled through the 2-cm lead walls in each of the four sides and through the lid.

Fungal cultures were exposed to these collimated beams of radioactivity as conidiospores and the direction of growth of germinating hyphae, in relation to the radiation source, was measured. Of the 27 fungal/radiation interactions, 18 (66.7%) showed positive stimulation of growth towards the radiation source (low mean return angle) and 8 showed no response. Figure 16.2 provides examples of oriented growth. With respect to particular radiation sources, 69% of fungi exposed to ^{109}Cd and 64% of fungi exposed to ^{32}P showed preferred growth towards the source of ionizing radiation. With respect to the isolation site of the tested fungi, 86% of the interactions involving fungal species isolated from radioactively contaminated sites showed positive directional growth to the radiation beam, whereas of the limited number of fungi isolated from less contaminated areas (e.g., *Cladosporium sphaerospermum* 3176, *Penicillium roseopurpureum* 100, and *Paecilomyces lilacinus* 101) only one instance of directed growth was observed.

In this study, directional growth occurred in the absence of any chemical influence of the source of radioactivity, because the source is physically separated from the fungal culture, and corresponds solely to response to clean beta and gamma emissions. Table 16.2 indicates the source of fungal isolates in relation to radiation levels at the source of isolation and their directional growth response. This response shows a hormetic character (Grodzinsky, 1989; Calabrese and Baldwin, 2000; Pelevina et al. 2003; Petin et al. 2003; Grodzinsky et al. 2005), inasmuch as directional growth is maximal at moderate irradiation levels and is lowest at no or

Table 16.3 Percent spore germination in a range of fungal species in the absence (control) and presence of ionizing radiation from two sources in relation to the radiation level at source of isolation of the fungal strain

Species (and Designation) of Isolate	Radioactivity of Substrate (mR h^{-1})	Percent Spore Germination with Ionizing Radiation Treatment		
		Control	^{121}Sn – γ	^{137}Cs – β+γ
Group I (0.1–100 mR h^{-1})				
Penicillium steckii (Pst2)	20	88.24	88.24	78.72
Penicillium hirsutum (Phir1)	20	10.32	41.89	1.48
Cladosporium cladosporioides (Ccl4)	30	26.67	50.53	30.5
Penicillium lanosum (Plan10)	20	5.75	9.38	6.25
Aspergillus versicolor (Aver43)	55	39.52	48.21	54.25
Aspergillus versicolor (Aver54)	98	2.17	1.83	5.61
Cladosporium sphaerospermum (Csph35)	35	22.75	33.04	26.86
Cladosporium sphaerospermum (Csph70)	40	47.56	62.74	75.0
Hormoconis resinae (Hres 30)	40	33.37	67.49	15.16
Group II (101–500 mR h^{-1})				
Aspergillus versicolor (Aver55)	250	8.06	2.73	9.44
Aspergillus versicolor (Aver57)	500	15.27	14.15	17.22
Hormoconis resinae (Hres52)	120	6.38	9.59	7.25
Hormoconis resinae (Hres61)	200	24.07	51.55	33.33
Hormoconis resinae (Hres21)	300	50.67	51.52	82.14
Penicillium aurantiogriseum (Paur 60)	380	32.0	57.38	25.74
Group III (501–5000 mR h^{-1}) and Group IV (above 5,000 mR h^{-1})				
Aspergillus versicolor (Aver101)	100,000	42.78	35.44	45.34
Aspergillus versicolor (Aver102)	100,000	9.54	5.42	6.42
Penicillium spinulosum (Pspin85)	30,000	32.11	40.0	33.8
Penicillium spinulosum (Pspin87)	100,000	38.89	13.11	48.50
Hormoconis resinae (Hres11)	1,500	12.7	9.17	10.18
Hormoconis resinae (Hres76)	10,000	86.7	82.2	89.3
Hormoconis resinae (Hres77)	30,000	70.3	51.62	51.03
Cladosporium herbarum (Cher80)	100,000	75.91	73.87	79.4
	Control			
Aspergillus versicolor (Aver222)	Control	12.17	7.69	3.23
Aspergillus versicolor (Aver429)	Control	38	24.11	20.18
Aspergillus versicolor (Aver432)	Control	9,33	8.06	3.61
Cladosporium sphaerospermum (Csph3925)	Control	64.54	33.48	13.08
Cladosporium sphaerospermum (Csph2538)	Control	43.4	12.7	8.6
Cladosporium sphaerospermum (Ccl4061)	Control	28.8	24.2	8

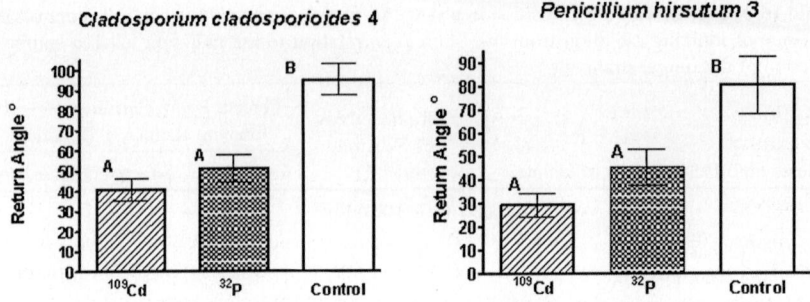

Fig. 16.2 Response of two fungal isolates to collimated beams of ionizing radiation showing directional growth of hyphae (as evidenced by low mean return angle) compared to nonirradiated control. (From Zhdanova et al. (2004.)

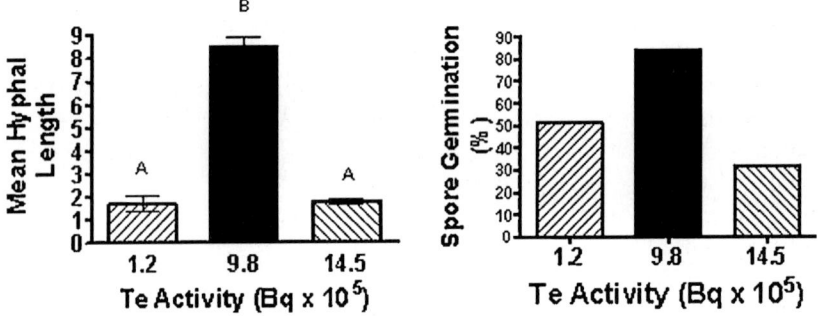

Fig. 16.3 Response of *Cladosporium cladosporioides* to mixed radiation at three levels of activity (left: hyphal growth, right: spore germination) showing hormesis, with maximal radio-stimulation at intermediate levels of activity

high levels at the site of isolation. Hormetic response to radionuclide level is also seen in the growth response and spore germination of *Cladosporium cladosporioides* (Fig. 16.3).

The nature of the ionizing radiation receptor system within the fungal cell remains unknown. However, it is possible that melanin or other natural quinine pigments in the fungal cell wall could act as a receptor. Melanin has been shown to protect lichen and epiphytic fungi from UV radiation (Gauslaa and Solhaug 2001), and has the capacity to change biochemical pathways in cells (Huselton and Hill 1990) and fungal spores (Durrell and Shields 1960) exposed to UV light. Thus, it is possible that melanin or quinines could have a protective effect against ionizing radiation and, by regulating changes in biochemical pathways, could act as a sensor triggering altered growth characteristics of fungal hyphae. This may explain why there is a greater proportion of melanized fungal species in soils subjected to long-term low levels of ionizing radiation (Zhdanova et al. 1995).

These studies suggest that the presence of ionizing radiation and long-term effects of exposure to radiation can alter the growth response of fungi. The ability of fungi to withstand high levels of radiation makes them candidates for regulating radionuclide movement in highly contaminated areas. Saprotrophic microfungi isolated from contaminated areas seem be attracted towards sources of ionizing radiation and to exhibit growth stimulation in response to these sources, which they are able to decompose. In a forest soil context, it is likely that these organisms play an important role in both mineralizing inorganically or organically bound radionuclides, allowing these to be available for immobilization by plants, leaching down the soil profile, or to be immobilized by fungi themselves and be retained in the upper soil horizons.

16.4 Conclusions and Perspectives

The various observations reported earlier in this review suggest that fungi are very resistant to radionuclides. It is possible that the presence of melanin pigment in the hyphae of mitosporic fungal species may provide some protection against ionizing radiation in the same way that it has been shown to protect against ultraviolet light. Resistance may also be related to absorption and retention of radionuclides within fungal hyphae (Bohac et al. 1989; Rafferty et al. 1997; Mahmoud 2004). Due to the long-lived and extensive hyphal network, fungi appear to be very efficient in absorbing radionuclides from the environment. The abundance of fungal biomass in upper soil horizons of forest ecosystems has been suggested as an important determinant of long-term accumulation of radionuclides (Avery 1996; Steinera et al. 2002; Avila and Moberg 1999). Indeed, biomass figures of $2\,t\,ha^{-1}$ of hyphae have been reported by Zvyagintsev (1987, 1999), such biomass forming a diffuse but extensive surface of adsorption and absorption.

This abundant fungal mycelium in soil has been proposed by Drissner et al. (1998) as a factor explaining why models of radionuclide adsorption/desorption rates by plants do not conform to observed patterns. The discrepancy between modeled and actual plant absorption rates would be attributable to mycorrhizal fungi acting as significant enhancers of soil-to-plant transfer rates. Internal translocation of radionuclides between sources and physiological sinks occurs in the same way as that of essential nutrients and accounts for the long-term retention of radionuclides within the fungal biomass. Adsorption of radionuclides onto ion exchange sites of fungal hyphal walls has been reported in the literature and this property of hyphal walls has been used in industrial effluent cleanup.

Many aspects of the interaction between fungi and radionuclides need further investigation. The intriguing concept of behavioral adaptations of fungi to evolve radiotropism needs further study to identify the triggers and the physiological mechanisms of the response. In terms of ecosystem processes, the potential for immobilization of a range of radionuclides into fungal tissue, the rates of translocation into harvestable fruitbodies, as well as the usefulness of fungi for environmental remediation need to be evaluated. Indeed, Entry et al. (1993)

compared the uptake of ^{137}Cs and ^{90}Sr by Ponderosa and Monterey pine seedlings as both are fast growing and potential candidate tree species for site remediation. They showed that Ponderosa and Monterey pines, respectively, accumulated 6.3 and 8.3% of the radiocesium and 1.5 and 4.5% of the radiostrontium present in an artificial growth medium during 4 weeks of growth. The possibility of enhancing this uptake by the addition of ectomycorrhizal symbionts was suggested in a subsequent study (Entry et al. 1994), where 3–5 times more ^{90}Sr was taken up in the ectomycorrhizal seedlings than in nonmycorrhizal controls. In practice, the combination of mycorrhizal-enhanced uptake of radionuclides by trees together with the harvesting of radionuclide-rich fruitbodies could prove an effective soil remediation technique.

More recently, models of radionuclide accumulation into fungal fruitbodies have been related to the depth distribution of the mycelia of different fungal species and their time-dependent activity of radionuclide absorption (Rühm et al. 1998), with models predicting an ecological half-life of 3–8 years for radionuclides. However, an experimental evaluation of a similar model adapted from Avila et al. (1999) suggests that this estimated half-life is a gross underestimation of fungal retention times of radionuclides in soil (Shaw et al. 2005).

Thus, a vast body of knowledge is now available on the accumulation and distribution of radionuclides in fungal components of the soil environment. Much of this information is related to mycorrhizal fungi, due to their importance in transfers to host plant material and, consequentially upwards through food chains (Barnett et al. 2001). However, more information and validation of existing models of fungal retention are needed. A detailed investigation of internal translocation processes and of soil to plant transfer rates is also required, to design procedures for fungal use in phyto- or mycoremediation (Roséna et al. 2005). Research along those lines should yield to significant practical advances, because currently the use of fungi in phytoremediation is often disregarded (Raskin and Enlsey 2000).

References

Aiking H, Tempest DW (1977) Rubidium as a probe for function and transport of potassium in the yeast *Candida utilis* NCYC-321 grown in chemostat culture. *Arch Microbiol* 115: 215–221

Avery SV (1996) Fate of caesium in the environment: Distribution between the abiotic and biotic components of aquatic and terrestrial ecosystems *J Environ Radioactiv* 30:139–171

Avila R, Johanson KJ, Bergström R (1999) Model of the seasonal variations of fungi ingestion and 137Cs activity concentrations in roe deer. *J Environ Radioactiv* 46:99–112

Avila R, Moberg L (1999) A systematic approach to the migration of ^{137}Cs in forest ecosystems using interaction matrices. *J Environ Radioactiv* 45:217–282

Baeza A, Guillén FJ, Hernández S (2002) Transfer of ^{134}Cs and ^{85}Sr to *Pleurotus eryngii* fruiting bodies under laboratory conditions: A compartmental model approach. *Bull Environ Contam Toxicol* 69:817–828

Barnett CL, Beresford NA, Frankland JC, Self PL, Howard BJ, Marriott JVR (2001) Radiocaesium intake in Great Britain as a consequence of the consumption of wild fungi. *Mycologist* 15:98–104

Berreck M, Hasselwandter K (2001) Effect of the arbuscular mycorrhizal symbiosis upon uptake of cesium and other cations by plants. *Mycorrhiza* 10:275–280

Bohac JD, Krivolutskii A, Antonova TB (1989) The role of fungi in the biogenous migration of elements and in the accumulation of radionuclides. *Agric Ecosyst Environ* 28:31–34

Burlakova EB, Michailov VF, Mazurik VK (2001) System of an oxidation-reduction homeostasis at the instability genome induced by radiation. *Rad Biol Radioecol* 41:489–499

Byrne AR (1988) Radioactivity in fungi in Slovenia, Yugoslavia, following the Chernobyl accident. *J Environ Radioactiv* 6:177–183

Calabrese EJ, Baldwin LA (2000). Radiation hormesis: Its historical foundations as a biological hypothesis. *Hum Exper Toxicol* 19:41–75

Cawse PA (1983) The accumulation of caesium-137 and plutonium-239 + 240 in soils of Great Britain, and transfer to vegetation. In: Coughtree PJ, Bell JNB, Roberts TM (eds) *Ecological Aspects of Radionuclide Release*. Blackwell Scientific, Oxford, pp 47–62

Clint GM, Dighton J (1992) Uptake and accumulation of radiocaesium by mycorrhizal and non-mycorrhizal heather plants. *New Phytol* 122:555–561

Clint GM, Dighton J, Rees S (1991) Influx of ^{137}Cs into hyphae of basidiomycete fungi. *Mycolog Res* 95:1047–1051

Connolly JH, Shortle WC, Jellison J (1998) Translocation and incorporation of strontium carbonate derived strontium into calcium oxalate crystals by the wood decay fungus *Resinicium bicolor. Can J Bot* 77:179–187

Coughtree PJ (ed) (1983) *Ecological Aspects of Radionuclide Release*. Blackwell Scientific, Oxford

Das J (1991) Influence of potassium in the agar medium on the growth pattern of the filamentous fungus *Fusarium solani. Appl Environ Microbiol* 57:3033

de Boulois HD, Delvaux B, Declerck S (2005) Effects of arbuscular mycorrhizal fungi on the root uptake and translocation of radiocaesium. *Environ Poll* 134:515–524

Denny HJ, Wilkins DA (1987a) Zinc tolerance in *Betula* spp. I. Effects of external concentration of zinc on growth and uptake. *New Phytol* 106:517–524

Denny HJ, Wilkins DA (1987b) Zinc tolerance in *Betula* spp. IV. The mechanism of ectomycorrhizal amelioration of zinc toxicity. *New Phytol* 106:545–553

Dighton J (2003) *Fungi in Ecosystem Processes*. Marcel Dekker, New York

Dighton J, Clint GM, Poskitt JM (1991) Uptake and accumulation of ^{137}Cs by upland grassland soil fungi: A potential pool of Cs immobilization. *Mycol Res* 95:1052–1056

Dighton J, Horrill AD (1988) Radiocaesium accumulation in the mycorrhizal fungi *Lactarius rufus* and *Inocybe longicystis*, in upland Britain. *Trans Brit Mycol Soc* 91:335–337

Dighton J, Terry GM (1996) Uptake and immobilization of caesium in UK grassland and forest soils by fungi following the Chernobyl accident. In: Frankland JC, Magan N and Gadd GM (eds) *Fungi and Environmental Change*. Cambridge University Press, Cambridge, pp 184–200

Drissner J, Bürmann W, Enslin F, Heider R, Klemt E, Miller R, Schick G, Zibold G (1998) Availability of caesium radionuclides to plants – Classification of soils and role of mycorrhiza *J Environ Radioactiv* 41:19–32

Durrell LW, Shields LA (1960) Fungi isolated in culture from soils of the Nevada test site. *Mycologia* 52:636–641

Eckl P, Hoffman W, Turk R (1986) Uptake of natural and man-made radionuclides by lichens and mushrooms. *Radiat Environ Biophys* 25:43–54

Elstner EF, Fink R, Holl W, Lengfelder E, Ziegler H (1987) Natural and Chernobyl-caused radioactivity in mushrooms, mosses and soil-samples of defined biotops in SW Bavaria. *Oecologia* 73: 553–558

Entry JA, Rygiewicz PT, Emmingham WH (1993) Accumulation of cesium-137 and strontium-90 in Ponderosa pine and Monterey pine seedlings. *J Environ Qual* 22:742–746

Entry JA, Rygiewicz PT, Emmingham WH (1994) ^{90}Sr uptake by *Pinus ponderosa* and *Pinus radiata* seedlings inoculated with ectomycorrhizal fungi. *Environ Poll* 86:201–206

Fogel R, Hunt G (1983) Contribution of mycorrhiza and soil fungi to nutrient cycling in a Douglas-fir ecosystem. *Can J For Res* 13:219–232

Gauslaa Y, Solhaug KA (2001) Fungal melanins as a sun screen for symbiotic green algae in the lichen *Lobaria pulmonaria. Oecologia* 126:462–471

Giovani C, Nimis PL, Land P, Padovani R (1990) Investigation of the performance of macromycetes as bioindicators of radioactive contamination. In: Desmet G, Nassimbeni P and Belli M (eds) *Transfer of Radionuclides in Natural and Semi-Natural Environments*. Elsevier Applied Science, London, pp 485–491

Gray SN (1998) Fungi as potential bioremediation agents in soil contaminated with heavy or radioactive metals. *Biochem Soc Trans* 26: 666–670

Gray SN, Dighton J, Jennings DH (1996) The physiology of basidiomycete linear organs III. Uptake and translocation of radiocaesium within differentiated mycelia of *Armillaria* spp. growing in microcosms and in the field. *New Phytol* 132:471–482

Gray SN, Dighton J, Olsson S, Jennings DH (1995) Real-time measurement of uptake and translocation of ^{137}Cs within mycelium of *Schizophyllum commune* Fr. by autoradiography followed by quantitative image analysis. *New Phytol* 129:449–465

Grodzinsky DM (1989) *Radiobiology of Plants*. Nauk. Dumka Press, Kiev

Grodzinsky DM, Shelina YuV, Meheev OM, Guscha NI (2005) Radiation hormesis retrospective review and contemporaneity. *Prob Nucl Power Plant Safe Chernobyl* 3:17–28

Guillitte O, Fraiture A, Lambinon J (1990) Soil-fungi radiocaesium transfers in forest ecosystems. In: Desmet G, Nassimbeni P and Belli M (eds) *Transfer of Radionuclides in Natural and Semi-Natural Environments*. Elsevier Applied Science, London

Guillitte O, Gasia MC, Lambinon J, Fraiture A, Colard J, Kirchmann R (1987) La radiocontamination des champignons sauvages en Belgique et au Grand-Duché de Luxembourg après l'accident nucléaire de Tchernobyl. *Mem Soc Roy Bot Belg* 9: 79–93

Guillitte O, Melin J, Wallberg L (1994) Biological pathways of radionuclides originating from the Chernoyl fallout in a boreal forest ecosystem. *Sci Total Environ* 157:207–215

Haselwandter K (1978) Accumulation of the radioactive nuclide ^{137}Cs in fruitbodies of basidiomycetes. *Health Phys* 34:713–715

Haselwandter K, Bereck M, Brunner P (1988) Fungi as bioindicators of radiocaesium contamination. Pre- and post Chernobyl activities. *Trans Br Mycol Soc* 90:171–176

Haselwandter K, Berreck M (1994) Accumulation of radionuclides in fungi. In: Winkelmann G and Winge DR (eds) *Metal Ions in Fungi*. Marcel Dekker, New York, pp 259–277

Heal OW, Horrill AD (1983) Terrestrial ecosystems: An ecological context for radionuclide research. In: Coughtree, PJ (ed.) *Ecological Aspects of Radionuclide Release*. Blackwell Scientific, Oxford, pp 31–46

Huselton CA, Hill HZ (1990) Melanin photosensitizes ultraviolet light (UVC) DNA damage in pigmented cells. *Environ Mol Mutagen* 16: 37–43

Jackson NE, Miller RH, Franklin RE (1973) The influence of vesicular-arbuscular mycorrhizae on uptake of ^{90}Sr from soil by soybeans. *Soil Biol Biochem* **5**:205–212

Jennings DH (1990) The ability of basidiomycete mycelium to move nutrients through the soil ecosystem. In: Harrison AF, Ineson P and Heal OW (eds) *Nutrient Cycling in Terrestrial Ecosystems: Field Methods, Applications and Interpretation*. Elsevier, Amsterdam, pp 233–245

Joner EJ, Roos P, Jansa J, Frossard E, Leyval C, Jakobsen I (2004) No significant contribution of arbuscular mycorrhizal fungi to transfer of radiocesium from soil to plants. *Appl Environ. Microbiol* 70:6512–6517

Kirchner G, Dalliant O (1998) Accumulation of ^{210}Pb, ^{226}Ra and radioactive cesium by fungi. *Sci Total Environ* 222: 63–70.

Mahmoud YA-G (2004) Uptake of radionuclides by some fungi. *Mycobiology* 32:110–114

Malinowska E, Szefer P, Bojanowski R (2006) Radionuclide content in *Xercomus badius* and other commercial mushrooms from several regions of Poland. *Food Chem* 97: 19–24

Mietelski JW, Jasinska M, Kubica B, Kozak K, Macharski P (1994) Radioactive contamination of Polish mushrooms. *Sci Total Environ* 157:217–226

Mironenko NV, Alekhina IA, Zhdanova NN, Bulat SA (2000) Intraspecific variation in gamma-radiation resistance and genomic structure in the filamentous fungus *Alternaria alternata*: A case study of strains inhabiting Chernobyl Reactor No. 4. *Ecotox Environ Safe* 45: 177–187

Muramatsu Y, Yoshida S, Sumia M (1991) Concentrations of radiocesium and potassium in basidiomycetes collected in Japan. *Sci Total Environ* 105:29–39

Olsen RA, Joner E, Bakken LR (1990) Soil fungi and the fate of radiocaesium in the soil ecosystem - a discussion of possible mechanisms involved in the radiocaesium accumulation in fungi, and the role of fungi as a Cs-sink in the soil. In: Desmet G, Nassimbeni P and Belli M (eds) *Transfer of Radionuclides in Natural and Semi-Natural Environments*. Elsevier Applied Science, London, pp 657–663

Olsson S (1995) Mycelial density profiles of fungi on heterogenous media and their interpretation in terms of nutrient reallocation patterns. *Mycol Res* 99:143–153

Olsson S, Jennings DH (1991) Evidence for diffusion being the mechanism of translocation in the hyphae of three moulds. *Exper Mycol* 15:302–309

Oolbekkink GT, Kuyper TW (1989) Radioactive caesium from Chernobyl in fungi. *Mycologist* 3:3–6

Osburn WS (1967) Ecological concentration of nuclear fallout in a Colorado mountain watershed. In: Aberg and Hungate (eds.) *Radiological Concentration Process*. Pergamon Press, New York, pp 675–709

Oughton DH (1989) The environmental chemistry of radiocaesium and other nuclides. PhD Thesis, University of Manchester, UK

Pelevina II, Aleschnko AV, Antoschina MM, Gotlib VJ, Kudriashova OV, Semenova LP, Serebryanyi AM (2003) The reaction of cell population to low level of irradiation. *Rad. Biol Radioecol* 43:161–166

Petin VG, Morozov II, Kabakova NM, Gorshkova TA (2003) Some effects of radiation hormesis for bacterial and yeast cells. *Rad Biol Radioecol* 43:176–178

Rafferty B, Dawson D, Kliashtorn A (1997) Decomposition in two pine forests: The mobilization of ^{137}Cs and K from forest litter. *Soil Biol Biochem* 29:1673–1681

Raskin I and Ensley BD (eds) (2000) *Phytoremediation of Toxic Metals: Using Plants to Clean Up the Environment*. Wiley, New York

Rayner ADM (1991) The challenge of the individualistic mycelium. *Mycologia* 83:48–71

Ritz K (1995) Growth responses of some fungi to spatially heterogeneous nutrients. *FEMS Microbiol Ecol* 16:269–280

Rommelt R, Hiersche L, Schaller G, Wirth E (1990) Influence of soil fungi (Basidiomycetes) on the migration of $Cs^{134+137}$ and SR^{90} in coniferous forest soils. In: Desmet G, Nassimbeni P and Belli M (eds) *Transfer of Radionuclides in Natural and Semi-Natural Environments*. Elsevier Applied Science, London, pp 152–160

Roséna K, Zhong Weiliang Z, Mårtensson A (2005) Arbuscular mycorrhizal fungi mediated uptake of ^{137}Cs in leek and ryegrass. *Sci Total Environ* 338:283–290

Rühm W, Steiner M, Kammerer L, Hiersche L, Wirth E (1998) Estimating future radiocaesium contamination of fungi on the basis of behavioural patterns derived from past instances of contamination. *J Environ Radioactiv* 39:129–147

Sanchez AL, Parekh NR, Dodd BA, Ineson P (2000) Microbial component of radiocaesium retention in highly organic soils. *Soil Biol Biochem* 32:2091–2094

Shaw G, Venter A, Avila R, Bergman R, Bulgakov A, Calmon P, Fesenko S, Frissel M, Goor F, Konoplev A, Linkov I, Mamikhin S, Moberg L, Orlov A, Rantavaara A, Spiridonov S, Thiry Y (2005) Radionuclide migration in forest ecosystems - Results of a model validation study *J Environ Radioactiv* 84:285–296

Singleton I, Tobin JM (1996) Fungal interactions with metals and radionuclides for environmental bioremediation. In: Frankland JC, Magan N and Gadd GM (eds.) *Fungi and Environmental Change*. Cambridge University Press, Cambridge, pp 282–298

Skladany GJ, Metting F (1992) Bioremediation of contaminated soil. *Soil Microb Ecol* 438–513

Smith SE, Read DJ (1997) *Mycorrhizal Symbiosis*. Academic Press, San Diego

Steinera M, Linkovb M, Yoshida S (2002) The role of fungi in the transfer and cycling of radionuclides in forest ecosystems *J Environ Radioactiv* 58: 217–241

Strandberg M, Johansson M (1998) ^{134}Cs in heather seed plants grown with and without mycorrhiza. *J Environ Radioactiv* 40:175–184

Tobin JM, Cooper DG, Neufeld RJ (1984) Uptake of metal ions by *Rhizopus arrhizus* biomass. *Appl Environ Microbiol* 47:821–824.

Tugay T, Zhdanova N, Zheltonozhsky V, Sadovnikov L (2006b) Influence of small dozes radiation on antioxidant activity of anamorphic fungal species *Aspergillus versicolor* and *Paecilomyces lilacinus*, having radio adaptive properties. Abstract of the *35th Annual Meeting of the European Radiation Research Society* 22–25 August, Kiev 2006, p 84

Tugay TI (2006) Features of evince of adaptable reactions at micromycetes, isolated from radioactively polluted territories. Abstract of *XII meeting of Ukrainian Botany Society*, 15–18 May, Odessa, p 26

Tugay TI, Zhdanova NN, Retchits TI, Zheltonozhsky VA, Sadovnikov LV (2003) Influence of low level ionizing irradiation on spread of radiotropism among fungi. *Sci Pap Inst Nuc Res* 2:72–79

Tugay TI, Zhdanova NN, Zheltonozhsky VA, Sadovnikov LV, Dighton J (2006a) The influence of ionizing radiation on spore germination and emergent hyphal growth response reactions of microfungi. *Mycologia* 98:521–527

Tugay TI, Zhdanova NN, Zheltonozhsky VA, Sadovnikov LV, Telichko MV (2005) Response reactions of the fungi, isolated from inner locations of "Ukryttya", which have different levels of radioactivity. *Sci Pap Inst Nuc Res* 1:128–136

Tugay TI, Zheltonozhsky VA, Sadovnikov LV (2004) Response reactions of fungi under exposure of ionizing irradiation. *Sci Pap Inst Nuc Res* 2:132–138

Vember VV, Zhdanova NN, Tugay TI (1999) Irradiation influence on the physiologo-biochemical properties of *Cladosporium cladosporioides* (Fres.) de Vries strains which differ in radiotropism sign. *Microbiologichny Zhurnal* 61:25–32

Vogt KA, Grier CC, Edmonds RL, Meier CE (1982) Mycorrhizal role in net primary production and nutrient cycling in *Abies amabilis* (Dougl.) Forbes ecosystems in western Washington. *Ecology* 63:370–380

Watling R, Laessoe T, Whalley AJS, Lepp NW (1993) Radioactive caesium in British mushrooms. *Bot J Scotl* 46:487–497

White C, Gadd GM (1990) Biosorption of radionuclides by fungal biomass. *J Chem Tech Biotech* 49:331–343

Witkamp M (1968) Accumulation of ^{137}Cs by *Trichoderma viride* relative to ^{137}Cs in soil organic matter and soil solution. *Soil Sci* 106:309–311

Witkamp M, Barzansky B (1968) Microbial immobilization of ^{137}Cs in forest litter. *Oikos* 19:392–395

Yoshida S, Muramatsu Y (1994) Accumulation of radiocesium in basidiomycetes collected from Japanese forests. *Sci Total Environ* 157:197–205

Zhdanova NN, Lashko TN, Redchitz TI, Vasiliveskaya AI, Bosisyuk LG, Sinyavskaya OI, Gavrilyuk VI, Muzalev PN (1991) Interaction of soil micromycetes with 'hot' particles in a model system. *Microbiologichny Zhurnal* 53:9–17

Zhdanova NN, Melezhik AV, Vasilevskaya AI, Pokhodenko VD (1978) Formation and disappearance of photo induced paramagnetic centers in melanin-containing fungi. *Herald Acad Sci USSR* 4:576–581

Zhdanova NN, Redchitz TI, Krendayaskova VG, Lacshko TN, Gavriluk VI, Muzalev PI, Sherbachenko AM (1994) Tropism under the influence of ionizing radiation. *Mikologia i Fitopatologiya* 28:8–13

Zhdanova NN, Tugay T, Dighton J, Zheltonozhsky V, McDermott P (2004) Ionizing radiation attracts fungi. *Mycol Res* 108:1089–1096

Zhdanova NN, Vasilevskaya AI, Artyshkova LV, Sadovnikov YuS, Gavrilyuk VI, Dighton J (1995) Changes in the micromycete communities in soil in response to pollution by long-lived radionuclides emitted in the Chernobyl accident. *Mycol Res* 98:789–795

Zhdanova NN, Vasilevskaya AA, Sadnovikov YuS, Artyshkova LA (1990) The dynamics of micromycete complexes contaminated with soil radionuclides. *Mikologia i Fitopatologiya* 24:504–512

Zhdanova NN, Zakharchenko VA, Haselwandter K (2005b) Radionuclides and fungal communities. In: Dighton J, White JF and Oudemans P (eds) *The Fungal Community: Its Organization and Role in the Ecosystem.* CRC Press, Baton Rouge, pp 759–768

Zhdanova NN, Zakharchenko VA, Tugay NI, Karpenko YV (2005a) Fungi lesion of inner locations object shelter. *Prob Nuc Power Plants' Safe Chernobyl*, 3:78–86

Zhdanova NN, Zakharchenko VA, Vember VV, Nakonechnaya LT (2000) Fungi from Chernobyl: Mycobiota of the inner regions of the containment structures of the damaged nuclear reactor. *Mycol Res* 104:1421–1426

Zvyagintsev DG (1987) *Soil and Microorganisms*. Moscow State University Press, Moscow

Zvyagintsev DG (1999) *Structure and Functioning of a Complex Soil Microorganisms/ Structurally Functional Role of Soil in Biosphere*. GEOS, Moscow, pp101–112

Index

A
Abiotic process
   for degradation of soil organic matter, 27, 122
   for oxygen consumption, 220
ACC (1-aminocyclopropane-1-carboxylic acid) deaminase, 320
Acetobacterium, 139
   tundrae, 145
Acetogens, 139
Acidification of drainage effluents, 312
Acidiphilium, 207
Acidithiobacillus, 17, 216
   ferrooxidans, 22, 206
Acidity
   in peatlands, 178
   tolerance, 208
Acidobacteria, 24, 252, 323
Acidobacteriales, 124
Acidobacterium, 258
Acidophiles, 22
   archaea, 22
Acidothermus, 311
Acidovorax wohlfahrtii, 198
Acinetobacter, 88, 101, 139, 258, 260, 278, 282, 284
Acremonium strictum, 342
Acrobeloides, 74
Actinobacter, 252
Actinobacteria, 17, 123, 141, 251, 311, 322
   actinomycetes, 19, 97, 275, 310
   nocardioforms, 278, 279, 282, 284
Actinomyces, 89
Actinophyris, 209
Actinopolyspora, 98
Adaptation
   homeophasic, 168
   homeoviscous, 168
   to environmental stress, 25
   to extreme conditions, 16
   to extreme pH, 22
   to extreme soils, 16, 57
   to heavy metals, 321
   to high temperatures, 21, 168
Adaptive landscape, 58–63
   composite topography, 63
   fitness peaks, 58–63
Afipia, 123
Agrobacterium, 251, 258
Alcaligenes, 88
Aleutian Islands, 194
Algae, 119, 145, 252
Alicante, Spain 88
Alicyclobacillus, 157, 158
Alkalibacillus
   haloalkaliphilus, 96
   salilacus, 96
Alkaliphiles, 22
   Bacillus, 95
   oil-utilizing, 283
Alkanes, 120, 254, 259, 277
Alps, 247
Alternaria alternata, 335, 342, 343, 345
Alyssum bertolonii, 322
Amanita fulva, 186
Amino acid transport, Na-dependent, 170
Ammonium chloride, 306
*amo*A gene, 311
Anabaena, 228
Andes, 119
Aneurinibacillus, 157, 158
   terranovensis, 157, 159, 166
Anoxybacillus, 157, 158
Antarctica, 89, 156, 159, 174, 237, 248
   Dry Valleys, 29, 31, 73, 122, 250
   Ellsworth Land, 75

Antarctica (*Cont.*)
  Jubany Station, 259, 261
  Marambio Station, 262
  Mount Erebus, 160
  Mount Melbourne, 160, 161
  Mount Rittmann, 160, 161
  Ross Desert, 29, 251
  Scott Base, 257
Antarctic Treaty, 249
Anti-freeze compounds, 20
Antofagasta, Chile, 119
Aphelenchus avenae, 74
Apoplast, 318
Arabian Gulf, 275, 276, 279
Aral Sea, 11
Archaea, 11, 67, 133, 141, 184, 251
Arctic, 133, 247
Arhodomonas aquaeoli, 105
Arid soils, 6
Armillaria, 336
Aromatic compounds *See* Hydrocarbons, polycyclic aromatic
  degradation by halophiles, 91, 104
Arsenic, 317
Arsenite resistance operon, 326
Arthrinium phaeospermum, 344
Arthrobacter, 89, 123, 139, 141, 250, 278, 279, 280, 290
Ascomycota, 145, 337
Aspergillus, 278, 279, 343
  niger, 335, 342
  pulverulens, 335
  versicolor, 341, 342, 347
Asticcacaulis, 123
Astrobiology, 134, 205, 220
Atacama Desert, Chile, 29, 75, 117, 227
  Yungay area, 120
Atmosphere, 228
Aureobasidium pullulans, 342, 343
Automated ribosomal intergenic spacer region analysis (ARISA), 35
Autotroph, bacterial, 310
Azospirillum, 197, 278

B

Bacillus, 30, 89, 102, 123, 139, 155, 156, 158, 185, 226, 251, 258, 278, 279, 283, 284, 322
  firmus, 17
  fumarioli, 156, 158, 165
  krulwichiae, 95
  licheniformis, 165, 169
  luciferensis, 168
  megaterium, 165, 169
  oshimensis, 95
  patagoniensis, 95
  pumilus, 123
  schlegelii, 157, 170
  shackletonii, 168
  subtilis, 50, 53, 127, 160
  thermantarcticus, 157, 164, 169
  tusciae, 157, 170
Bacteria, 133
  hydrocarbon-degrading, 248, 256, 259
  hydrocarbon-degrading in pristine soils, 257
  in Antarctic soils, 249
  in desert soils, 275
  in hydrocarbon-contaminated polar soils, 256
  moderately halophilic, 92
  susceptibility to soil fires, 310
Bacteroides, 90
Bacteroidetes, 141, 197, 252, 258, 320, 322
Basalt, 225, 226, 237
Basidiomycota, 145, 337
Beijerinckia indica, 197
Biodiversity, 71, 182
Biofilms *see* microbial mats
  genetic diversification, 53
  in desert areas, 124
  in the Río Tinto ecosystem, 209
Biogeochemical cycle, 146, 302
Biogeography, 10, 78, 79
Bioindicator, of radioactive contamination, 344
Bioleaching, 209, 312
Biological interactions, 60
Biological soil crust, 30
Biomarkers, 218
Biomass
  fungal, 349
  microbial, 308
Bioremediation, 248
  by bioaugmentation, 262, 285, 289
  by biostimulation, 262, 289
  mycoremediation, 339, 350
  of hydrocarbon-contaminated coastal regions, 291
  of hydrocarbon-contaminated cold soils, 261
  of oil-contaminated desert soils, 289
  of radionuclide-contaminated soils, 334
  phytoremediation, 292, 321, 335
  surfactant-enhanced, 289
  using genetically engineered micro-organisms, 326
Biotechnology, 101, 149, 312
Blastococcus, 252

Bodo, 208
Borehole fluids analysis, 213
Bradyrhizobium, 123
Brevibacillus
  levickii, 123, 157, 158, 159, 166, 167
Brevibacterium, 89, 250, 278
Burkholderia, 197, 260
  cepacia, 326

C

Cactus, 119
Cadmium, 25
Caldalkalibacillus, 157
Candida utilis, 335
Candidatus Nitrotoga arctica, 145
Candlemas Island, 163
Capillary electrophoresis, 236
Carbon
  cycle, 147, 182
  monoxide, 228, 230
  reservoir, 147, 194
  translocation, 336
Carbonates, 213
Carnobacterium pleistocenium, 145
Carotenoids, 104
Caves, 231, 238
Cellular fatty acid profile, 164
Cellulomonas, 123, 139, 250, 283
  flavigena, 280
Centralia, Pennsylvania, USA, 300
Centro de Astrobiología, Spain, 210
Ceratobasidium, 187
Cercomonas, 208
Chaetomium aureum, 344
Chaperones, chemical, 20
Chemolithotrophs, 28, 205, 232
Chernobyl, 333, 339
Chernobyl Atomic Electric Station, 340
Chlamydomonas, 208
Chlorella, 208
Chloroflexi, 141
Chlorophyta, 208
Chromium, 25, 318
Chromohalobacter, 103, 105
Chromosomal rearrangements, 53
Chroococcidiopsis, 124, 228
Chytridiomycota, 145
Citrate synthase, 326
Cladosporium, 278, 343
  chlorocephalum, 342
  cladosporioides, 336, 341, 342, 346, 348
  herbarum, 341, 342, 343, 347
  sphaerospermum, 341–343, 345–347

Clostridium, 250, 311
  algoriphilum, 144
Coal, 304
Coastal regions, contamination with oil, 276
Cobalt, 106, 318, 336
Collembola, 73
Collimonas fungivorans, 198
Colonization
  cryptoendolithic, 30, 124
  endoevaporitic, 126
  endolithic, 124, 251
  epilithic, 282, 284
  hypolithic, 119, 124, 125
  of peatlands, 184
Comamonas testosteronii, 197
Community 321 *see* Diversity
  bacterial, 247
  fungal, 344–345
  heterogeneity, 126
  impact of fire on microbial, 300
  microbial, 133, 139–146, 218, 229, 308
  of aquatic environments, 251
  structure, 4, 6, 32–37
  study by culture-dependent methods, 33, 88, 122, 216, 251, 300, 320
  study by culture-independent methods, 90, 122, 212, 216, 250, 301, 320
  study by molecular methods and in-situ analysis, 33
Compatible solutes, 19, 20, 102
  ectoines, 20, 103
  glutamate, glutamine, 103
Competence, for transformation, 54
Competition, 49, 193, 325
Composite selection pressure, 47
Conidia, 335, 345–346
Consortia
  cyanobacteria-dominated, 124
  hydrocarbon-degrading, 260, 282, 283
  in the rhizosphere of desert plants, 284
  use in bioremediation, 262
Containment, 235
Contingency genes, 52
Copiapite, 209
Copper, 25, 318
Coquimbite, 209
Corynebacterium, 89, 250, 278
Cosmic rays, 234
Crenarchaeota, 90, 141
Cropping, for soil bioremediation, 291
Crude oil *see* Hydrocarbons
  Arabian Gulf spill, 276
  composition, 277
  contamination of desert soils, 276

Cryolithosphere, 138
Cryopedogenesis, 138
Cryosol, 138
Cryptobiosis, 74
   anhydrobiosis, 74
   cryobiosis, 74
   osmobiosis, 74
Cryptogam Ridge, 161
Cupriavidus metallidurans, 326
Cyanidium, 208
Cyanobacteria, 6, 29, 72, 171, 228, 231, 251, 278, 280
   hydrocarbon oxidation by, 282
   hypolithic, 121, 125
   in desert soils, 275
Cystoderma amianthinum, 340
Cytokinin, 320
Cytophaga, 197
Cytophagales, 252

D

Dead Sea, 100
Death Valley, California, USA, 98
Deception Island, 163
Deinococcus, 252
   radiodurans, 6, 17, 23, 55, 228
Deinococcus/Thermus phylum, 30, 252, 258
Dematiaceae, 335
Denaturing gradient gel electrophoresis (DGGE), 35, 123, 193, 301
Denitrification, 181, 198
Desiccation resistance, 6, 19, 55, 128
Desulfobacter halotolerans, 97
Desulfotomaculum, 250
Desulfovibrio, 250
Detection of life, 235
Dew, in the Atacama Desert, 126
Dibenzofurans, 308
Dispersal, of micro-organisms, 9, 11, 57
Distribution
   bacterial,
   islands of fertility in the Atacama Desert, 125
   of *Bacillus*, 89
   of liquid water, 125
   of microbes and contaminants in Antarctic soils, 262
   of microbial populations and root exudates, 319
   of species in extreme soils, 77
   of subsurface micro-organisms, 28, 216
   patchy, 28, 49, 128
Diversification, 48, 53, 57
Diversity *see* Community

definition of, 32
eukaryotic, 208
of Antarctic soil bacteria, 251
of Arctic micro-organisms, 134
of halophilic soil micro-organisms, 88
of marine bacteria, 4
of micro-organisms in arid soils, 30
of micro-organisms in fire-impacted soils, 308
of micro-organisms in heavy-metal contaminated soils, 35
of micro-organisms in hydrocarbon-contaminated soils, 36, 257
of micro-organisms in Río Tinto, 206
of micro-organisms in soils, 27, 32
of micro-organisms in tundra soils, 31, 36, 252
DNA reassociation rate measurements, 33
DNA repair, 17, 23, 52, 54
Dormancy, 19, 21, 54, 146
Downeyite, 306
Drilling, 234
Dunaliella, 208

E

Ecosystem, 75
   service, 72
Electron
   acceptors, 220
   donors, 220
   transport chain, 236
Ellesmere Island, Canada, 141, 148
Endolithic bacteria, 231
Endophytes, 317, 319–321
Endospore formers
   from Antarctica, 164–168, 171
   from geothermal sources, 156
   in fire-impacted soils, 310
   in permafrost, 144
   potential for survival in Mars, 226
   taxonomy, 155
Endospores, 54, 96, 155, 286, 310
Energy source, 229
Enterobacter cancerogenes, 325
Enterobacteriaceae, 258
Environment
   extreme, 15
   hypersaline, 87
   impact of fire on, 301, 304
   unstable, 15
   variables affecting growth of oil-degrading bacteria, 284
Enzymes, halophilic, 101
   amylase, 101
   lipase, 102

protease, 102
pullulanase, 102
Epifluorescence, 140
Epiphytic bacteria, 197
Epistasis, 58
Error-prone DNA polymerase, 50, 52
Erwinia, 144
Escherichia coli, 50, 52
Eubacterium, 250
Eudorylaimus antarcticus, 75, 78
Euglena, 208
Eukarya, 133
Euplotes, 208
Europa, 230
Europium, 336
Euryarchaeota, 90, 141
  group III, 90
Evaporites, 126, 232
Evolution, 59
  macroevolution, 46
  Markovian dependency, 60
  microevolution, 46
  of prokaryotes, 46
  of soil bacteria, 48
  tradeoff, 59, 60
Evolutionary layer, 48, 56
Exiguobacterium, 141, 145
Exobiology, 128
Exopolymers, 325
Exopolysaccharides, 19, 103, 127
Extant life, 225, 235
Extremophiles, 16
  oil-degrading, 282
Extremotrophs, 16
Exudates, 319

F

Ferroplasma, 17, 22, 206
Fibrobacter, 258
Filobacillus milolensis, 96
Fires, 299–316
  duration, 305
  intensity, 301, 305
  migration rate, 309
  soil sterilization by, 308
  surface, 299
  underground, 299
Firmicutes, 90, 123, 141, 252, 320
Flavobacterium, 88, 139, 141, 197, 251, 258, 278
Flexibacter, 197
Fluorescence in situ hybridization (FISH), 141, 144, 207, 212, 217

Food chain, 145, 338, 350
Food web, 72, 75
Frigoribacterium, 144
Fruitbodies, of fungi, 336, 339
Fumarole, 160, 161, 311
Functional groups, 75
Fungi, 184, 333–355
  bacteria with antagonist activity against, 196
  ectomycorrhizal, 186, 310, 338
  endomycorrhizal, 338
  hydrocarbon-degrading, 278
  in biological soil crusts, 30
  in desert soils, 275
  in permafrost, 145
  in Río Tinto, 208
  mitosporic, 349
  surface area, 334
  susceptibility to soil fires, 310
  tolerance to water stress, 18
Fusarium, 278, 279
  solani, 59
  verticilloides, 335

G

Galerina, 185
Gas chromatography, 213
Gelatin, 156, 166
Gemmatimonadetes, 123, 141
Gene amplification, 53
Genetic engineering, 326
Genetic instability, 50, 51
Genome stability, 47
Geobacillus, 156, 158, 311
  stearothermophilus, 156, 279
  thermodenitrificans, 156, 158, 168
Geochemistry, 237
Geodermatophilaceae, 123
Geomonhystera, 78
Geomyces, 145
Geothermal environments, 311
Global changes, 77, 79, 133
Glomeromycota, 145, 337
Glomus mosseae, 337
Glutamic acid uptake, 171
Goethite, 209
Great Salt Lake, Utah, USA, 100
Greenhouse gases, 312
Greenhouse warming, 10
Guild, 32
Gypsum, 124, 209, 306

H

Hafnia, 197
Halanaerobacter salinarius, 100
Halanaerobiales, 100
Halanaerobium, 100
Halite, 124
Haloarchaea, 90
Haloarcula, 91
  japonica, 17
Halobacillus, 89, 96, 102, 103
Halobacteriaceae, 91
Halobacteriales, 90
Halobacterium, 89
Haloferax, 91, 105
Halomonas, 88, 98, 103, 104, 105
Halophiles, 19
  archaea, 20, 91
  in soils, 91
  oil-utilizing, 283
  physiological groups, 87
Halophilic methanogens, 90
Halorubrum, 90, 91
Haloterrigena, 90, 91
Halothermothrix, 100, 101
Halotolerant micro-organisms, 19, 88
  in soils, 91
Heavy metals, 335
  binding by bacterial exopolymers, 322
  bioavailability, 24, 321
  biotransformation, 25
  chelation, 25
  detoxification, 318
  foliar concentrations, 318
  hyperaccumulating plants, 209, 317
  impact on microbial diversity, 321
  in oil-polluted soils, 288
  in Río Tinto, 206
  mobilization, 319, 324
  phytoextraction of, 325
  tolerance, 25, 208, 317, 324
  tolerance in halophiles, 105
  toxicity, 25
Hematite, 209
Herbaspirillum seropedicae, 326
Herbicide, 29, 61, 104
Heterotroph, 122, 139, 236
  aerobic, 256
  bacterial, 122, 123, 216, 281, 310
  halophilic and halotolerant, 91
Holophaga/Acidobacterium group, 322, 323
Horizontal gene transfer, 9, 27, 46, 54, 261, 321
Hormoconis resinae, 341, 342, 343, 346, 347
Hot spring, 157, 158

Humboldt Current, 117
Humic acids, 182
Hydrocarbons *See* Crude oil
  abiotic loss from soil, 261
  anaerobic degradation, 250, 287
  as inclusions in bacteria and fungi, 278
  bioavailability, 262
  biodegradation, 26
  contamination, 26
  contamination of desert soils, 275
  contamination of polar soils, 247, 248, 252
  contamination, acute, 255
  contamination, chronic, 255
  degradation by a dioxygenase, 278
  degradation by a monooxygenase, 278
  degradation by fungi, 278
  degradation by halophiles, 105
  degradation in cold soils, 37, 247
  -degrading bacteria, 37
  effect on bacterial populations, 257
  influence of moisture and aeration on degradation of, 287
  influence of pH on degradation of, 286
  polycyclic aromatic (PAHs), 120, 253, 254, 256, 308
  utilization by bacteria associated with fish, 284
  utilization by *Geobacillus,* 170
Hydrogen, 215, 228
Hydronium jarosite, 209
Hydrothermal activity, 229
Hydrothermal vents, 158
Hygrocybe, 185
Hymenobacter, 123
Hymenoscyphus ericae, 188
Hypermutability, 51, 53
Hypermutation, 52
Hypersaline environment, 231

I

Iberian Pyritic Belt, 206
Ice, 230
Indole-3-acetic acid, 320
Infrared spectroscopy, 121
Initial land colonists, 56
Insurance hypothesis, 29
International Commision on Radiological Protection, 333
Intestinal pathogens, survival in soils, 61
Invertebrates, 72
Ion exchange, in fungal hyphal walls, 349
Ionizing radiation, 344, 345

Iron, 232
    bioformations, 209
    cycle, 207, 208
    ferric, 206
    ferrous, 206
    oxidation, 207
    reduction, 207
Iron-oxidizing micro-organisms, 206, 214
Isolation, of microbial populations, 58

J
Jarosite, 225
Jet Propulsion Laboratory, 129
Jharia Coalfield, India, 300

K
Karaj region, Iran, 102
Klebsormidium, 208
Kocuria, 123
Kurthia, 250
Kuwaiti Desert, 276, 277, 279

L
Labyrinthula, 208
Lactarius tabidus, 186
Lake Chaka, China, 90
Laphamite, 306
Laptev Sea, 148
Last Universal Common Ancestor, 45
Leaching, 30
Lead, 25, 318
Leaf litter, 186, 334, 336
Leccinum holopus, 186
Lena Delta, Siberia, 137, 138
Lentibacillus, 96
Leptodontidium orchidicola, 188
Leptospirillum, 206
Leucosporidium, 145
Lichens, 30, 119, 252, 339
Life signature, 238
Life strategies
    L-strategy, 18
    r- and K-selection, 16, 24, 261, 322
Limit, to life, 16
Linkage disequilibrium, 60

M
Magnetic field, 238
Manganese, 318
Manned mission, to Mars, 234

Marine fog, in the Atacama Desert, 119, 126
Marinobacter, 99, 102
Marinococcus, 89, 96, 103
Mars, 8, 117, 150, 220, 225
MARTE project, 205, 210, 234
Mass spectrometry, 235
Melanin, 335, 342, 349
Membrane *See* Temperature, 18
    adaptation to acid, 22
    adaptation to cold, 20
Mercury resistance, 7, 25
Mesophile, 96
Mesorhizobium, 123
Metagenome, 34
Metal toxicity, 25, 61
Metallic sulfides, mechanisms of oxidation, 206
Metallothioneines, 318
Metarhizium anisopliae, 344
Methane, 10, 134, 181, 218, 220, 228, 230
    monooxygenase, 193
    oxidation pathway, 195
    oxidation, anaerobic, 218
Methanobacteriaceae, 142
Methanobacterium
    bryantii, 189
    espanolae, 190
    uliginosum, 190
Methanococcus, 189
Methanogenesis, 142, 184, 189, 216
    influence of environmental factors on, 191
Methanogenium organophilium, 189
Methanogens, 142, 188, 214, 218, 230
    acidophilic, 189
    psychrophilic, 191
Methanol dehydrogenase, 195
Methanomicrobiaceae, 142
Methanopyrus kandleri, 189
Methanosaeta, 189
Methanosaetaceae, 142
Methanosarcina, 139, 142, 189
    acetivorans, 230
    barkeri, 189
Methanosarcinaceae, 142
Methanospirillum hungatei, 189
Methanotrophs, 142, 192
    type I, 192
    type II, 192
    type X, 192
Methylobacter, 139, 144, 192, 194
    psychrophilus, 144
    tundripaludum, 144

Methylobacterium, 322
   extorquens, 322
   goesingense, 323
   mesophilicum, 198, 322
Methylocaldum, 192
Methylocapsa, 192
Methylocella, 192
   palustris, 194
Methylococcus, 192, 194
Methylocystis, 144, 192, 194
   rosea, 144
Methylohalobius, 192
Methylomicrobium, 144, 192
Methylomonas, 144, 192, 194
Methylosarcina, 144, 192
Methylosinus, 144, 192, 194
Methylosphaera, 144, 192
Microaerophilic micro-organisms, 220
Microarthropods, 73
Microbacteriaceae, 141
Microbacterium, 144
   halotolerans, 97
   saperdae, 325
Microbial mats *See* Biofilms
   in Arabian Gulf coasts, 280
   on the Kuwaiti coast, 281
Microbiota, 15, 335
   of Antarctic soils, 262
   of cold peat soils, 251
   of contaminated soils, 248
   of hypersaline soils, 92
   of permafrost, 139–142
Micrococcus, 89, 139, 258, 278, 279, 282, 283, 284
Microcoleus, 278, 280
Micro-organisms
   oil-utilizing, 277, 280
   responses to salt, 87
Microscopy, 212, 217, 219, 237
Microsphaera, 252
Mismatch repair, 50, 51
Mobile genetic elements, 54
Modularity, 5
Moniliaceae, 335
Moraxella, 258
Mrakia, 145
Mucor, 278, 279
Muller's ratchet, 59
Mullite, 306
Mushrooms, 185, 338
Mutation rate, 9, 50, 51
Mutator phenotypes, 51
Mutinus caninus, 336
Mycelial cords, 335

Mycelium, 98, 334, 335
Mycena
   epipterygia, 185
   polygramma, 340
   sanguinolenta, 340
Mycorrhizae *See* Fungi
   arbuscular, 337
   ectomycorrhizae, 186, 337
   endomycorrhizae, 337
   ericoid, 187, 337
   orchid mycorrhizae, 187
   role in radionuclide uptake, 337
Myrothecium roridum, 344
Myxotrichum arcticum, 188

N

Naegleria, 209
Naphthalene dioxygenase (*ndo*) genes, 261
NASA, 210
Natrinema, 90
Natroniella acetica, 100
Natronobacterium, 17
Natronococcus, 90
Natronojarosite, 209
Natronorubrum, 90
ncc-nre nickel resistance system, 326
Nematodes, 73
Nesterenkonia, 89, 97, 101
Neurospora, 310
Nevada Test Site, 340, 342
Nickel, 318
Nitrate deposits, 117
Nitrification, 198
Nitrobacter, 139, 310
Nitrogen *See* Nutrients
   cycle, 148
   fixation, 30, 168, 193, 229
   nitrate, 24, 213
   nitrite, 198, 213
   NOx gases, 24, 148, 218
   source, 229, 284
Nitrogen-cycling bacteria, 310
   ammonia oxidizers, 148, 311
   denitrifiers, 185, 214, 218, 311
   nitrifiers, 49, 139, 310
   nitrogen fixers, 139, 282, 284, 291, 310
Nitrosomonas, 139
Nitrosospira, 310
Nitrous oxide, 198
Nocardia, 89
Nocardiopsis, 97

Nostoc, 124
Nuclear magnetic resonance, 120
Nuclear weapons, 333
Nunavut, Canada, 234
Nutrients
  availability, 300
  carbon, 303
  cycling, 30
  effect of soil fire on, 302, 306
  inorganic, 288
  low supply, 23
  nitrogen, 303, 306
  phosphorus, 303
  starvation for, 275
  sulfur, 304, 307
  uptake, 24

O

Ochroomonas, 208
Oidiodendron maius, 188
Okibacterium, 322
Oligotrophs, 23
Olivine, 226
Orenia, 100
Organophosphonates, degradation by halophiles, 105
Orpiment, 306
Oscillatoria, 278
Osmoadaptation, 103
Osmoregulation, 19
Oxidative damage, 23
Oxygen, 218
  in hydrocarbon degradation, 284, 287
  in peatlands, 177
  in soil fires, 305
Oxytrichia, 209

P

Paecilomyces lilacinus, 344
Paenibacillus, 144
  cineris, 168
  cookii, 168
  wynnii, 168
Paleo-archives, 181
Palleronia marisminoris, 99, 104
Paludification, 180
Panagrolaimus, 74
Pantoea, 197
Pathogens, human, 10
Peat, 177
Peatland, 177
  acrotelm, 178
  bog, 23, 178
  catotelm, 178
  fen, 178
  minerotrophic, 178
  ombrotrophic, 178
Peña de Hierro, Spain, 210
Penicillium, 278, 279
  aurantiogriseum, 341, 342, 347
  chrysogenum, 335
  italicum, 335
  roseopurpureum, 336
  spinulosum, 341, 342, 347
Permafrost, 21, 31, 133, 135, 161, 227, 234, 238, 249, 250
  active layer, 31, 133, 135, 146, 250, 256
  cryopeg, 32, 138
  deep sediments, 135
  distribution, 134
  ground ice, 136
  ice complex, 138
  ice wedge, 136
  polygon structures, 31, 136
  submarine, 138
  table, 136
  talik, 138
  upper sediments, 135
Permanganate oxidation, for analysis of total organics, 121
Pesticides, 26
  fungicides, 26, 61
  triazines, 26
pH limits for life, 22
Phallus impudicus, 336
Phase variation, 46
Phenotypic switching, 51
Phialocephala fortinii, 188
Phormidium, 124, 278
Phospholipids fatty acid analysis, 25, 144, 301
Phosphorus, 26
Photoautotrophy, 56, 285
Photodegradation, 122
Photorhabdus luminescens, 198
Photosynthesis, 232
Phototrophic bacteria, 278
Phragmites, 321
Phytochelatin synthase, 326
Phytochelatins, 318
Picrophilus, 17
  oshimae, 22
  torridus, 22
Planctomycetales, 252
Planctomycetes, 123, 141, 323
Planococcus, 89

Planomicrobium, 141
Plant growth-promoting bacteria, 291, 320
Plants, carnivorous, 182
Plectonema, 228
Plectus, 77
Pleurotus eringii, 336
Poikilotrophic organisms, 16
Polychlorinated dibenzo-p-dioxins, 308
Polymers, intracellular, 24
Porphyrins, 237
Potential *See* Water
    matric, 18
    osmotic, 18
Predation, 3, 27, 49
Programmed cell death, 55
Proteobacteria, 123, 141, 197, 258, 320
    alpha, 30, 123, 124, 141, 252, 311
    beta, 123, 141, 311
    delta, 141, 311
    gamma, 98, 124, 141, 192, 252
Proton translocation, 22
Pseudomonas, 88, 139, 197, 251, 258, 260, 278, 279, 322
    aeruginosa, 53, 198, 326
    aureofaciens, 326
    grimontii, 198
    monteilii, 325
    putida, 17, 54, 261
Psychrobacter, 17, 141, 144
Psychrophiles, 21, 248
Psychrotolerance, of Antarctic soil bacteria, 259
Psychrotrophs, 21, 248
Pyrite, 206
Pyrobaculum aerophilum, 51
Pyrolobus fumarii, 17
Pyrolysis-gas chromatography-mass spectrometry, 120

Q
Qinghai-Tibet Plateau, 23
Quartz, 124
Quinines, 348
Quinones, 237

R
Radiation, 23
    beta, 345
    gamma, 345
    ionizing, 228
    light, 23
    resistance to, 6, 17, 55, 344
    ultraviolet, 227

Radioactive debris, carbon-based, 336
Radiocarbon, 124
Radiocesium, 350
Radionuclide, 333
    absorption, 350
    adsorption/desorption, 349
    binding by ion exchange, 335, 349
    biosorption, 335
    concentration factor, 334
    immobilization, 336
    translocation, 335, 338
Radiorespirometry, 236
Radiothorium, 335
Radiotropism, 341, 347
    hormetic character of, 347
Rahnella, 260
Rainfall, 124
Ralstonia, 17
RecA, 50
Redox compounds, as life signatures, 236
Redundancy, 76
    functional, 6, 29
Resinicium bicolor, 335
Respiration, 207
    aerobic, 207
    anaerobic, 207, 213
Restriction fragment length polymorphism (RFLP), 252
Rhizomorphs, 336
Rhizoplane, 319
Rhizopus arrhizus, 335
Rhizoscyphus ericae, 188
Rhizosphere
    bacteria, 317, 319, 324
    hydrocarbon-degrading bacteria, 291
    of desert plants, 279
Rhodococcus, 139, 144, 252, 258, 259, 278, 279, 284, 322
    alkane-degrading, 259
    erythropolis, 198, 280
Rhodoferax fermentans, 197
Rhodopseudomonas, 278
    acetophilia, 197
Rhodospirillum, 197, 278
Ribosomal DNA nucleotide sequence analysis, 35, 197, 212, 301, 323
Río Tinto, Spain, 206
Rock leachate analysis, 213
Rocks, 121, 227, 237, 238
Rotylenchulus reniformis, 74
RpoS, 56
Rubrivivax gelatinosus, 197
Rubrobacter, 17, 252

Rubrobacteria, 17, 323
Russula, 187
    betularum, 186
    claroflava, 186

S
Saccharomonospora, 98
Sahara, 29
Salinicoccus hispanicus, 96
Salinity, 19, 88
    influence on oil degradation, 288
Salinivibrio, 88
Salipiger, 99, 104
Sampling, 234
Saprospira, 197
Saprotrophic fungi, 334
Schizophyllum commune, 336
Schwermannite, 209
Scottnema lindsayae, 76
Scuttellonema brachyurum, 74
Scytonemin, 127
Sebacina, 187
Selection, 25, 47, 248
    biotic and abiotic in soil, 25, 47, 248, 220
Selenium, 324
Self-organization, of microbial communities, 50
Serratia, 197
Sewage sludge, 25, 317
Shewanella, 258
Siderophores, 320, 325
Sigma factor, 50
Single strand conformational polymorphism analysis (SSCP), 35
Smallpox virus, 11, 45
Sodium chloride, 288
Soil
    acid deposition in, 24
    adaptation to, 59, 63
    aggregates, 27, 49
    Alpine, 252
    Antarctic, 249
    Arctic, 250
    biogeography, 79
    biopolymers, 26
    calamine, 318
    clay colloids, 27
    cold, 247, 248
    colonization by micro-organisms, 26–28, 49
    containing Cu and Co, 318
    contamination, 24
    definition of, 26
    desert, 29, 275
    enzymes, 25, 27
    extreme, 28–32, 71
    extremization, 10
    fell-field, 31
    geothermal, 155, 156, 159–163
    heavy-metal containing, 317
    heterogeneity, 27, 29, 77, 88, 138, 247
    hypersaline, 87
    hyphal biomass, 334
    indicators, 25
    Mars-like, 117, 124
    metalliferous, 317
    microbial populations, 9, 122
    microhabitats, 27
    mineral content, 162
    moisture, 302, 306
    oligotrophic conditions in, 264
    organic matter, 100, 119, 147, 177, 285
    oxidizing agents, 119, 129, 225, 236
    pH, 22
    pit, 123
    pristine, 257, 279
    selenium-rich, 318
    self-cleaning capacity, 280
    serpentine, 317, 318
    structure, 26, 49, 72, 77, 178
    temperature, 32
    texture, 26, 265
    thermal, 21
    transfer of energy from fire, 302, 305
    tundra, 31
    ultramafic, 318
Solute transport, 22
SOS response, 50, 63
Spacecraft, 228, 234
Specific affinity value, 255
Sphagnum, 177
Sphingobacterium, 197
Sphingomonas, 123, 141, 252, 258, 323
Spore, 31
    germination, 19, 346, 347, 348
    survival, 344
Sporichthya, 252
Sporohalobacter lortetii, 97
Staphylococcus, 89, 197
Starvation, 133
Stationary phase, 102, 104
Stemphylium, 336
Stenotrophomonas, 260
Stramenopiles, 208
Streptomonospora, 98
Streptomyces, 127, 139, 278, 279, 322

Stress *See* Water
  biotic and abiotic, 321
  effect on community structure, 36
  environmental, 227
  hyperosmotic, 103
  in desert soils, 276
  response, 49–56
  tolerance, 18, 325
Stringent response, 276
Strontium, 335
Subsurface, 123, 227, 238, 309
  ecosystems, 205
  geomicrobiology, 205, 210
Subtercola, 144
Sulfate
  reducers, 97, 148, 185, 216, 218
  reduction, 181
Sulfobacillus, 157, 170
Sulfolobulus acidoalcarius, 51
Sulfur, 306
  cycle, 148, 208
  dioxide, 24
  distribution in fire-affected soils, 307
  oxidation, 207
  source of, 229
  sulfate, 24, 206
  sulfide, 25, 148, 206
Sulfur-cycling bacteria, 311
  thiosulfate oxidizers, 216
Sulfur-oxidizing micro-organisms, 148, 206
Sunlight, 127
Surface adsorption, 601
Surface area, 324, 337
Surfactants, produced by oil-utilizing
    micro-organisms, 289
Survival mechanisms, 275
Svalbard, Norway, 36, 144
Synechococcus, 228, 279
Synechocystis, 279

T
Tataouine sand dunes, Tunisia, 30
Taxonomy
  of *Bacillus* and related genera, 155–158
  of bacteria colonizing Antarctic soils, 249
Temperature
  effect on membranes, 20, 21, 168
  limits for microbial activity, 20
Tenuibacillus multivorans, 96
Terminal restriction fragment length
    polymorphism (T-RFLP), 35, 193, 301
Terrestrialization, 180
Thalassobacillus, 96, 105
Thermoanaerobacter thermohydrosulfuricus, 164

Thermobaculum terrenum, 17
Thermomicrobia, 123
Thermophiles, 21, 155, 158
  endospore forming, 164
  halophilic, 100
  hydrocarbon-utilizing, 279, 286
  in cold environments, 158
  in fire-affected soils, 310
  nutrition, 170
  sulfur metabolizers, 311
Thermoplasmales, 91
Thermotrophs, 21
Thermus thermophilus, 51, 54
Thlaspi
  caerulescens, 323
  goesingense, 322
Transcription-associated mutagenesis, 53
Translational stress-induced mutagenesis, 53
Trichoderma viride, 334
Trophic level, 75
Tulasnella, 187
Turgor pressure, 18

U
Ultramicrobacteria, 275
Unalaska Island, 194
United Nations Convention on Biodiversity, 71
United Nations Convention to Combat
    Desertification, 79
UV light, 127, 344

V
Vacuole, 318
Valhkampfia, 209
Variovorax, 258
  paradoxus, 197
Varnish, mineral, 232
Vent, 305
Verrucomicrobia, 141, 252, 322
Vibrio cholerae, 54
Viking missions, 119, 129, 225
Virgibacillus, 96
Volcanic regions, 238
Vozrozhdeniye Island, 11
Vulcanibacillus, 157, 158

W
Water
  activity (a_w), 18
  athalassohaline, 87
  availability for life, 29, 124, 127
  hypersaline, 87

    in hydrocarbon degradation, 288
    in soil, 18
    potential, 18
    presence on Mars, 228, 230, 232, 233
    stress, 19
    thalassohaline, 87
Water table, 178, 213

X
Xanthomonadaceae, 141
Xanthomonas, 258
Xerophiles, 19

Y
Yeasts, 145
Yersinia, 197
    pestis, 11

Z
Zinc, 25, 318, 338
Zygnemopsis, 208
Zygomycota, 145